P9-CCT-568

Boundary Value Problems

Fourth Edition

Boundary Value Problems

Fourth Edition

David L. Powers
Clarkson University

San Diego London Boston
New York Sydney Tokyo Toronto

This book is printed on acid free paper. ∞

Copyright © 1999, 1987, 1979, 1972 by Academic Press

All rights reserved.
No part of this publication may be reproduced or transmitted in any form or by any means, electronic or mechanical, including photocopy, recording, or any information storage and retrieval system, without permission in writing from the publisher.

Cover Image
Frank Stella
Pergusa Three
woodcut, relief on paper
66-3/8"x 51-7/8"
Printed and published by Tyler Graphics Ltd, 1983
© Frank Stella/Tyler Graphics Ltd./Artists Rights Society (ARS), New York
Photo: Steven Sloman

Academic Press
a division of Harcourt Brace & Company
525 B Street, Suite 1900, San Diego, CA 92101-4495, USA
http://www.apnet.com

Academic Press
24–28 Oval Road, London NW1 7DX, UK
http://www.hbuk.co.uk/ap/

Harcourt/Academic Press
200 Wheeler Road, Burlington, MA 01803, USA
http//www.harcourt-ap.com

Library of Congress Catalog Card Number: 98-89676

International Standard Book Number: 0-12-563734-9

Printed in the United States of America
99 00 01 02 03 IP 9 8 7 6 5 4 3 2 1

Preface

This text is designed for a one-semester or two-quarter course in partial differential equations given to third- and fourth-year students of engineering and science. It can also be used as the basis for an introductory course for graduate students. Mathematical prerequisites have been kept to a minimum — calculus and differential equations. Vector calculus is used for only one derivation, and necessary linear algebra is limited to determinants of order two. A reader needs enough background in physics to follow the derivations of the heat and wave equations.

The principal objective of the book is solving boundary value problems involving partial differential equations. Separation of variables receives the greatest attention because it is widely used in applications and because it provides a uniform method for solving important cases of the heat, wave, and potential equations. One technique is not enough, of course. D'Alembert's solution of the wave equation is developed in parallel with the series solution, and the distributed-source solution is constructed for the heat equation. In addition, there are chapters on Laplace transform techniques and on numerical methods.

The second objective is to tie together the mathematics developed and the student's physical intuition. This is accomplished by deriving the mathematical model in a number of cases, by using physical reasoning in the mathematical development, by interpreting mathematical results in physical terms, and by studying the heat, wave, and potential equations separately.

In the service of both objectives, there are many fully worked examples and now about 850 exercises, including miscellaneous exercises at the end of each chapter. The level of difficulty ranges from drill and verification of details to development of new material. Answers to odd-numbered exercises are in the back of the book. An Instructor's Manual is available with the answers to the even-numbered problems.

There are many ways of choosing and arranging topics from the book to provide an interesting and meaningful course. The following sections form the core, requiring at least 14 hours of lecture: Sections 1.1-1.3, 2.1-2.5, 3.1-3.3, 4.1, 4.2, and 4.4. These cover the basics of Fourier series and the solutions of heat, wave, and potential equations in finite regions. My choice for the next most important block of material is the Fourier integral and the solution of problems on unbounded regions: Sections 1.9, 2.10-2.12, 3.6 and 4.3. These require at least six more lectures.

The tastes of the instructor and the needs of the audience will govern the choice of further material. A rather theoretical flavor results from including: Sections 1.4-1.7 on convergence of Fourier series; Sections 2.7-2.9 on Sturm-Liouville problems, and the sequel, Section 3.4; and the more difficult parts of Chapter 5, Sections 5.5-5.10 on Bessel functions and Legendre polynomials. On the other hand, inclusion of numerical methods in Sections 1.8, 3.3 and Chapter 7 gives a very applied flavor.

Chapter 0 reviews solution techniques and theory of ordinary differential equations and boundary value problems. Equilibrium forms of the heat and wave equations are derived also. This material belongs in an elementary differential equations course and is strictly optional. However, many students have either forgotten it or never seen it.

For this Fourth Edition I have rewritten several sections to improve clarity and provide a little more detail. Sections 5.10, Applications of Legendre Polynomials, and 2.12, The Error Function, are both new. Throughout the book, I have added almost 100 new exercises; over 20 of these are based on current engineering literature, showing actual applications with authentic parameter values.

The new Appendix to Chapter 7 explains the use of spreadsheet programs for numerical methods. I believe that spreadsheets are superior to programming as a teaching device for numerical methods. Furthermore, engineers and engineering students all have and use spreadsheet programs, while few are competent programmers or even take advantage of computer algebra programs such as Maple and Mathematica.

I wish to acknowledge the skillful work of Cindy Smith, who was the Latex compositor and corrected many of my mistakes, the help of Academic Press editors and consultants, and the guidance of reviewers for this edition:

Linda Allen, Texas Tech University
Ilya Bakelman, Texas A&M University
Herman Gollwitzer, Drexel University

James Herod, Georgia Institute of Technology
Robert Hunt, Humboldt State University
Mohammad Khavanin, the University of North Dakota
Jeff Morgan, Texas A&M University
Jim Mueller, California Polytechnic State University
Ron Perline, Drexel University
William Royalty, the University of Idaho
Lawrence Schovanec, Texas Tech University
Al Shenk, the University of California at San Diego
Michael Smiley, Iowa State University
Monty Strauss, Texas Tech University
Kathie Yerion, Gonzaga University

Contents

Chapter 0

Ordinary Differential Equations

0.1 HOMOGENEOUS LINEAR EQUATIONS

The subject of most of this book is partial differential equations: their physical meaning, problems in which they appear, and their solutions. Our principal solution technique will involve separating a partial differential equation into ordinary differential equations. Therefore, we begin by reviewing some facts about ordinary differential equations and their solutions.

We are interested mainly in linear differential equations of first and second orders, as shown here:

$$\frac{du}{dt} = k(t)u + f(t), \tag{1}$$

$$\frac{d^2u}{dt^2} + k(t)\frac{du}{dt} + p(t)u = f(t). \tag{2}$$

In either equation, if $f(t)$ is 0, the equation is homogeneous. (Another test: if the constant function $u(t) \equiv 0$ is a solution, the equation is homogeneous.) In the rest of this section, we review homogeneous linear equations.

A. First-Order Equations

The most general first-order linear homogeneous equation has the form

$$\frac{du}{dt} = k(t)u. \tag{3}$$

This equation can be solved by isolating u on one side and then integrating:

$$\frac{1}{u}\frac{du}{dt} = k(t)$$

$$\ln|u| = \int k(t)dt + C$$

$$u(t) = \pm e^{C} e^{\int k(t)dt} = c e^{\int k(t)dt}. \tag{4}$$

It is easy to check directly that the last expression is a solution of the differential equation for any value of c. That is, c is an arbitrary constant and can be used to satisfy an initial condition if one has been specified. For example, let us solve the homogeneous differential equation

$$\frac{du}{dt} = -tu.$$

The procedure outlined here gives the general solution

$$u(t) = ce^{-t^2/2}$$

for any c. If an initial condition such as $u(0) = 5$ is specified, then c must be chosen to satisfy it ($c = 5$).

The most common case of this differential equation has $k(t) = k$ constant. The differential equation and its general solution are

$$\frac{du}{dt} = ku, \qquad u(t) = ce^{kt}. \tag{5}$$

If k is negative, then $u(t)$ approaches 0 as t increases. If k is positive, then $u(t)$ increases rapidly in magnitude with t. This kind of exponential growth often signals disaster in physical situations, as it cannot be sustained indefinitely.

B. Second-Order Equations

It is not possible to give a solution method for the general second-order linear homogeneous equation,

$$\frac{d^2u}{dt^2} + k(t)\frac{du}{dt} + p(t)u = 0. \tag{6}$$

Nevertheless, we can solve some important cases that we detail in what follows. The most important point in the general theory is the following.

Principle of Superposition.
If $u_1(t)$ and $u_2(t)$ are solutions of the same linear homogeneous equation (6), then so is any linear combination of them: $u(t) = c_1u_1(t) + c_2u_2(t)$.

This theorem, which is very easy to prove, merits the name of *principle* because it applies, with only superficial changes, to many other kinds of linear, homogeneous equations. Later, we will be using the same principle on partial differential equations. To be able to satisfy an unrestricted initial condition, we need two linearly independent solutions of a second-order equation. Two solutions are *linearly independent* on an interval if the only linear combination of them (with constant coefficients) that is identically 0 is the combination with 0 for its coefficients. There is an alternative test.

Two solutions of the same linear homogeneous equation (6) are independent on an interval if and only if their *Wronskian*

$$W(u_1, u_2) = \begin{vmatrix} u_1(t) & u_2(t) \\ u_1'(t) & u_2'(t) \end{vmatrix} \tag{7}$$

is nonzero on that interval.

If we have two independent solutions $u_1(t)$, $u_2(t)$ of a linear second-order homogeneous equation, then the linear combination $u(t) = c_1u_1(t) + c_2u_2(t)$ is a general solution of the equation: given any initial conditions, c_1 and c_2 can be chosen so that $u(t)$ satisfies them.

1. Constant coefficients.
The most important type of second-order linear differential equation that can be solved in closed form is the one with constant coefficients,

$$\frac{d^2u}{dt^2} + k\frac{du}{dt} + pu = 0 \quad (k, p \text{ are constants}). \tag{8}$$

There is always at least one solution of the form $u(t) = e^{mt}$ for an appropriate constant m. To find m, substitute the proposed solution into the differential equation, obtaining

$$m^2e^{mt} + kme^{mt} + pe^{mt} = 0,$$

or

$$m^2 + km + p = 0 \tag{9}$$

(since e^{mt} is never 0). This is called the *characteristic equation* of the differential equation (8). There are three cases for the roots of the characteristic equation (9), which determine the nature of the general solution of Eq. (8). These are summarized in Table 1.

Table 1: Solutions of $\frac{d^2u}{dt^2} + k\frac{du}{dt} + pu = 0$

Roots of Characteristic Polynomial	General Solution of Differential Equation
Real, distinct: $m_1 \neq m_2$	$u(t) = c_1 e^{m_1 t} + c_2 e^{m_2 t}$
Real, double: $m_1 = m_2$	$u(t) = c_1 e^{m_1 t} + c_2 t e^{m_1 t}$
Conjugate complex: $m_1 = \alpha + i\beta, m_2 = \alpha - i\beta$	$u(t) = c_1 e^{\alpha t}\cos(\beta t) + c_2 e^{\alpha t}\sin(\beta t)$

This method of assuming an exponential form for the solution works for linear homogeneous equations of any order with constant coefficients. In all cases, a pair of complex conjugate roots $m = \alpha \pm i\beta$ leads to a pair of complex solutions

$$e^{\alpha t}e^{i\beta t}, \qquad e^{\alpha t}e^{-i\beta t} \tag{10}$$

which can be traded for the pair of real solutions

$$e^{\alpha t}\cos(\beta t), \qquad e^{\alpha t}\sin(\beta t). \tag{11}$$

We include two important examples. First, consider the differential equation

$$\frac{d^2u}{dt^2} + \lambda^2 u = 0 \tag{12}$$

where λ is constant. The characteristic equation is $m^2 + \lambda^2 = 0$, with roots $m = \pm i\lambda$. The third case of Table 1 applies if $\lambda \neq 0$; the general solution of the differential equation is

$$u(t) = c_1\cos(\lambda t) + c_2\sin(\lambda t). \tag{13}$$

Second, consider the similar differential equation

$$\frac{d^2u}{dt^2} - \lambda^2 u = 0. \tag{14}$$

The characteristic equation now is $m^2 - \lambda^2 = 0$, with roots $m = \pm\lambda$. If $\lambda \neq 0$, the first case of Table 1 applies, and the general solution is

$$u(t) = c_1 e^{\lambda t} + c_2 e^{-\lambda t}. \tag{15}$$

It is sometimes helpful to write the solution in another form. The hyperbolic sine and cosine are defined by

$$\sinh(A) = \frac{1}{2}\left(e^A - e^{-A}\right), \quad \cosh(A) = \frac{1}{2}\left(e^A + e^{-A}\right). \tag{16}$$

Thus, $\sinh(\lambda t)$ and $\cosh(\lambda t)$ are linear combinations of $e^{\lambda t}$ and $e^{-\lambda t}$. By the Principle of Superposition, they too are solutions of Eq. (14). The Wronskian test shows them to be independent. Therefore, we may equally well write

$$u(t) = c_1' \cosh(\lambda t) + c_2' \sinh(\lambda t)$$

as the general solution of Eq. (14), where c_1' and c_2' are arbitrary constants.

2. Cauchy-Euler equation.

One of the few equations with variable coefficients that can be solved in complete generality is the Cauchy-Euler equation:

$$t^2 \frac{d^2u}{dt^2} + kt\frac{du}{dt} + pu = 0. \tag{17}$$

The distinguishing feature of this equation is that the coefficient of the nth derivative is the nth power of t, multiplied by a constant. The style of solution for this equation is quite similar to the preceding: assume that a solution has the form $u(t) = t^m$, then find m. Substituting u in this form into Eq. (17) leads to

$$t^2 m(m-1)t^{m-2} + ktmt^{m-1} + pt^m = 0, \quad \text{or}$$

$$m(m-1) + km + p = 0 \quad (k, p \text{ are constants}). \tag{18}$$

This is the characteristic equation for Eq. (17), and the nature of its roots determines the solution as summarized in Table 2.

One important example of the Cauchy-Euler equation is

$$t^2 \frac{d^2u}{dt^2} + t\frac{du}{dt} - \lambda^2 u = 0 \tag{19}$$

where $\lambda > 0$. The characteristic equation is $m(m-1)+m-\lambda^2 = m^2 - \lambda^2 = 0$. The roots are $m = \pm\lambda$, so the first case of Table 2 applies, and

$$u(t) = c_1 t^\lambda + c_2 t^{-\lambda} \tag{20}$$

is the general solution of Eq. (19).

For the general linear equation

$$\frac{d^2u}{dt^2} + k(t)\frac{du}{dt} + p(t)u = 0,$$

Table 2: Solutions of $t^2\frac{d^2u}{dt^2}+kt\frac{du}{dt}+pu=0$

Roots of Characteristic Polynomial	General Solution of Differential Equation
Real, distinct roots: $m_1 \neq m_2$	$u(t) = c_1t^{m_1} + c_2t^{m_2}$
Real, double root: $m_1 = m_2$	$u(t) = c_1t^{m_1} + c_2(\ln t)t^{m_1}$
Conjugate complex roots: $m_1 = \alpha + i\beta, m_2 = \alpha - i\beta$	$u(t) = c_1t^{\alpha t}\cos(\beta \ln t) + c_2t^{\alpha t}\sin(\beta \ln t)$

any point where $k(t)$ or $p(t)$ fails to be continuous is a *singular point* of the differential equation. At such a point, solutions may break down in various ways. However, if t_0 is a singular point where both of the functions

$$(t-t_0)k(t) \quad \text{and} \quad (t-t_0)^2p(t) \tag{21}$$

have Taylor series expansions, then t_0 is called a *regular singular point*. The Cauchy-Euler equation is an example of an important differential equation having a regular singular point (at $t_0 = 0$). The behavior of its solution near that point provides a model for more general equations.

3. Other equations.

Other second-order equations may be solved by power series, by change of variable to a kind already solved, or by sheer luck. For example, the equation

$$t^4\frac{d^2u}{dt^2} + \lambda^2u = 0, \tag{22}$$

which occurs in the theory of beams, can be solved by the change of variables

$$t = \frac{1}{z}, \qquad u(t) = \frac{1}{z}v(z).$$

In terms of the new variables, the differential equation (22) becomes

$$\frac{d^2v}{dz^2} + \lambda^2v = 0.$$

This equation is easily solved, and the solution of the original is then found by reversing the change of variables:

$$u(t) = t(c_1\cos(\lambda/t) + c_2\sin(\lambda/t)). \tag{23}$$

C. Second Independent Solution

Although it is not generally possible to solve a second-order linear homogeneous equation with variable coefficients, we can always find a second independent solution if one solution is known. This method is called "reduction of order."

Suppose $u_1(t)$ is a solution of the general equation

$$\frac{d^2u}{dt^2} + k(t)\frac{du}{dt} + p(t)u = 0. \tag{24}$$

Assume that $u_2(t) = v(t)u_1(t)$ is a solution. We wish to find $v(t)$ so that u_2 is indeed a solution. However, $v(t)$ must not be constant, as that would not supply an independent solution. A straightforward substitution of $u_2 = vu_1$ into the differential equation leads to

$$v''u_1 + 2v'u_1' + vu_1'' + k(t)(v'u_1 + vu_1') + p(t)vu_1 = 0.$$

Now collect terms in the derivatives of v. The preceding equation becomes

$$u_1v'' + (2u_1' + k(t)u_1)v' + (u_1'' + k(t)u_1' + p(t)u_1)v = 0.$$

However, u_1 is a solution of Eq. (24), so the coefficient of v is 0. This leaves

$$u_1v'' + (2u_1' + k(t)u_1)v' = 0, \tag{25}$$

which is a first-order linear equation for v'. Thus, a nonconstant v can be found, at least in terms of some integrals.

For example, consider the equation

$$(1 - t^2)u'' - 2tu' + 2u = 0, \qquad -1 < t < 1,$$

which has $u_1(t) = t$ as a solution. By assuming that $u_2 = v \cdot t$ and substituting, we obtain

$$(1 - t^2)(v''t + 2v') - 2t(v't + v) + 2vt = 0.$$

After collecting terms, we have

$$(1 - t^2)tv'' + (2 - 4t^2)v' = 0.$$

From here, it is fairly easy to find

$$\frac{v''}{v'} = \frac{4t^2 - 2}{t(1 - t^2)} = \frac{-2}{t} + \frac{1}{1 - t} - \frac{1}{1 + t}$$

(using partial fractions), then

$$\ln v' = -2\ln(t) - \ln(1-t) - \ln(1+t).$$

Finally, each side is exponentiated to obtain

$$v' = \frac{1}{t^2(1-t^2)} = \frac{1}{t^2} + \frac{1/2}{1-t} + \frac{1/2}{1+t}$$

$$v = -\frac{1}{t} + \frac{1}{2}\ln\left|\frac{1+t}{1-t}\right|.$$

D. Higher-Order Equations

Linear homogeneous equations of order higher than 2 — especially order 4 — occur frequently in elasticity and fluid mechanics. A general, nth-order homogeneous linear equation may be written

$$u^{(n)} + k_1(t)u^{(n-1)} + \cdots + k_{n-1}(t)u^{(1)} + k_n(t)u = 0, \tag{26}$$

in which the coefficients $k_1(t)$, $k_2(t)$, etc., are given functions of t. The techniques of solution are analogous to those for second-order equations. In particular, they depend on the Principle of Superposition, which remains valid for this equation. That Principle allows us to say that the general solution of Eq. (26) has the form of a linear combination of n independent solutions $u_1(t), u_2(t), \ldots, u_n(t)$ with arbitrary constant coefficients,

$$u(t) = c_1u_1(t) + c_2u_2(t) + \cdots + c_nu_n(t).$$

Of course, we cannot solve the general nth-order equation (26), but we can indeed solve any homogeneous linear equation with *constant* coefficients,

$$u^{(n)} + k_1u^{(n-1)} + \cdots + k_{n-1}u^{(1)} + k_nu = 0. \tag{27}$$

We must now find n independent solutions of this equation. As in the second-order case, we assume that a solution has the form $u(t) = e^{mt}$, and find values of m for which this is true. That is, we substitute e^{mt} for u in the differential equation (27) and divide out the common factor of e^{mt}. The result is the polynomial equation

$$m^n + k_1m^{n-1} + \cdots + k_{n-1}m + k_n = 0, \tag{28}$$

called the *characteristic equation* of the differential equation (27).

Each distinct root of the characteristic equation contributes as many independent solutions as its multiplicity, which might be as high as n.

Table 3: Contributions to general solution

Root	Multiplicity	Contribution
m (real)	1	ce^{mt}
m (real)	k	$(c_1 + c_2t + \cdots c_kt^{k-1})e^{mt}$
$m, \bar{m}$ (complex) $m = \alpha + i\beta$	1	$(a\cos(\beta t) + b\sin(\beta t))e^{\alpha t}$
$m, \bar{m}$ (complex)	k	$(a_1 + a_2t + \cdots + a_kt^{k-1})\cos(\beta t)e^{\alpha t} +$ $(b_1 + b_2t + \cdots + b_kt^{k-1})\sin(\beta t)e^{\alpha t}$

Recall also that the polynomial equation (28) may have complex roots, which will occur in conjugate pairs if — as we assume — the coefficients k_1, k_2, etc., are real. When this happens, we prefer to have real solutions, in the form of an exponential times sine or cosine, instead of complex exponentials. The contribution of each root or pair of conjugate roots of Eq. (28) is summarized in Table 3. Since the sum of the multiplicities of the roots of Eq. (28) is n, the sum of the contributions produces a solution with n terms, which can be shown to be the general solution.

As an example, let us find the general solution of this fourth-order equation,

$$u^{(4)} + 3u^{(2)} - 4u = 0.$$

The characteristic equation is $m^4 + 3m^2 - 4 = 0$, which is easy to solve because it is a biquadratic. We find that $m^2 = -4$ or 1, and thus the roots are $m = \pm 2i$, ± 1, all with multiplicity 1. From Table 3 we find that $a\cos(2t) + b\sin(2t)$ corresponds to the complex conjugate pair, $m = \pm 2i$, while e^t and e^{-t} correspond to $m = 1$ and $m = -1$. Thus we build up the general solution,

$$u(t) = a\cos(2t) + b\sin(2t) + c_1e^t + c_2e^{-t}.$$

As another example, we find the general solution of the fourth-order equation

$$u^{(4)} - 2u^{(2)} + u = 0.$$

The characteristic polynomial is $m^4 - 2m^2 + 1 = 0$, whose roots, found as in the preceding, are ± 1, both with multiplicity 2. From Table 3 we find that each of the roots contributes a first-degree polynomial times an exponential. Thus, we assemble the general solution as

$$u(t) = (c_1 + c_2t)e^t + (c_3 + c_4t)e^{-t}.$$

With $\sinh(t) = (e^t - e^{-t})/2$ and $\cosh(t) = (e^t + e^{-t})/2$, the terms of the combination above can be rearranged to give the general solution in a different form,

$$u(t) = (C_1 + C_2 t)\cosh(t) + (C_3 + C_4 t)\sinh(t).$$

Summary.
Some important equations and their solutions follow.

1. $\dfrac{du}{dt} = ku \quad (k \text{ is constant})$

$u(t) = ce^{kt}$

2. $\dfrac{d^2u}{dt^2} + \lambda^2 u = 0$

$u(t) = a\cos(\lambda t) + b\sin(\lambda t)$

3. $\dfrac{d^2u}{dt^2} - \lambda^2 u = 0$

$u(t) = a\cosh(\lambda t) + b\sinh(\lambda t) \quad \text{or} \quad u(t) = c_1 e^{\lambda t} + c_2 e^{-\lambda t}$

4. $t^2 u'' + tu' - \lambda^2 u = 0$

$u(t) = c_1 t^{\lambda} + c_2 t^{-\lambda}$

Exercises

In Exercises 1-6, find the general solution of the differential equation. Be careful to identify the dependent and independent variables.

1. $\dfrac{d^2\phi}{dx^2} + \lambda^2\phi = 0$

2. $\dfrac{d^2\phi}{dx^2} - \mu^2\phi = 0$

3. $\dfrac{d^2u}{dt^2} = 0$

4. $\dfrac{dT}{dt} = -\lambda^2 kT$

5. $\dfrac{1}{r}\dfrac{d}{dr}\left(r\dfrac{dw}{dr}\right) - \dfrac{\lambda^2}{r^2}w = 0$

6. $\rho^2\dfrac{d^2R}{d\rho^2} + 2\rho\dfrac{dR}{d\rho} - n(n+1)R = 0$

In Exercises 7-11, find the general solution. In some cases, it is helpful to carry out the indicated differentiation, in others it is not.

7. $\dfrac{d}{dx}\left((h+kx)\dfrac{dv}{dx}\right) = 0$ (h, k are constants) **8.** $(e^x\phi')' + \lambda^2 e^x \phi = 0$

9. $\dfrac{d}{dx}\left(x^3\dfrac{du}{dx}\right) = 0$ **10.** $r^2\dfrac{d^2u}{dr^2} + r\dfrac{du}{dr} + \lambda^2 u = 0$

11. $\dfrac{1}{r}\dfrac{d}{dr}\left(r\dfrac{du}{dr}\right) = 0$

12. Compare and contrast the form of the solutions of these three differential equations and their behavior as $t \to \infty$.

a. $\dfrac{d^2u}{dt^2} + u = 0$ **b.** $\dfrac{d^2u}{dt^2} = 0$ **c.** $\dfrac{d^2u}{dt^2} - u = 0$

In Exercises 13-15, use the "exponential guess" method to find the general solution of the differential equations (λ is constant).

13. $\dfrac{d^4u}{dx^4} + \lambda^4 u = 0$

14. $\dfrac{d^4u}{dx^4} - \lambda^4 u = 0$

15. $\dfrac{d^4u}{dx^4} + 2\lambda^2\dfrac{d^2u}{dx^2} + \lambda^4 u = 0$

In Exercises 16-18, one solution of the differential equation is given. Find a second independent solution.

16. $\dfrac{d^2u}{dt^2} + 2a\dfrac{du}{dt} + a^2u = 0, \qquad u_1(t) = e^{-at}$

17. $t^2\dfrac{d^2u}{dt^2} + (1-2b)t\dfrac{du}{dt} + b^2u = 0, \qquad u_1(t) = t^b$

18. $\dfrac{d}{dx}\left(x\dfrac{du}{dx}\right) + \dfrac{4x^2-1}{4x}u = 0, \qquad u_1(x) = \dfrac{\cos(x)}{\sqrt{x}}$

In Exercises 19-21, use the indicated change of variable to solve the differential equation.

19. $\dfrac{d}{d\rho}\left(\rho^2\dfrac{dR}{d\rho}\right) + \lambda^2\rho^2 R = 0, \qquad R(\rho) = u(\rho)/\rho$

20. $\dfrac{d}{d\rho}\left(\rho\dfrac{d\phi}{d\rho}\right)+\dfrac{4\lambda^2\rho^2-1}{4\rho}\phi=0, \qquad \phi(\rho)=v(\rho)/\sqrt{\rho}$

21. $t^2\dfrac{d^2u}{dt^2}+kt\dfrac{du}{dt}+pu=0, \qquad x=\ln t, \qquad u(t)=v(x)$

22. The displacement $u(t)$ of a mass in a mass-spring-damper system (Fig. 1) is described by the initial value problem

$$\frac{d^2u}{dt^2}+b\frac{du}{dt}+\omega^2u=0,$$

$$u(0)=u_0, \qquad \frac{du}{dt}(0)=v_0.$$

(The coefficients b and ω^2 are proportional to the characteristic constants of the damper and the spring, respectively.) Solve the initial value problem for each of the parameter ranges given here, and explain why these ranges might have been chosen:

(i) $b=0$, (ii) $0<b<2\omega$, (iii) $b=2\omega$, (iv) $b>2\omega$.

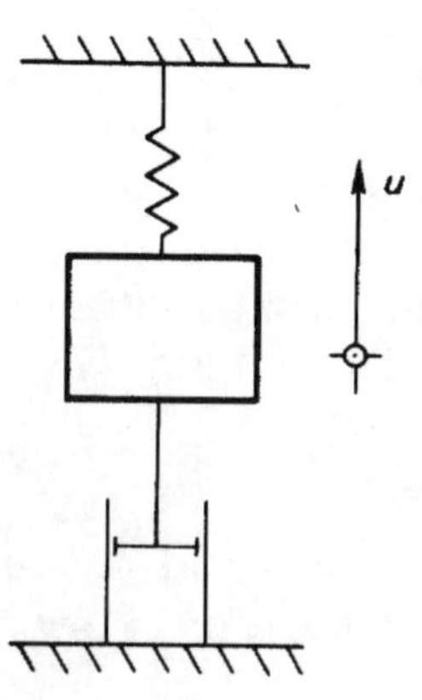

Figure 1: Mass-spring-damper system.

23. Suppose that $u(t)$ is a function, not identically 0, for which

$$\frac{u''}{u}=\text{constant}>0.$$

Show that this relation is a differential equation and solve it. (Call the constant p^2.) Prove that exactly one of the following three possibilities holds:

(i) $u(t) = 0$ for one value of t and $u'(t)$ is never 0;

(ii) $u'(t) = 0$ for one value of t and $u(t)$ is never 0;

(iii) neither $u(t)$ nor $u'(t)$ is ever 0.

0.2 NONHOMOGENEOUS LINEAR EQUATIONS

In this section, we will review methods for solving nonhomogeneous linear equations of first and second orders,

$$\frac{du}{dt} = k(t)u + f(t)$$

$$\frac{d^2u}{dt^2} + k(t)\frac{du}{dt} + p(t)u = f(t).$$

Of course, we assume that the inhomogeneity $f(t)$ is not identically 0. The simplest nonhomogeneous equation is

$$\frac{du}{dt} = f(t). \tag{1}$$

This can be solved in complete generality by one integration:

$$u(t) = \int f(t)dt + c. \tag{2}$$

We have used an indefinite integral and have written c as a reminder that there is an arbitrary additive constant in the general solution of Eq. (1). A more precise way to write the solution is

$$u(t) = \int_{t_0}^{t} f(z)dz + c. \tag{3}$$

Here we have replaced the indefinite integral by a definite integral with variable upper limit. The lower limit of integration is usually an initial time. Note that the name of the integration variable is changed from t to something else (here, z) to avoid confusing the limit with the dummy variable of integration. The simple second-order equation

$$\frac{d^2u}{dt^2} = f(t) \tag{4}$$

can be solved by two successive integrations.

The two theorems that follow summarize some properties of linear equations that are useful in constructing solutions.

Theorem 1

The general solution of a nonhomogeneous linear equation has the form $u(t) = u_p(t) + u_c(t)$, $u_p(t)$ is any particular solution of the nonhomogeneous equation and $u_c(t)$ is the general solution of the corresponding homogeneous equation.

Theorem 2

If $u_{p1}(t)$ and $u_{p2}(t)$ are particular solutions of a differential equation with inhomogeneities $f_1(t)$ and $f_2(t)$, respectively, then $k_1 u_{p1}(t) + k_2 u_{p2}$ is a particular solution of the differential equation with inhomogeneity $k_1 f_1(t)$ $+k_2 f_2(t)$ (k_1, k_2 are constants).

To illustrate the application of these theorems, we will put together the solution of the differential equation

$$\frac{d^2u}{dt^2} + u = 1 - e^{-t}.$$

The corresponding homogeneous equation is

$$\frac{d^2u}{dt^2} + u = 0,$$

with general solution $u_c(t) = c_1 \cos(t) + c_2 \sin(t)$ (found in Section 1). A particular solution of the equation with the inhomogeneity $f_1(t) = 1$, that is, of the equation

$$\frac{d^2u}{dt^2} + u = 1,$$

is $u_{p1}(t) = 1$. A particular solution of the equation

$$\frac{d^2u}{dt^2} + u = e^{-t}$$

is $u_{p2}(t) = \frac{1}{2}e^{-t}$. (Later in this section, we will review methods for constructing these particular solutions.) Then, by Theorem 2, a particular solution of the given nonhomogeneous eq. (5) is $u_p(t) = 1 - \frac{1}{2}e^{-t}$. Finally, by Theorem 1, the general solution of the given equation is

$$u(t) = 1 - \frac{1}{2}e^{-t} + c_1 \cos(t) + c_2 \sin(t).$$

If two initial conditions are given, then c_1 and c_2 are available to satisfy them. Of course, an initial condition applies to the entire solution of the given differential equation, not just to $u_c(t)$.

Now we turn our attention to methods for finding particular solutions of nonhomogeneous linear differential equations.

Table 4: Undetermined coefficients

Inhomogeneity, $f(t)$	Form of Trial Solution, $u_p(t)$
$(a_0 t^n + a_1 t^{n-1} + \cdots + a_n)e^{\alpha t}$	$(A_0 t^n + A_1 t^{n-1} + \cdots + A_n)e^{\alpha t}$
$(a_0 t^n + \cdots + a_n)e^{\alpha t}\cos(\beta t) +$ $(b_0 t^n + \cdots + b_n)e^{\alpha t}\sin(\beta t)$	$(A_0 t^n + \cdots + A_n)e^{\alpha t}\cos(\beta t) +$ $(B_0 t^n + \cdots + B_n)e^{\alpha t}\sin(\beta t)$

A. Undetermined Coefficients

This method involves guessing the form of a trial solution and then finding the appropriate coefficients. Naturally, it is limited to the cases in which we can guess successfully: when the equation has constant coefficients and the inhomogeneity is simple in form. Table 4 offers a summary of admissible inhomogeneities and the corresponding forms for particular solution. The Table compresses several cases. For instance, $f(t)$ in line 1 is a polynomial if $\alpha = 0$, or an exponential if $n = 0$ and $\alpha \neq 0$. In line 2, both sine and cosine must be included in the trial solution even if one is absent from $f(t)$; but $\alpha = 0$ is allowed, and so is $n = 0$.

Example.

To find a particular solution of

$$\frac{d^2u}{dt^2} + 5u = te^{-t},$$

we use line 1 of Table 4. Evidently, $n = 1$ and $\alpha = -1$. The appropriate form for the trial solution is

$$u_p(t) = (A_0 t + A_1)e^{-t}.$$

When we substitute this form into the differential equation, we obtain

$$(A_0 t + A_1 - 2A_0)e^{-t} + 5(A_0 t + A_1)e^{-t} = te^{-t}.$$

Now, equating coefficients of like terms gives these two equations for the coefficients

$$6A_0 = 1 \quad \text{(coefficient of } te^{-t}\text{)}$$

$$6A_1 - 2A_0 = 0 \quad \text{(coefficient of } e^{-t}\text{)}.$$

These we solve easily to find $A_0 = 1/6$, $A_1 = 1/18$. Finally, a particular solution is

$$u_p(t) = \left(\frac{1}{6}t + \frac{1}{18}\right)e^{-t}.$$

A trial solution from Table 4 will not work if it contains any term that is a solution of the homogeneous differential equation. In that case, the trial solution has to be revised by this rule: *Multiply by the lowest positive integral power of t such that no term in the trial solution satisfies the corresponding homogeneous equation.*

Example.
Table 4 suggests the trial solution $u_p(t) = (A_0t+A_1)e^{-t}$ for the differential equation

$$\frac{d^2u}{dt^2} - u = te^{-t}.$$

However, we know that the solution of the corresponding homogeneous equation, $u'' - u = 0$, is

$$u_c(t) = c_1e^t + c_2e^{-t}.$$

The trial solution contains a term (A_1e^{-t}) that is a solution of the homogeneous equation. Multiplying the trial solution by t eliminates the problem. Thus, the trial solution is

$$u_p(t) = t(A_0t + A_1)e^{-t} = (A_0t^2 + A_1t)e^{-t}.$$

Similarly, the trial solution for the differential equation

$$\frac{d^2u}{dt^2} + 2\frac{du}{dt} + u = te^{-t}$$

has to be revised. The solution of the corresponding homogeneous equation is $u_c(t) = c_1e^{-t} + c_2te^{-t}$. The trial solution from the table has to be multiplied by t^2 to eliminate solutions of the homogeneous equation.

B. Variation of Parameters
Generally, if a linear homogeneous differential equation can be solved, the corresponding nonhomogeneous equation can also be solved, at least in terms of integrals.

1. First-order equations.
Suppose that $u_c(t)$ is a solution of the homogeneous equation

$$\frac{du}{dt} = k(t)u. \tag{5}$$

Then to find a particular solution of the nonhomogeneous equation

$$\frac{du}{dt} = k(t)u + f(t), \tag{6}$$

we assume that $u_p(t) = v(t)u_c(t)$. Substituting u_p in this form into the differential equation (6) we have

$$\frac{dv}{dt}u_c + v\frac{du_c}{dt} = k(t)vu_c + f(t). \tag{7}$$

However, $u_c' = k(t)u_c$, so one term on the left cancels a term on the right, leaving

$$\frac{dv}{dt}u_c = f(t), \quad \text{or} \quad \frac{dv}{dt} = \frac{f(t)}{u_c(t)}. \tag{8}$$

The latter is a nonhomogeneous equation of simplest type, which can be solved for $v(t)$ in one integration.

Example.
By this method, we would seek a solution of the homogeneous equation

$$\frac{du}{dt} = 5u + t$$

in the form $u_p(t) = v(t) \cdot e^{5t}$, as e^{5t} is a solution of $u' = 5u$. Substituting the preceding form for u_p, we find

$$\frac{dv}{dt} \cdot e^{5t} + v \cdot 5e^{5t} = 5ve^{5t} + t,$$

or, after canceling $5ve^{5t}$ from both sides and simplifying, we find

$$\frac{dv}{dt} = e^{-5t}t.$$

This equation is integrated once (by parts) to find

$$v(t) = \left(-\frac{t}{5} - \frac{1}{25}\right)e^{-5t}.$$

From here, we obtain $u_p(t) = v(t) \cdot e^{5t} = -\left(\frac{1}{5}t + \frac{1}{25}\right)$.

2. Second-order equations.
To find a particular solution of the nonhomogeneous second-order equation

$$\frac{d^2u}{dt^2} + k(t)\frac{du}{dt} + p(t)u = f(t), \tag{9}$$

we need two independent solutions, $u_1(t)$ and $u_2(t)$, of the corresponding homogeneous equation

$$\frac{d^2u}{dt^2} + k(t)\frac{du}{dt} + p(t)u = 0. \tag{10}$$

Then we assume that our particular solution has the form

$$u_p(t) = v_1(t)u_1(t) + v_2(t)u_2(t), \tag{11}$$

where v_1 and v_2 are functions to be found. If we simply insert u_p in this form into Eq. (9), we obtain one complicated second-order equation in two unknown functions. However, if we impose the extra requirement that

$$\frac{dv_1}{dt}u_1 + \frac{dv_2}{dt}u_2 = 0, \tag{12}$$

then we find that

$$u_p' = v_1'u_1 + v_2'u_2 + v_1u_1' + v_2u_2' = v_1u_1' + v_2u_2' \tag{13}$$

$$u_p'' = v_1'u_1' + v_2'u_2' + v_1u_1'' + v_2u_2'', \tag{14}$$

and the equation that results from substituting Eq. (11) into Eq. (9) becomes

$$v_1'u_1' + v_2'u_2' + v_1\left(u_1'' + k(t)u_1' + p(t)u_1\right) + v_2\left(u_2'' + k(t)u_2' + p(t)u_2\right) = f(t).$$

This simplifies further: the multipliers of v_1 and v_2 are both 0, because u_1 and u_2 satisfy the homogeneous Eq. (10).

Thus, we are left with a pair of simultaneous equations

$$v_1'u_1 + v_2'u_2 = 0 \tag{12'}$$

$$v_1'u_1' + v_2'u_2' = f(t) \tag{15}$$

in the unknowns v_1' and v_2'. The determinant of this system is

$$\begin{vmatrix} u_1 & u_2 \\ u_1' & u_2' \end{vmatrix} = W(t), \tag{16}$$

the Wronskian of u_1 and u_2. Since these were to be independent solutions of Eq. (10), their Wronskian is nonzero, and we may solve for $v_1'(t)$, and $v_2'(t)$, and hence for v_1 and v_2.

Example.
To solve the nonhomogeneous equation

$$\frac{d^2u}{dt^2} + u = \cos(\omega t),$$

we assume a solution in the form

$$u_p(t) = v_1 \cos(t) + v_2 \sin(t),$$

because $\sin(t)$ and $\cos(t)$ are independent solutions of the corresponding homogeneous equation $u'' + u = 0$. The assumption of Eq. (12) is

$$v_1' \cos(t) + v_2' \sin(t) = 0. \tag{17}$$

Then our equation reduces to the following, corresponding to Eq. (15):

$$-v_1' \sin(t) + v_2' \cos(t) = \cos(\omega t). \tag{18}$$

Now we solve Eqs. (17) and (18) simultaneously to find

$$v_1' = -\sin(t)\cos(\omega t), \qquad v_2' = \cos(t)\cos(\omega t). \tag{19}$$

These equations are to be integrated to find v_1 and v_2, and then $u_p(t)$.

Finally, we note that $v_1(t)$ and $v_2(t)$ can be found from Eqs. (12) and (15) in general:

$$v_1' = -\frac{u_2 f}{W}, \qquad v_2' = \frac{u_1 f}{W}. \tag{20}$$

Integrating these two equations, we find that

$$v_1(t) = -\int \frac{u_2(t)f(t)}{W(t)}dt, \qquad v_2(t) = \int \frac{u_1(t)f(t)}{W(t)}. \tag{21}$$

Now, Eq. (11) may be used to form a particular solution of the nonhomogeneous equation (9).

We may also obtain v_1 and v_2 by using definite integrals with variable upper limit:

$$v_1(t) = -\int_{t_0}^{t} \frac{u_2(z)f(z)}{W(z)}dz, \qquad v_2(t) = \int_{t_0}^{t} \frac{u_1(z)f(z)}{W(z)}dz. \tag{22}$$

The lower limit is usually the initial value of t, but may be any convenient value. The particular solution can now be written as

$$u_p(t) = -u_1(t)\int_{t_0}^{t} \frac{u_2(z)f(z)}{W(z)}dz + u_2(t)\int_{t_0}^{t} \frac{u_1(z)f(z)}{W(z)}dz.$$

Furthermore, the factors $u_1(t)$ and $u_2(t)$ can be inside the integrals (which are *not* with respect to t) and these can be combined to give a tidy formula, as follows.

Theorem 3
Let $u_1(t)$ and $u_2(t)$ be independent solutions of

$$\frac{d^2u}{dt^2} + k(t)\frac{du}{dt} + p(t)u = 0 \qquad (H)$$

with Wronskian $W(t) = u_1(t)u_2'(t) - u_2(t)u_1'(t)$. Then

$$u_p(t) = \int_{t_0}^{t} G(t,z)f(z)dz$$

is a particular solution of the nonhomogeneous equation

$$\frac{d^2u}{dt^2} + k(t)\frac{du}{dt} + p(t)u = f(t), \qquad (NH)$$

where G is the Green's function defined by

$$G(t,z) = \frac{u_1(z)u_2(t) - u_2(z)u_1(t)}{W(z)}. \qquad (23)$$

Exercises
In Exercises 1-10, find the general solution of the differential equation.

1. $\frac{du}{dt} + a(u - T) = 0$

2. $\frac{du}{dt} + au = e^{at}$

3. $\frac{du}{dt} + au = e^{-at}$

4. $\frac{d^2u}{dt^2} + u = \cos(\omega t) \quad (\omega \neq 1)$

5. $\frac{d^2u}{dt^2} + u = \cos(t)$

6. $\frac{d^2u}{dx^2} - \gamma^2(u - U) = 0$
(U, γ^2 are constants)

7. $\frac{d^2u}{dt^2} + 3\frac{du}{dt} + 2u = \cosh(t)$

8. $\frac{1}{r}\frac{d}{dr}\left(r\frac{du}{dr}\right) = -1$

9. $\frac{1}{\rho^2}\frac{d}{d\rho}\left(\rho^2\frac{du}{d\rho}\right) = -1$

10. $\frac{d^2u}{dt^2} = -1$

11. Let $h(t)$ be the height of a parachutist above the surface of the earth. Consideration of forces on his body leads to the initial value problem for h:

$$M\frac{d^2h}{dt^2} + K\frac{dh}{dt} = -Mg$$

$$h(0) = h_0, \quad \frac{dh}{dt}(0) = 0.$$

(M = mass, g = acceleration of gravity, K = parachute constant) Solve the problem, taking $g = 32$ ft/s^2 and $K/M = 0.1$/s.

12. The displacement of $u(t)$ of a mass in a mass-spring-damper system with an external force (Fig. 2) is described by this initial value problem

$$\frac{d^2u}{dt^2} + b\frac{du}{dt} + \omega^2 u = f_0 \cos(\mu t),$$

$$u(0) = 0, \quad \frac{du}{dt}(0) = 0.$$

(See Exercise 22, Section 1. The coefficient f_0 is proportional to the amplitude of the external force.) Solve the problem for these three cases: (i) $b = 0$, $\mu \neq \omega$; (ii) $b = 0$, $\mu = \omega$; (iii) $b > 0$.

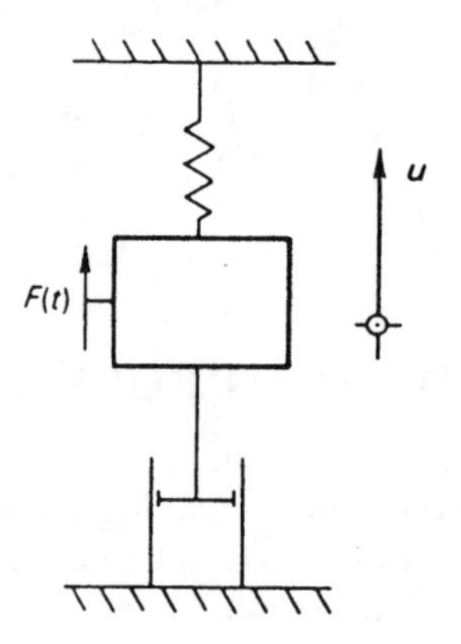

Figure 2: Mass-spring-damper system with an external force.

In Exercises 13-19, use variation of parameters to find a particular solution of the differential equation. Be sure that the differential equation is in the correct form.

13. $\dfrac{du}{dt} + au = e^{-at}; \quad u_c(t) = e^{-at}$

14. $t\dfrac{du}{dt} = -1; \quad u_c(t) = 1$

15. $\dfrac{d^2y}{dx^2} + y = \tan(x); \quad y_1(x) = \cos(x), \quad y_2(x) = \sin(x)$

16. $\dfrac{d^2y}{dx^2} + y = \sin(x); \quad y_1(x) = \cos(x), \quad y_2(x) = \sin(x)$

17. $\dfrac{d^2u}{dt^2} = -1; \quad u_1(t) = 1, u_2(t) = t$

18. $\dfrac{1}{r}\dfrac{d}{dr}\left(r\dfrac{du}{dr}\right) = -1; \quad u_1(r) = 1, u_2(r) = \ln(r)$

19. $t^2\dfrac{d^2u}{dt^2} + t\dfrac{du}{dt} - u = 1; \quad u_1(t) = t, u_2(t) = 1/t$

In Exercises 20-22, use Theorem 3 to develop the formula shown for a particular solution of the differential equation.

20. $\dfrac{d^2u}{dt^2} + \gamma^2 u = f(t), \qquad u_p(t) = \dfrac{1}{\gamma}\displaystyle\int_0^t \sin(\gamma(t-z))f(z)dz$

21. $\dfrac{du}{dt} + au = f(t), \qquad u_p(t) = \displaystyle\int_0^t e^{a(t-z)}f(z)dz$

22. $\dfrac{d^2u}{dt^2} - \gamma^2 u = f(t), \qquad u_p(t) = \dfrac{1}{\gamma}\displaystyle\int_0^t \sinh(\gamma(t-z))f(z)dz$

0.3 BOUNDARY VALUE PROBLEMS

A *boundary value problem* in one dimension is an ordinary differential equation together with conditions involving values of the solution and/or its derivatives at two or more points. The number of conditions imposed is equal to the order of the differential equation. Usually, boundary value problems of any physical relevance have these characteristics: (1) the conditions are imposed at two different points; (2) the solution is of interest only between those two points; and (3) the independent variable is a space variable, which we shall represent as x. In addition, we are primarily concerned with cases where the differential equation is linear and of second order. However, problems in elasticity often involve fourth-order equations.

In contrast to initial value problems, even the most innocent looking boundary value problem may have exactly one solution, no solution, or an infinite number of solutions. Exercise 1 illustrates these cases.

When the differential equation in a boundary value problem has a known general solution, we use the two boundary conditions to supply two equations that are to be satisfied by the two constants in the general solution. If the differential equation is linear, these are two linear equations and can be easily solved, if there is a solution.

In the rest of this section we examine some physical examples that are naturally associated with boundary value problems.

Example — Hanging Cable.
First we consider the problem of finding the shape of a cable that is fastened at each end and carries a distributed load. The cables of a suspension bridge provide an important example. Let $u(x)$ denote the position of the centerline of the cable, measured upward from the x-axis, which we assume to be horizontal. (See Fig. 3.) Our objective is to find the function $u(x)$.

The shape of the cable is determined by the forces acting on it. In our analysis, we consider the forces that hold a small segment of the cable in place. (See Fig. 4.) The key assumption is that the cable is perfectly flexible. This means that force inside the cable is always a tension and its direction at every point is the direction tangent to the centerline.

We suppose that the cable is not moving. Then by Newton's second law, the sum of the horizontal components of the forces on the segment is 0, and likewise for the vertical components. If $T(x)$ and $T(x+\Delta x)$ are the magnitudes of the tensions at the ends on the segment, we have these two equations:

$$T(x+\Delta x)\cos(\phi(x+\Delta x)) - T(x)\cos(\phi(x)) = 0 \quad \text{(Horizontal)} \qquad (1)$$

$$T(x+\Delta x)\sin(\phi(x+\Delta x)) - T(x)\sin(\phi(x)) - f(x)\Delta x = 0 \quad \text{(Vertical)} \quad (2)$$

In the second equation, $f(x)$ is the intensity of the distributed load, measured in force per unit of horizontal length, so $f(x)\Delta x$ is the load borne by the small segment.

From Eq. (1) we see that the horizontal component of the tension is the same at both ends of the segment. In fact, the horizontal component of tension has the same value — call it T — at every point, including the endpoints where the cable is attached to solid supports. By simple algebra

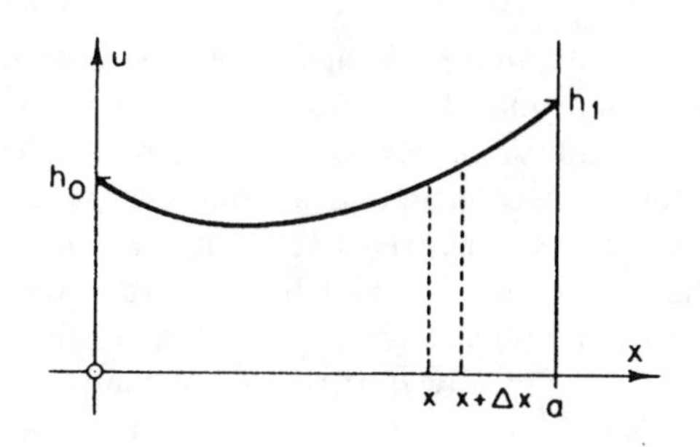

Figure 3: The hanging cable.

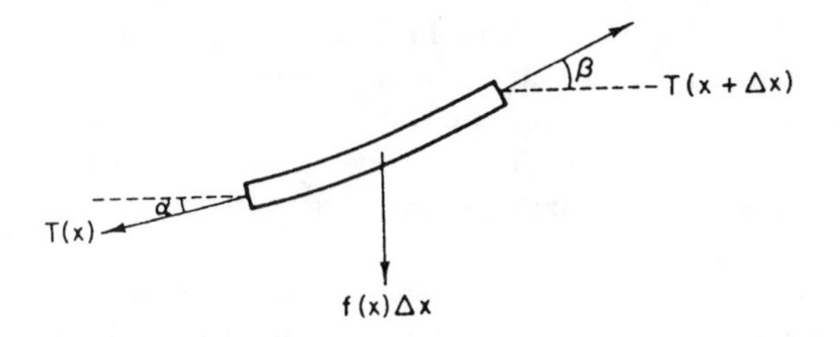

Figure 4: Section of cable showing forces acting on it.

we can now find the tension in the cable at the ends of our segment,

$$T(x + \Delta x) = \frac{T}{\cos(\phi(x + \Delta x))}, \qquad T(x) = \frac{T}{\cos(\phi(x))},$$

and substitute these into Eq. (2), which becomes

$$\frac{T}{\cos(\phi(x + \Delta x))} \sin(\phi(x + \Delta x)) - \frac{T}{\cos(\phi(x))} \sin(\phi(x)) - f(x)\Delta x = 0$$

or

$$T(\tan(\phi(x + \Delta x)) - \tan(\phi(x))) - f(x)\Delta x = 0.$$

Before going further we should note (Fig. 4) that $\phi(x)$ measures the angle between the tangent to the centerline of the cable and the horizontal. As the position of the centerline is given by $u(x)$, $\tan(\phi(x))$ is just the slope of the cable at x. From elementary calculus we know

$$\tan(\phi(x)) = \frac{du}{dx}(x).$$

Substituting the derivative for the slope and making some algebraic adjustments, we obtain

$$T(u'(x+\Delta x) - u'(x)) = f(x)\Delta x.$$

Dividing through by Δx yields

$$T\frac{u'(x+\Delta x) - u'(x)}{\Delta x} = f(x).$$

In the limit, as Δx approaches 0, the difference quotient in the left member becomes the second derivative of u, and the result is the equation

$$T\frac{d^2u}{dx^2} = f(x), \tag{3}$$

which is valid for x in the range $0 < x < a$, where the cable is located. In addition, $u(x)$ must satisfy the boundary conditions

$$u(0) = h_0, \qquad u(a) = h_1. \tag{4}$$

Regarding the load $f(x)$, we might first assume that the cable is hanging under its own weight of w units of weight per unit length of cable. Then in Eq. (2), we should put

$$f(x)\Delta x = w\frac{\Delta s}{\Delta x}\Delta x$$

where s represents arclength along the cable. In the limit, as Δx approaches 0, $\Delta s/\Delta x$ has the limit

$$\lim_{\Delta x \to 0} \frac{\Delta s}{\Delta x} = \sqrt{1 + \left(\frac{du}{dx}\right)^2}.$$

Therefore, with this assumption, the boundary value problem that determines the shape of the cable is

$$\frac{d^2u}{dx^2} = \frac{w}{T}\sqrt{1 + \left(\frac{du}{dx}\right)^2}, \qquad 0 < x < a \tag{5}$$

$$u(0) = h_0, \qquad u(a) = h_1. \tag{6}$$

Notice that the differential equation is nonlinear. Nevertheless, we can find its general solution in closed form and satisfy the boundary conditions by appropriate choice of the arbitrary constants that appear. (See Exercises 4 and 5.)

Another case arises when the cable supports a load uniformly distributed in the horizontal direction, as given by

$$f(x)\Delta x = w\Delta x.$$

This is approximately true for a suspension bridge. The boundary value problem to be solved is then

$$\frac{d^2u}{dx^2} = \frac{w}{T}, \qquad 0 < x < a$$

$$u(0) = h_0, \qquad u(a) = h_1. \tag{7}$$

Then general solution of the differential equation (7) can be found by the procedures of Sections 1 and 2. It is

$$u(x) = \left(\frac{w}{2T}\right)x^2 + c_1x + c_2,$$

where c_1 and c_2 are arbitrary. The two boundary conditions require

$$u(0) = h_0 : \qquad c_2 = h_0$$

$$u(a) = h_1 : \qquad \left(\frac{w}{2T}\right)a^2 + c_1a + c_2 = h_1.$$

These two are solved for c_1 and c_2 in terms of given parameters. The result, after some beautifying algebra, is

$$u(x) = \frac{w}{2T}\left(x^2 - ax\right) + \frac{h_1 - h_0}{a}x + h_0. \tag{8}$$

Clearly, this function specifies the cable's shape as part of a parabola opening upward.

Example — Heat Conduction in a Rod.
A long rod of uniform material and cross section conducts heat along its axial direction (see Fig. 5). We assume that the temperature in the rod, $u(x)$, does not change in time. A heat balance ("what goes in must come out") applied to a slice of the rod between x and $x + \Delta x$ (Fig. 6) shows that the heat flow rate q, measured in units of heat per unit time per unit area, obeys the equation

$$q(x)A + g(x)A\Delta x = q(x + \Delta x)A \tag{9}$$

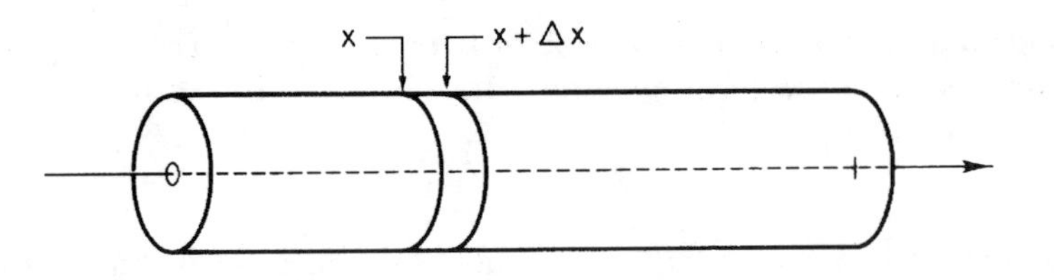

Figure 5: Cylinder of heat-conducting material.

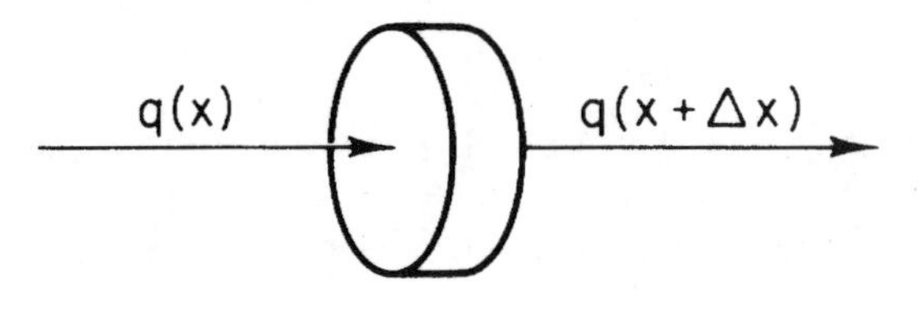

Figure 6: Section cut from heat-conducting cylinder showing heat flow.

in which A is the cross-sectional area and g is the rate at which heat enters the slice by means other than conduction through the two faces. For instance, if heat is generated in the slice by an electric current I, we might have

$$g(x)A\Delta x = I^2 R\Delta x \tag{10}$$

where R is the resistance of the rod per unit length. If heat is lost through the cylindrical surface of the rod by convection to a surrounding medium at temperature T, then $g(x)$ would be given by "Newton's law of cooling,"

$$g(x)A\Delta x = -h(u(x) - T)C\Delta x, \tag{11}$$

where C is the circumference of the rod and h is the heat transfer coefficient. (This minus sign appears because, if $u(x) > T$, heat actually leaves the rod.)

Equation (9) may be altered algebraically to read

$$\frac{q(x + \Delta x) - q(x)}{\Delta x} = g(x),$$

and application of the limiting process leaves

$$\frac{dq}{dx} = g(x). \tag{12}$$

The unknown function $u(x)$ does not appear in Eq. (12). However, a well-known experimental law (Fourier's law) says that the heat flow rate through a unit area of material is directly proportional to the temperature difference and inversely proportional to thickness. In the limit, this law takes the form

$$q = -\kappa \frac{du}{dx}. \tag{13}$$

The minus sign expresses the fact that heat moves from hotter toward cooler regions.

Combining Eqs. (12) and (13) gives the differential equation

$$-\kappa \frac{d^2u}{dx^2} = g(x), \qquad 0 < x < a, \tag{14}$$

where a is the length of the rod and the conductivity κ is assumed to be constant.

If the two ends of the rod are held at constant temperature, the boundary conditions on u would be

$$u(0) = T_0, \qquad u(a) = T_1. \tag{15}$$

On the other hand, if heat were supplied at $x = 0$ (by a heating coil, for instance), the boundary condition there would be

$$-\kappa A \frac{du}{dx}(0) = H, \tag{16}$$

where H is measured in units of heat per unit time.

As an example, we solve the problem

$$-\kappa \frac{d^2u}{dx^2} = -hu(x)\frac{C}{A}, \qquad 0 < x < a \tag{17}$$

$$u(0) = T_0, \qquad u(a) = T_0. \tag{18}$$

(Physically, the rod is losing heat to a surrounding medium at temperature 0, while both ends are held at the same temperature T_0.) If we designate $\mu^2 = hC/\kappa A$, the differential equation becomes

$$\frac{d^2u}{dx^2} - \mu^2 u = 0, \qquad 0 < x < a,$$

with general solution

$$u(x) = c_1 \cosh(\mu x) + c_2 \sinh(\mu x).$$

Application of the boundary condition at $x = 0$ gives $c_1 = T_0$; the second boundary condition requires that

$$u(a) = T_0 : T_0 = T_0 \cosh(\mu a) + c_2 \sinh(\mu a).$$

Thus $c_2 = T_0(1 - \cosh(\mu a))/\sinh(\mu a)$ and

$$u(x) = T_0\left(\cosh(\mu x) + \frac{1 - \cosh(\mu a)}{\sinh(\mu a)}\sinh(\mu x)\right).$$

It should be clear now that solving a boundary value problem is not substantially different from solving an initial value problem. The procedure is (1) find the general solution of the differential equation, which must contain some arbitrary constants, and (2) apply the boundary conditions to determine values for the arbitrary constants. In our examples the differential equations have been of second order, causing the appearance of two arbitrary constants, which are to be determined by the boundary conditions.

The next example is somewhat different in spirit from the others. Instead of just finding the solution of a boundary value problem, we will be looking for parameter values that permit the existence of solutions of special form.

Example — Buckling of a Column.
A long, slender column whose bottom end is hinged carries an axial load as shown in Fig. 7. The upper end of the column can move up or down but not sideways. The displacement of the column's centerline from a vertical reference line is given by $u(x)$. If the column were cut at any point x, an upward force P and a clockwise moment $Pu(x)$ would have to be applied to the upper part to keep it in equilibrium (see Fig. 8). This force and moment must be supplied by the lower part of the column.

It is known that the internal bending moment (positive when counterclockwise) in a column is given by the product

$$EI\frac{d^2u}{dx^2}$$

where E is Young's modulus and I is the moment of inertia of the cross-sectional area. (The moment $I = b^4/12$ for a column whose cross section is a square of side b.) Thus equating the external moment to the internal moment gives the differential equation

$$EI\frac{d^2u}{dx^2} = -Pu, \qquad 0 < x < a \tag{19}$$

which, together with the boundary conditions

$$u(0) = 0, \qquad u(a) = 0, \tag{20}$$

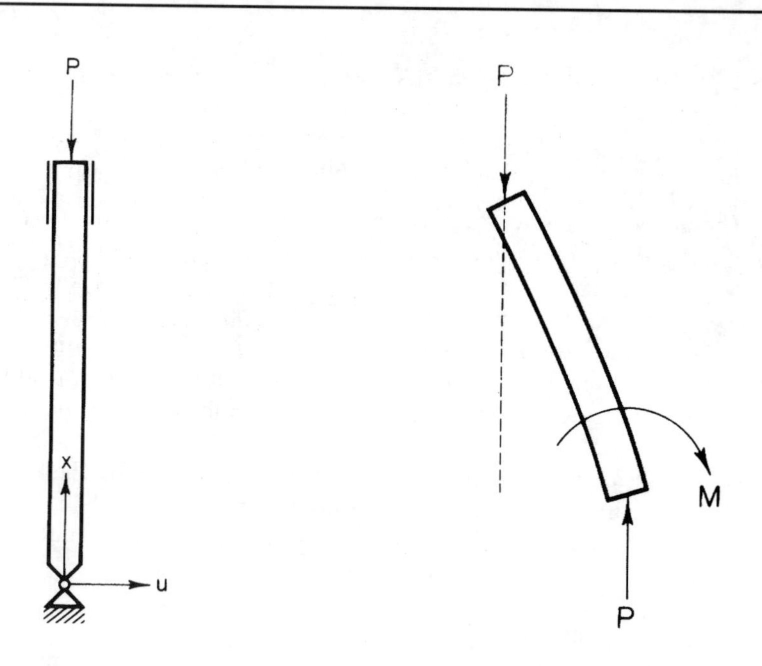

Figure 7: Column carrying load P.

Figure 8: Section of column showing forces and moments.

determines the function $u(x)$.

In order to study this problem more conveniently, we set

$$\frac{P}{EI} = \lambda^2,$$

so that the differential equation becomes

$$\frac{d^2u}{dx^2} + \lambda^2 u = 0, \qquad 0 < x < a. \tag{21}$$

Now, the general solution of this differential equation is

$$u(x) = c_1 \cos(\lambda x) + c_2 \sin(\lambda x).$$

As $u(0) = 0$, we must choose $c_1 = 0$, leaving $u(x) = c_2 \sin(\lambda x)$. The second boundary condition requires that

$$u(a) = 0 : c_2 \sin(\lambda a) = 0.$$

If $\sin(\lambda a)$ is not 0, the only possibility is that $c_2 = 0$. In this case we find that the solution is

$$u(x) \equiv 0, \qquad 0 < x < a.$$

Physically, this means that the column stands straight and transmits the load to its support, as it was probably intended to do.

Something quite different happens if $\sin(\lambda a) = 0$, for then any choice of c_2 gives a solution. The physical manifestation of this case is that the column assumes a sinusoidal shape and may then collapse, or buckle, under the axial load. Mathematically, the condition $\sin(\lambda a) = 0$ means that λa is an integer multiple of π, since $\sin(\pi) = 0$, $\sin(2\pi) = 0$, etc., and integer multiples of π are the only arguments for which the sine function is 0. The equation $\lambda a = \pi$, in terms of the original parameters, is

$$\sqrt{\frac{P}{EI}}a = \pi.$$

It is reasonable to think of E, I, and a as given quantities; thus it is the force

$$P = EI\left(\frac{\pi}{a}\right)^2,$$

called the critical or Euler load, that causes the buckling. The higher critical loads, corresponding to $\lambda a = 2\pi$, $\lambda a = 3\pi$, etc., are so unstable as to be of no physical interest in this problem.

The buckling example is one instance of an *eigenvalue* problem. The general setting is a homogeneous differential equation containing a parameter λ and accompanied by homogeneous boundary conditions. Because both differential equations and boundary conditions are homogeneous, the constant function 0 is always a solution. The question to be answered is: What values of the parameter λ allow the existence of nonzero solutions? Eigenvalue problems often are employed to find the dividing line between stable and unstable behavior. We will see them frequently in later chapters.

Exercises

1. Of these three boundary value problems, one has no solution, one has exactly one solution, and one has an infinite number of solutions. Which is which?

a. $\dfrac{d^2u}{dx^2} + u = 0, \qquad u(0) = 0, \qquad u(\pi) = 0$

b. $\dfrac{d^2u}{dx^2} + u = 1, \qquad u(0) = 0, \qquad u(1) = 0$

c. $\dfrac{d^2u}{dx^2} + u = 0, \qquad u(0) = 0, \qquad u(\pi) = 1$

2. Find the Euler buckling load of a steel column with a 2 in. × 3 in. rectangular cross section. The parameters are $E = 30 \times 10^6$ lb/in^2, $I = 2$ in^4, $a = 10$ ft.

3. Find all values of the parameter λ for which these homogeneous boundary value problems have a solution other than $u(x) \equiv 0$.

a. $\dfrac{d^2u}{dx^2} + \lambda^2 u = 0, \qquad u(0) = 0, \qquad \dfrac{du}{dx}(a) = 0$

b. $\dfrac{d^2u}{dx^2} + \lambda^2 u = 0, \qquad \dfrac{du}{dx}(0) = 0, \qquad u(a) = 0$

c. $\dfrac{d^2u}{dx^2} + \lambda^2 u = 0, \qquad \dfrac{du}{dx}(0) = 0, \qquad \dfrac{du}{dx}(a) = 0$

4. Verify, by differentiating and substituting, that

$$u(x) = c' + \frac{1}{\mu}\cosh(\mu(x + c))$$

is the general solution of the differential equation (5). (Here $\mu = w/T$. The graph of $u(x)$ is called a *catenary*.)

5. Find the values of c and c' for which the function $u(x)$ in Exercise 4 satisfies the conditions

$$u(0) = h, \qquad u(a) = h.$$

6. A beam that is simply supported at its ends carries a distributed lateral load of uniform intensity w (force/length) and an axial tension load T (force/length^2). The displacement $u(x)$ of its centerline (positive down) satisfies the boundary value problem here. Find $u(x)$.

$$\frac{d^2u}{dx^2} - \frac{T}{EI}u = -\frac{w}{EI}\frac{Lx - x^2}{2}, \qquad 0 < x < L$$

$$u(0) = 0, \qquad u(L) = 0$$

7. The temperature $u(x)$ in a cooling fin satisfies the differential equation

$$\frac{d^2u}{dx^2} = \frac{hC}{\kappa A}(u - T), \qquad 0 < x < a$$

and boundary conditions

$$u(0) = T_0, \qquad -\kappa\frac{du}{dx}(a) = h(u(a) - T).$$

That is, the temperature at the left end is held at $T_0 > T$, while the surface of the rod and its right end exchange heat with a surrounding medium at temperature T. Find $u(x)$.

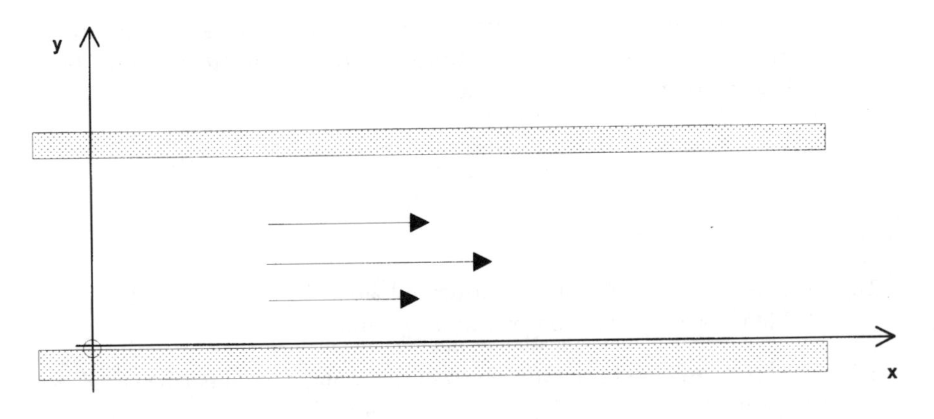

Figure 9: Poiseuille flow.

8. Calculate the limit as a tends to infinity of $u(x)$, the solution of the problem in Exercise 7. Is the result physically reasonable?

9. In an electrical heating element, the temperature $u(x)$ satisfies the boundary value problem that follows. Find $u(x)$.

$$\frac{d^2u}{dx^2} = \frac{hC}{\kappa A}(u - T) - \frac{I^2 R}{\kappa A}, \qquad 0 < x < a$$

$$u(0) = T, \qquad u(a) = T$$

10. Verify that the solution of the problem given in Eqs. (17) and (18) can also be written

$$u(x) = T_0 \frac{\cosh(\mu(x - a/2))}{\cosh(\mu a/2)}.$$

11. (Poiseuille flow) A viscous fluid flows steadily between two large parallel plates so that its velocity is parallel to the x-axis. (See Fig. 9.) The x-component of velocity of the fluid at any point (x, y) is a function of y only. It can be shown that this component $u(x)$ satisfies the differential equation

$$\frac{d^2u}{dy^2} = \frac{g}{\mu}, \qquad 0 < y < L,$$

where μ is the viscosity and g is a constant, negative pressure gradient. Find $u(y)$, subject to the "no-slip" boundary conditions, $u(0) = 0$, $u(L) = 0$.

12. If the beam mentioned in Exercise 6 is subjected to axial compression, instead of tension, the boundary value problem for $u(x)$ becomes the one here. Solve for $u(x)$.

$$\frac{d^2u}{dx^2} + \frac{P}{EI}u = -\frac{w}{EI}\frac{Lx - x^2}{2}, \qquad 0 < x < L,$$

$$u(0) = 0, \qquad u(L) = 0$$

13. For what value(s) of the compressive load P in Exercise 12 does the problem have no solution or infinitely many solutions?

14. The pressure $p(x)$ in the lubricant under a plane pad bearing satisfies the problem

$$\frac{d}{dx}\left(x^3\frac{dp}{dx}\right) = -K, \qquad a < x < b,$$

$$p(a) = 0, \qquad p(b) = 0.$$

Find $p(x)$ in terms of a, b and K (constant). Hint: the differential equation can be solved by integration.

15. In a nuclear fuel rod, nuclear reaction constantly generates heat. If we treat a rod as a one-dimensional object, the temperature $u(x)$ in the rod might satisfy the boundary-value problem

$$\frac{d^2u}{dx^2} + \frac{g}{\kappa} = \frac{hC}{\kappa A}(u - T), \qquad 0 < x < a,$$

$$u(0) = T, \qquad u(a) = T.$$

Here, g is the heat generation rate or power density, and the terms on the right-hand side represent heat transfer by convection to a surrounding medium, usually pressurized water. Find $u(x)$.

16. Sketch the solution of the problem above and determine the maximum temperature encountered. Typical values for the parameters are g = 300 W/cm^3, T = 325 °C, κ = 0.01 cal/cm s °C, a = 2.9 m, C/A = 4/cm, h = 0.035 cal/cm^2 s °C. It will be useful to know that 1 W = 0.239 cal/s.

17. An assembly of nuclear fuel rods is housed in a pressure vessel that is shaped roughly like a cylinder with flat or hemispherical ends. The temperature in the thick steel wall of the vessel affects its strength and thus must be studied for design and safety. Treating the vessel as a long cylinder (that is, ignoring the effects of the ends), it is easy to derive this differential equation in cylindrical coordinates for the temperature $u(r)$ in the wall:

$$\frac{1}{r}\frac{d}{dr}\frac{(rdu)}{dr} = 0, \qquad a < r < b$$

where a and b are the inner and outer radii. The boundary conditions both involve convection, with hot pressurized water at the inner radius and with air at the outer radius:

$$-\kappa u'(a) = h_0 \left(T_w - u(a)\right)$$
$$\kappa u'(b) = h_1 \left(T_a - u(b)\right).$$

Find $u(r)$ in terms of the parameters, carefully checking the dimensions.

18. Find the outer surface temperature $u(b)$ for the preceding exercise, using these values for the parameters: $T_w = 550$ °F, $T_a = 100$ °F, $a = 40$ in, $b = 48$ in, $\kappa = 24$ Btu/h ft °F, $h_0 = 250$ Btu/h ft^2 °F, $h_1 = 10$ Btu/h ft^2 °F.

0.4 SINGULAR BOUNDARY VALUE PROBLEMS

A boundary value problem can be singular in two different ways. In one case, an end point of the interval of interest is a singular point of the differential equation. In the other, the interval is infinitely long.

Regular Singular Point.
Recall that a point x_0 is a (regular) singular point of the differential equation

$$u'' + k(x)u' + p(x)u = f(x)$$

if the products

$$(x - x_0)k(x), \qquad (x - x_0)^2 p(x)$$

both have Taylor series expansions centered at x_0, but either $k(x)$ or $p(x)$ or both become infinite as $x \to x_0$. For example, the point $x_0 = 1$ is a regular singular point of the differential equation

$$(1-x)u'' + u' + xu = 0.$$

In standard form, the equation is

$$u'' + \frac{1}{1-x}u' + \frac{x}{1-x}u = 0.$$

Since both

$$k(x) = \frac{1}{1-x} \quad \text{and} \quad p(x) = \frac{x}{1-x}$$

become infinite at $x = 1$, but $(x-1)k(x)$ and $(x-1)^2 p(x)$ both have Taylor series expansions about the center $x = 1$, the point $x_0 = 1$ is a regular singular point. Another convenient example is provided by the Cauchy-Euler equation of Section 1, which has a regular singular point at the origin.

This situation typically arises when a boundary point is a mathematical boundary without being a physical boundary. For instance, a circular disk of radius c may be described in polar (r, θ) coordinates as occupying the region $0 \le r \le c$. The origin, at $r = 0$, is a mathematical boundary, yet physically this point is in the interior of the disk.

At a singular point, one cannot specify a value for $u(x_0)$, the solution of the differential equation, or for its derivative. However, it is usually necessary to require that both $u(x_0)$ and $u'(x_0)$ be finite, or bounded. Tacitly, we *always* require that the solution and its derivative be finite at *every* point of the interval where we are solving a differential equation. But when a singular point is a boundary point of that interval, we enforce the condition explicitly. In the example that follows we shall see how these conditions act so as to make the solution of a boundary value problem unique.

Example — Radial Heat Flow.
Suppose a long cylindrical bar, surrounded by a medium at temperature T, carries an electrical current. If heat flows in the radial direction much faster than in the axial direction, the temperature $u(r)$ in the rod may be described by the problem

$$\frac{1}{r}\frac{d}{dr}\left(r\frac{du}{dr}\right) = -H, \qquad 0 \le r < c, \tag{1}$$

$$u(c) = T. \tag{2}$$

Here, c is the radius of the rod, r is a polar coordinate, and H (constant) is proportional to the electrical power being converted into heat.

In this problem, only the physical boundary condition has been noted. The mathematical boundary $r = 0$ is a singular point, as is clear from the differential equation in the form

$$\frac{d^2u}{dr^2} + \frac{1}{r}\frac{du}{dr} = -H.$$

Thus, at this point we will require that u and du/dr be finite:

$$u(0), \quad u'(0) \text{ finite.} \tag{3}$$

Now the differential equation (1) is easy to solve. Multiply through by r and integrate once to find that

$$r\frac{du}{dr} = -H\frac{r^2}{2} + c_1.$$

Divide through this equation by r and integrate once more to determine that

$$u(r) = -H\frac{r^2}{4} + c_1 \ln(r) + c_2.$$

Application of the special condition, that $u(0)$ and $u'(0)$ be finite, immediately tells us that $c_1 = 0$; for both $\ln(r)$ and its derivative $1/r$ become infinite as r approaches 0.

The physical boundary condition, Eq. (2), says that

$$u(c) = -H\frac{c^2}{4} + c_2 = T.$$

Hence, $c_2 = Hc^2/4 + T$, and the complete solution is

$$u(r) = H\frac{(c^2 - r^2)}{4} + T. \tag{4}$$

From this example, it is clear that the "artificial" boundary condition, boundedness of $u(r)$ at the singular point $r = 0$, works just the way an ordinary boundary condition works at an ordinary (not singular) point. It gives one condition to be fulfilled by the unknown constants c_1 and c_2, which are then completely determined by the second boundary condition.

Semi-Infinite and Infinite Intervals.
Another type of singular boundary value problem is one for which the interval of interest is infinite. (Of course, this is always a mathematical abstraction that cannot be realized physically.) For instance, on the interval $0 < x < \infty$, sometimes called a *semi-infinite interval*, as it does have one finite endpoint, a boundary condition would normally be imposed at $x = 0$. At the other "end," no boundary condition is imposed, because no boundary exists. However, we normally require that both $u(x)$ and $u'(x)$ remain bounded as x increases. In precise terms, we require that there exist constants M and M' for which

$$|u(x)| \leq M \quad \text{and} \quad |u'(x)| \leq M'$$

are both satisfied for all x, no matter how large. We never identify M or M', and the entire condition is usually written

$$u(x) \ \text{ and } \ u'(x) \ \text{ bounded as } \ x \to \infty.$$

Example — Cooling Fin.
A long cooling fin has one end held at a constant temperature T_0 and exchanges heat with a medium at temperature T through convection. The temperature $u(x)$ in the fin satisfies the requirements

$$\frac{d^2u}{dx^2} = \frac{hC}{\kappa A}(u - T), \qquad 0 < x \tag{5}$$

$$u(0) = T_0 \tag{6}$$

(see Section 3). As the problem has been posed for a semi-infinite interval (because the fin is very long and, perhaps, to mask our ignorance of what is happening at the other physical end), we must also impose the condition

$$u(x), \quad u'(x) \ \text{ bounded as } \ x \to \infty. \tag{7}$$

Now, the general solution of the differential equation (5) is

$$u(x) = T + c_1 \cosh(\mu x) + c_2 \sinh(\mu x),$$

where $\mu = \sqrt{hC/\kappa A}$. The boundary condition at $x = 0$ requires that

$$u(0) = T_0 : T + c_1 = T_0.$$

The boundedness condition, Eq. (7), requires that

$$c_2 = -c_1.$$

The reason for this is that of all the linear combinations of cosh and sinh, the only one that is bounded as $x \to \infty$ is

$$\cosh(\mu x) - \sinh(\mu x) = e^{-\mu x},$$

and its constant multiples. The final solution is easily found to be

$$u(x) = T + (T_0 - T)(\cosh(\mu x) - \sinh(\mu x)).$$

Satisfying the boundedness condition would have been simpler if we had expressed the general solution of the differential equation (5) as

$$u(x) = T + c_1' e^{\mu x} + c_2' e^{-\mu x}.$$

We would have seen immediately that choosing $c_1' = 0$ is the only way to satisfy the boundedness condition. We summarize the observation as a rule of thumb: the solution of

$$\frac{d^2u}{dx^2} - \mu^2 u = 0$$

on an interval I is best expressed as

$$u(x) = \begin{cases} c_1 \cosh(\mu x) + c_2 \sinh(\mu x), & \text{if } I \text{ is finite,} \\ c_1 e^{\mu x} + c_2 e^{-\mu x}, & \text{if } I \text{ is infinite.} \end{cases}$$

Exercises

1. Put each of the following equations in the form

$$u'' + ku' + pu = f$$

and identify the singular point(s).

a. $\frac{1}{r}\frac{d}{dr}\left(r\frac{du}{dr}\right) = u$ **b.** $\frac{d}{dx}\left((1-x^2)\frac{du}{dx}\right) = 0$

c. $\frac{d}{d\phi}\left(\sin(\phi)\frac{du}{d\phi}\right) = \sin(\phi)u$ **d.** $\frac{1}{\rho^2}\frac{d}{d\rho}\left(\rho^2\frac{du}{d\rho}\right) = -\lambda^2 u$

2. The temperature u in a large object having a hole of radius c in the middle may be said to obey the equations

$$\frac{1}{r}\frac{d}{dr}\left(r\frac{du}{dr}\right) = 0, \qquad r > c$$

$$u(c) = T.$$

Solve the problem, adding the appropriate boundedness condition.

3. Compact kryptonite produces heat at a rate of H cal/s cm^3. If a sphere (radius c) of this material transfers heat by convection to a surrounding medium at temperature T, the temperature $u(\rho)$ in the sphere satisfies the boundary value problem

$$\frac{1}{\rho^2}\frac{d}{d\rho}\left(\rho^2\frac{du}{d\rho}\right) = \frac{-H}{\kappa}, \qquad 0 < \rho < c$$

$$-\kappa\frac{du}{d\rho}(c) = h(u(c) - T).$$

Supply the proper boundedness condition and solve. What is the temperature at the center of the sphere?

4. (Critical radius) The neutron flux u in a sphere of uranium obeys the differential equation

$$\frac{\lambda}{3}\frac{1}{\rho^2}\frac{d}{d\rho}\left(\rho^2\frac{du}{d\rho}\right) + (k-1)Au = 0$$

in the range $0 < \rho < a$, where λ is the effective distance traveled by a neutron between collisions, A is called the absorption cross section, and k is the number of neutrons produced by a collision during fission. In addition, the neutron flux at the boundary of the sphere is 0. Make the substitution $u = v/\rho$ and $3(k-1)A/\lambda = \mu^2$, and determine the differential equation satisfied by $v(\rho)$.

5. Solve the equation found in Exercise 4 and then find $u(\rho)$ that satisfies the boundary value problem (with boundedness condition) stated in Exercise 4. For what radius a is the solution not identically 0?

6. Inside a nuclear fuel rod, heat is constantly produced by nuclear reaction. A typical rod is about 3 m long and about 1 cm in diameter, so temperature variation along the length is much less than along a radius. Thus, we treat the temperature in such a rod as a function of the radial variable alone. Find this temperature $u(r)$, which is the solution of the boundary value problem

$$\frac{1}{r}\frac{d}{dr}\left(r\frac{du}{dr}\right) = -\frac{g}{\kappa}, \qquad 0 < r < a$$

$$u(a) = T_0.$$

7. For the problem of Exercise 6, find the temperature at the center of the rod, $u(0)$, using these values for the parameters: $a = 0.5$ cm, the power density $g = 418$ W/cm^3 = 100 cal/s cm^3, conductivity $\kappa = 0.01$ cal/s cm °C, and the surface temperature $T_0 = 325$ °C.

8. A model for microwave heating of food uses this equation for the temperature $u(x)$ in a large solid object:

$$\frac{d^2u}{dx^2} = -Ae^{-x/L}, \qquad 0 < x.$$

Here, A is a constant representing the strength of the radiation and properties of the object, and L is a characteristic length, known as penetration depth, that depends on frequency of the radiation and properties of the object. (Typically, L is about 12 cm in frozen raw beef or 2 cm thawed.) Show that the boundary condition $u'(0) = 0$ is incompatible with the condition that $u(x)$ be bounded as x goes to infinity. [See C.J. Coleman, The microwave heating of frozen substances, Applied Math. Modeling, Vol. 14, 1990, pp. 439-443.]

9. Solve the differential equation in Exercise 8 subject to the conditions

$$u(0) = T_0, \quad u(x) \text{ bounded.}$$

0.5 GREEN'S FUNCTIONS

The most important features of the solution of the boundary value problem[1]

$$\frac{d^2u}{dx^2} + k(x)\frac{du}{dx} + p(x)u = f(x), \qquad l < x < r \tag{1}$$

$$\alpha u(l) - \alpha' u'(l) = 0 \tag{2}$$

$$\beta u(r) + \beta' u'(r) = 0 \tag{3}$$

can be developed by using the variation-of-parameters solution of the differential equation (1), as presented in Section 2. To begin, we need to have two independent solutions of the homogeneous equation

$$\frac{d^2u}{dx^2} + k(x)\frac{du}{dx} + p(x)u = 0, \qquad l < x < r. \tag{4}$$

[1]The primes on the constants α', β' are not to indicate differentiation, of course, but to show that they are coefficients of derivatives.

Let us designate these two solutions as $u_1(x)$ and $u_2(x)$. It will simplify algebra later if we require that u_1 satisfy the boundary condition at $x = l$ and u_2 the condition at $x = r$;

$$\alpha u_1(l) - \alpha' u_1'(l) = 0, \tag{5}$$

$$\beta u_2(r) + \beta' u_2'(r) = 0. \tag{6}$$

According to Theorem 3 of Section 2, the general solution of the differential equation (1) can be written as

$$u(x) = c_1 u_1(x) + c_2 u_2(x) + \int_l^x (u_1(z)u_2(x) - u_2(z)u_1(x)) \frac{f(z)}{W(z)} dz. \tag{7}$$

Recall that in the denominator of the integrand, we have the Wronskian of u_1 and u_2,

$$W(z) = \begin{vmatrix} u_1(z) & u_2(z) \\ u_1'(z) & u_2'(z) \end{vmatrix}, \tag{8}$$

which is nonzero because u_1 and u_2 are independent. We will need to know the following derivative of the function in Eq. (7):

$$\frac{du}{dx} = c_1 u_1'(x) + c_2 u_2'(x) + \int_l^x (u_1(z)u_2'(x) - u_2(z)u_1'(x)) \frac{f(z)}{W(z)} dz.$$

(See Leibniz's rule in the Appendix.)

Now, let us apply the boundary condition, Eq. (2), to the general solution $u(x)$. First, at $x = l$ we have

$$\alpha u(l) - \alpha' u'(l) = c_1(\alpha u_1(l) - \alpha' u_1'(l)) + c_2(\alpha u_2(l) - \alpha' u_2'(l)) = 0. \tag{9}$$

Note that the integrals in u and u' are both 0 at $x = l$. Because of the boundary condition (5) imposed on u_1, Eq. (9) reduces to

$$c_2(\alpha u_2(l) - \alpha' u_2'(l)) = 0, \tag{10}$$

and we conclude that $c_2 = 0$.

Second, the boundary condition at $x = r$ becomes

$$\begin{aligned} \beta u(r) + \beta' u'(r) &= c_1 \left(\beta u_1(r) + \beta' u'(r)\right) \\ &+ \int_l^r \left[u_1(z)\left(\beta u_2(r) + \beta' u_2'(r)\right) - u_2(z)\left(\beta u_1(r) + \beta' u_1'(r)\right)\right] \frac{f(z)}{W(z)} dz = 0. \end{aligned} \tag{11}$$

Now, the boundary condition (6) on u_2 at $x = r$ eliminates one term of the integrand, leaving

$$c_1 \left(\beta u_1(r) + \beta' u_1'(r)\right) - \int_l^r u_2(z) \left(\beta u_1(r) + \beta' u_1'(r)\right) \frac{f(z)}{W(z)} dz = 0. \quad (12)$$

The common factor of $\beta u_1(r) + \beta' u_1'(r)$ can be canceled from both terms, and we then find

$$c_1 = \int_l^r u_2(z) \frac{f(z)}{W(z)} dz. \quad (13)$$

Now we have found c_1 and c_2 so that $u(x)$ in Eq. (7) satisfies both boundary conditions. If we use the values of c_1 and c_2 as found, we have

$$u(x) = u_1(x) \int_l^r u_2(z) \frac{f(z)}{W(z)} dz + \int_l^x \left(u_1(z)u_2(x) - u_2(z)u_1(x)\right) \frac{f(z)}{W(z)} dz. \quad (14)$$

The solution becomes more compact if we break the interval of integration at x in the first integral, making it

$$\int_l^r u_2(z) \frac{f(z)}{W(z)} dz = \int_l^x u_2(z) \frac{f(z)}{W(z)} dz + \int_x^r u_2(z) \frac{f(z)}{W(z)} dz. \quad (15)$$

When the integrals on the range l to x are combined, there is some cancellation, and our solution becomes

$$u(x) = \int_l^x u_1(z)u_2(x) \frac{f(z)}{W(z)} dz + \int_x^r u_1(x)u_2(z) \frac{f(z)}{W(z)} dz. \quad (16)$$

Finally, these two integrals can be combined into one. We first define the *Green's function* for the problem (1), (2), (3) as

$$G(x, z) = \begin{cases} \dfrac{u_1(z)u_2(x)}{W(z)}, & l < z \leq x, \\ \dfrac{u_1(x)u_2(z)}{W(z)}, & x \leq z < r, \end{cases} \quad (17)$$

Then the formula given in Eq. (16) for u simplifies to

$$u(x) = \int_l^r G(x, z) f(z) dz. \quad (18)$$

Example.
We solve the problem that follows by constructing the Green's function.

$$\frac{d^2u}{dx^2} - u = -1, \quad 0 < x < 1$$

$$u(0) = 0, \quad u(1) = 0$$

First, we must find two independent solutions of the homogeneous differential equation $u'' - u = 0$ that satisfy the boundary conditions as required. The general solution of the homogeneous differential equation is

$$u(x) = c_1 \cosh(x) + c_2 \sinh(x).$$

As $u_1(x)$ is required to satisfy the condition at the left, $u_1(0) = 0$, we take $c_1 = 0$, $c_2 = 1$ and conclude $u_1(x) = \sinh(x)$. The second solution is to satisfy $u_2(1) = 0$. We may take

$$u_2(x) = \sinh(1)\cosh(x) - \cosh(1)\sinh(x) = \sinh(1 - x).$$

The Wronskian of the two solutions is

$$W(x) = \begin{vmatrix} \sinh(x) & \sinh(1-x) \\ \cosh(x) & -\cosh(1-x) \end{vmatrix} = -\sinh(1).$$

Now, by Eq. (17), the Green's function for this problem is

$$G(x, z) = \begin{cases} \dfrac{\sinh(z)\sinh(1-x)}{-\sinh(1)}, & 0 < z \leq x \\ \\ \dfrac{\sinh(x)\sinh(1-z)}{-\sinh(1)}, & x \leq z < 1. \end{cases}$$

Furthermore, since $f(x) = -1$, the solution, by Eq. (18), is the integral

$$u(x) = \int_0^1 -G(x, z)dz.$$

To actually carry out the integration, we must break the interval of integration at x, thus reverting in effect to Eq. (16). The result:

$$\begin{aligned}
u(x) &= \int_0^x \frac{\sinh(z)\sinh(1-x)}{\sinh(1)}dz + \int_x^1 \frac{\sinh(x)\sinh(1-z)}{\sinh(1)}dz \\
&= \frac{\sinh(1-x)}{\sinh(1)}\cosh(z)\Big|_0^x + \frac{\sinh(x)}{\sinh(1)}(-\cosh(1-z))\Big|_x^1 \\
&= \frac{\sinh(1-x)}{\sinh(1)}(\cosh(x)-1) + \frac{\sinh(x)}{\sinh(1)}(\cosh(1-x)-1) \\
&= \frac{\sinh(1-x)\cosh(x)+\sinh(x)\cosh(1-x)}{\sinh(1)} - \frac{\sinh(1-x)+\sinh(x)}{\sinh(1)} \\
&= 1 - \frac{\sinh(1-x)+\sinh(x)}{\sinh(1)}.
\end{aligned}$$

This, finally, is easily seen to be the correct solution. In this instance, there are much quicker ways to arrive at the same result. The advantage of the Green's function is that it shows how the solution of the problem depends on the inhomogeneity $f(x)$. It is an efficient way to obtain the solution in some cases.

Now, let us look back over the calculations and see if there is some place that they might fail. Aside from the possibility that the coefficients $k(x)$ or $p(x)$ in the differential equation might not be continuous, it seems that division by 0 is the only possibility of failure. Quantities cancelled or divided by were

$$W(x) = \begin{vmatrix} u_1(x) & u_2(x) \\ u_1'(x) & u_2'(x) \end{vmatrix}$$

$$\alpha u_2(l) - \alpha' u_2'(l)$$

$$\beta u_1(r) + \beta' u_1'(r)$$

in Eqs. (7), (10), and (12) respectively. It can be shown that all three of these are 0 if any one of them is 0, and, in that case, $u_1(x)$ and $u_2(x)$ are proportional. We summarize in a theorem.

Theorem
Let $k(x)$, $p(x)$ and $f(x)$ be continuous, $l \le x \le r$. The boundary value problem

$$\frac{d^2u}{dx^2} + k(x)\frac{du}{dx} + p(x)u = f(x), \qquad l < x < r$$

$$\alpha u(l) - \alpha' u'(l) = 0 \tag{i}$$

$$\beta u(r) + \beta' u'(r) = 0 \tag{ii}$$

has one and only one solution, unless there is a nontrivial solution of

$$\frac{d^2u}{dx^2} + k(x)\frac{du}{dx} + p(x)u = 0, \qquad l < x < r$$

that satisfies (i) and (ii).

When a unique solution exists, it is given by Eqs. (17) and (18).

Example.
The boundary value problem

$$\frac{d^2u}{dx^2} + u = -1, \qquad 0 < x < \pi$$

$$u(0) = 0, \qquad u(\pi) = 0$$

does not have a unique solution, according to the Theorem, because $u(x) = \sin(x)$ is a nontrivial solution of the problem

$$\frac{d^2u}{dx^2} + u = 0, \qquad 0 < x < \pi$$

$$u(0) = 0, \qquad u(\pi) = 0.$$

Indeed, if we try to follow through the construction, we find that $u_1(x) = \sin(x)$ and also $u_2(x) = \sin(x)$ (or a multiple thereof), and so all three quantities in Eq. (19) are 0.

On the other hand, suppose we try to obtain a solution by the usual method. The general solution of the differential equation is

$$u(x) = -1 + c_1 \cos(x) + c_2 \sin(x).$$

However, application of the boundary conditions leads to the contradictory requirements

$$-1 + c_1 = 0 \ \text{ and } \ -1 - c_1 = 0.$$

Thus, in this case, there simply is no solution to the problem stated.

If the differential equation (1) has a singular point at $x = l$ or $x = r$ (or both), a Green's function may still be constructed. The boundary condition (2) or (3) would be replaced by a boundedness condition, which would also apply to u_1 or u_2 as the case may be.

For example, consider the problem

$$\frac{1}{x}\frac{d}{dx}\left(x\frac{du}{dx}\right) = f(x), \qquad 0 < x < 1$$

$$u(0) \text{ bounded}, \qquad u(1) = 0.$$

The general solution of the corresponding homogeneous equation is $u(x) = c_1 + c_2 \ln(x)$. Thus, we would choose

$$u_1(x) = 1, \qquad u_2(x) = \ln(x)$$

so that $u_1(x)$ is bounded at $x = 0$ and $u_2(x)$ is 0 at $x = 1$. The Green's function is thus

$$G(x, z) = \begin{cases} z\ln(x), & 0 < z \le x, \\ z\ln(z), & x \le z < 1. \end{cases}$$

A similar procedure is followed if the interval $l < x < r$ is infinite in length.

Exercises

In Exercises 1-8, find the Green's function for the problem stated.

1. $\dfrac{d^2u}{dx^2} = f(x), \qquad 0 < x < a$

$u(0) = 0,\ u(a) = 0.$

2. $\dfrac{d^2u}{dx^2} = f(x), \qquad 0 < x < a$

$u(0) = 0, \quad \dfrac{du}{dx}(a) = 0.$

3. $\dfrac{d^2u}{dx^2} - \gamma^2 u = f(x), \qquad 0 < x < a$

$\dfrac{du}{dx}(0) = 0, \quad u(a) = 0.$

4. $\dfrac{1}{r}\dfrac{d}{dr}\left(r\dfrac{du}{dr}\right) = f(r), \quad 0 \le r < c$

$u(c) = 0,\ u(r)$ bounded at $r = 0.$

5. $\dfrac{1}{\rho^2}\dfrac{d}{d\rho}\left(\rho^2\dfrac{du}{d\rho}\right) = f(\rho), \quad 0 \le \rho < c$
$u(c) = 0$, $u(\rho)$ bounded at $\rho = 0$.

6. $\dfrac{d^2u}{dx^2} + \dfrac{1}{x}\dfrac{du}{dx} - \dfrac{1}{4x^2}u = f(x), \quad 0 \le x < a$
$u(a) = 0$, $u(x)$ bounded at $x = 0$.

7. $\dfrac{d^2u}{dx^2} - \gamma^2 u = f(x), \quad 0 < x$
$u(0) = 0$, $u(x)$ bounded as $x \to \infty$.

8. $\dfrac{d^2u}{dx^2} - \gamma^2 u = f(x), \quad -\infty < x < \infty,$
$u(x)$ bounded as $x \to \pm\infty$.

9. Use the Green's function of Exercise 5 to solve the problem

$$\frac{1}{\rho^2}\frac{d}{d\rho}\left(\rho^2\frac{du}{d\rho}\right) = 1, \quad 0 \le \rho < c,$$
$$u(c) = 0,$$

and compare with the solution found by integrating the equation directly.

10. Use the Green's function of Exercise 8 to solve the problem

$$\frac{d^2u}{dx^2} - \gamma^2 u = -\gamma^2, \qquad -\infty < x < \infty$$
$$u(x) \text{ bounded as } x \to \pm\infty$$

and compare with the result found directly.

11. Use the Green's function of Exercise 1 to solve the problem stated there, if

$$f(x) = \begin{cases} 0, & 0 < x < a/2, \\ 1, & a/2 < x < a \end{cases}$$

12. In confirmation of the Theorem, show that: the homogeneous problem **a.** has a nontrivial solution; problem **b.** has no solution (existence fails); and problem **c.** has infinitely many solutions (uniqueness fails).

a. $u'' + u = 0, \quad u(0) = 0, \quad u(\pi) = 0,$

b. $u'' + u = -1, \quad u(0) = 0, \quad u(\pi) = 0,$

c. $u'' + u = \pi - 2x, \quad u(0) = 0, \quad u(\pi) = 0.$

13. Considering z to be a parameter ($l < z < r$), define the function $v(x) = G(x, z)$ with G as in Eq. (17). Show that v has these four properties, which are sometimes used to define the Green's function.

(i) v satisfies the boundary conditions, Eqs. (2) and (3), at $x = l$ and r.

(ii) v is continuous, $l < x < r$. (The point $x = z$ needs to be checked.)

(iii) v' is discontinuous at $x = z$, and

$$\lim_{h \to 0} (v'(z + h) - v'(z - h)) = 1.$$

(iv) v satisfies the differential equation $v'' + k(x)v' + p(x)v = 0$ for $l < x < z$ and $z < x < r$.

14. Show that the boundary value problem

$$\frac{d^2u}{dx^2} + \lambda^2 u = f(x), \qquad 0 < x < a,$$

$$u(0) = 0, \qquad u(a) = 0,$$

will have no solution or infinitely many solutions, if λ is an eigenvalue of

$$\frac{d^2u}{dx^2} + \lambda^2 u = 0,$$

$$u(0) = 0, \qquad u(a) = 0.$$

MISCELLANEOUS EXERCISES

In Exercises 1 to 15, solve the given boundary value problem, supplying boundedness conditions where necessary.

1. $\dfrac{d^2u}{dx^2} - \gamma^2 u = 0, \quad 0 < x < a$

$u(0) = T_0, \quad u(a) = T_1$

2. $\dfrac{d^2u}{dx^2} - r = 0, \quad 0 < x < a \quad (r \text{ is constant})$

$u(0) = T_0, \quad \dfrac{du}{dx}(a) = 0$

3. $\dfrac{d^2u}{dx^2} = 0, \quad 0 < x < a$

$u(0) = T_0, \quad \dfrac{du}{dx}(a) = 0$

4. $\dfrac{d^2u}{dx^2} - \gamma^2 u = 0, \quad 0 < x < a$

$\dfrac{du}{dx}(0) = 0, \quad u(a) = T_1$

5. $\dfrac{1}{r}\dfrac{d}{dr}\left(r\dfrac{du}{dr}\right) = -p, \quad 0 < r < a$

$u(a) = 0$

6. $\dfrac{1}{r}\dfrac{d}{dr}\left(r\dfrac{du}{dr}\right) = 0, \quad a < r < b$

$u(a) = T_0, \quad u(b) = T_1$

7. $\dfrac{1}{\rho^2}\dfrac{d}{d\rho}\left(\rho^2\dfrac{du}{d\rho}\right) = -H, \quad 0 < \rho < a$

$u(a) = T_0$

8. $\dfrac{1}{r}\dfrac{d}{dr}\left(r\dfrac{du}{dr}\right) = 0, \quad a < r < \infty$

$u(a) = T$

9. $\dfrac{d^2u}{dx^2} - \gamma^2(u - T) = 0, \quad 0 < x < a$

$\dfrac{du}{dx}(0) = 0, \quad u(a) = T_1$

10. $\dfrac{d^2u}{dx^2} - \gamma^2 u = 0, \quad 0 < x < \infty$

$u(0) = T$

11. $\dfrac{d^2u}{dx^2} = \gamma^2(u - T_0), \quad 0 < x < \infty$

$u(0) = T$

12. $\dfrac{d}{dx}\left(x^3\dfrac{du}{dx}\right) = -k, \quad a < x < b \quad (k \text{ is constant})$

$u(a) = 0, \quad u(b) = 0 \quad (\text{Note}: \ 0 < a.)$

13. In this problem, h is the groundwater level between two trenches in which water is held at constant levels. Note that the equation is nonlinear.

$$\frac{d}{dx}\left(h\frac{dh}{dx}\right) + e = 0, \qquad 0 < x < a$$

$$h(0) = h_0, \qquad h(a) = h_1$$

14. $\dfrac{d^4u}{dx^4} = w, \quad 0 < x < a \quad (w \text{ is constant})$

$u(0) = 0, \quad u(a) = 0, \quad \dfrac{d^2u}{dx^2}(0) = 0, \quad \dfrac{d^2u}{dx^2}(a) = 0$

15. $\dfrac{d^4u}{dx^4} + \dfrac{k}{EI}u = w, \quad 0 < x < \infty \quad (w \text{ is constant})$

$u(0) = 0, \quad \dfrac{d^2u}{dx^2}(0) = 0$

16. Show that any two of the four functions $\sinh(\lambda x)$, $\sinh(\lambda(a - x))$, $\cosh(\lambda x)$, $\cosh(\lambda(a-x))$ are independent solutions of the differential equation

$$\phi'' - \lambda^2\phi = 0.$$

17. In this problem, u is the temperature in a wall composed of two substances. Find $u(x)$.

$$\frac{d^2u}{dx^2} = 0, \qquad 0 < x < \alpha a \ \text{ and } \ \alpha a < x < a$$

$$u(0) = T_0, \qquad u(a) = T_1$$

$$\kappa_1\frac{du}{dx}(\alpha a-) = \kappa_2\frac{du}{dx}(\alpha a+)$$

$$u(\alpha a-) = u(\alpha a+)$$

The last two conditions say that the heat flow rate and the temperature are both continuous across the interface at $x = \alpha a$.

18. Find the general solution of the differential equation

$$\frac{1}{x^2}\frac{d}{dx}\left(x^2\frac{du}{dx}\right) + ku = 0$$

for the cases $k = \lambda^2$ and $k = -p^2$. (Hint: let $u(x) = v(x)/x$ and find the equation that $v(x)$ satisfies.)

19. Find the solution of the boundary value problem

$$e^x\frac{d}{dx}\left(e^x\frac{du}{dx}\right) = -1, \qquad 0 < x < a$$

$$u(0) = 0, \qquad u(a) = 0.$$

20. Solve the boundary value problem

$$\frac{1}{r}\frac{d}{dr}\left(r\frac{du}{dr}\right) = -r^k, \qquad 0 < r < a$$

$$u(0) \text{ bounded and } u(a) = 0.$$

21. Solve the differential equation

$$\frac{d^2u}{dx^2} = p^2u, \qquad 0 < x < a$$

subject to the following sets of boundary conditions.

a. $u(0) = 0, \quad u(a) = 1$

b. $u(0) = 1, \quad u(a) = 0$

c. $u'(0) = 0, \quad u(a) = 1$

d. $u(0) = 1, \quad u'(a) = 0$

e. $u'(0) = 1, \quad u'(a) = 0$

f. $u'(0) = 0, \quad u'(a) = 1$

22. Solve the integro-differential boundary value problem

$$\frac{d^2u}{dx^2} = \gamma^2\left(u - \int_0^1 u(x)dx\right), \qquad 0 < x < 1$$

$$\frac{du}{dx}(0) = 0, \qquad u(1) = T.$$

Hint: Look for a solution in the form

$$u(x) = A\cosh(\gamma x) + B\sinh(\gamma x) + C.$$

23. Use a variation of parameters to find a second independent solution of each of the following differential equations. In each case, one solution is given in parentheses.

a. $\dfrac{d^2u}{dx^2} - \dfrac{2x}{1-x^2}\dfrac{du}{dx} + \dfrac{2}{1-x^2}u = 0 \quad (u = x)$

b. $\dfrac{d^2u}{dx^2} + \dfrac{1-x}{x}\dfrac{du}{dx} + \dfrac{1}{x}u = 0 \quad (u = 1-x)$

24. By applying the method of variation of parameters, derive this formula for a particular solution of the differential equation

$$\frac{d^2u}{dx^2} - \gamma^2 u = f(x)$$

$$u(x) = \int_0^x f(x')\frac{\sinh \gamma(x-x')}{\gamma}dx'.$$

25. The absolute temperature $u(x)$ in a cooling fin that radiates heat to a medium at absolute temperature T obeys the differential equation $u'' = \gamma^2(u^4 - T^4)$. Solve the special version in the boundary value problem that follows, which can be done in closed form.

$$\frac{d^2u}{dx^2} = \gamma^2 u^4, \qquad 0 < x,$$

$$u(0) = U, \qquad \lim_{x\to\infty} u(x) = 0$$

26. If a beam of uniform cross section is simply supported at its ends and carries a distributed load $w(x)$ along its length, then the displacement $u(x)$ of its centerline satisfies the boundary-value problem

$$\frac{d^4u}{dx^4} = \frac{w(x)}{EI}, \qquad 0 < x < a$$

$$u(0) = 0, \qquad u''(0) = 0, \qquad u(a) = 0, \qquad u''(a) = 0.$$

(Here, E is Young's modulus and I is the second moment of the cross section.) Solve this problem if $w(x) = w_0$, constant.

27. If the beam of Exercise 26 is built into a wall at the left end and is unsupported at the right end, the boundary conditions become

$$u(0) = 0, \qquad u'(0) = 0, \qquad u''(a) = 0, \qquad u'''(a) = 0.$$

Solve the same differential equation subject to these conditions.

28. A uniform, straight shaft exhibits violent behavior at certain frequencies of rotation. Let the x-axis between 0 and a represent the undeflected centerline of the shaft, and let $u(x)$ be the displacement of the actual centerline of the shaft measured from the x-axis. Centrifugal force provides a transverse loading on the shaft when u is not identically equal to zero. The equation for the displacement is

$$\frac{d^4u}{dx^4} - \frac{\omega^2 w}{EIg}u = 0, \qquad 0 < x < a,$$

where w is the weight per unit length of the shaft, g is the acceleration of gravity, E is Young's modulus, I is the second moment of the cross-sectional area of the shaft, and ω is the angular velocity. If the shaft is held in narrow bearings at the ends, these can be interpreted as simple supports, leading to boundary conditions

$$u(0) = 0, \qquad u''(0) = 0, \qquad u(a) = 0, \qquad u''(a) = 0.$$

Find a formula for those values of angular velocity (critical values or whirling speeds) that permit the existence of nonzero solutions to this boundary-value problem.

29. Find the lowest critical value for the angular velocity of a steel shaft with these specifications: diameter, 1.5 in; length, 48 in; $w = 0.5$ lb/in; $E = 30 \times 10^6$ lb/in^2, $I = 0.5$ in^4.

30. Sulphur dioxide (SO_2) is a common air pollutant that reacts with water to form sulphuric acid. If the water is airborne, the result is acid rain; if the water is in snow, the result is acid runoff when the snow melts. Use an analysis similar to that of Section 3 to obtain a boundary value problem for the concentration $u(x)$ (in units of mass per unit volume) of sulphur dioxide in the air included in a layer of snow. Introduce $q(x)$, the flow rate of sulphur dioxide (in units of mass per unit time per unit of cross-sectional area.) There are two important physical facts: (1) Diffusion is governed by Fick's law (similar to Fourier's law)

$$q(x) = -D\frac{du}{dx},$$

where D is the *diffusion constant*; and (2) when the sulphur dioxide reacts with water, it "disappears" at a rate proportional to its concentration, say $ku(x)$ (in units of mass per unit time per unit volume).

31. The sulphur dioxide concentration in the air in a deep layer of snow satisfies this boundary value problem in equilibrium conditions:

$$\frac{d^2u}{dx^2} - a^2u = 0, \qquad 0 < x$$

$$u(0) = C_0.$$

Here, C_0 is the concentration in freely circulating air. Add an appropriate boundedness condition and solve for $u(x)$.

Chapter 1

Fourier Series and Integrals

1.1 PERIODIC FUNCTIONS AND FOURIER SERIES

A function f is said to be *periodic with period* $p > 0$ if: (1) $f(x)$ has been defined for all x; and (2) $f(x+p) = f(x)$ for all x. The familiar functions $\sin(x)$ and $\cos(x)$ are simple examples of periodic functions with period 2π, and the functions $\sin(2\pi x/p)$ and $\cos(2\pi x/p)$ are periodic with period p.

A periodic function has many periods, for if $f(x) = f(x+p)$ then also

$$f(x) = f(x+p) = f(x+2p) = \cdots = f(x+np),$$

where n is any integer. Thus $\sin(x)$ has periods $2\pi, 4\pi, 6\pi, \cdot,\ n \cdots 2\pi$. The period of a periodic function is generally taken to be positive, but the periodicity condition holds for negative as well as positive changes in the argument. That is to say, $f(x-p) = f(x)$ for all x, since $f(x) = f(x-p+p) = f(x-p)$. Also

$$f(x) = f(x-p) = f(x-2p) = \cdots = f(x-np).$$

The definition of periodic says essentially that functional values repeat themselves. This implies that the graph of a periodic function can be drawn for all x by making a template of the graph on any interval of length p and then copying the graph from the template up and down the x axis (see Fig. 1).

Many of the functions that occur in engineering and physics are periodic in space or time — for example, acoustic waves — and in order to

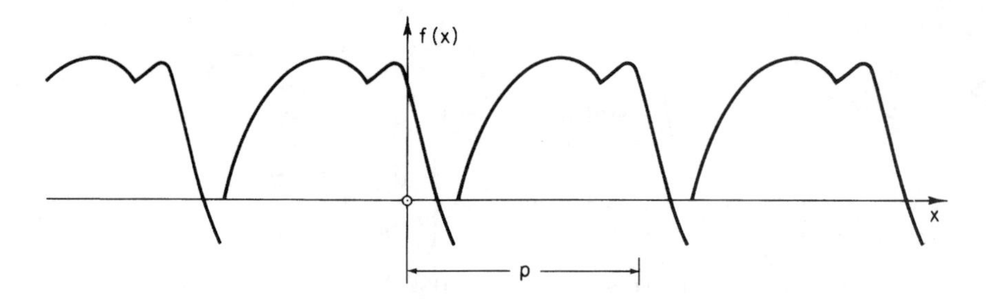

Figure 1: A periodic function of period p.

understand them better it is often desirable to represent them in terms of the very simple periodic functions 1, $\sin(x)$, $\cos(x)$, $\sin(2x)$, $\cos(2x)$, and so forth. *All* of these functions have the common period 2π, although each has other periods as well.

If f is periodic with period 2π, then we attempt to represent f in the form of an infinite series

$$f(x) = a_0 + \sum_{n=1}^{\infty} (a_n \cos(nx) + b_n \sin(nx)). \tag{1}$$

Each term of the series has period 2π, so if the sum of the series exists, it will be a function of period 2π. There are two questions to be answered: (a) What values must a_0, a_n, b_n have; and (b) if the appropriate values are assigned to the coefficients, does the series actually represent the given function $f(x)$?

On the face of it, the first question is tremendously difficult, for Eq. (1) represents an equation in an infinite number of unknowns. But a reasonable answer can be found easily by using the following *orthogonality*[1]

[1]The word "orthogonality" should not be thought of in the geometric sense.

relations:

$$
\begin{aligned}
\int_{-\pi}^{\pi} \sin(nx)dx &= 0 \\
\int_{-\pi}^{\pi} \cos(nx)dx &= \begin{cases} 0, & n \neq 0 \\ 2\pi, & n = 0 \end{cases} \\
\int_{-\pi}^{\pi} \sin(nx)\cos(mx)dx &= 0 \\
\int_{-\pi}^{\pi} \sin(nx)\sin(mx)dx &= \begin{cases} 0, & n \neq m \\ \pi, & n = m \end{cases} \\
\int_{-\pi}^{\pi} \cos(nx)\cos(mx)dx &= \begin{cases} 0, & n \neq m \\ \pi, & n = m \neq 0. \end{cases}
\end{aligned}
\tag{2}
$$

We may summarize these relations by saying: The definite integral (over the interval $-\pi$ to π) of the product of any two different functions from the series in Eq. (1) is zero.

The fundamental idea is that if the equality proposed in Eq. (1) is to be a real equality, then both sides must give the same result after the same operation. The orthogonality relations then suggest operations that simplify the right-hand side of Eq. (1). Namely, we multiply both sides of the proposed equation by one of the functions that appears there and integrate from $-\pi$ to π. (We must assume that the integration of the series can be carried out term-by-term. This is sometimes difficult to justify, but we do it nonetheless.)

Multiplying both sides of Eq. (1) by the constant 1 ($= \cos(0x)$) and integrating from $-\pi$ to π, we find

$$\int_{-\pi}^{\pi} f(x)dx = \int_{-\pi}^{\pi} a_0 dx + \sum_{n=1}^{\infty} \int_{-\pi}^{\pi} (a_n \cos(nx) + b_n \sin(nx))dx.$$

Each of the terms in the integrated series is zero, so the right-hand side of this equation reduces to $2\pi \cdot a_0$, giving

$$a_0 = \frac{1}{2\pi} \int_{-\pi}^{\pi} f(x)dx.$$

(In words, a_0 is the mean value of $f(x)$ over one period.)

Now multiplying each side of Eq. (1) by $\sin(mx)$, where m is a fixed integer, and integrating from $-\pi$ to π, we find

$$\int_{-\pi}^{\pi} f(x)\sin(mx)dx = \int_{-\pi}^{\pi} a_0 \sin(mx)dx + \sum_{n=1}^{\infty}\int_{-\pi}^{\pi} a_n \cos(nx)\sin(mx)dx$$
$$+\sum_{n=1}^{\infty}\int_{-\pi}^{\pi} b_n \sin(nx)\sin(mx)dx.$$

All terms containing a_0 or a_n disappear, according to Eq. (2). Furthermore, of all those containing a b_n, the only one that is not zero is the one in which $n = m$. (Notice that n is a summation index and runs through all the integers $1, 2, \cdots$. We chose m to be a fixed integer, so $n = m$ once.) We now have the formula

$$b_m = \frac{1}{\pi}\int_{-\pi}^{\pi} f(x)\sin(mx)dx.$$

By multiplying both sides of Eq. (1) by $\cos(mx)$ (m is a fixed integer) and integrating, we also find

$$a_m = \frac{1}{\pi}\int_{-\pi}^{\pi} f(x)\cos(mx)dx.$$

We can now summarize our results. In order for the proposed equality

$$f(x) = a_0 + \sum_{n=1}^{\infty}(a_n\cos(nx) + b_n\sin(nx)) \qquad (1')$$

to hold, the a's and b's must be chosen according to the formulas

$$a_0 = \frac{1}{2\pi}\int_{-\pi}^{\pi} f(x)dx \qquad (3)$$

$$a_n = \frac{1}{\pi}\int_{-\pi}^{\pi} f(x)\cos(nx)dx \qquad (4)$$

$$b_n = \frac{1}{\pi}\int_{-\pi}^{\pi} f(x)\sin(nx)dx. \qquad (5)$$

When the coefficients are chosen this way, the right-hand side of Eq. (1) is called the *Fourier series* of f. The a's and b's are called *Fourier coefficients*. We have not yet answered question (b) about equality, so we write

$$f(x) \sim a_0 + \sum_{n=1}^{\infty}(a_n\cos(nx) + b_n\sin(nx))$$

to indicate that the Fourier series *corresponds* to $f(x)$.

Example.
Suppose that $f(x)$ is periodic with period 2π and is given by the formula $f(x) = x$ in the interval $-\pi < x < \pi$ (see Fig. 2). According to our formulas,

$$a_0 = \frac{1}{2\pi}\int_{-\pi}^{\pi} f(x)dx = \frac{1}{2\pi}\int_{-\pi}^{\pi} xdx = 0$$

$$\begin{aligned} a_n &= \frac{1}{\pi}\int_{-\pi}^{\pi} f(x)\cos(nx)dx = \frac{1}{\pi}\int_{-\pi}^{\pi} x\cos nx \ \ dx \\ &= \frac{1}{\pi}\left[\frac{\cos(nx)}{n^2} + \frac{x\sin(nx)}{n}\right|_{-\pi}^{\pi} = 0 \\ b_n &= \frac{1}{\pi}\int_{-\pi}^{\pi} f(x)\sin(nx)dx = \frac{1}{\pi}\int_{-\pi}^{\pi} x\sin(nx) \ \ dx \\ &= \frac{1}{\pi}\left[\frac{\sin(nx)}{n^2} - \frac{x\cos(nx)}{n}\right|_{-\pi}^{\pi} \\ &= \frac{1}{\pi}\frac{(-2\pi)\cos n\pi}{n} = \frac{2}{n}(-1)^{n+1}. \end{aligned}$$

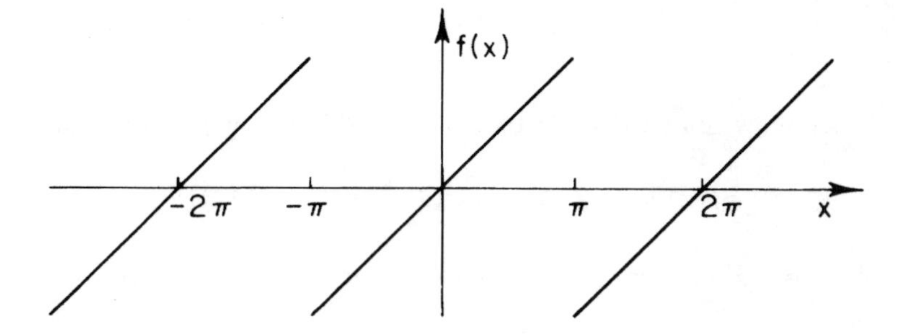

Figure 2: $f(x) = x$, $-\pi < x < \pi$, f periodic with period 2π.

Thus, for this function, we have

$$\begin{aligned} f(x) &\sim \sum_{n=1}^{\infty} \frac{2(-1)^{n+1}}{n}\sin(nx) \\ &\sim 2(\sin(x) - \frac{1}{2}\sin(2x) + \frac{1}{3}\sin(3x) - + \cdots). \end{aligned}$$

The Appendix contains some integration formulas that are convenient for finding Fourier coefficients. It is also useful to know these special values of sines and cosines that come up frequently in Fourier series.

$$\sin(n\pi) = 0, \qquad \cos(n\pi) = (-1)^n, \quad \text{for } n = 0, \pm 1, \pm 2, \cdots$$

$$\sin\left(\frac{(2n-1)\pi}{2}\right) = (-1)^{n+1}, \quad \cos\left(\frac{(2n-1)\pi}{2}\right) = 0, \quad \text{for } n = 0, \pm 1, \pm 2, \cdots.$$

Note that the second line involves only *odd* multiples of $\pi/2$. Even multiples of $\pi/2$ are included in the first line.

Exercises

1. Find the Fourier coefficients of the functions given in what follows. All are supposed to be periodic with period 2π. Sketch the graph of the function.

a. $f(x) = x, \quad -\pi < x < \pi$

b. $f(x) = |x|, \quad -\pi < x < \pi$

c. $f(x) = \begin{cases} 0, & -\pi < x < 0 \\ 1, & 0 < x < \pi \end{cases}$

d. $f(x) = |\sin x|$

2. Sketch for at least two periods the graphs of the functions defined by:

a. $f(x) = x, \quad -1 < x \leq 1, \quad f(x+2) = f(x)$

b. $f(x) = \begin{cases} 0, & -1 < x \leq 0, \\ x, & 0 < x < 1, \end{cases} \quad f(x+2) = f(x)$

c. $f(x) = \begin{cases} 0, & -\pi < x \leq 0, \\ 1, & 0 < x \leq 2\pi, \end{cases} \quad f(x+3\pi) = f(x)$

d. $f(x) = \begin{cases} 0, & -\pi < x \leq 0, \\ \sin x, & 0 < x \leq \pi, \end{cases} \quad f(x+2\pi) = f(x)$

3. Show that the constant function $f(x) = 1$ is periodic with every possible period $p > 0$.

4. Carry out the details of deriving the equation for a_m.

5. Suppose $f(x)$ has period p. Show that for any c,

$$\int_c^{c+p} f(x)dx = \int_0^p f(x)dx.$$

6. Suppose $f(x)$, $g(x)$ are periodic with a common period p. Show that $af(x) + bg(x)$ and $f(x) \cdot g(x)$ also are periodic with period p (a, b are constants).

7. Find the Fourier series of each of the following periodic functions. Integration is not necessary.

a. $f(x) = \cos^2(x)$

b. $f(x) = \sin(x - \pi/6)$

c. $f(x) = \sin(x)\cos(2x)$

8. Verify that $\sin(2\pi x/p)$ and $\cos(2\pi x/p)$ are periodic with period p.

1.2 ARBITRARY PERIOD AND HALF-RANGE EXPANSIONS

In Section 1 we found a way to represent a periodic function of period 2π with a Fourier series. It is not necessary to restrict ourselves to this period. In fact, we may broaden the idea of Fourier series to include functions of any period by a simple rescaling of the variables. Let us suppose that a function f is periodic with period $2a$. (We use $2a$ in place of p for later convenience.) Then we may relate f to a series of the functions 1, $\sin(\pi x/a)$, $\cos(\pi x/a)$, $\sin(2\pi x/a)$, $\cos(2\pi x/a)$, $\cdots$, all having period $2a$, in the form

$$f(x) \sim a_0 + \sum_{n=1}^{\infty} a_n \cos\left(\frac{n\pi x}{a}\right) + b_n \sin\left(\frac{n\pi x}{a}\right).$$

The coefficients of this Fourier series may be determined either by scaling from the formulas of Section 1 or through the concept of orthogonality. In either case, the coefficients are

$$\begin{aligned} a_0 &= \frac{1}{2a}\int_{-a}^{a} f(x)dx, \qquad a_n = \frac{1}{a}\int_{-a}^{a} f(x)\cos\left(\frac{n\pi x}{a}\right)dx \\ b_n &= \frac{1}{a}\int_{-a}^{a} f(x)\sin\left(\frac{n\pi x}{a}\right)dx. \end{aligned} \tag{1}$$

Example.
Let $f(x) = |\sin(\pi x)|$, which is periodic with period 1. The Fourier coefficients of f are thus ($a = \frac{1}{2}$)

$$a_0 = \int_{-1/2}^{1/2} |\sin(\pi x)| dx = \frac{2}{\pi}$$

$$a_n = 2\int_{-1/2}^{1/2} |\sin(\pi x)| \cos(2n\pi x) dx = -\frac{4}{\pi}\frac{1}{4n^2-1}$$

$$b_n = 2\int_{-1/2}^{1/2} |\sin(\pi x)| \sin(2n\pi x) dx = 0.$$

Consequently, the Fourier series of the function is

$$|\sin \pi x| \sim \frac{2}{\pi} - \frac{4}{\pi}\sum_{n=1}^{\infty} \frac{1}{4n^2-1} \cos 2n\pi x.$$

It is often necessary to represent by a Fourier series a function that has been defined only in a finite interval. We can justify such a representation by making the given function part of a periodic function. If the given function f is defined on the interval $-a < x < a$, we may construct $\bar{f}$, the *periodic extension* of period $2a$, by using the following definitions:

$$\begin{aligned} \bar{f}(x) &= f(x), & -a < x < a \\ \bar{f}(x) &= f(x+2a), & -3a < x < -a \\ \bar{f}(x) &= f(x-2a), & a < x < 3a \end{aligned}$$

and so on, up and down the x-axis. Notice that the argument of f on the right-hand side always falls in the interval $-a < x < a$, where f was originally given. Graphically, this kind of extension amounts to making a template of the graph of f on $-a < x < a$ and then copying from the template in abutting intervals of length $2a$.

For the extended function with period $2a$, the formulas for the Fourier coefficients become

$$a_0 = \frac{1}{2a}\int_{-a}^{a} \bar{f}(x)dx,$$

$$a_n = \frac{1}{a}\int_{-a}^{a} \bar{f}(x)\cos\left(\frac{n\pi x}{a}\right)dx \tag{2}$$

$$b_n = \frac{1}{a}\int_{-a}^{a} \bar{f}(x)\sin\left(\frac{n\pi x}{a}\right)dx.$$

If we are concerned with $f(x)$ only in the interval $-a < x < a$ where it was originally given, the process of periodic extension is strictly formal, because the formulas for the coefficients involve f only on the original interval. Thus, we may write

$$f(x) \sim a_0 + \sum_{n=1}^{\infty} a_n \cos\left(\frac{n\pi x}{a}\right) + b_n \sin\left(\frac{n\pi x}{a}\right), \qquad -a < x < a.$$

The inequality for x draws attention to the fact that f was defined only on the interval $-a$ to a.

Example.
Suppose $f(x) = x$ in the interval $-1 < x < 1$. The graph of its periodic extension (with period 2) is seen in Fig. 3, and the Fourier coefficients are

$$a_0 = 0, \quad a_n = 0, \quad b_n = \int_{-1}^{1} x\sin(n\pi x)dx = -\frac{2\cos(n\pi)}{n\pi} = \frac{2}{\pi}\frac{(-1)^{n+1}}{n}.$$

The sine and cosine functions that appear in a Fourier series have some special symmetry properties that are useful in evaluating the coefficients. The graph of the cosine function is symmetric about the vertical axis, and that of the sine is antisymmetric. We formalize these properties with a definition.

Definition
A function $g(x)$ is *even* if $g(-x) = g(x)$; $h(x)$ is *odd* if $h(-x) = -h(x)$. Note that a function must be defined on a symmetric interval, say $-c < x < c$ (where c might be ∞), in order to qualify as even or odd.

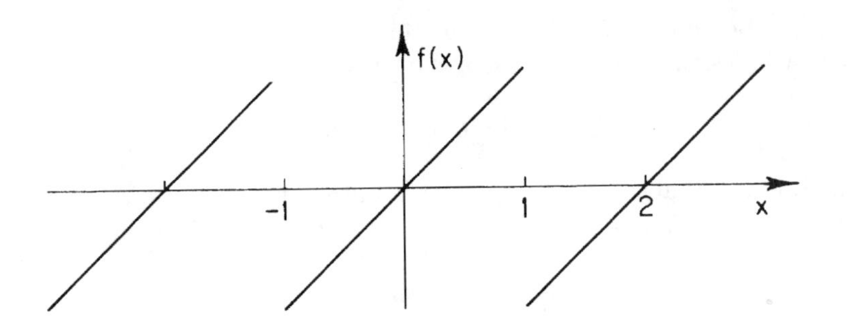

Figure 3: $f(x) = x$, $-1 < x < 1$, f periodic with period 2.

An even function is often said to be symmetric about the vertical axis, and an odd function is said to be symmetric in the origin. Many familiar functions are either even or odd. For example, $\sin(kx)$, x, x^3 and any other odd power of x are all odd functions defined on the interval $-\infty < x < \infty$. Similarly, $\cos(kx)$, $|x|$, $1(= x^0)$, x^2, and any other even power of x are even functions over the same interval. Most functions are neither even nor odd, but any function that is defined on a symmetric interval can be written as a sum of an even and an odd function:

$$f(x) = \frac{1}{2}(f(x) + f(-x)) + \frac{1}{2}(f(x) - f(-x)).$$

It is easy to show that the first term is an even function and the second is odd.

Even and odd functions preserve their symmetries in some algebraic operations, as summarized here:

$$\text{even} + \text{even} = \text{even}, \quad \text{odd} + \text{odd} = \text{odd}$$

$$\text{even} \times \text{even} = \text{even}, \quad \text{odd} \times \text{odd} = \text{even}, \quad \text{odd} \times \text{even} = \text{odd}.$$

We are also concerned with definite integrals of even and odd functions over symmetric intervals. The symmetry properties lead to important simplifications in our calculations.

Theorem 1

Let $g(x)$ be an even function defined in a symmetric interval $-c < x < c$ that includes the interval $-a < x < a$. Then

$$\int_{-a}^{a} g(x)dx = 2\int_{0}^{a} g(x)dx.$$

Let $h(x)$ be an odd function defined in a symmetric interval $-c < x < c$ that includes the interval $-a < x < a$. Then

$$\int_{-a}^{a} h(x)dx = 0.$$

Suppose now that g is an even function in the interval $-a < x < a$. Since the sine function is odd, and the product $g(x)\sin(n\pi x/a)$ is odd,

$$b_n = \frac{1}{a}\int_{-a}^{a} g(x)\sin\frac{n\pi x}{a}dx = 0.$$

That is, all the sine coefficients are zero. Also, since the cosine is even, so is $g(x)\cos(n\pi x/a)$, and then

$$a_n = \frac{1}{a}\int_{-a}^{a} g(x)\cos\left(\frac{n\pi x}{a}\right)dx = \frac{2}{a}\int_{0}^{a} g(x)\cos\left(\frac{n\pi x}{a}\right)dx.$$

Thus the cosine coefficients can be computed from an integral over the interval from 0 to a.

Parallel results hold for odd functions: the cosine coefficients are all zero and the sine coefficients can be simplified. We summarize the results.

Theorem 2
If $g(x)$ is even on the interval $-a < x < a$ $(g(-x) = g(x))$, then

$$g(x) \sim a_0 + \sum_{n=1}^{\infty} a_n \cos\left(\frac{n\pi x}{a}\right), \qquad -a < x < a$$

where

$$a_0 = \frac{1}{a}\int_{0}^{a} g(x)dx, \qquad a_n = \frac{2}{a}\int_{0}^{a} g(x)\cos\left(\frac{n\pi x}{a}\right)dx.$$

If $h(x)$ is odd on the interval $-a < x < a$ $(h(-x) = -h(x))$, then

$$h(x) \sim \sum_{n=1}^{\infty} b_n \sin\left(\frac{n\pi x}{a}\right), \qquad -a < x < a$$

where

$$b_n = \frac{2}{a}\int_{0}^{a} h(x)\sin\left(\frac{n\pi x}{a}\right)dx.$$

Very frequently, a function given in an interval $0 < x < a$ must be represented in the form of a Fourier series. There are infinitely many ways of doing this, but two ways are especially simple and useful: extending the given function to one defined on a symmetric interval $-a < x < a$ by making the extended function either odd or even.

Definition
Let $f(x)$ be given for $0 < x < a$. The *odd extension* of f is defined by

$$f_o(x) = \begin{cases} f(x), & 0 < x < a \\ -f(-x), & -a < x < 0. \end{cases}$$

The *even extension* of f is defined by

$$f_e(x) = \begin{cases} f(x), & 0 < x < a \\ f(-x), & -a < x < 0. \end{cases}$$

Notice that if $-a < x < 0$, then $0 < -x < a$, so the functional values on the right are known from the given functions.

Graphically, the even extension is made by reflecting the graph in the vertical axis. The odd extension is made by reflecting first in the vertical then in the horizontal axis (see Fig. 4).

Now the Fourier series of either extension may be calculated from the formulas above. Since f_e is even and f_o is odd, we have

$$f_e(x) \sim a_0 + \sum_{n=1}^{\infty} a_n \cos\left(\frac{n\pi x}{a}\right), \qquad -a < x < a$$

$$f_o(x) \sim \sum_{n=1}^{\infty} b_n \sin\left(\frac{n\pi x}{a}\right), \qquad -a < x < a.$$

If the series on the right converge, they actually represent periodic functions with period $2a$. The cosine series would represent the *even periodic* extension of f — the periodic extension of f_e; and the sine series would represent the *odd periodic* extension of f.

When the problem at hand is to represent the function $f(x)$ in the interval $0 < x < a$ where it was originally given, we may use either the Fourier sine series or the cosine series because both f_e and f_o coincide with f in the interval.

Thus we may summarize by saying: If $f(x)$ is given for $0 < x < a$, then

$$f(x) \sim a_0 + \sum_{n=1}^{\infty} a_n \cos\left(\frac{n\pi x}{a}\right), \qquad 0 < x < a$$

$$a_0 = \frac{1}{a}\int_0^a f(x)dx, \qquad a_n = \frac{2}{a}\int_0^a f(x)\cos\left(\frac{n\pi x}{a}\right)dx$$

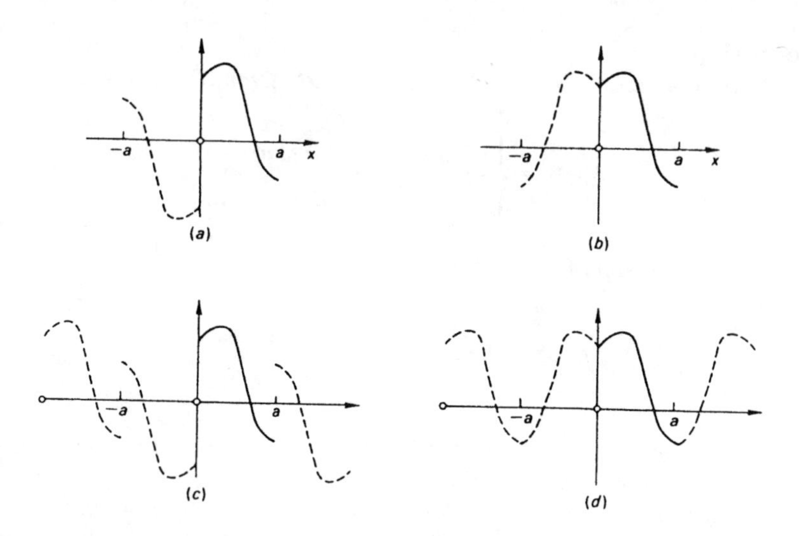

Figure 4: A function is given in the interval $0 < x < a$ (heavy curve). The figure shows: (a) the odd extension; (b) the even extension; (c) the odd periodic extension; and (d) the even periodic extension.

and

$$f(x) \sim \sum_{n=1}^{\infty} b_n \sin\left(\frac{n\pi x}{a}\right), \qquad 0 < x < a$$

$$b_n = \frac{2}{a}\int_0^a f(x)\sin\left(\frac{n\pi x}{a}\right) dx.$$

These two representations are called *half-range expansions*. We shall need these, more than any other kind of Fourier series, in the applications we make later in this book.

Example.
Let us suppose that the function f has the formula

$$f(x) = x, \qquad 0 < x < 1.$$

Then the odd periodic extension of f is as shown in Fig. 5, and the Fourier sine coefficients of f are

$$b_n = 2\int_0^1 x\sin(n\pi x)\ dx = -\frac{2}{n\pi}\cos(n\pi)$$

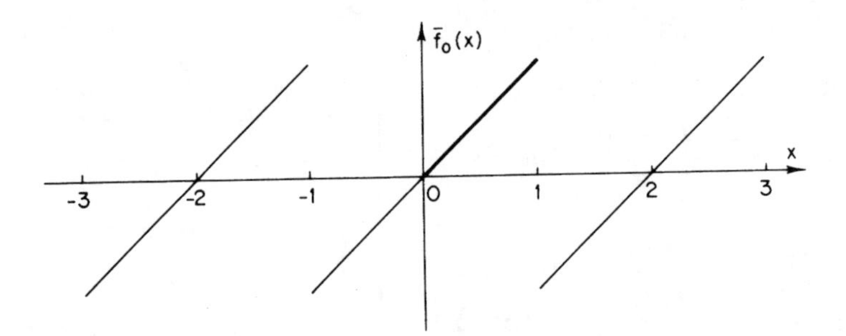

Figure 5: Odd periodic extension (period 2) of $f(x) = x$, $0 < x < 1$.

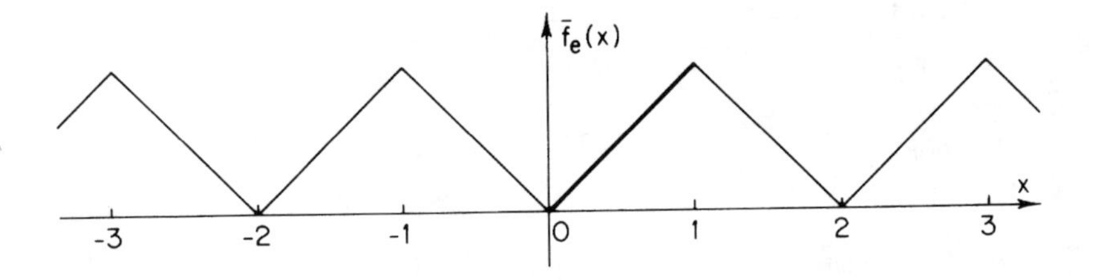

Figure 6: Even periodic extension (period 2) of $f(x) = x$, $0 < x < 1$.

The even periodic extension of f is shown in Fig. 6. The Fourier cosine coefficients are

$$
\begin{aligned}
a_0 &= \int_0^1 x dx = \frac{1}{2} \\
a_n &= 2\int_0^1 x\cos(n\pi x)dx = -\frac{2}{n^2\pi^2}(1-\cos(n\pi)).
\end{aligned}
$$

The following six correspondences (we will later show them to be equalities) follow from the ideas of this section. Note that the inequalities showing the applicable range of x are crucial.

$$\sum_{n=1}^{\infty} \frac{-2\cos(n\pi)}{n\pi} \sin(n\pi x) \sim \begin{cases} f(x) = x, & 0 < x < 1 \\ f_o(x) = x, & -1 < x < 1 \\ \bar{f}_o(x), & -\infty < x < \infty \end{cases}$$

$$\frac{1}{2} - \sum_{n=1}^{\infty} \frac{2(1-\cos(n\pi))}{n^2\pi^2} \cos(n\pi x) \sim \begin{cases} f(x) = x, & 0 < x < 1 \\ f_e(x) = |x|, & -1 < x < 1 \\ \bar{f}_e(x), & -\infty < x < \infty \end{cases}$$

Exercises

1. Find the Fourier series of each of the following functions. Sketch the graph of the periodic extension of f for at least two periods.

a. $f(x) = |x|, \quad -1 < x < 1$

b. $f(x) = \begin{cases} -1, & -2 < x < 0 \\ 1, & 0 < x < 2 \end{cases}$

c. $f(x) = x^2, \quad -\frac{1}{2} < x < \frac{1}{2}$

2. Show that the functions $\cos(n\pi x/a)$ and $\sin(n\pi x/a)$ satisfy orthogonality relations similar to those given in Section 1.

3. Suppose a Fourier series is needed for a function defined in the interval $0 < x < 2a$. Show how to construct a periodic extension with period $2a$ and give formulas for the Fourier coefficients which use only integrals from 0 to $2a$. (Hint: see Exercise 5, Section 1.)

4. Show that the formula

$$e^x = \cosh(x) + \sinh(x)$$

gives the decomposition of the function e^x into a sum of an even and an odd function.

5. Identify each of the following as being even, odd, or neither. Sketch.

a. $f(x) = x$ **b.** $f(x) = |x|$

c. $f(x) = |\cos(x)|$ **d.** $f(x) = \arcsin(x)$

e. $f(x) = x\cos(x)$ **f.** $f(x) = x + \cos(x+1)$

6. If $f(x)$ is given in the interval $0 < x < a$, what other ways are there to extend it to a function on $-a < x < a$?

7. Find the Fourier series of the functions:

a. $f(x) = x, \quad -1 < x < 1$ **b.** $f(x) = 1, \quad -2 < x < 2$

c. $f(x) = \begin{cases} x, & -\frac{1}{2} < x < \frac{1}{2} \\ 1 - x, & \frac{1}{2} < x < \frac{3}{2} \end{cases}$

8. Is it true, if all the sine coefficients of a function f defined on $-a < x < a$ are zero, then f is even?

9. We know that if $f(x)$ is odd on the interval $-a < x < a$, its Fourier series is composed only of sines. What additional symmetry condition on f will make the sine coefficients with even indices be zero? Give an example.

10. Sketch both the even and odd extensions of the functions:

a. $f(x) = 1, \quad 0 < x < a$ **b.** $f(x) = x, \quad 0 < x < a$

c. $f(x) = \sin(x), \quad 0 < x < 1$ **d.** $f(x) = \sin(x), \quad 0 < x < \pi$

11. Find the Fourier sine series and cosine series for the functions given in Exercise 10. Sketch the even and odd periodic extensions for several periods.

12. Prove the orthogonality relations

$$\int_0^a \sin\left(\frac{n\pi x}{a}\right)\sin\left(\frac{m\pi x}{a}\right)dx = \begin{cases} 0, & n \neq m \\ a/2, & n = m \end{cases}$$

$$\int_0^a \cos\left(\frac{m\pi x}{a}\right)\cos\left(\frac{n\pi x}{a}\right)dx = \begin{cases} 0, & n \neq m \\ a/2, & n = m \neq 0 \\ a, & n = m = 0. \end{cases}$$

13. If $f(x)$ is continuous on the interval $0 < x < a$, is its even periodic extension continuous? What about the odd periodic extension? Check especially at $x = 0$ and $\pm a$.

14. Justify Theorem 1 by considering the integral as a sum of signed areas. See Fig. 4 for typical even and odd functions.

15. Justify or prove these statements.

a. If $h(x)$ is an odd function, then $|h(x)|$ is an even function.

b. If $f(x)$ is defined for all positive x, then $f(|x|)$ is an even function.

c. If $f(x)$ is defined for all x and $g(x)$ is any even function, then $f(g(x))$ is even.

d. If $h(x)$ is an odd function, $g(x)$ is even, and $g(x)$ is defined for all x, then $g(h(x))$ is an even function.

1.3 CONVERGENCE OF FOURIER SERIES

Now we are ready to take up the second question of Section 1: Does the Fourier series of a function actually represent that function? The word *represent* has many interpretations, but for most practical purposes, we really want to know the answer to this question:

If a value of x is chosen, the numbers $\cos(n\pi x/a)$ and $\sin(n\pi x/a)$ are computed for each n and inserted into the Fourier series of f, and the sum of the series is calculated, is that sum equal to the functional value $f(x)$?

In this section we shall state, without proof, some theorems that answer the question (a proof of the convergence theorem is given in Section 7). But first we need a few definitions about limits and continuity.

The ordinary limit $\lim_{x \to x_0} f(x)$ can be rewritten as $\lim_{h \to 0} f(x_0 + h)$. Here h may approach zero in any manner. But if h is required to be positive only, we get what is called the *right-hand limit* of f at x_0, defined by

$$f(x_0+) = \lim_{h \to 0+} f(x_0 + h) = \lim_{\substack{h \to 0 \\ h > 0}} f(x_0 + h).$$

The *left-hand limit* is defined similarly:

$$f(x_0-) = \lim_{h \to 0-} f(x_0 + h) = \lim_{\substack{h \to 0 \\ h < 0}} f(x_0 + h) = \lim_{h \to 0+} f(x_0 - h).$$

Note that $f(x_0+)$ and $f(x_0-)$ need not be values of the function f.

If both left- and right-hand limits exist and are equal, the ordinary limit exists and is equal to the one-handed limits. It is quite possible that the left- and right-handed limits exist but are different. This happens, for instance, at $x = 0$ for the function

$$f(x) = \begin{cases} 1, & 0 < x < \pi \\ -1, & -\pi < x < 0. \end{cases}$$

In this case, the left-hand limit at $x_0 = 0$ is -1, whereas the right-hand limit is $+1$. A discontinuity at which the one-handed limits exist, but do not agree, is called a *jump discontinuity*.

It is also possible that at some point both limits exist and agree, but that the function is not defined at that point. In such a case, a function is said to have a *removable discontinuity*. If the value of the function at the troublesome point is redefined to be equal to the limit, the function will become continuous. For example, the function $f(x) = \sin(x)/x$ has a removable discontinuity at $x = 0$. The discontinuity is eliminated by redefining $f(x) = \sin(x)/x \quad (x \neq 0)$, $f(0) = 1$. Removable discontinuities are so simple that we may assume they have been removed from any function under discussion.

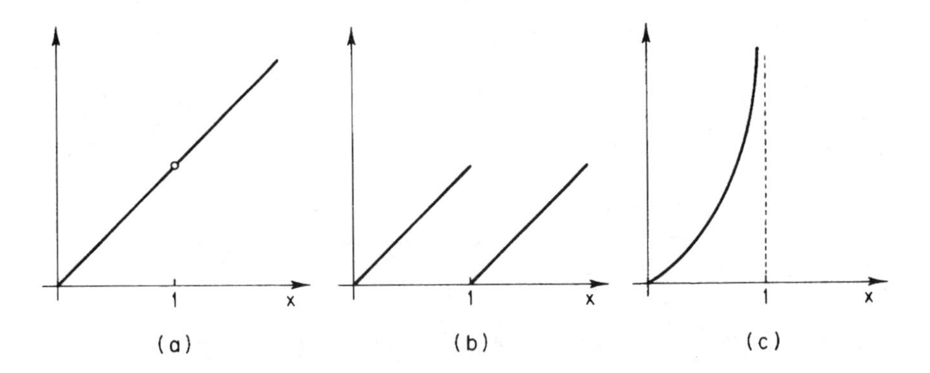

Figure 7: Three functions with different kinds of discontinuities at $x = 1$. (*a*) $f(x) = (x - x^2)/(1 - x)$ has a removable discontinuity. (*b*) $f(x) = x$ for $0 < x < 1$ and $f(x) = x - 1$ for $1 < x$; this function has a jump discontinuity. (*c*) $f(x) = -\ln(|1 - x|)$ has a "bad" discontinuity.

Other discontinuities are more serious. They occur if one or both of the one-handed limits fail to exist. Each of the functions $\sin(1/x)$, $e^{1/x}$, $1/x$ has a discontinuity at $x = 0$ that is neither removable nor a jump (see Fig. 7). Table 1 summarizes continuity behavior at a point.

We shall say that a function is *sectionally continuous* on an interval $a < x < b$ if it is bounded and continuous, except possibly for a finite number of jumps and removable discontinuities. (See Fig. 8.) A function is sectionally continuous (without qualification) if it is sectionally continuous on every interval of finite length. For instance, if a periodic function is sectionally continuous on any interval whose length is one period or more, then it is sectionally continuous.

Table 1: Types of continuity behavior at x_0

Name	Criterion
Continuity	$f(x_0+) = f(x_0-) = f(x_0)$
Removable discontinuity	$f(x_0+) = f(x_0-) \neq f(x_0)$
Jump discontinuity	$f(x_0+) \neq f(x_0-)$
"Bad" discontinuity	$f(x_0+)$ or $f(x_0-)$ or both fail to exist

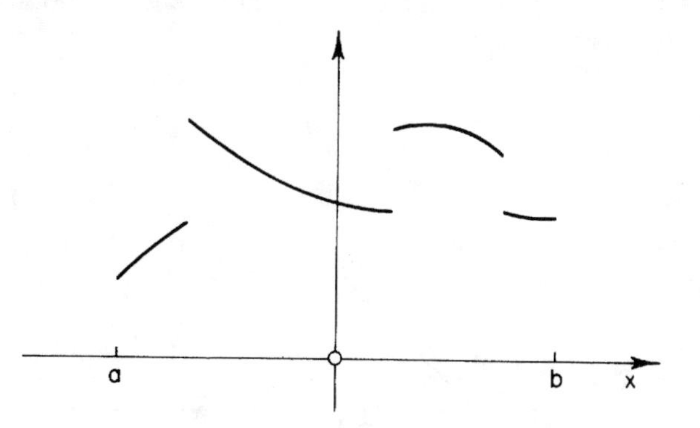

Figure 8: Typical sectionally continuous function made up of four continuous "sections."

Examples.

1. The *square wave*, defined by

$$f(x) = \begin{cases} 1, & 0 < x < a, \\ -1, & -a < x < 0, \end{cases} \qquad f(x+2a) = f(x)$$

is sectionally continuous. There are jump discontinuities at $x = 0$, $\pm a$, $\pm 2a$, etc.

2. The function $f(x) = 1/x$ cannot be sectionally continuous on any interval that contains 0 or even has 0 as an endpoint, because the function is not bounded at $x = 0$.

3. If $f(x) = x$, $-1 < x < 1$, then f is continuous on that interval. Its periodic extension is *sectionally continuous* but not continuous.

The examples clarify a couple of facts about the meaning of sectional continuity. Most important is that a sectionally continuous function must not "blow up" at any point — even an endpoint — of an interval. Note also that a function need not be defined at every point in order to qualify as sectionally continuous. No value was given for the square wave function at $x = 0, \pm a$, but the function remains sectionally continuous, no matter what values are assigned for these points.

A function is *sectionally smooth* in an interval $a < x < b$ if: f is sectionally continuous; $f'(x)$ exists, except perhaps at a finite number of points; and $f'(x)$ is sectionally continuous. The graph of a sectionally smooth function then has a finite number of removable discontinuities, jumps, and corners. (The derivative will not exist at these points.) *Between* these points, the graph will be continuous, with a continuous derivative. No vertical tangents are allowed, for these indicate that the derivative is infinite.

Examples.

1. $f(x) = |x|^{1/2}$ is continuous, but not sectionally smooth in any interval that contains 0, because $|f'(x)| \to \infty$ as $x \to 0$.

2. The square wave is sectionally smooth, but not continuous.

Most of the functions useful in mathematical modeling are sectionally smooth. Fortunately we can also give a positive statement about the Fourier series of such functions.

Theorem
If $f(x)$ is sectionally smooth and periodic with period $2a$, then at each point x the Fourier series corresponding to f converges, and its sum is

$$a_0 + \sum_{n=1}^{\infty} a_n \cos\left(\frac{n\pi x}{a}\right) + b_n \sin\left(\frac{n\pi x}{a}\right) = \frac{f(x+) + f(x-)}{2}.$$

This theorem gives an answer to the question at the beginning of the section. Recall that a sectionally smooth function has only a finite number of jumps and no bad discontinuities in every finite number of jumps and no bad discontinuities in every finite interval. Hence,

$$f(x-) = f(x+) = \frac{1}{2}\left(f(x+) + f(x-)\right) = f(x),$$

except perhaps at a finite number of points on any finite interval. For this reason, if f satisfies the hypotheses of the theorem, we write f *equal* to its Fourier series, even though the equality may fail at jumps.

In constructing the periodic extension of a function, we never defined the values of $f(x)$ at the endpoints. Since the Fourier coefficients are given by integrals, the value assigned to $f(x)$ at one point cannot influence them; in that sense, the value of f at $x = +a$ is unimportant. But because of the averaging features of the Fourier series, it is reasonable to define

$$f(a) = f(-a) = \frac{1}{2}\left(f(a-) + f(-a+)\right).$$

That is, the value of f at the endpoints is the average of the one-handed limits at the endpoints, each limit taken from the interior. For instance, if $f(x) = 1 + x$, $0 < x < 1$, and $f(x) = 0$, $-1 < x < 0$, then $f(\pm 1)$ should be taken to be 1, and $f(0)$ should be $\frac{1}{2}$.

Examples.

1. The square wave function

$$f(x) = \begin{cases} 1, & 0 < x < 1 \\ -1, & -1 < x < 0 \end{cases}$$

is sectionally smooth; therefore the corresponding Fourier series converges to

$$\begin{cases} 1, & \text{for} \quad 0 < x < 1 \\ -1, & \text{for} \quad -1 < x < 0 \\ 0, & \text{for} \quad x = 0, 1, -1, \end{cases}$$

and is periodic with period 2.

2. For the function $f(x) = |x|^{1/2}$, $-\pi < x < \pi$, $f(x + 2\pi) = f(x)$, the preceding theorem does not guarantee convergence of the Fourier series at any point, even though the function is continuous. Nevertheless, the series does converge at any point x! This shows that the conditions in the theorem above are perhaps too strong. (But they are useful.)

Exercises

1. In each part that follows, a function is equated to its Fourier series as justified by the Theorem of this section. By evaluating both sides of the equality at an appropriate value of x, derive the second equality.

a. $$|x| = \frac{1}{2} - \frac{4}{\pi^2}\sum_{k=0}^{\infty}\frac{1}{(2k+1)^2}\cos((2k+1)\pi x), \qquad -1 < x < 1$$

$$\frac{\pi^2}{8} = 1 + \frac{1}{9} + \frac{1}{25} + \cdots$$

b. $$\frac{4}{\pi}\sum_{k=0}^{\infty}\frac{1}{2k+1}\sin((2k+1)\pi x) = \begin{cases} 1, & 0 < x < 1 \\ -1, & -1 < x < 0 \end{cases}$$

$$\frac{\pi}{4} = 1 - \frac{1}{3} + \frac{1}{3} + \frac{1}{5} - \frac{1}{7} + \cdots$$

c. $$|\sin(x)| = \frac{2}{\pi} - \frac{4}{\pi}\sum_{n=1}^{\infty}\frac{1}{4n^2-1}\cos(2nx)$$

2. Check each function described in what follows to see whether it is sectionally smooth. If it is, state the value to which its Fourier series converges at each point x in the interval and at the end points. Sketch.

a. $f(x) = |x| + x, \quad -1 < x < 1$

b. $f(x) = x\cos(x), \quad -\frac{\pi}{2} < x < \frac{\pi}{2}$

c. $f(x) = x\cos(x), \quad -1 < x < 1$

d. $$f(x) = \begin{cases} 0, & 1 < x < 3 \\ 1, & -1 < x < 1 \\ x, & -3 < x < -1 \end{cases}$$

3. To what value does the Fourier series of f converge if f is a *continuous*, sectionally smooth, periodic function?

4. State convergence theorems for the Fourier sine and cosine series that arise from half-range expansions.

1.4 UNIFORM CONVERGENCE

The Theorem of the preceding section treats convergence at individual points of an interval. A stronger kind of convergence is uniform convergence in an interval. Let

$$S_N(x) = a_0 + \sum_{n=1}^{N} a_n \cos\left(\frac{n\pi x}{a}\right) + b_n \sin\left(\frac{n\pi x}{a}\right)$$

be the partial sum of the Fourier series of a function f. The maximum deviation between the graphs of $S_N(x)$ and $f(x)$ is

$$\delta_N = \max|f(x) - S_N(x)|, \qquad -a \le x \le a,$$

where the maximum[2] is taken over all x in the interval, including the end points. If the maximum deviation tends to zero as N increases, we say that the series *converges uniformly* in the interval $-a \le x \le a$.

Roughly speaking, if a Fourier series converges uniformly, then the sum of a finite number N of terms gives a good approximation — to within $\pm\delta_N$ — of the value of $f(x)$ at *any* and *every* point of the interval. Furthermore, by taking a large enough N, one can make the error as small as necessary.

There are two important facts about uniform convergence. If a Fourier series converges uniformly in a period interval, then (1) it must converge to a continuous function, and (2) it must converge to the (continuous) function that generates the series. Thus, a function that has a nonremovable discontinuity *cannot* have a uniformly convergent Fourier series. (And not all continuous functions have uniformly convergent Fourier series.)

Figure 9 presents graphs of some partial sums of a square-wave function. It is easy to see that for every N there are points near $x = 0$ and $x = \pm\pi$ where $|f(x) - S_N(x)|$ is nearly equal to 1, so convergence is *not* uniform. (Incidentally, the graphs in Fig. 9 also show the partial sums of $f(x)$ overshooting their mark near $x = 0$. This feature of Fourier series is called *Gibbs' phenomenon* and always occurs near a jump.) On the other hand, Fig. 10 shows graphs of a "sawtooth" function and the partial sums of its Fourier series. The maximum deviation always occurs at $x = 0$, and the convergence is uniform.

One of the ways of proving uniform convergence is by examining the coefficients.

---

[2]If f is not continuous, the maximum must be replaced by a supremum, or least upper bound.

Theorem 1

If the series $\sum_{n=1}^{\infty}(|a_n| + |b_n|)$ converges, then the Fourier series

$$a_0 + \sum_{n=1}^{\infty} a_n \cos\left(\frac{n\pi x}{a}\right) + b_n \sin\left(\frac{n\pi x}{a}\right)$$

converges uniformly in the interval $-a \leq x \leq a$, and, in fact, on the whole interval $-\infty < x < \infty$.

Example.

For the function

$$f(x) = |x|, \qquad -\pi < x < \pi$$

the Fourier coefficients are

$$a_0 = \frac{\pi}{2}, \qquad a_n = \frac{2}{\pi}\frac{\cos(n\pi) - 1}{n^2}, \qquad b_n = 0.$$

Since the series $\sum_{n=1}^{\infty} 1/n^2$ converges, the series of absolute values of the coefficients converges, and so the Fourier series converges uniformly on the interval $-\pi \leq x \leq \pi$ to $|x|$. The Fourier series converges uniformly to the periodic extension of $f(x)$ on the whole real line (see Fig. 10).

Another way of proving uniform convergence of a Fourier series is by examining the function f that generates it.

Theorem 2

If f is periodic, continuous, and has a sectionally continuous derivative, then the Fourier series corresponding to f converges uniformly to $f(x)$ on the entire real axis.

While this theorem is stated for a periodic function, it may be adapted to a function $f(x)$ given on the interval $-a < x < a$. If the *periodic extension* of f satisfies the conditions of the theorem, then the Fourier series of f converges uniformly on the interval $-a \leq x \leq a$.

Example.

Consider the function

$$f(x) = x, \qquad -1 < x < 1.$$

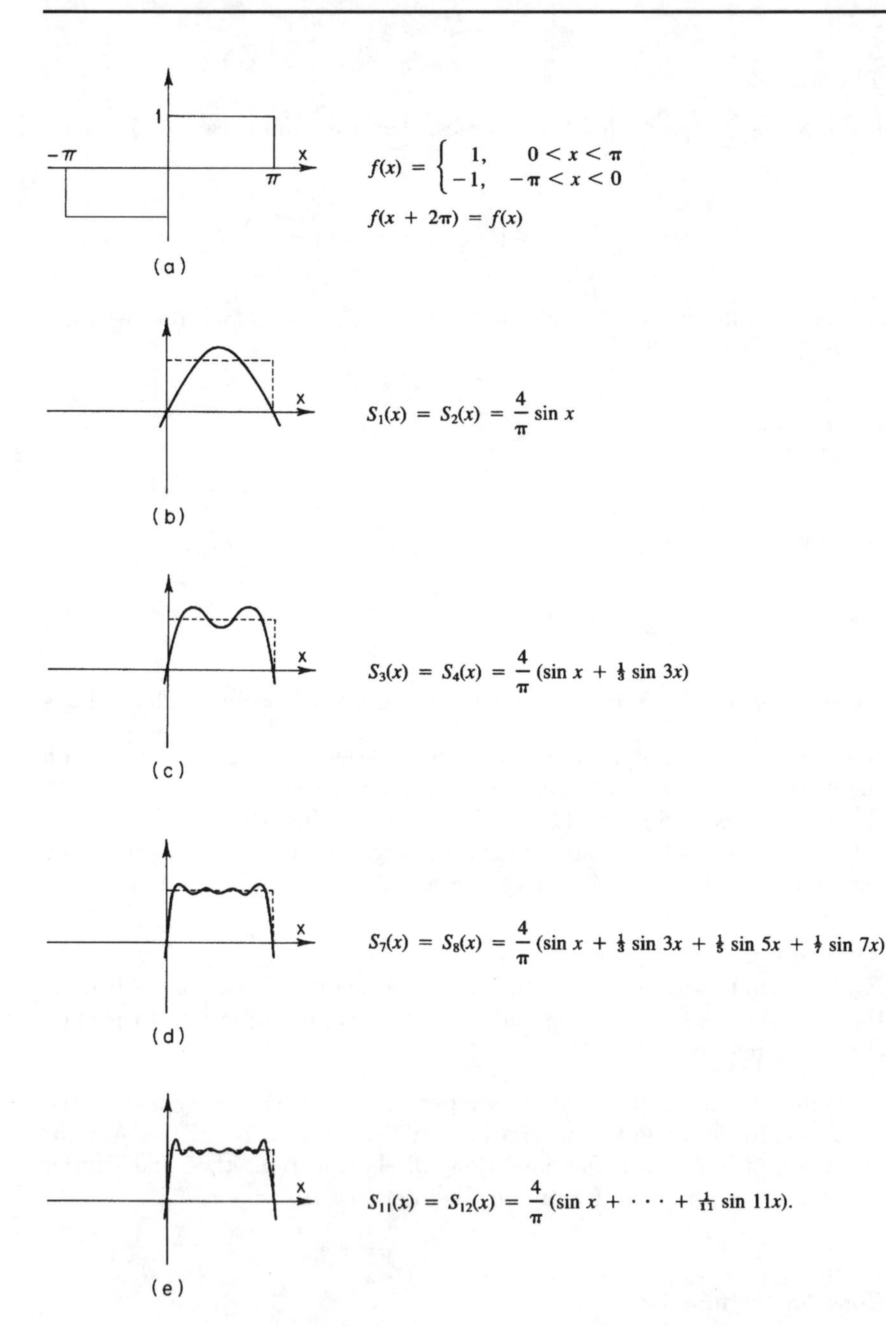

Figure 9: Partial sums of the square wave function. Convergence is *not* uniform.

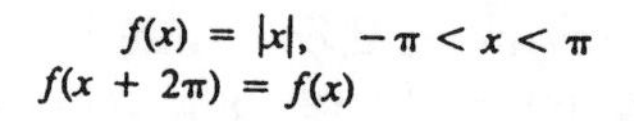

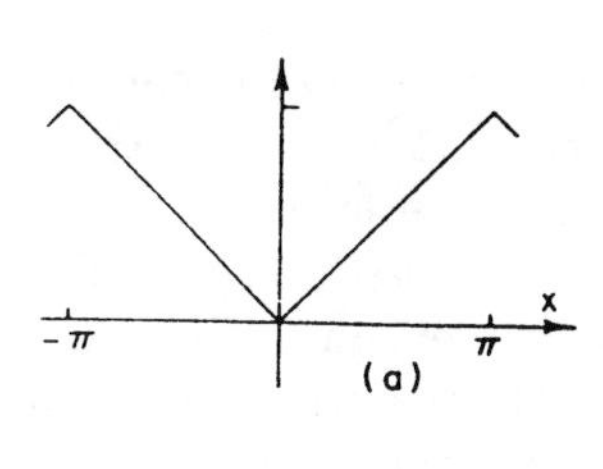

$$S_0(x) = \frac{\pi}{2}$$

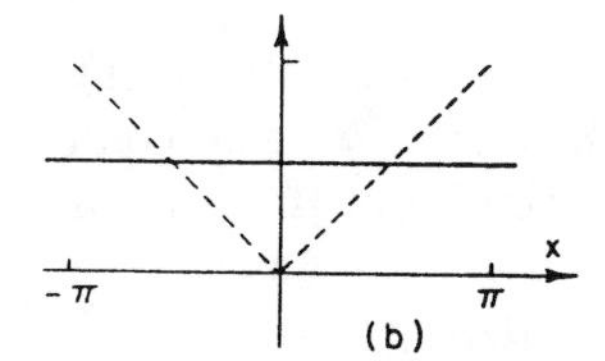

$$S_2(x) = S_1(x) = \frac{\pi}{2} - \frac{4}{\pi}\cos x$$

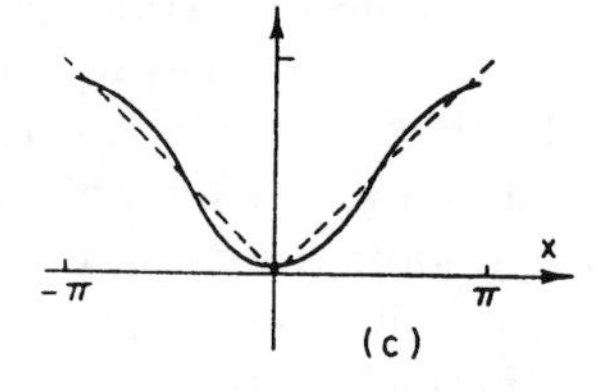

$$S_3(x) = S_4(x) = \frac{\pi}{2} - \frac{4}{\pi}\left(\cos x + \tfrac{1}{9}\cos 3x\right)$$

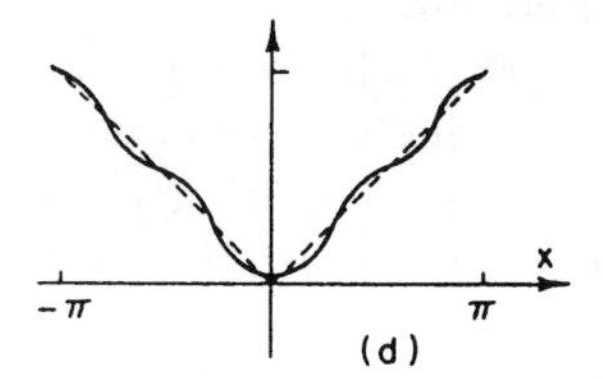

$$S_5(x) = S_6(x) = \frac{\pi}{2} - \frac{4}{\pi}\left(\cos x + \tfrac{1}{9}\cos 3x + \tfrac{1}{25}\cos 5x\right)$$

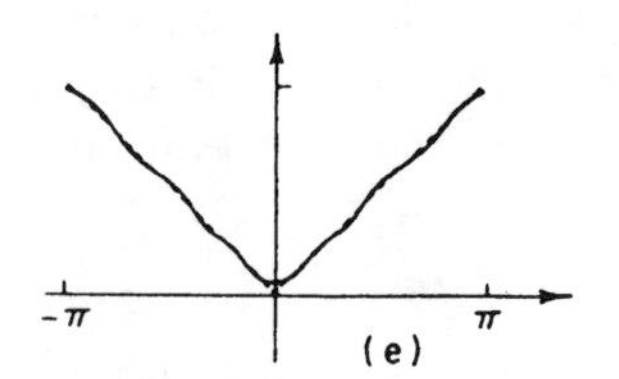

Figure 10: Partial sums of a sawtooth function. Convergence is uniform.

Although $f(x)$ is continuous and has a continuous derivative in the interval $-1 < x < 1$, the periodic extension of f is *not* continuous. The Fourier series cannot converge uniformly in any interval containing 1 or -1 because the periodic extension of f has jumps there, but uniform convergence must produce a continuous function.

On the other hand, the function $f(x) = |\sin(x)|$, periodic with period 2π, is continuous and has a sectionally continuous derivative. Therefore, its Fourier series converges uniformly to $f(x)$ everywhere.

Here is a restatement of Theorem 2 for a function given on the interval $-a < x < a$. The condition at the endpoints replaces the condition of continuity of the periodic extension of f.

Theorem 3
If $f(x)$ is given on $-a < x < a$, if f is continuous and bounded and has a sectionally continuous derivative, and if $f(-a+) = f(a-)$, then the Fourier series of f converges uniformly to f on the interval $-a \leq x \leq a$. (The series converges to $f(a-) = f(-a+)$ at $x = \pm a$.)

If an odd periodic function is to be continuous, it must have value 0 at $x = 0$ and at the endpoints of the symmetric period-interval. Thus, the odd periodic extension of a function given in $0 < x < a$ may have jump discontinuities even though it is continuous where originally given. The even periodic extension causes no such difficulty, however.

Theorem 4
If $f(x)$ is given on $0 < x < a$, if f is continuous and bounded and has a sectionally continuous derivative, and if $f(0+) = f(a-) = 0$, then the Fourier sine series of f converges uniformly to f in the interval $0 \leq x \leq a$. (The series converges to 0 at $x = 0$ and $x = a$.)

Theorem 5
If $f(x)$ is given on $0 < x < a$, and if f is continuous and bounded and has a sectionally continuous derivative, then the Fourier cosine series of f converges uniformly to f in the interval $0 \leq x \leq a$. (The series converges to $f(0+)$ at $x = 0$ and to $f(a-)$ at $x = a$.)

Exercises

1. Determine whether the series of the following functions converge uniformly or not. Sketch each function.

   a. $f(x) = e^x, \qquad -1 < x < 1$

b. $f(x) = \sinh(x), \qquad -\pi < x < \pi$

c. $f(x) = \sin(x), \qquad -\pi < x < \pi$

d. $f(x) = \sin(x) + |\sin(x)|, \qquad -\pi < x < \pi$

e. $f(x) = x + |x|, \qquad -\pi < x < \pi$

f. $f(x) = x(x^2 - 1), \qquad -1 < x < 1$

g. $f(x) = 1 + 2x - 2x^3, \qquad -1 < x < 1$

2. The Fourier series of the function

$$f(x) = \frac{\sin(x)}{x}, \qquad -\pi < x < \pi$$

converges at every point. To what value does the series converge at $x = 0$? At $x = \pi$? The convergence is uniform. Why?

3. Determine whether the sine and cosine series of the following functions converge uniformly. Sketch.

a. $f(x) = \sinh(x), \qquad 0 < x < \pi$

b. $f(x) = \sin(x), \qquad 0 < x < \pi$

c. $f(x) = \sin(\pi x), \qquad 0 < x < \frac{1}{2}$

d. $f(x) = 1/(1 + x), \qquad 0 < x < 1$

e. $f(x) = 1/(1 + x^2), \qquad 0 < x < 2$

4. If a_n and b_n tend to zero as n tends to infinity, show that the series

$$a_0 + \sum_{n=1}^{\infty} e^{-\alpha n} \left(a_n \cos(nx) + b_n \sin(nx)\right)$$

converges uniformly ($\alpha > 0$).

5. Determine which of the series of Exercise 1 of Section 1.3 converge uniformly.

1.5 OPERATIONS ON FOURIER SERIES

In the course of this book we shall have to perform certain operations on Fourier series. The purpose of this section is to find conditions under which they are legitimate. Two things must be noted, however. First, the theorems stated here are not the best possible: There are theorems with weaker hypotheses and the same conclusions. Second, in applying mathematics, we often carry out operations formally, legitimate or not. The results must then be checked for correctness.

Throughout this section we shall state results about functions and Fourier series with period 2π, for typographic convenience. The results remain true when the period is $2a$ instead. For functions defined only on a finite interval, the periodic extension must fulfill the hypotheses. We shall refer to a function $f(x)$ with the series shown.

$$f(x) \sim a_0 + \sum_{n=1}^{\infty} a_n \cos(nx) + b_n \sin(nx) \tag{1}$$

Theorem 1
The Fourier series of the function $cf(x)$ has coefficients ca_0, ca_n, and cb_n (c is constant).

This theorem is a simple consequence of the fact that a constant passes through an integral. The fact that the integral of a sum is the sum of the integrals leads to the following.

Theorem 2
The Fourier coefficients of the sum $f(x) + g(x)$ are the sums of the corresponding coefficients of $f(x)$ and $g(x)$.

These two theorems are so natural that the reader has probably used them already without thinking about it. The theorems that follow are much more difficult to prove, but extremely important.

Theorem 3
If $f(x)$ is periodic and sectionally continuous, then the Fourier series of f may be integrated term by term:

$$\int_a^b f(x)dx = \int_a^b a_0 dx + \sum_{n=1}^{\infty} \int_a^b (a_n \cos(nx) + b_n \sin(nx))\, dx. \tag{2}$$

Theorem 4
If $f(x)$ is periodic and sectionally continuous, and if $g(x)$ is sectionally continuous for $a \leq x \leq b$, then

$$\int_a^b f(x)g(x)dx = \int_a^b a_0 g(x)dx + \sum_{n=1}^{\infty} \int_a^b \left(a_n \cos(nx) + b_n \sin(nx)\right) g(x)dx. \tag{3}$$

In Theorems 3 and 4, the function $f(x)$ is only required to be sectionally continuous. It is not necessary that the Fourier series of $f(x)$ converge at all. Nevertheless, the theorems guarantee that the series on the right converges and equals the integral on the left in Eqs. (2) and (3).

One important application of Theorem 4 was the derivation of the formulas for the Fourier coefficients in Section 1. An application of Theorems 3 and 4 is given in what follows.

Example.
The periodic function $g(x)$ whose formula in the interval $0 < x < 2\pi$ is

$$g(x) = x, \qquad 0 < x < 2\pi$$

has the Fourier series

$$g(x) \sim \pi - 2\sum_{n=1}^{\infty} \frac{\sin(nx)}{n}.$$

By applying Theorems 1 and 2, we find that the function $f(x)$ defined by $f(x) = [\pi - g(x)]/2$ has the series

$$f(x) \sim \sum_{n=1}^{\infty} \frac{\sin(nx)}{n}.$$

This manipulation would be simple algebra if the correspondence $\sim$ were an equality.

The function $f(x)$ satisfies the hypotheses of Theorem 3. Thus we may integrate the preceding series from 0 to b to obtain

$$\int_0^b f(x)dx = \sum_{n=1}^{\infty} \frac{1 - \cos(nb)}{n^2}.$$

Theorem 3 guarantees that this equality holds for any b. In the interval from 0 to 2π we have the formula $f(x) = (\pi x)/2$. Hence

$$\int_0^b f(x)dx = \frac{\pi b}{2} - \frac{b^2}{4} = \sum_{n=1}^{\infty} \frac{1 - \cos(nb)}{n^2}, \quad 0 \leq b \leq 2\pi.$$

Now, replacing b by x, we have

$$\frac{x(2\pi - x)}{4} = \sum_{n=1}^{\infty} \frac{1}{n^2} - \sum_{n=1}^{\infty} \frac{\cos(nx)}{n^2}, \quad 0 \leq x \leq 2\pi. \tag{4}$$

Outside the indicated interval, the periodic extension of the function on the left equals the series on the right.

It is worthwhile to mention that the series on the right of Eq. (4) is the Fourier series of the function on the left. That is to say,

$$\frac{1}{2\pi} \int_0^{2\pi} \frac{x(2\pi - x)}{4} dx = \sum_{n=1}^{\infty} \frac{1}{n^2} \tag{5}$$

$$\frac{1}{\pi} \int_0^{2\pi} \frac{x(2\pi - x)}{4} \cos(nx) dx = \frac{-1}{n^2} \tag{6}$$

$$\frac{1}{\pi} \int_0^{2\pi} \frac{x(2\pi - x)}{4} \sin(nx) dx = 0. \tag{7}$$

Equations (6) and (7) can be verified directly, of course, but Theorem 4, together with the orthogonality relations of Section 1, also guarantees them. In addition, Eq. (5) gives us a way to evaluate the series on the right.

Although the uniqueness property stated in this theorem is so very natural that we tend to assume it is true without checking, it really is a consequence of Theorem 4.

Theorem 5
If $f(x)$ is periodic and sectionally continuous, its Fourier series is unique.

That is to say, only one series can correspond to $f(x)$. We often make use of uniqueness in this way: if two Fourier series are equal (or correspond to the same function), then the coefficients of like terms must match.

The last operation to be discussed is differentiation, one that plays a principal role in applications.

Theorem 6
If $f(x)$ is periodic, continuous, and sectionally smooth, then the differentiated Fourier series of $f(x)$ converges to $f'(x)$ at every point x where $f''(x)$ exists:

$$f'(x) = \sum_{n=1}^{\infty} (-na_n \sin(nx) + nb_n \cos(nx)). \tag{8}$$

The hypotheses on $f(x)$ itself imply (see Section 4) that the Fourier series of $f(x)$ converges uniformly. If $f(x)$ (or its periodic extension) fails to be continuous, it is certain that the differentiated series of $f(x)$ will fail to converge, at some points at least.

As an example, take the function that is periodic with period 2π and has the formula

$$f(x) = |x|, \qquad -\pi < x < \pi.$$

This function is indeed continuous and sectionally smooth and is equal to its Fourier series,

$$f(x) = \frac{\pi}{2} - \frac{4}{\pi}\left(\cos(x) + \frac{\cos(3x)}{9} + \frac{\cos(5x)}{25} + \cdots\right).$$

According to Theorem 5, the differentiated series

$$\frac{4}{\pi}\left(\sin(x) + \frac{\sin(3x)}{3} + \frac{\sin(5x)}{5} + \cdots\right)$$

converges to $f'(x)$ at any point x where $f''(x)$ exists. Now, the derivative of the sawtooth function $f(x)$ (see Fig. 10) is the square-wave function

$$f'(x) = \begin{cases} 1, & 0 < x < \pi \\ -1, & -\pi < x < 0 \end{cases} \tag{9}$$

(see Fig. 9). Moreover, we know that the foregoing sine series is the Fourier series of the square wave $f'(x)$ and that it converges to the values given by Eq. (9), except at the points where $f'(x)$ has a jump. These are precisely the points where $f''(x)$ does not exist.

Later on, it will frequently happen that we know a function only through its Fourier series. Thus, it will be important to obtain properties of the function by examining its coefficients, as the next theorem does.

Theorem 7
If f is periodic, with Fourier coefficients a_n, b_n, and if the series

$$\sum_{n=1}^{\infty}(|n^k a_n| + |n^k b_n|)$$

converges for some integer $k \geq 1$, then f has continuous derivatives $f', \cdots, f^{(k)}$ whose Fourier series are differentiated series of f.

For example, consider the function defined by the series

$$f(x) = \sum_{n=1}^{\infty} e^{-n\alpha} \cos(nx),$$

in which α is a positive parameter. For this function we have $a_0 = 0$, $a_n = e^{-n\alpha}$, $b_n = 0$. By the integral test, the series $\sum n^k e^{-n\alpha}$ converges for any k. Therefore f has derivatives of all orders. The Fourier series of f' and f'' are

$$f'(x) = \sum_{n=1}^{\infty} -ne^{-n\alpha} \sin(nx),$$

$$f''(x) = \sum_{n=1}^{\infty} -n^2 e^{-n\alpha} \cos(nx).$$

Exercises

1. Evaluate the sum of the series $\sum_{n=1}^{\infty} 1/n^2$ by performing the integration indicated in Eq. (5).

2. Sketch the graphs of the periodic extension of the function

$$f(x) = \frac{\pi - x}{2}, \quad 0 < x < 2\pi$$

and of its derivative $f'(x)$ and of

$$F(x) = \int_0^x f(t)dt.$$

3. Suppose that a function has the formula $f(x) = x$, $0 < x < \pi$. What is its derivative? Can the Fourier sine series of f be differentiated term by term? What about the cosine series?

4. Verify Eqs. (6) and (7) by integration.

5. Suppose that a function $f(x)$ is continuous and sectionally smooth in the interval $0 < x < a$. What additional conditions must $f(x)$ satisfy in order to guarantee that its sine series can be differentiated term by term? The cosine series?

6. Is the derivative of a periodic function periodic? Is the integral of a periodic function periodic?

7. It is known that the equality

$$\ln\left(\left|2\cos\left(\frac{x}{2}\right)\right|\right) = \sum_{n=1}^{\infty} \frac{(-1)^{n+1}}{n} \cos(nx)$$

is valid except when x is an odd multiple of π. Can the Fourier series be differentiated term by term?

8. Use the series that follows, together with integration or differentiation, to find a Fourier series for the function $p(x) = x(\pi - x)$, $0 < x < \pi$.

$$x = 2\sum_{n=1}^{\infty} \frac{(-1)^{n+1}}{n} \sin(nx), \quad 0 < x < \pi$$

9. Let $f(x)$ be an odd, periodic, sectionally smooth function with Fourier sine coefficients $b_1, b_2, \cdots$. Show that the function defined by

$$u(x,t) = \sum_{n=1}^{\infty} b_n e^{-n^2 t} \sin(nx), \quad t \geq 0,$$

has the following properties:

a. $\dfrac{\partial^2 u}{\partial x^2} = \displaystyle\sum_{n=1}^{\infty} -n^2 b_n e^{-n^2 t} \sin(nx), \quad t > 0,$

b. $u(0,t) = 0, \quad u(\pi,t) = 0, \quad t > 0,$

c. $u(x,0) = \dfrac{1}{2}\left(f(x+) + f(x-)\right).$

10. Let f be as in Exercise 9, but define $u(x,y)$ by

$$u(x,y) = \sum_{n=1}^{\infty} b_n e^{-ny} \sin(nx), \quad y > 0.$$

Show that $u(x,y)$ has these properties:

a. $\dfrac{\partial^2 u}{\partial x^2} = \displaystyle\sum_{n=1}^{\infty} -n^2 b_n e^{-ny} \sin(nx), \quad y > 0,$

b. $u(0,y) = 0, \quad u(\pi,y) = 0, \quad y > 0,$

c. $u(x,0) = \dfrac{1}{2}\left(f(x+) + f(x-)\right).$

1.6 MEAN ERROR AND CONVERGENCE IN MEAN

While we can study the behavior of infinite series, we must almost always use finite series in practice. Fortunately, Fourier series have some properties that make them very useful in this setting. Before going on to these properties, we shall develop a useful formula.

Suppose f is a function defined in the interval $-a < x < a$, for which

$$\int_{-a}^{a} (f(x))^2 dx$$

is a finite number. Let

$$f(x) \sim a_0 + \sum_{n=1}^{\infty} a_n \cos\left(\frac{n\pi x}{a}\right) + b_n \sin\left(\frac{n\pi x}{a}\right)$$

and let $g(x)$ have a finite Fourier series

$$g(x) = A_0 + \sum_{1}^{N} A_n \cos\left(\frac{n\pi x}{a}\right) + B_n \sin\left(\frac{n\pi x}{a}\right).$$

Then we may perform the following operations:

$$\int_{-a}^{a} f(x)g(x)dx = \int_{-a}^{a} f(x)\left[A_0 + \sum_{1}^{N} A_n \cos\left(\frac{n\pi x}{a}\right) + B_n \sin\left(\frac{n\pi x}{a}\right)\right] dx$$

$$= A_0 \int_{-a}^{a} f(x)dx + \sum_{1}^{N} A_n \int_{-a}^{a} f(x)\cos\left(\frac{n\pi x}{a}\right) dx$$

$$+ \sum_{1}^{N} B_n \int_{-a}^{a} f(x)\sin\left(\frac{n\pi x}{a}\right) dx.$$

We recognize the integrals as multiples of the Fourier coefficients of f, and rewrite

$$\frac{1}{a}\int_{-a}^{a} f(x)g(x)dx = 2a_0A_0 + \sum_{1}^{N}(a_nA_n + b_nB_n). \qquad (1)$$

Now suppose we wish to approximate $f(x)$ by a *finite* Fourier series. The difficulty here is deciding what "approximate" means. Of the many ways we can measure approximation, the one that is easiest to use is the following:

$$E_N = \int_{-a}^{a} (f(x) - g(x))^2 dx. \tag{2}$$

(Here g is the function with a Fourier series containing terms up to and including $\cos(N\pi x/a)$.) Clearly, E_N can never be negative, and if f and g are "close," then E_N will be small. Thus our problem is to choose the coefficients of g so as to minimize E_N. (We assume N fixed.)

To compute E_N, we first expand the integrand:

$$E_N = \int_{-a}^{a} f^2(x)dx - 2\int_{-a}^{a} f(x)g(x)dx + \int_{-a}^{a} g^2(x)dx. \tag{3}$$

The first integral has nothing to do with g; the other two integrals clearly depend on the choice of g and can be manipulated so as to minimize E_N. We already have an expression for the middle integral. The last one can be found by replacing f with g in Eq. (1)

$$\int_{-a}^{a} g^2(x)dx = a\left[2A_0^2 + \sum_1^N A_n^2 + B_n^2\right]. \tag{4}$$

Now we have a formula for E_N in terms of the variables A_0, A_n, B_n:

$$E_N = \int_{-a}^{a} f^2(x)dx - 2a\left[2A_0a_0 + \sum_1^N A_na_n + B_nb_n\right]$$

$$+a\left[2A_0^2 + \sum_1^N A_n^2 + B_n^2\right]. \tag{5}$$

The error E_N takes it minimum value when all of the partial derivatives with respect to the variables are zero. We must then solve the equations

$$\frac{\partial E_N}{\partial A_0} = -4aa_0 + 4aA_0 = 0$$

$$\frac{\partial E_N}{\partial A_n} = -2aa_n + 2aA_n = 0$$

$$\frac{\partial E_N}{\partial B_n} = -2ab_n + 2aB_n = 0.$$

These equations require that $A_0 = a_0$, $A_n = a_n$, $B_n = b_n$. Thus g should be chosen to be the *truncated* Fourier series of f,

$$g(x) = a_0 + \sum_{n=1}^{N} a_n \cos\left(\frac{n\pi x}{a}\right) + b_n \sin\left(\frac{n\pi x}{a}\right)$$

in order to minimize E_N.

Now that we know which choice of A's and B's minimizes E_N, we can compute that minimum value. After some algebra, we see that

$$\min(E_N) = \int_{-a}^{a} f^2(x)dx - a\left[2a_0^2 + \sum_{1}^{N} a_n^2 + b_n^2\right]. \tag{6}$$

Even this minimum error must be greater than or equal to zero, and thus we have the *Bessel inequality*

$$\frac{1}{a}\int_{-a}^{a} f^2(x)dx \geq 2a_0^2 + \sum_{1}^{N} a_n^2 + b_n^2. \tag{7}$$

This inequality is valid for any N and therefore is also valid in the limit as N tends to infinity. The actual fact is that, in the limit, the inequality becomes *Parseval's equality:*

$$\frac{1}{a}\int_{-a}^{a} f^2(x)dx = 2a_0^2 + \sum_{1}^{\infty} a_n^2 + b_n^2. \tag{8}$$

Another very important consequence of Bessel's inequality is that the two series $\sum a_n^2$ and $\sum b_n^2$ must converge if the left-hand side of Eqs. (7) and (8) is finite. Thus, the numbers a_n and b_n must tend to 0 as n tends to infinity.

By comparing Eqs. (6) and (8), we get a different expression for the minimum error

$$\min(E_N) = a\sum_{N+1}^{\infty} a_n^2 + b_n^2.$$

This quantity decreases steadily to zero as N increases. Since $\min(E_N)$ is, according to Eq. (2), a mean deviation between f and the truncated Fourier series of f, we often say "the Fourier series of f converges to f in the mean." (Another kind of convergence!)

Summary

If $f(x)$ has been defined in the interval $-a < x < a$, and if

$$\int_{-a}^{a} f^2(x)dx$$

is finite, then:

1. Among all finite series of the form

$$g(x) = A_0 + \sum_{1}^{N} A_n \cos\left(\frac{n\pi x}{a}\right) + B_n \sin\left(\frac{n\pi x}{a}\right)$$

the one that best approximates f in the sense of the error described by Eq. (2) is the truncated Fourier series of f:

$$a_0 + \sum_{1}^{N} a_n \cos\left(\frac{n\pi x}{a}\right) + b_n \sin\left(\frac{n\pi x}{a}\right).$$

2. $\displaystyle\frac{1}{a}\int_{-a}^{a} f^2(x)dx = 2a_0^2 + \sum_{1}^{\infty} a_n^2 + b_n^2,$

3. $\displaystyle a_n = \frac{1}{a}\int_{-a}^{a} f(x)\cos\left(\frac{n\pi x}{a}\right)dx \to 0 \quad \text{as } n \to \infty$

$$b_n = \frac{1}{a}\int_{-a}^{a} f(x)\sin\left(\frac{n\pi x}{a}\right)dx \to 0 \quad \text{as } n \to \infty$$

4. The Fourier series of f converges to f in the sense of the mean.

Properties 2 and 3 are very useful for checking computed values of Fourier coefficients.

Exercises

1. Use properties of Fourier series to evaluate the definite integral

$$\frac{1}{\pi}\int_{-\pi}^{\pi}\left(\ln\left|2\cos\left(\frac{x}{2}\right)\right|\right)^2 dx.$$

(Hint: see Section 10, Eq. (4) and Section 5, Eq. (5).)

2. Verify Parseval's equality for these functions:

a. $f(x) = x, \quad -1 < x < 1$ **b.** $f(x) = \sin(x), \quad -\pi < x < \pi.$

3. What can be said about the behavior of the Fourier coefficients of the following functions as $n \to \infty$?

a. $f(x) = |x|^{1/2}, \quad -1 < x < 1$ **b.** $f(x) = |x|^{-1/2}, \quad -1 < x < 1$

4. How do we know that E_N has a minimum and not a maximum?

5. If a function f defined on the interval $-a < x < a$ has Fourier coefficients

$$a_n = 0, \quad b_n = \frac{1}{\sqrt{n}}$$

what can you say about

$$\int_{-a}^{a} f^2(x)dx?$$

6. Show that, as $n \to \infty$, the Fourier sine coefficients of the function

$$f(x) = \frac{1}{x}, \quad -\pi < x < \pi$$

tend to a nonzero constant. (Since this is an odd function, we can take the cosine coefficients to be zero, although strictly speaking they do not exist.) Use the fact that

$$\int_0^{\infty} \frac{\sin(t)}{t} dt = \frac{\pi}{2}.$$

1.7 PROOF OF CONVERGENCE

In this section we prove the Fourier convergence theorem stated in Section 3. Most of the proof requires nothing more than simple calculus, but there are three technical points that we state here.

Lemma 1.
For all $N = 1, 2, \cdots,$

$$\frac{1}{\pi} \int_{-\pi}^{\pi} \left(\frac{1}{2} + \sum_{n=1}^{N} \cos(ny) \right) dy = 1.$$

Lemma 2.
For all $N = 1, 2, \cdots$,

$$\frac{1}{2} + \sum_{n=1}^{N} \cos(ny) = \frac{\sin\left((N+\frac{1}{2})y\right)}{2\sin(\frac{1}{2}y)}.$$

Lemma 3.
If $\phi(y)$ is sectionally continuous, $-\pi < y < \pi$, then its Fourier coefficients tend to 0 with n:

$$\lim_{n\to\infty} \frac{1}{\pi} \int_{-\pi}^{\pi} \phi(y)\cos(ny)dy = 0,$$

$$\lim_{n\to\infty} \frac{1}{\pi} \int_{-\pi}^{\pi} \phi(y)\sin(ny)dy = 0.$$

In Exercises 1 and 2 of this section, you are asked to verify Lemmas 1 and 2 (also see Miscellaneous Exercise 17 at the end of this chapter). Lemma 3 was proved in Section 6.

The theorem we are going to prove is restated here for easy reference. Period 2π is used for typographic convenience; we have seen that any other period can be obtained by a simple change of variables.

Theorem
If $f(x)$ is sectionally smooth and periodic with period 2π, then the Fourier series corresponding to f converges at every x, and the sum of the series is

$$a_0 + \sum_{n=1}^{\infty} a_n \cos(nx) + b_n \sin(nx) = \frac{1}{2}(f(x+) + f(x-)). \tag{1}$$

Proof.
Let the point x be chosen; it is to remain fixed. To begin with, we assume that f is *continuous* at x, so that the sum of the series should be $f(x)$. Another way to say this is that

$$\lim_{N\to\infty} S_N(x) - f(x) = 0,$$

where S_N is the partial sum of the Fourier series of f,

$$S_N(x) = a_0 + \sum_{n=1}^{N} a_n \cos(nx) + b_n \sin(nx). \tag{2}$$

Of course, the a's and b's are the Fourier coefficients of f,

$$\begin{aligned} a_0 &= \frac{1}{2\pi}\int_{-\pi}^{\pi} f(z)dz, \qquad a_n = \frac{1}{\pi}\int_{-\pi}^{\pi} f(z)\cos(nz)dz, \\ b_n &= \frac{1}{2\pi}\int_{-\pi}^{\pi} f(z)\sin(nz)dz. \end{aligned} \tag{3}$$

The integrals have z as their variable of integration, but that does not affect their value.

Part 1. Transformation of $S_N(x)$.
In order to show a relationship between $S_N(x)$ and f, we replace the coefficients in Eq. (2) by the integrals that define them and use elementary algebra on the results.

$$\begin{aligned} S_N(x) &= \frac{1}{2\pi}\int_{-\pi}^{\pi} f(z)dx + \sum_{n=1}^{N}\left[\frac{1}{\pi}\int_{-\pi}^{\pi} f(z)\cos(nz)dz\cos(nx) \right. \\ &\qquad \left. +\frac{1}{\pi}\int_{-\pi}^{\pi} f(z)\sin(nz)dz\sin(nx)\right] && (4) \\ &= \frac{1}{2\pi}\int_{-\pi}^{\pi} f(z)dx + \sum_{n=1}^{N}\left[\frac{1}{\pi}\int_{-\pi}^{\pi} f(z)\cos(nz)\cos(nx)dz \right. \\ &\qquad \left. +\frac{1}{\pi}\int_{-\pi}^{\pi} f(z)\sin(nz)\sin(nx)dz\right] && (5) \\ &= \frac{1}{2\pi}\int_{-\pi}^{\pi} f(z)dx + \sum_{n=1}^{N}\left[\frac{1}{\pi}\int_{-\pi}^{\pi} f(z)(\cos(nz)\cos(nx) \right. \\ &\qquad \left. +\sin(nz)\sin(nx))dz\right] && (6) \\ &= \frac{1}{\pi}\int_{-\pi}^{\pi} f(z)\left(\frac{1}{2}+\sum_{n=1}^{N}\cos(nz)\cos(nx)+\sin(nz)\sin(nx)\right)dz && (7) \\ &= \frac{1}{\pi}\int_{-\pi}^{\pi} f(z)\left(\frac{1}{2}+\sum_{n=1}^{N}\cos(n(z-x))\right)dz. && (8) \end{aligned}$$

In this very compact formula for $S_N(x)$, we now change the variable of integration from z to $y = z - x$:

$$S_N(x) = \frac{1}{\pi}\int_{-\pi+x}^{\pi+x} f(x+y)\left(\frac{1}{2}+\sum_{n=1}^{N}\cos(ny)\right)dy. \tag{9}$$

Note that both factors in the integrand are periodic with period 2π. The interval of integration can be any interval of length 2π with no change in the result. (See Exercise 5 of Section 1.) Therefore,

$$S_N(x) = \frac{1}{\pi}\int_{-\pi}^{\pi} f(x+y)\left(\frac{1}{2} + \sum_{n=1}^{N} \cos(ny)\right) dy. \tag{10}$$

Part 2. Expression for $S_N(x) - f(x)$.
Since we must show that the difference $S_N(x) - f(x)$ goes to 0, we need to have $f(x)$ in a form compatible with that for $S_N(x)$. Recall that x is fixed (although arbitrary), so $f(x)$ is to be thought of as a number. Lemma 1 suggests the appropriate form,

$$\begin{aligned} f(x) &= f(x)\cdot\frac{1}{\pi}\int_{-\pi}^{\pi}\left(\frac{1}{2} + \sum_{n=1}^{N}\cos(ny)\right) dy \\ &= \frac{1}{\pi}\int_{-\pi}^{\pi} f(x)\left(\frac{1}{2} + \sum_{n=1}^{N}\cos(ny)\right) dy \end{aligned} \tag{11}$$

Now, using Eq. (10) to represent $S_N(x)$, we have

$$S_N(x) - f(x) = \frac{1}{\pi}\int_{-\pi}^{\pi} (f(x+y) - f(x))\left(\frac{1}{2} + \sum_{n=1}^{N}\cos(ny)\right) dy. \tag{12}$$

Part 3. The limit.
The next step is to use Lemma 2 to replace the sum in Eq. (12). The result is

$$S_N(x) - f(x) = \frac{1}{\pi}\int_{-\pi}^{\pi} (f(x+y) - f(x))\frac{\sin\left((N+\frac{1}{2})y\right)}{2\sin(\frac{1}{2}y)} dy. \tag{13}$$

The addition formula for sines gives the equality

$$\sin\left(\left(N+\tfrac{1}{2}\right)y\right) = \cos(Ny)\sin\left(\tfrac{1}{2}y\right) + \sin(Ny)\cos\left(\tfrac{1}{2}y\right).$$

Substituting it in Eq. (13) and using simple properties of integrals, we obtain

$$\begin{aligned} S_N(x) - f(x) = {} & \frac{1}{\pi}\int_{-\pi}^{\pi} (f(x+y) - f(x))\frac{1}{2}\cos(Ny)dy \\ & + \frac{1}{\pi}\int_{-\pi}^{\pi} (f(x+y) - f(x))\frac{\cos(\frac{1}{2}y)}{2\sin(\frac{1}{2}y)}\sin(Ny)dy. \end{aligned} \tag{14}$$

The first integral in Eq. (14) can be recognized as the Fourier cosine coefficient of the function

$$\psi(y) = \frac{1}{2}\left(f(x+y) - f(x)\right). \tag{15}$$

Since f is a sectionally smooth function, so is ψ, and the first integral has limit 0 as N increases, by Lemma 3.

The second integral in Eq. (14) can also be recognized, as the Fourier sine coefficient of the function

$$\phi(y) = \frac{f(x+y) - f(x)}{2\sin(\frac{1}{2}y)} \cos\left(\tfrac{1}{2}y\right). \tag{16}$$

To proceed as before, we must show that $\phi(y)$ is at least sectionally continuous, $-\pi \le y \le \pi$. The only difficulty is to show that the apparent division by 0 at $y = 0$ does not cause $\phi(y)$ to have a bad discontinuity there.

First, if f is continuous and differentiable near x, then $f(x+y) - f(x)$ is continuous and differentiable near $y = 0$. Then L'Hôpital's rule gives

$$\lim_{y\to 0} \frac{f(x+y) - f(x)}{2\sin(\frac{1}{2}y)} = \lim_{y\to 0} \frac{f'(x+y)}{\cos(\frac{1}{2}y)} = f'(x). \tag{17}$$

Under these conditions, the function $\phi(y)$ of Eq. (16) has a removable discontinuity at $y = 0$ and thus is sectionally continuous.

Second, if f is continuous at x but has a corner there, then $f(x+y) - f(x)$ is continuous with a corner at $y = 0$. In this case, L'Hôpital's rule applies with the one-sided limits, which show

$$\lim_{y\to 0+} \frac{f(x+y) - f(x)}{2\sin(\frac{1}{2}y)} = \lim_{y\to 0+} \frac{f'(x+y)}{\cos(\frac{1}{2}y)} = f'(x+) \tag{18}$$

$$\lim_{y\to 0-} \frac{f(x+y) - f(x)}{2\sin(\frac{1}{2}y)} = \lim_{y\to 0-} \frac{f'(x+y)}{\cos(\frac{1}{2}y)} = f'(x-). \tag{19}$$

Under these conditions, the function $\phi(y)$ of Eq. (16) has a jump discontinuity at $y = 0$, and again is sectionally continuous.

In either case, we see that the second integral in Eq. (14) is the Fourier sine coefficient of a sectionally continuous function. By Lemma 3, then, it too has limit 0 as N increases, and the proof is complete for every x where f is continuous.

Part 4. If f is not continuous at x.
Now, let us suppose that f has a jump discontinuity at x. In this case, we must return to Part 2 and express the proposed sum of the series as

$$\frac{1}{2}\left(f(x+)+f(x-)\right)=\frac{1}{\pi}\int_0^\pi f(x+)\left(\frac{1}{2}+\sum_{n=1}^N\cos(ny)\right)dy$$

$$+\frac{1}{\pi}\int_{-\pi}^0 f(x-)\left(\frac{1}{2}+\sum_{n=1}^N\cos(ny)\right)dy. \tag{20}$$

Here, we have used the evenness of the integrand in Lemma 1 to write

$$\frac{1}{\pi}\int_0^\pi\left(\frac{1}{2}+\sum_{n=1}^N\cos(ny)\right)dy=\frac{1}{\pi}\int_{-\pi}^0\left(\frac{1}{2}+\sum_{n=1}^N\cos(ny)\right)dy=\frac{1}{2}. \tag{21}$$

Next, we have a convenient way to write the quantity to be limited:

$$S_N(x)-\frac{1}{2}\left(f(x+)+f(x-)\right)$$

$$=\frac{1}{\pi}\int_0^\pi\left(f(x+y)-f(x+)\right)\left(\frac{1}{2}+\sum_{n=1}^N\cos(ny)\right)dy \tag{22}$$

$$+\frac{1}{\pi}\int_{-\pi}^0\left(f(x+y)-f(x-)\right)\left(\frac{1}{2}+\sum_{n=1}^N\cos(ny)\right)dy.$$

The interval of integration for $S_N(x)$ as shown in Eq. (10) has been split in half to conform to the integrals in Eq. (20).

The last step is to show that each of the integrals in Eq. (22) approaches 0 as N increases. Since the technique is the same as in Part 3, this is left as an exercise.

Let us emphasize that the crux of the proof is to show that the function from Eq. (16),

$$\phi(y)=\frac{f(x+y)-f(x)}{2\sin(\frac{1}{2}y)}\cos\left(\tfrac{1}{2}y\right) \tag{23}$$

(or a similar function that arises from the integrands in Eq. (22)) does not have a bad discontinuity at $y=0$.

Exercises

1. Verify Lemma 2. Multiply through by $2\sin(\frac{1}{2}y)$. Use the identity

$$\sin\left(\tfrac{1}{2}y\right)\cos(ny) = \tfrac{1}{2}\left(\sin\left((n+\tfrac{1}{2})y\right) - \sin\left((n-\tfrac{1}{2})y\right)\right).$$

Note that most of the series then disappears. (To see this, write out the result for $N = 3$.)

2. Verify Lemma 1 by integrating the sum term by term.

3. Let $f(x) = f(x + 2\pi)$ and $f(x) = |x|$ for $-\pi < x < \pi$. Note that f is continuous and has a corner at $x = 0$. Sketch the function $\phi(y)$ as defined in Eq. (16) if $x = 0$. Find $\phi(0+)$ and $\phi(0-)$.

4. Let f be the odd periodic extension of the function whose formula is $\pi - x$ for $0 < x < \pi$. In this case, f has a jump discontinuity at $x = 0$. Taking $x = 0$, sketch the functions

$$\phi_R(y) = \frac{f(x+y) - f(x+)}{2\sin\left(\frac{1}{2}y\right)}\cos\left(\tfrac{1}{2}y\right) \quad (y > 0)$$

$$\phi_L(y) = \frac{f(x+y) - f(x-)}{2\sin\left(\frac{1}{2}y\right)}\cos\left(\tfrac{1}{2}y\right) \quad (y < 0).$$

(These functions appear if the integrands in Eq. (22) are developed as in Part 3 of the proof.)

5. Consider the function f that is periodic with period 2π and has the formula $f(x) = |x|^{3/4}$ for $-\pi < x < \pi$.

a. Show that f is continuous at $x = 0$ but is not sectionally smooth.

b. Show that the function $\phi(y)$ (from Eq. (16), with $x = 0$) is sectionally continuous, $-\pi < x < \pi$, except for a bad discontinuity at $y = 0$.

c. Show that the Fourier coefficients of $\phi(y)$ tend to 0 as n increases despite the bad discontinuity.

1.8 NUMERICAL DETERMINATION OF FOURIER COEFFICIENTS

There are many functions whose Fourier coefficients cannot be determined analytically because the integrals involved are not known in terms of easily evaluated functions. Also, it may happen that a function is not known explicitly, but its value can be found at some points. In either case, if a Fourier series is to be found for the function, some numerical technique must be employed to approximate the integrals that give the Fourier coefficients. It turns out that one of the crudest numerical integration techniques is the best.

We shall assume that the function whose coefficients are to be approximated is continuous, periodic, and sectionally smooth. If this is not the case, a sectionally smooth function can be modified to make it fit this description by following the procedure illustrated in Fig. 11.

Suppose then that $f(x)$ is continuous, sectionally smooth, and periodic with period $2a$. Let the interval $-a \leq x \leq a$ be divided into r equal subintervals whose endpoints are $-a = x_0, x_1, \cdots, x_r = a$ (or, in general, $x_i = -a + i\Delta x$, and $\Delta x = 2a/r$). Then the approximate Fourier coefficients of f are (carets indicate approximate values):

$$\hat{a}_0 = \frac{1}{2a}\left(f(x_1) + f(x_2) + \cdots + f(x_r)\right)\Delta x$$

$$\hat{a}_n = \frac{\left(f(x_1)\cos(n\pi/a)x_1 + \cdots + f(x_r)\cos(n\pi/a)x_r\right)\Delta x}{\left(\cos^2(n\pi/a)x_1 + \cdots + \cos^2(n\pi/a)x_r\right)\Delta x} \tag{1}$$

$$\hat{a}_n = \frac{\left(f(x_1)\sin(n\pi/a)x_1 + \cdots + f(x_r)\cos(n\pi/a)x_r\right)\Delta x}{\left(\sin^2(n\pi/a)x_1 + \cdots + \sin^2(n\pi/a)x_r\right)\Delta x}.$$

The coefficient $\hat{a}_0$ is an approximate mean value of f. For the other equations, we start from the formulas for the coefficients that were obtained by orthogonality:

$$a_n = \frac{\displaystyle\int_{-1}^{a} f(x)\cos(n\pi x/a)dx}{\displaystyle\int_{-a}^{a} \cos^2(n\pi x/a)dx}.$$

Then, the integrals in both numerator and denominator are approximated by a Riemann sum

$$\int_{-a}^{a} f(x)\cos\left(\frac{n\pi x}{a}\right)dx$$
$$\cong \left(f(x_1)\cos\left(\frac{n\pi x_1}{a}\right) + \cdots + f(x_r)\cos\left(\frac{n\pi x_r}{a}\right)\right)\Delta x.$$

In all cases, $\Delta x = 2a/r$.

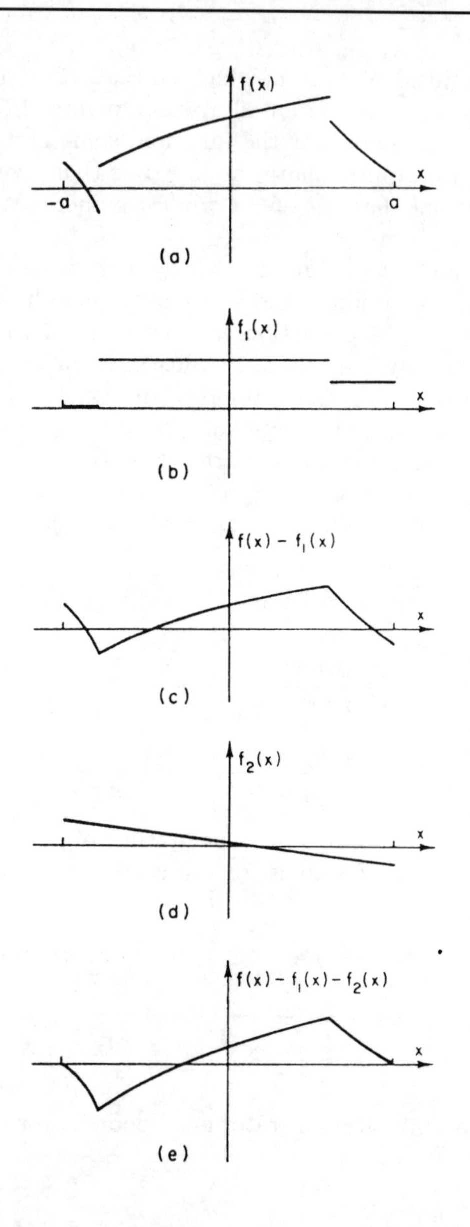

Figure 11: Preparation of a function for numerical integration of Fourier coefficients. (a) Graph of sectionally smooth function $f(x)$ given on $-a < x < a$. (b) Graph of $f_1(x)$, which has jumps of the same magnitude and position as $f(x)$. Coefficients can be found analytically. (c) Graph of $f(x) - f_1(x)$. This function has no jumps in $-a < x < a$. (d) Graph of $f_2(x)$. The periodic extensions of $f_2(x)$ and of $f(x) - f_1(x)$ have jumps of the same magnitude at $x = \pm a$, and so forth. The coefficients of f_2 can be found analytically. (e) Graph of $f_3(x) = f(x) - f_1(x) - f_2(x)$. The Fourier series of $f_3(x)$ converges uniformly (the coefficients tend to zero rapidly).

The denominators of the preceding expressions can be found by some tedious calculations:

$$\cos^2\left(\frac{n\pi x_1}{a}\right) + \cdots + \cos^2\left(\frac{n\pi x_r}{a}\right) = \begin{cases} r, & \text{if } n = 0, \frac{r}{2}, r, \cdots \\ \frac{r}{2}, & \text{otherwise} \end{cases} \tag{2}$$

$$\sin^2\left(\frac{n\pi x_1}{a}\right) + \cdots + \sin^2\left(\frac{n\pi x_r}{a}\right) = \begin{cases} 0, & \text{if } n = 0, \frac{r}{2}, r, \cdots \\ \frac{r}{2}, & \text{otherwise.} \end{cases} \tag{3}$$

Of course, n can equal $r/2$ only if r is even.

Now we can simplify our formulas. Taking into account the results already provided here, we have

$$\begin{aligned} \hat{a}_0 &= \frac{1}{r}\left(f(x_1) + \cdots + f(x_r)\right) \\ \hat{a}_n &= \frac{2}{r}\left(f(x_1)\cos\left(\frac{n\pi x_1}{a}\right) + \cdots + f(x_r)\cos\left(\frac{n\pi x_r}{a}\right)\right) \\ \hat{b}_n &= \frac{2}{r}\left(f(x_1)\sin\left(\frac{n\pi x_1}{a}\right) + \cdots + f(x_r)\sin\left(\frac{n\pi x_r}{a}\right)\right). \end{aligned} \tag{4}$$

These are valid for $n < r/2$. If r is even ($r = 2q$), then the sine coefficient $\hat{b}_q$ is undefined, and the cosine coefficient is

$$\hat{a}_q = \frac{1}{r}\left(f(x_1)\cos\left(\frac{q\pi x_1}{a}\right) + \cdots + f(x_r)\cos\left(\frac{q\pi x_r}{a}\right)\right). \tag{5}$$

How many coefficients calculated numerically are reasonably accurate? Since the values of the function at $x_1, \cdots, x_r$ represent r pieces of information, we may expect r coefficients to be reasonably accurate; this is indeed the case. Thus, if r is an even number, say, $r = 2q$, we can find

$$\hat{a}_0, \hat{a}_1, \hat{b}_1, \cdots, \hat{a}_q$$

but not $\hat{b}_q$, which is undefined. (Remember that $\hat{a}_q$ is found from the special formula.) If r is an odd number, say $r = 2q + 1$, then we can find

$$\hat{a}_0, \hat{a}_1, \hat{b}_1, \cdots, \hat{a}_q, \hat{b}_q.$$

The formulas in Eq. (4) were derived for the case in which x_0, x_1, $\cdots, x_r$ are equally spaced points in the interval $-a \leq x \leq a$. However, they remain valid for equally spaced points on the interval $0 \le x \le 2a$. That is,

$$x_0 = 0, \quad x_1 = \frac{2a}{r}, \quad x_2 = \frac{4a}{r}, \cdots, x_r = 2a. \tag{6}$$

Note also that when $f(x)$ is given in the interval $0 \le x \le a$, and the sine or cosine coefficients are to be determined, the formulas may be derived from those already given here. Let the interval be divided into s equal subintervals with endpoints $0 = x_0, x_1, \cdots, x_s = a$ (in general, $x_i = ia/s$). Then the approximate Fourier cosine coefficients for f or its even extension are

$$\hat{a}_0 = \frac{1}{s}\left(\frac{1}{2}f(x_0) + f(x_1) + \cdots + f(x_{s-1}) + \frac{1}{2}f(x_s)\right)$$

$$\hat{a}_n = \frac{2}{s}\left(\frac{1}{2}f(x_0) + f(x_1)\cos\left(\frac{n\pi x_1}{a}\right) + \cdots + \frac{1}{2}f(x_s)\cos\left(\frac{n\pi x_s}{a}\right)\right), \quad n = 1, \cdots, s-1$$

$$\hat{a}_s = \frac{1}{s}\left(\frac{1}{2}f(x_0) + f(x_1)\cos\left(\frac{s\pi x_1}{a}\right) + \cdots + \frac{1}{2}f(x_s)\cos\left(\frac{s\pi x_s}{a}\right)\right) \tag{7}$$

Similarly, the approximate Fourier sine coefficients for f or its odd extension are

$$\hat{b}_n = \frac{2}{s}\left(f(x_1)\sin\left(\frac{n\pi x_1}{a}\right) + \cdots + f(x_{s-1})\sin\left(\frac{n\pi x_{s-1}}{a}\right)\right), \quad n = 1, 2, \cdots, s. \tag{8}$$

In the formulas of Eq. (7) the endpoints acquire factors of $\frac{1}{2}$. This occurs because, if the even extension of f is used in Eq. (4), the values of f at 0 and a are counted once whereas the values of f at points between are counted twice. In Eq. (8), values at the endpoints do not enter, because the sine function is zero there.

An important feature of the approximate Fourier coefficients is this: If

$$F(x) = \hat{a}_0 + \hat{a}_1\cos\left(\frac{\pi x}{a}\right) + \hat{b}_1\sin\left(\frac{\pi x}{a}\right) + \cdots$$

is a finite Fourier series using a total of r approximate coefficients calculated from Eqs. (4) and (5), then $F(x)$ actually interpolates the function $f(x)$ at $x_1, x_2, \cdots, x_r$. That is,

$$F(x_i) = f(x_i), \qquad i = 1, 2, \cdots, r.$$

Thus the graph of $F(x)$ cuts the graph of $f(x)$ at the points x_i, $i = 1, 2, \cdots, r$.

Example.
Calculate the approximate Fourier coefficients of $f(x) = \sin(x)/x$ in $-\pi < x < \pi$.

Since f is even, it will have a cosine series. We simplify computation by using the half-range formulas and making s even. We take $s = 6$, $x_0 = 0$, $x_1 = \pi/6$, $\cdots, x_5 = 5\pi/6$, $x_6 = \pi$. The numerical information is given in Table 2.

Table 2: Numerical information

i	x_i	$\cos x_i$	$\cos 2x_i$	$\cos 3x_i$	$\sin(x_i)/x_i$
0	0	1.0	1.0	1.0	1.0
1	$\frac{\pi}{6}$	0.86603	0.5	0	0.95493
2	$\frac{\pi}{3}$	0.5	−0.5	−1.0	0.82699
3	$\frac{\pi}{2}$	0	−1.0	0	0.63662
4	$\frac{2\pi}{3}$	−0.5	−0.5	1.0	0.41350
5	$\frac{5\pi}{6}$	−0.86603	0.5	0	0.19099
6	π	−1.0	1.0	−1.0	0.0

The results of the calculation are given in Table 3. On the left are the approximate coefficients calculated from the table. On the right are the correct values (to five decimals), obtained with the aid of a table of the sine integral (see Exercise 2). Figure 12 shows $f(x)$, $F(x)$ (the sum of the Fourier series using the approximate coefficients through $\hat{a}_6$), and the difference between them.

For hand calculation, choosing s to be a multiple of 4 makes many of the cosines "easy" numbers such as 1 or 0.5. When the calculation is done by digital computer, this is not a consideration.

Table 3: Approximate coefficients of $\sin(x)/x$

n	$\hat{a}_n$	a_n	error
0	0.58717	0.58949	0.00232
1	0.45611	0.45141	0.00470
2	−0.06130	−0.05640	0.00490
3	0.02884	0.02356	0.00528

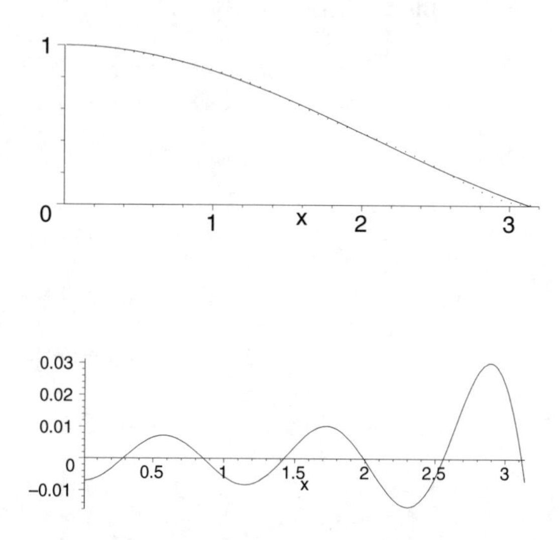

Figure 12: (a) Graphs over the interval $0 \le x \le \pi$ of the functions $f(x) = (\sin(x))/x$ (solid curve) and $F(x)$ (dashed), the sum of the Fourier cosine series using the approximate coefficients through $\hat{a}_6$. (b) Graph of $y = f(x) - F(x)$. Notice that the difference is zero at six places in the interval, confirming the statement about the interpolation property of the series with approximate coefficients.

Exercises

1. Since the table in the example gives $\sin(x)/x$ for seven points, seven cosine coefficients can be calculated. Find $\hat{a}_6$.

2. Express the Fourier cosine coefficients of the example in terms of integrals of the form

$$\mathrm{Si}((n+1)\pi) = \int_0^{(n+1)\pi} \frac{\sin(t)}{t}\,dt.$$

This is the sine integral function and is tabulated in many books, especially *Handbook of Mathematical Functions*, Abramowitz and Stegun, 1972.

3. Each entry in the list that follows represents the depth of the water in Lake Ontario (minus the low-water datum of 242.8 feet) on the first of the corresponding month. Assuming that the water level is a periodic function of period one year, and that the observations are taken at equal intervals, compute the Fourier coefficients $\hat{a}_0$, $\hat{a}_1$, $\hat{b}_1$, $\hat{a}_2$, $\hat{b}_2$, thus identifying the mean level, and fluctuations of period 12 months, 6 months, 4 months, and so forth. Take x_0 as January,$\cdots$, x_{11} as December, and x_{12} as January again.

Jan.	0.75	July	2.35
Feb.	0.60	Aug.	2.15
Mar.	0.65	Sept.	1.75
Apr.	1.15	Oct.	1.05
May	1.80	Nov.	1.00
June	2.25	Dec.	0.90

4. The numbers in the table that follows represent the monthly precipitation (in inches of water) in Lake Placid NY averaged over the period 1950-1959. Find the approximate Fourier coefficients $\hat{a}_0, \cdots, \hat{a}_6$ and $\hat{b}_1, \cdots, \hat{b}_5$.

Jan.	2.751	July	3.861
Feb.	2.004	Aug.	4.088
Mar.	3.166	Sept.	4.093
Apr.	2.909	Oct.	3.434
May	3.215	Nov.	2.902
June	3.767	Dec.	3.011

1.9 FOURIER INTEGRAL

In Sections 1 and 2 of this chapter we developed the representation of a periodic function in terms of sines and cosines with the same period. Then, by means of periodic extension, we obtained series representations for functions defined only on a finite interval. Now we must deal with non-periodic functions defined for x between $-\infty$ and ∞. Can such functions also be represented in terms of sines and cosines?

That such a representation may exist for some nonperiodic functions is evident from some simple integral formulas. For instance, it is known that the following definite integral depends on the parameter x as shown.

$$\int_0^\infty \frac{\sin(\lambda x)}{\lambda} d\lambda = \begin{cases} \frac{\pi}{2}, & x < 0 \\ 0, & x = 0 \\ -\frac{\pi}{2}, & x < 0 \end{cases} \tag{1}$$

From this point it is not difficult to determine (See Exercise 10 of this section) that the integral

$$K(x, h) = \frac{2}{\pi} \int_0^\infty \frac{\sin(\lambda h)}{\lambda} \cos(\lambda x) d\lambda \tag{2}$$

depends on the two parameters x and h, in this way (assuming that $h > 0$):

$$K(x, h) = \begin{cases} 1, & |x| < h \\ 0, & |x| > h. \end{cases} \tag{3}$$

The fact that this function (whose graph as a function of x is a rectangular pulse of width $2h$ and height 1) can be represented in terms of $\cos(\lambda x)$ is the key to representing other functions.

Suppose that a function $f(x)$ is defined for all x and is sectionally continuous in every finite interval. Then the mean value theorem guarantees that the equality

$$f(x) = \lim_{h \to 0+} \frac{1}{2h} \int_{x-h}^{x+h} f(x')dx' \tag{4}$$

is valid at every point x where f is continuous. We incorporate sines and cosines into this equality by rewriting the right-hand side of Eq. (4) in this way:

$$f(x) = \lim_{h \to 0+} \frac{1}{2h} \int_{-\infty}^{\infty} f(x')K(x - x', h)dx'. \tag{5}$$

This is exactly the same as Eq. (4), because $K(x-x',h)$ is 1 for x' between $x-h$ and $x+h$ and is 0 everywhere else (see Fig. 13). Thus we have

$$f(x) = \lim_{h\to 0+} \frac{1}{2h} \int_{-\infty}^{\infty} f(x') \frac{2}{\pi} \int_0^{\infty} \frac{\sin(\lambda h)}{\lambda} \cos(\lambda(x-x'))d\lambda \ dx'. \quad (6)$$

If $f(x)$ is such that the order of integration may be reversed, we may progress to the form

$$f(x) = \lim_{h\to 0+} \int_0^{\infty} \frac{\sin(\lambda h)}{\lambda h} \frac{1}{\pi} \int_{-\infty}^{\infty} f(x') \cos(\lambda(x-x'))dx'd\lambda. \quad (7)$$

In the limit as h approaches 0, $\sin(\lambda h)/\lambda h$ approaches 1. Thus we expect to obtain the representation

$$f(x) = \int_0^{\infty} \frac{1}{\pi} \int_{-\infty}^{\infty} f(x') \cos(\lambda(x-x'))dx'd\lambda \quad (8)$$

for appropriate functions $f(x)$. When $\cos(\lambda(x-x'))$ is expanded by the formula for the cosine of a difference, Eq. (8) can be written in the following form, which we shall use almost exclusively from now on:

$$f(x) = \int_0^{\infty} \left(A(\lambda)\cos(\lambda x) + B(\lambda)\sin(\lambda x)\right) d\lambda,$$

$$A(\lambda) = \frac{1}{\pi}\int_{-\infty}^{\infty} f(x)\cos(\lambda x)dx, \qquad B(\lambda) = \frac{1}{\pi}\int_{-\infty}^{\infty} f(x)\sin(\lambda x)dx.$$

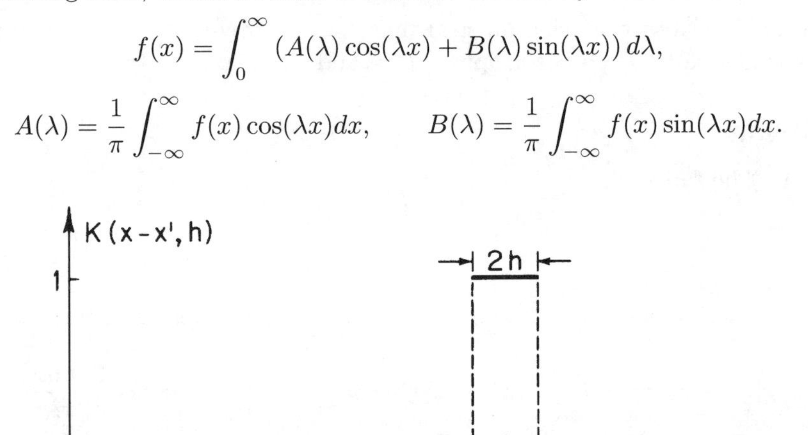

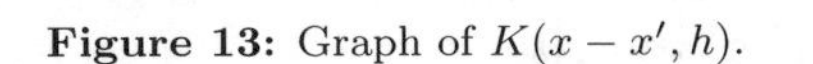
Figure 13: Graph of $K(x-x',h)$.

This is called the *Fourier integral representation* of the function f; $A(\lambda)$ and $B(\lambda)$ are called the *Fourier integral coefficient functions* of f. The following theorem gives some conditions under which the representation is valid.

Theorem

Let $f(x)$ be sectionally smooth on every finite interval, and let $\int_{-\infty}^{\infty} |f(x)|dx$ be finite. Then at every point x,

$$\int_0^{\infty} (A(\lambda)\cos(\lambda x) + B(\lambda)\sin(\lambda x))\, d\lambda = \frac{1}{2}\left(f(x+) + f(x-)\right)$$

where

$$A(\lambda) = \frac{1}{\pi}\int_{-\infty}^{\infty} f(x)\cos(\lambda x)dx, \qquad B(\lambda) = \frac{1}{\pi}\int_{-\infty}^{\infty} f(x)\sin(\lambda x)dx.$$

We have by no means proved the theorem in our derivation, but we have some indication of the roles played by the various hypotheses. Incidentally, the hypotheses of the theorem eliminate periodic functions from consideration; no periodic function, other than 0, can meet the condition that

$$\int_{-\infty}^{\infty} |f(x)|dx$$

be finite. Note that this integral may be interpreted as the total (unsigned) area between the graph of $f(x)$ and the x-axis.

Examples.

1. The function

$$f(x) = \begin{cases} 1, & |x| < 1 \\ 0, & |x| > 1, \end{cases}$$

which is also $K(x, 1)$, has the Fourier integral coefficient functions

$$\begin{aligned} A(\lambda) &= \frac{1}{\pi}\int_{-\infty}^{\infty} f(x)\cos(\lambda x)dx = \frac{1}{\pi}\int_{-1}^{1}\cos(\lambda x)dx = \frac{2\sin(\lambda)}{\pi\lambda} \\ B(\lambda) &= 0. \end{aligned}$$

Since $f(x)$ is sectionally smooth, the Fourier integral representation is legitimate, and we write

$$f(x) = \int_0^{\infty} \frac{2\sin(\lambda)}{\pi\lambda}\cos(\lambda)d\lambda.$$

(Actually the integral here equals $\frac{1}{2}$ at $x = \pm 1$, so equality fails at these two points.)

2. Let $f(x) = \exp(-|x|)$. Then direct integration gives

$$\begin{aligned} A(\lambda) &= \frac{1}{\pi}\int_{-\infty}^{\infty} \exp(-|x|)\cos(\lambda x)dx \\ &= \frac{2}{\pi}\int_0^{\infty} e^{-x}\cos(\lambda x)dx \\ &= \frac{2}{\pi}\left.\frac{e^{-x}(-\cos(\lambda x) + \lambda\sin(\lambda x))}{1+\lambda^2}\right|_0^{\infty} = \frac{2}{\pi}\frac{1}{1+\lambda^2}. \end{aligned}$$

$B(\lambda) = 0$, because $\exp(-|x|)$ is even. Since $\exp(-|x|)$ is continuous and sectionally smooth, we may write

$$\exp(-|x|) = \frac{2}{\pi}\int_0^{\infty} \frac{\cos(\lambda x)}{1+\lambda^2}d\lambda, \quad -\infty < x < \infty.$$

These two examples illustrate the fact that, in general, one cannot evaluate the integral in the Fourier integral representation. It is the theorem stated in the preceding that allows us to write the equality between a suitable function and its Fourier integral.

If $f(x)$ is defined only in the interval $0 < x < \infty$, one can construct an even or odd extension whose Fourier integral contains only $\cos(\lambda x)$ or $\sin(\lambda x)$. These are called the *Fourier cosine and sine integral representations* of f, respectively. Thus we may write either

$$f(x) = \int_0^{\infty} A(\lambda)\cos(\lambda x)d\lambda, \qquad 0 < x < \infty$$

with

$$A(\lambda) = \frac{2}{\pi}\int_0^{\infty} f(x)\cos(\lambda x)dx,$$

or

$$f(x) = \int_0^{\infty} B(\lambda)\sin(\lambda x)d\lambda, \qquad 0 < x < \infty$$

with

$$B(\lambda) = \frac{2}{\pi}\int_0^{\infty} f(x)\sin(\lambda x)dx.$$

Example.
Let $f(x)$ be given for $0 < x$ by

$$f(x) = \begin{cases} \sin(x), & 0 < x < \pi \\ 0, & \pi < x. \end{cases}$$

Since $f(x) = 0$ for $x > \pi$, the integral for $B(\lambda)$ reduces to one over the interval $0 < x < \pi$:

$$\begin{aligned} B(\lambda) &= \frac{2}{\pi}\int_0^\infty f(x)\sin(\lambda x)dx = \frac{2}{\pi}\int_0^\infty \sin(x)\sin(\lambda x)dx \\ &= \frac{2}{\pi}\left[\frac{\sin((\lambda-1)x)}{2(\lambda-1)} - \frac{\sin((\lambda+1)x)}{2(\lambda+1)}\right]_0^\pi \\ &= \frac{2}{\pi}\left[\frac{\sin((\lambda-1)\pi)}{2(\lambda-1)} - \frac{\sin((\lambda+1)\pi)}{2(\lambda+1)}\right]. \end{aligned}$$

This expression can be simplified by using the fact that

$$\sin((\lambda \pm 1)\pi) = \sin(\lambda\pi \pm \pi) = -\sin(\lambda\pi).$$

Then, creating a common denominator, we obtain

$$B(\lambda) = \frac{-2\sin(\lambda\pi)}{\pi(\lambda^2 - 1)}.$$

Hence the Fourier sine integral representation of $f(x)$ is

$$f(x) = \int_0^\infty \frac{-2\sin(\lambda\pi)}{\pi(\lambda^2-1)}\sin(\lambda x)d\lambda, \qquad 0 < x.$$

Since $f(x)$ is continuous for $0 < x$, the equality holds at every point.

Similarly, we can compute the cosine coefficient function

$$A(\lambda) = \frac{-2(1+\cos(\lambda\pi))}{\pi(\lambda^2-1)},$$

and the cosine integral representation of $f(x)$ is

$$f(x) = \int_0^\infty \frac{-2(1+\cos(\lambda\pi))}{\pi(\lambda^2-1)}\cos(\lambda x)d\lambda, \qquad 0 < x.$$

Note that both $A(\lambda)$ and $B(\lambda)$ have removable discontinuities at $\lambda = 1$.

It seems to be a rule of thumb that if the Fourier coefficient functions $A(\lambda)$ and $B(\lambda)$ can be found in closed form for some function $f(x)$, then the integral in the Fourier integral representation cannot be carried out by elementary means, and vice versa. (See Exercise 3.)

Exercises

1. Sketch the even and odd extensions of each of the following functions, and find the Fourier cosine and sine integrals for f. Each function is given in the interval $0 < x < \infty$.

a. $f(x) = e^{-x}$

b. $f(x) = \begin{cases} 1, & 0 < x < 1 \\ 0, & 1 < x \end{cases}$

c. $f(x) = \begin{cases} \pi - x, & 0 < x < \pi \\ 0, & \pi < x \end{cases}$

2. Change the variable of integration in the formulas for A and B and justify each step of the following string of equalities. (Do not worry about changing order of integration.)

$$\begin{aligned} f(x) &= \frac{1}{\pi}\int_0^\infty \int_{-\infty}^\infty f(t)\left(\cos(\lambda t)\cos(\lambda x) + \sin(\lambda t)\sin(\lambda x)\right) dt \;\; d\lambda \\ &= \frac{1}{\pi}\int_{-\infty}^\infty f(t)\int_0^\infty \cos(\lambda(t-x))d\lambda dt \\ &= \frac{1}{\pi}\int_{-\infty}^\infty f(t)\left[\lim_{\omega\to\infty}\frac{\sin(\omega(t-x))}{t-x}\right] dt \\ &= \lim_{\omega\to\infty}\frac{1}{\pi}\int_{-\infty}^\infty f(t)\frac{\sin(\omega(t-x))}{t-x}dt \end{aligned}$$

The last integral is called *Fourier's single integral.* Sketch the function $\sin(\omega v)/v$ as a function of v for several values of ω. What happens near $v = 0$? Sometimes notation is compressed and, instead of the last line, we write

$$f(x) = \int_{-\infty}^\infty f(t)\delta(t-x)dt.$$

Although δ is not, strictly speaking, a function, it is called "Dirac's delta function."

3. Find the Fourier integral representation of each of the following functions.

a. $f(x) = \dfrac{1}{1+x^2}$ **b.** $f(x) = \dfrac{\sin(x)}{x}$

(Hint: To evaluate the Fourier integral coefficient functions, consult the Fourier integral representations found in the Examples.)

4. In Exercise 3b, the integral $\int_{-\infty}^{\infty} |f(x)|dx$ is not finite. Nevertheless, $A(\lambda)$ and $B(\lambda)$ do exist ($B(\lambda) = 0$). Find a rationale in the convergence theorem for saying that this function can be represented by its Fourier integral. (Hint: See Example 1.)

5. Suppose that $f(x)$ is a continuous, differentiable function with Fourier integral coefficient functions $A(\lambda)$ and $B(\lambda)$. How are the Fourier integral coefficient functions of the derivative $f'(x)$ related to A and B?

6. Show that if k and K are positive, then the following are true:

a. $\displaystyle\int_0^{\infty} e^{-kx}\sin(x)dx = \frac{1}{1+k^2}$

b. $\displaystyle\int_0^{\infty} \frac{1-e^{-Kx}}{x}\sin(x)dx = \tan^{-1}(K)$

c. $\displaystyle\int_0^{\infty} \frac{\sin(x)}{x}dx = \frac{\pi}{2}$

(Part (a) by direct integration, (b) by integration of (a) with respect to k over the interval 0 to k, (c) by limit of (b) as $K \to \infty$.)

7. Starting from Exercise 6c, derive Eq. (1).

8. Use trigonometric identities for the product $\sin(\lambda h)\cos(\lambda x)$ to derive Eq. (3) from Eq. (1).

9. Find the Fourier integral representation of each of these functions:

a. $f(x) = \begin{cases} \sin(x), & -\pi < x < \pi \\ 0, & |x| > \pi \end{cases}$

b. $f(x) = \begin{cases} \sin(x), & 0 < x < \pi \\ 0, & \text{otherwise} \end{cases}$

c. $f(x) = \begin{cases} |\sin(x)|, & -\pi < x < \pi \\ 0, & \text{otherwise.} \end{cases}$

10. Rewrite Eq. (1) as

$$\int_0^\infty \frac{\sin(\lambda x)}{\lambda} d\lambda = \frac{\pi}{2}\mathrm{sgn}(x),$$

where $\mathrm{sgn}(x)$ is $+1, 0$ or -1 as x is positive, zero, or negative. Then, show that the function of Eq. (3) is

$$K(x,h) = \frac{1}{2}\left(\mathrm{sgn}(x+h) - \mathrm{sgn}(x-h)\right).$$

Finally, obtain Eq. (2) using Eq. (1) and the identity

$$\sin(\lambda(x+h)) - \sin(\lambda(x-h)) = 2\sin(\lambda h)\cos(\lambda x).$$

11. The Fourier integral can be obtained as a kind of limit of the Fourier series. Suppose $f(x)$ is defined for all x and sectionally smooth. It can be represented on any interval $-a < x < a$ by a Fourier series

$$f(x) = a_0 + \sum_{n=1}^\infty a_n \cos\left(\frac{n\pi x}{a}\right) + b_n \sin\left(\frac{n\pi x}{a}\right), \qquad -a < x < a$$

$$a_0 = \frac{1}{2a}\int_{-a}^{a} f(x)dx, \qquad a_n = \frac{1}{a}\int_{-a}^{a} f(x)\cos\left(\frac{n\pi x}{a}\right)dx,$$

$$b_n = \frac{1}{a}\int_{-a}^{a} f(x)\sin\left(\frac{n\pi x}{a}\right)dx.$$

Let $\lambda_n = n\pi/a$ and define two functions,

$$A_a(\lambda_n) = \frac{1}{\pi}\int_{-a}^{a} f(x)\cos(\lambda_n x)dx, \qquad B_a(\lambda_n) = \frac{1}{\pi}\int_{-a}^{a} f(x)\sin(\lambda_n)dx.$$

The Fourier series for f now becomes

$$f(x) = a_0 + \sum_{n=1}^\infty \left(A_a(\lambda_n)\cos(\lambda_n x) + B_a(\lambda_n)\sin(\lambda_n x)\right)\Delta\lambda \qquad (*)$$

where $\Delta\lambda = \lambda_{n+1} - \lambda_n = \pi/a$.

Show that the following limits hold as $a \to \infty$, if $\int_{-\infty}^{\infty} |f(x)|dx$ is finite:

$$A_a(\lambda) \to \frac{1}{\pi}\int_{-\infty}^{\infty} f(x)\cos(\lambda x)dx = A(\lambda)$$

$$B_a(\lambda) \to \frac{1}{\pi}\int_{-\infty}^{\infty} f(x)\sin(\lambda x)dx = B(\lambda)$$

$$A_0 \to 0.$$

12. Does the Fourier series in the equaton, designated by (*) here, approach the Fourier integral for f as $a \to \infty$?

1.10 COMPLEX METHODS

Suppose that a function $f(x)$ corresponds to the Fourier series

$$f(x) \sim a_0 + \sum_{n=1}^{\infty} a_n \cos(nx) + b_n \sin(nx).$$

(We use period 2π for simplicity only.) A famous formula of Euler states that

$$e^{i\theta} = \cos(\theta) + i\sin(\theta), \qquad \text{where } i^2 = -1.$$

Some simple algebra then gives the exponential definitions of the sine and cosine:

$$\cos(\theta) = \frac{1}{2}\left(e^{i\theta} + e^{-i\theta}\right), \qquad \sin(\theta) = \frac{1}{2i}\left(e^{i\theta} - e^{-i\theta}\right).$$

By substituting the exponential forms into the Fourier series of f we arrive at the alternate form

$$\begin{aligned} f(x) &\sim a_0 + \frac{1}{2}\sum_{n=1}^{\infty} a_n\left(e^{inx} + e^{-inx}\right) - ib_n\left(e^{inx} - e^{-inx}\right) \\ &\sim a_0 + \frac{1}{2}\sum_{n=1}^{\infty} \left(a_n - ib_n\right)e^{inx} + \left(a_n + ib_n\right)e^{-inx}. \end{aligned}$$

We are now led to define *complex Fourier coefficients* for f:

$$c_0 = a_0, \quad c_n = \frac{1}{2}(a_n - ib_n), \quad c_{-n} = \frac{1}{2}(a_n + ib_n), \quad n = 1, 2, 3, \cdots.$$

In terms of these two coefficients, we have

$$f(x) \sim c_0 + \sum_{n=1}^{\infty}\left(c_n e^{inx} + c_{-n}e^{-inx}\right) \sim \sum_{-\infty}^{\infty} c_n e^{inx}. \tag{1}$$

This is the complex form of the Fourier series for f. It is easy to derive the universal formula

$$c_n = \frac{1}{2\pi}\int_{-\pi}^{\pi} f(x)e^{-inx}\,dx, \tag{2}$$

which is valid for all integers n, positive, negative, or zero. The complex form is used especially in physics and electrical engineering.

Sometimes the function corresponding to a Fourier series can be recognized by use of the complex form. For instance, the series

$$\sum_{n=1}^{\infty} \frac{(-1)^{n+1}}{n} \cos(nx)$$

may be considered the real part of

$$\sum_{n=1}^{\infty} \frac{(-1)^{n+1}}{n} e^{inx} = \sum_{1}^{\infty} \frac{(-1)^{n+1}}{n} (e^{ix})^n \tag{3}$$

because the real part of $e^{i\theta}$ is $\cos(\theta)$. The series on the right in Eq. (3) is recognized as a Taylor series

$$\sum_{n=1}^{\infty} \frac{(-1)^{n+1}}{n} (e^{ix})^n = \ln(1 + e^{ix}).$$

Some manipulations yield

$$1 + e^{ix} = e^{ix/2}\left(e^{ix/2} + e^{-ix/2}\right) = 2e^{ix/2} \cos\left(\frac{x}{2}\right)$$

$$\ln(1 + e^{ix}) = \frac{ix}{2} + \ln\left(2\cos\left(\frac{x}{2}\right)\right).$$

The real part of $\ln(1 + e^{ix})$ is $\ln(2\cos(x/2))$ when $-\pi < x < \pi$. Thus, we derive the relation

$$\ln\left(2\cos\left(\frac{x}{2}\right)\right) \sim \sum_{n=1}^{\infty} \frac{(-1)^{n+1}}{n} \cos(nx), \quad -\pi < x < \pi. \tag{4}$$

(The series actually converges except at $x = \pm\pi, \pm 3\pi, \cdots$.)

The Fourier integral of a function $f(x)$ defined in the entire interval $-\infty < x < \infty$ can also be cast in complex form:

$$f(x) = \int_{-\infty}^{\infty} C(\lambda) e^{i\lambda x} d\lambda. \tag{5}$$

The complex Fourier integral coefficient function is given by

$$C(\lambda) = \frac{1}{2\pi} \int_{-\infty}^{\infty} f(x) e^{-i\lambda x} dx. \tag{6}$$

It is simple to show that

$$C(\lambda) = \frac{1}{2}\left(A(\lambda) - iB(\lambda)\right), \tag{7}$$

where A and B are the usual Fourier integral coefficients. The complex Fourier integral coefficient is often called the *Fourier transform* of the function $f(x)$.

Exercises

1. Relate the functions and series that follow by using complex form and Taylor series.

   a. $$1 + \sum_{n=1}^{\infty} r^n \cos(nx) = \frac{1 - r\cos(x)}{1 - 2r\cos(x) + r^2}, \qquad 0 \leq r < 1$$

   b. $$\sum_{n=1}^{\infty} \frac{\sin(nx)}{n!} = e^{\cos(x)} \sin(\sin(x)).$$

2. Show by integrating that

   $$\int_{-\pi}^{\pi} e^{inx} e^{-imx} dx = \begin{cases} 0, & n \neq m \\ 2\pi, & n = m \end{cases}$$

   and develop the formula for the complex Fourier coefficients using this idea of orthogonality.

3. Use the complex form

   $$a_n - ib_n = \frac{1}{\pi}\int_{-\pi}^{\pi} f(x)e^{-inx} dx, \qquad n \neq 0$$

   to find the Fourier series of the function

   $$f(x) = e^{\alpha x}, \qquad -\pi < x < \pi.$$

4. Develop formula (2) for the complex Fourier coefficients c_n from the formulas for a_n and b_n.

5. Find the complex Fourier integral representation of the following functions:

   a. $$f(x) = \begin{cases} e^{-x}, & x > 0 \\ 0, & x < 0 \end{cases}$$

b. $f(x) = \begin{cases} \sin(x), & 0 < x < \pi \\ 0, & \text{elsewhere.} \end{cases}$

1.11 APPLICATIONS OF FOURIER SERIES AND INTEGRALS

Fourier series and integrals are among the most basic tools of applied mathematics. In what follows, give just a few applications that do not fall within the scope of the rest of this book.

A. Nonhomogeneous Differential Equation

Many mechanical and electrical systems may be described by the differential equation

$$\ddot{y} + \alpha\dot{y} + \beta y = f(t).$$

The function $f(t)$ is called the "forcing function," βy the "restoring term," and $\alpha\dot{y}$ the "damping term."

If $f(t)$ is periodic with period 2π, let its Fourier series be

$$f(t) = a_0 + \sum_{n=1}^{\infty} a_n \cos(nt) + b_n \sin(nt).$$

If $y(t)$ is periodic with period 2π, it and its derivatives have Fourier series:

$$y(t) = A_0 + \sum_{n=1}^{\infty} A_n \cos(nt) + B_n \sin(nt)$$

$$\dot{y}(t) = \sum_{n=1}^{\infty} -nA_n \sin(nt) + nB_n \cos(nt)$$

$$\ddot{y}(t) = \sum_{n=1}^{\infty} -n^2 A_n \cos(nt) - n^2 B_n \sin(nt).$$

Then the differential equation can be written in the form:

$$\beta A_0 + \sum_{n=1}^{\infty} \left(-n^2 A_n + \alpha n B_n + \beta A_n\right) \cos(nt)$$
$$+ \sum_{n=1}^{\infty} \left(-n^2 B_n - \alpha n A_n + \beta B_n\right) \sin(nt) = a_0 + \sum_{n=1}^{\infty} a_n \cos(nt) + b_n \sin(nt).$$

The A's and B's are now determined by matching coefficients

$$\beta A_0 = a_0$$
$$(\beta - n^2)A_n + \alpha n B_n = a_n$$
$$-\alpha n A_n + (\beta - n^2)B_n = b_n.$$

When these equations are solved for the A's and B's, we find

$$A_n = \frac{(\beta - n^2)a_n - \alpha n b_n}{\Delta}, \qquad B_n = \frac{(\beta - n^2)b_n + \alpha n a_n}{\Delta}$$

where

$$\Delta = (\beta - n^2)^2 + \alpha^2 n^2.$$

Now, given the function f, the a's and b's can be determined, thus giving the A's and B's. The function $y(t)$ represented by the series found is the periodic part of the response. Depending on the initial conditions, there may also be a transient response, which dies out as t increases.

B. Boundary Value Problems

By way of introduction to the next chapter, we apply the idea of Fourier series to the solution of the boundary value problem

$$\frac{d^2u}{dx^2} + pu = f(x), \qquad 0 < x < a$$

$$u(0) = 0, \qquad u(a) = 0.$$

First, we will assume that $f(x)$ is equal to its Fourier sine series,

$$f(x) = \sum_{n=1}^{\infty} b_n \sin\left(\frac{n\pi x}{a}\right), \qquad 0 < x < a.$$

And second, we will assume that the solution $u(x)$, which we are seeking, equals its Fourier sine series,

$$u(x) = \sum_{n=1}^{\infty} B_n \sin\left(\frac{n\pi x}{a}\right), \qquad 0 < x < a$$

and that this series may be differentiated twice to give

$$\frac{d^2u}{dx^2} = \sum_{n=1}^{\infty} -\left(\frac{n^2\pi^2}{a^2} B_n\right) \sin\left(\frac{n\pi x}{a}\right), \qquad 0 < x < a.$$

When we insert the series forms for u, u'', and $f(x)$ into the differential equation, we find that

$$\sum_{n=1}^{\infty}\left(-\frac{n^2\pi^2}{a^2}B_n + pB_n\right)\sin\left(\frac{n\pi x}{a}\right) = \sum_{n=1}^{\infty} b_n \sin\left(\frac{n\pi x}{a}\right), \quad 0 < x < a.$$

Since the coefficients of like terms in the two series must match, we may conclude that

$$\left(p - \frac{n^2\pi^2}{a^2}\right)B_n = b_n, \qquad n = 1, 2, 3, \cdots.$$

If it should happen that $p = m^2\pi^2/a^2$ for some positive integer m, there is no value of B_m that satisfies

$$\left(p - \frac{m^2\pi^2}{a^2}\right)B_m = b_m$$

unless that $b_m = 0$ also, in which case any value of B_m is satisfactory. In summary, we may say that

$$B_n = \frac{b_n}{\left(p - \dfrac{n^2\pi^2}{a^2}\right)}$$

and

$$u(x) = \sum_{n=1}^{\infty} \frac{a^2 b_n}{a^2 p - n^2\pi^2}\sin\left(\frac{n\pi x}{a}\right)$$

with the agreement that a zero denominator must be handled separately.

As an example, consider the boundary value problem

$$\frac{d^2u}{dx^2} - u = -x, \qquad 0 < x < 1$$

$$u(0) = 0, \qquad u(1) = 0.$$

It may readily be verified that

$$-x = \sum_{n=1}^{\infty} \frac{2(-1)^n}{\pi n}\sin(n\pi x), \qquad 0 < x < 1.$$

Thus, by the preceding development, the solution must be

$$u(x) = \sum_{n=1}^{\infty} \frac{2}{\pi}\frac{(-1)^{n+1}}{n(n^2\pi^2 + 1)}\sin(n\pi x), \qquad 0 < x < 1.$$

Although this particular series belongs to a known function, one would not, in general, know any formula for the solution $u(x)$ other than its Fourier sine series.

C. The Sampling Theorem

One of the most important results of information theory is the sampling theorem, which is based on a combination of the Fourier series and the Fourier integral in their complex forms. What the electrical engineer calls a signal is just a function $f(t)$ defined for all t. If the function is integrable, there is a Fourier integral representation for it:

$$f(t) = \int_{-\infty}^{\infty} C(\omega) \exp(i\omega t) d\omega$$

$$C(\omega) = \frac{1}{2\pi} \int_{-\infty}^{\infty} f(t) \exp(-i\omega t) dt.$$

A signal is called *band limited* if its Fourier transform is zero except in a finite interval; that is, if

$$C(\omega) = 0, \qquad \mathit{for}\ \ |\omega| > \Omega.$$

Then Ω is called the cutoff frequency. If f is band limited, we can write it in the form

$$f(t) = \int_{-\Omega}^{\Omega} C(\omega) \exp(i\omega t) d\omega \tag{1}$$

because $C(\omega)$ is zero outside the interval $-\Omega < \omega < \Omega$. We focus our attention on this interval by writing $C(\omega)$ as a Fourier series:

$$C(\omega) = \sum_{-\infty}^{\infty} c_n \exp\left(\frac{in\pi\omega}{\Omega}\right), \qquad -\Omega < \omega < \Omega. \tag{2}$$

The (complex) coefficients are

$$c_n = \frac{1}{2\Omega} \int_{-\Omega}^{\Omega} C(\omega) \exp\left(\frac{-in\pi\omega}{\Omega}\right) d\omega.$$

The point of the sampling theorem is to observe that the integral for c_n actually is a value of $f(t)$ at a particular time. In fact, from the integral Eq. (1), we see that

$$c_n = \frac{1}{2\Omega} f\left(\frac{-n\pi}{\Omega}\right).$$

Thus there is an easy way of finding the Fourier transform of a band-limited function. We have

$$\begin{aligned} C(\omega) &= \frac{1}{2\Omega} \sum_{-\infty}^{\infty} f\left(\frac{-n\pi}{\Omega}\right) \exp\left(\frac{in\pi\omega}{\Omega}\right) \\ &= \frac{1}{2\Omega} \sum_{-\infty}^{\infty} f\left(\frac{n\pi}{\Omega}\right) \exp\left(\frac{-in\pi\omega}{\Omega}\right), \qquad -\Omega < \omega < \Omega. \end{aligned}$$

By utilizing Eq. (1) again, we can reconstruct $f(t)$:

$$\begin{aligned} f(t) &= \int_{-\Omega}^{\Omega} C(\omega) \exp(i\omega t) d\omega \\ &= \frac{1}{2\Omega} \sum_{-\infty}^{\infty} f\left(\frac{n\pi}{\Omega}\right) \int_{-\Omega}^{\Omega} \exp\left(\frac{-in\pi\omega}{\Omega}\right) \exp(i\omega t) d\omega. \end{aligned}$$

Carrying out the integration and using the identity

$$\sin(\theta) = \frac{(e^{i\theta} - e^{-i\theta})}{2i},$$

we find

$$f(t) = \sum_{-\infty}^{\infty} f\left(\frac{n\pi}{\Omega}\right) \frac{\sin(\Omega t - n\pi)}{\Omega t - n\pi}.$$

This is the main result of the sampling theorem. It says that the band-limited function $f(t)$ may be reconstructed from the samples of f at $t = 0$, $\pm\pi/\Omega, \cdots$. The sampling theorem can be realized physically.

1.12 COMMENTS AND REFERENCES

The first use of trigonometric series occurred in the middle of the eighteenth century. Euler seems to have originated the use of orthogonality for the determination of coefficients. In the early nineteenth century Fourier made extensive use of trigonometric series in studying problems of heat conduction (see Chapter 2). His claim, that an arbitrary function could be represented as a trigonometric series, led to an extensive reexamination of the foundations of calculus. Fourier seems to have been among the first to recognize that a function might have different analytical expressions in different places.

Dirichlet established sufficient conditions (similar to those of our convergence theorem) for the convergence of Fourier series around 1830. Later, Riemann was led to redefine the integral as part of his attempt to discover conditions on a function necessary and sufficient for the convergence of its Fourier series. This problem has never been solved. Many other great mathematicians have founded important theories (the theory of sets, for one) in the course of studying Fourier series, and they continue to be a subject of active research. An entertaining and readable account of the history and uses of Fourier series is in *The Mathematical Experience*, by Davis, Hersh and Marchisotto. (See the Bibliography.)

Historical interest aside, Fourier series and integrals are extremely important in applied mathematics, physics and engineering, and they merit further study. A superbly written and organized book is Tolstov's *Fourier Series.* Its mathematical prerequisites are not too high. *Fourier Series and Boundary Value Problems* by Churchill and Brown is a standard text for some engineering applications.

About 1960 it became clear that the numerical computation of Fourier coefficients could be rearranged to achieve dramatic reductions in the amount of arithmetic required. The result, called the *fast Fourier transform* or FFT, has revolutionized the use of Fourier series in applications. See *The Fast Fourier Transform and its Applications*, by E.O. Brigham. The *Scientific American* article, "The Fourier Transform" by R.N. Bracewell (June 1989, pages 86-95) gives an interesting and well-illustrated exposition of the Fourier series and related topics.

The sampling theorem mentioned in the last section has become bread and butter in communications engineering. For extensive information on this, as well as the FFT, see *Integral and Discrete Transforms with Applications and Error Analysis*, by A.J. Jerri.

MISCELLANEOUS EXERCISES

1. Find the Fourier sine series of the trapezoidal function given for $0 < x < \pi$ by

$$f(x) = \begin{cases} x/\alpha, & 0 < x < \alpha \\ 1, & \alpha < x < \pi - \alpha \\ (\pi - x)/\alpha, & \pi - \alpha < x < \pi. \end{cases}$$

2. Show that the series found in Exercise 1 converges uniformly.

3. When α approaches 0, the function of Exercise 1 approaches a square wave. Do the sine coefficients found in Exercise 1 approach those of a square wave?

4. Find the Fourier cosine series of the function

$$F(x) = \int_0^x f(t)dt,$$

where f denotes the function in Exercise 1. Sketch.

5. Find the Fourier sine series of the function given in the interval $0 < x < a$ by the formula (α is a parameter between 0 and 1)

$$f(x) = \begin{cases} \dfrac{hx}{\alpha a}, & 0 < x < \alpha a \\ \dfrac{h(a-x)}{(1-\alpha)a}, & \alpha a < x < a. \end{cases}$$

6. Sketch the function of Exercise 5. To what does its Fourier sine series converge at $x = 0$? at $x = \alpha a$? at $x = a$?

7. Suppose that $f(x) = 1$, $0 < x < a$. Sketch and find the Fourier series of the following extensions of $f(x)$:

a. even extension
b. odd extension
c. periodic extension (period a)
d. even periodic extension
e. odd periodic extension
f. the one corresponding to $f(x) = x$, $-a < x < 0$.

8. Perform the same task as in Exercise 7, but $f(x) = 0$, $0 < x < a$.

9. Find the Fourier series of the function given by

$$f(x) = \begin{cases} 0, & -a < x < 0 \\ 2x, & 0 < x < a. \end{cases}$$

Sketch the graph of $f(x)$ and its periodic extension. To what values does the series converge at $x = -a$, $x = -a/2$, $x = 0$, $x = a$, and $x = 2a$?

10. Sketch the odd periodic extension and find the Fourier sine series of the function given by

$$f(x) = \begin{cases} 1, & 0 < x < \dfrac{\pi}{2} \\ \dfrac{1}{2}, & \dfrac{\pi}{2} < x < \pi. \end{cases}$$

To what values does the series converge at $x = 0$, $x = \pi/2$, $x = \pi$, $x = 3\pi/2$, and $x = 2\pi$?

11. Sketch the even periodic extension of the function given in Exercise 10. Find its Fourier cosine series. To what values does the series converge at $x = 0$, $x = \pi/2$, $x = \pi$, $x = 3\pi/2$, and $x = 2\pi$?

12. Find the Fourier cosine series of the function

$$g(x) = \begin{cases} 1 - x, & 0 < x < 1 \\ 0, & 1 < x < 2. \end{cases}$$

Sketch the graph of the sum of the cosine series.

13. Find the Fourier sine series of the function defined by $f(x) = 1-2x$, $0 < x < 1$. Sketch the graph of the odd periodic extension of $f(x)$, and determine the sum of the sine series at points where the graph has a jump.

14. Following the same requirements as in Exercise 13, use the cosine series and the even periodic extension.

15. Find the Fourier series of the function given by

$$f(x) = \begin{cases} 0, & -\pi < x < -\frac{\pi}{2} \\ \sin(2x), & -\frac{\pi}{2} < x < \frac{\pi}{2} \\ 0, & \frac{\pi}{2} < x < \pi. \end{cases}$$

Sketch the graph of the function.

16. Show that the function given by the formula $f(x) = (\pi - x)/2$, $0 < x < 2\pi$, has the Fourier series

$$f(x) = \sum_{1}^{\infty} \frac{\sin(nx)}{n}, \qquad 0 < x < 2\pi.$$

Sketch $f(x)$ and its periodic extension.

17. Use complex methods and a finite geometric series to show that

$$\sum_{n=1}^{N} \cos(nx) = \frac{\sin((N + \frac{1}{2})x) - \sin(\frac{1}{2}x)}{2\sin(\frac{1}{2}x)}.$$

Then use trigonometric identities to identify

$$\sum_{n=1}^{N} \cos(nx) = \frac{\sin(\frac{1}{2}Nx)\cos(\frac{1}{2}(N+1)x)}{\sin(\frac{1}{2}x)}.$$

18. Identify the partial sums of the Fourier series in Exercise 16 as

$$S_N(x) = \sum_{n=1}^{N} \frac{\sin(nx)}{n}.$$

The series of Exercise 17 is $S'_N(x)$. Use this information to locate the maxima and minima of $S_N(x)$ in the interval $0 \le x \le \pi$. Find the value of $S_N(x)$ at the first point in the interval $0 < x < \pi$ where $S'_N(x) = 0$ for $N = 5$. Compare to $(\pi - x)/2$ at that point.

19. Find the Fourier sine series of the function given by

$$f(x) = \begin{cases} \sin\left(\frac{\pi x}{a}\right), & 0 < x < a \\ 0, & a < x < \pi \end{cases}$$

assuming that $0 < a < \pi$.

20. Find the Fourier cosine series of the function given in Exercise 19.

21. Find the Fourier integral representation of the function given by

$$f(x) = \begin{cases} 1, & 0 < x < a \\ 0, & x < 0 \text{ or } x > a. \end{cases}$$

22. Find the Fourier sine and cosine integral representations of the function given by

$$f(x) = \begin{cases} \frac{a - x}{a}, & 0 < x < a \\ 0, & a < x. \end{cases}$$

23. Find the Fourier sine integral representation of the function

$$f(x) = \begin{cases} \sin(x), & 0 < x < \pi \\ 0, & \pi < x. \end{cases}$$

24. Find the Fourier integral representation of the function

$$f(x) = \begin{cases} 1/\epsilon, & \alpha < x < \alpha + \epsilon \\ 0, & \text{elsewhere.} \end{cases}$$

25. Use integration by parts to establish the equality

$$\int_0^\infty e^{-\lambda} \cos(\lambda x) d\lambda = \frac{1}{1+x^2}.$$

26. The equation in Exercise 25 is valid for all x. Explain why its validity implies that

$$\frac{2}{\pi} \int_0^\infty \frac{\cos(\lambda x)}{1+x^2} dx = e^{-\lambda}, \qquad \lambda > 0.$$

27. Integrate both sides of the equality in Exercise 25 from 0 to t to derive the equality

$$\int_0^\infty \frac{e^{-\lambda} \sin(\lambda t)}{\lambda} d\lambda = \tan^{-1}(t).$$

28. Does the equality in Exercise 27 imply that

$$\frac{2}{\pi} \int_0^\infty \tan^{-1}(t) \sin(\lambda t) dt = \frac{e^{-\lambda}}{\lambda}?$$

29. From Exercise 27 derive the equality

$$\int_0^\infty \frac{1 - e^{-\lambda}}{\lambda} \sin(\lambda x) d\lambda = \frac{\pi}{2} - \tan^{-1}(x), \qquad x > 0.$$

30. Without using integration, obtain the Fourier series (period 2π) of each of the following functions:

a. $2 + 4\sin(50x) - 12\cos(41x)$ **b.** $\sin^2(5x)$

c. $\sin(4x+2)$ **d.** $\sin(3x)\cos(5x)$

e. $\cos^3(x)$ **f.** $\cos(2x + \frac{1}{3}\pi)$

31. Let the function $f(x)$ be given in the interval $0 < x < 1$ by the formula

$$f(x) = 1 - x.$$

Find (a) a sine series, (b) a cosine series, (c) a sine integral, and (d) a cosine integral that equals the given function for $0 < x < 1$. In each case, sketch the function to which the series or integral converges in the interval $-2 < x < 2$.

32. Verify the Fourier integral

$$\int_0^\infty \cos(\lambda q)\exp(-\lambda^2 t)d\lambda = \sqrt{\frac{\pi}{4t}}\exp\left(-\frac{q^2}{4t}\right), \qquad (t>0)$$

by transforming the left-hand side according to these steps: (a) convert to an integral from $-\infty$ to ∞ by using the evenness of the integrand; (b) replace $\cos(\lambda q)$ by $\exp(i\lambda q)$ (justify this step); (c) complete the square in the exponent; (d) change the variable of integration; (e) use the equality

$$\int_{-\infty}^\infty \exp(-u^2)du = \sqrt{\pi}.$$

33. Approximate the first seven cosine coefficients $(\hat{a}_0, \hat{a}_1, \cdots, \hat{a}_6)$ of the function

$$f(x) = \frac{1}{1+x^2}, \qquad 0 < x < 1.$$

34. Use Fourier sine series representations of $u(x)$ and of the function $f(x) = x$, $0 < x < a$, to solve the boundary value problem

$$\frac{d^2u}{dx^2} - \gamma^2 u = -x, \qquad 0 < x < a,$$

$$u(0) = 0, \qquad u(a) = 0.$$

35. - 43. For each of these exercises,

a. find the Fourier cosine series of the function;

b. determine the value to which the series converges at the given values of x;

c. sketch the even periodic extension of the given function for at least two periods.

44. - 52. For each of these exercises,

a. find the Fourier sine series of the function;

b. determine the value to which the series converges at the given values of x;

c. sketch the odd periodic extension of the given function for at least two periods.

35. & **44.** $f(x) = \begin{cases} 0, & 0 < x < \dfrac{a}{3} \\ x - \dfrac{a}{3}, & \dfrac{a}{3} < x < \dfrac{2a}{3}, \\ \dfrac{a}{3}, & \dfrac{2a}{3} < x < a \end{cases} \qquad x = 0, \dfrac{a}{3}, a, -\dfrac{a}{2}$

36. & **45.** $f(x) = \begin{cases} \dfrac{1}{2}, & 0 < x < \dfrac{a}{2} \\ x = \dfrac{a}{2}, & 2a, 0, -a \\ 1, & \dfrac{a}{2} < x < a \end{cases}$

37. & **46.** $f(x) = \begin{cases} \dfrac{2x}{a}, & 0 < x < \dfrac{a}{2} \\ x = 0, & \dfrac{a}{2}, a, \dfrac{3a}{2} \\ \dfrac{(3a - 2x)}{2a}, & \dfrac{a}{2} < x < a \end{cases}$

38. & **47.** $f(x) = \begin{cases} x, & 0 < x < \dfrac{a}{2} \\ x = 0, & a, -\dfrac{a}{2} \\ \dfrac{a}{2}, & \dfrac{a}{2} < x < a \end{cases}$

39. & **48.** $f(x) = \dfrac{(a - x)}{a}, \qquad 0 < x < a, \quad x = 0, a, -\dfrac{a}{2}$

40. & **49.** $f(x) = \begin{cases} 0, & 0 < x < \dfrac{a}{4} \\ 1, & \dfrac{a}{4} < x < \dfrac{3a}{4}, \\ 0, & \dfrac{3a}{4} < x < a \end{cases} \quad x = 0, \dfrac{a}{4}, \dfrac{a}{2}, a, -\dfrac{3a}{4}$

41. & **50.** $f(x) = x(a - x), \ \ 0 < x < a, \ \ x = 0, -a, -\dfrac{a}{2}$

42. & **51.** $f(x) = e^{kx}, \ \ 0 < x < a, \ \ x = 0, \dfrac{a}{2}, a, -a$

43. & 52. $f(x) = \begin{cases} 0, & 0 < x < \frac{a}{2} \\ x = -a, \frac{a}{2}, a & \\ 1, & \frac{a}{2} < x < a \end{cases}$

53. - 58. For each of these exercises,

a. find the Fourier cosine integral representation of the function;

b. sketch the even extension of the function.

59. - 64. For each of these exercises,

a. find the Fourier sine integral representation of the function;

b. sketch the odd extension of the function.

53. & 59. $f(x) = e^{-x}$, $0 < x$

54. & 60. $f(x) = \begin{cases} e^{-x}, & 0 < x < a \\ 0, & a < x \end{cases}$

55. & 61. $f(x) = \begin{cases} 1, & 0 < x < b \\ 0, & b < x \end{cases}$

56. & 62. $f(x) = \begin{cases} \cos(x), & 0 < x < \pi \\ 0, & \pi < x \end{cases}$

57. & 63. $f(x) = \begin{cases} 1 - x, & 0 < x < 1 \\ 0, & 1 < x \end{cases}$

58. & 64. $f(x) = \begin{cases} 1, & 0 < x < 1 \\ 2 - x, & 1 < x < 2 \\ 0, & 2 < x \end{cases}$

65. Let $f(x)$ be a periodic function with period 2π whose Fourier coefficients are $a_0, a_1, b_1, \cdots$. Then, the partial sum

$$S_N(x) = a_0 + \sum_{n=1}^{N} a_n \cos(nx) + b_n \sin(nx)$$

is an approximation to $f(x)$ if f is sectionally smooth and N is large enough. The average of these approximations is

$$\sigma_N(x) = \frac{1}{N}\left(S_1(x) + \cdots + S_N(x)\right).$$

It is known that $\sigma_N(x)$ converges uniformly to $f(x)$ if f is continuous. Show that

$$\sigma_N(x) = a_0 + \sum_{n=1}^{N} \frac{N+1-n}{N}(a_n \cos(nx) + b_n \sin(nx)).$$

66. In analogy to Lemma 2 of Section 7, prove that

$$\sum_{n=0}^{N-1} \sin\left(\left(n+\tfrac{1}{2}\right)y\right) = \frac{\sin^2(\frac{1}{2}Ny)}{\sin(\frac{1}{2}y)}.$$

67. Following the lines of Section 7, show that

$$\sigma_N(x) - f(x) = \frac{1}{2N\pi}\int_{-\pi}^{\pi} [f(x+y) - f(x)] \left(\frac{\sin(\frac{1}{2}Ny)}{\sin(\frac{1}{2}y)}\right)^2 dy.$$

This equality is the key to the proof of uniform convergence mentioned in Exercise 65.

68. In a study of river freezing, E.P. Foltyn and H.T. Shen (St. Lawrence River freeze-up forecast, Journal of Waterway, Port, Coastal and Ocean Engineering, vol. 112, 1986, pp. 467-481.) use data spanning 33 years to find this Fourier series representation of the air temperature in Massena NY.

$$T(t) = a_0 + \sum_{n=0}^{\infty} a_n \cos(2n\pi t) + b_n \sin(2n\pi t).$$

Here T is temperature in °C, t is time in years, and the origin is Oct. 1. The first coefficients were found to be

$$a_0 = 6.638, a_1 = 5.870, b_1 = -13.094, a_2 = 0.166, b_2 = 0.583,$$

and the remaining coefficients were all less than 0.3 in absolute value. The authors decided to exclude all the terms from a_2 and b_2 up, so their approximation could be written

$$T(t) \cong a_0 + A\sin(2\pi t + \theta).$$

a. Find the average temperature in Massena.

b. Find A, the amplitude of the annual variation, and the phase angle θ.

c. Find the approximate date when the minimum temperature occurs.

d. Find the dates when the approximate temperature passes through 0.

e. Discuss the effect on the answer to part **d** if the next two terms of the series were included.

Chapter 2

The Heat Equation

2.1 DERIVATION AND BOUNDARY CONDITIONS

As the first example of the derivation of a partial differential equation, we consider the problem of describing the temperature in a rod of conducting material. In order to simplify the problem as much as possible, we shall assume that the rod has a uniform cross section (like an extrusion) and that the temperature does not vary from point to point on a section. Thus, if we use a coordinate system as suggested in Fig. 1, we may say that the temperature depends only on position x and time t.

The basic idea in developing the partial differential equation is to apply the laws of physics to a small piece of the rod. Specifically, we apply the law of conservation of energy to a slice of the rod that lies between x and $x + \Delta x$ (Fig. 2).

The law of conservation of energy states that the amount of heat that enters a region plus what is generated inside is equal to the amount of heat that leaves plus the amount stored. The law is equally valid in terms of rates per unit time instead of amounts.

Now let $q(x, t)$ be the heat flux at point x and time t. The dimensions of q are[1] $[q] = H/tL^2$, and q is taken to be positive when heat flows to the right. The rate at which heat is entering the slice through the surface at x is $Aq(x, t)$, where A is the area of a cross section. The rate at which heat is leaving the slice through the surface at $x + \Delta x$ is $Aq(x + \Delta x, t)$.

The rate of heat storage in the slice of material is proportional to the rate of change of temperature. Thus, if ρ is the density and c is the heat capacity per unit mass ($[c] = H/mT$), we may approximate the rate of

[1] Square brackets are used to symbolize "dimension of." H = heat energy, t = time, T = temperature, L = length, m = mass, and so forth.

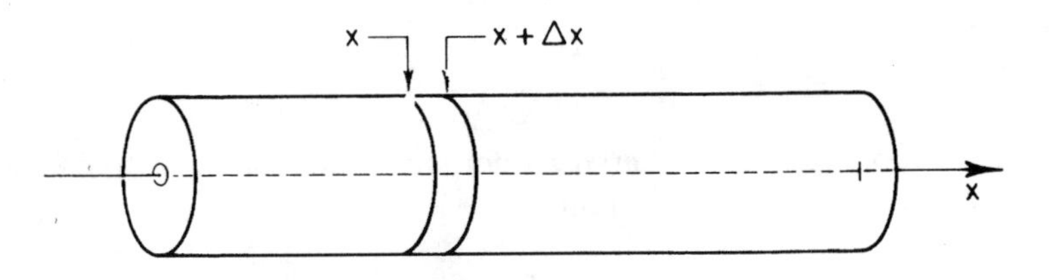

Figure 1:

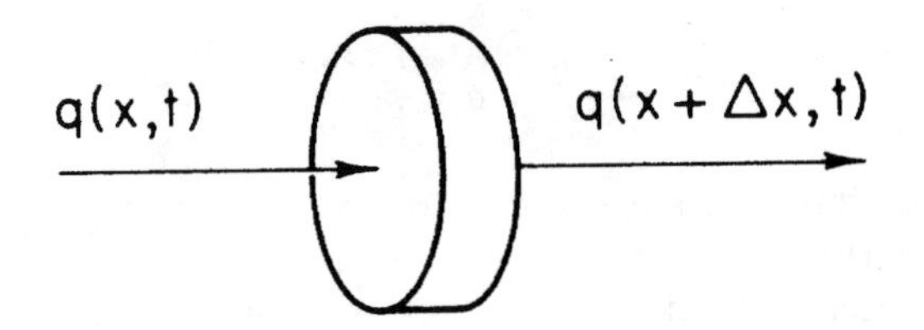

Figure 2:

heat storage in the slice by

$$\rho c A \Delta x \frac{\partial u}{\partial t}(x,t)$$

where $u(x,t)$ is the temperature.

There are other ways in which heat may enter (or leave) the section of rod we are looking at. One possibility is that heat is transferred by radiation or convection from (or to) a surrounding medium. Another is that heat is converted from another form of energy — for instance, by resistance to an electrical current or by chemical or nuclear reaction. All of these possibilities we lump together in a "generation rate." If the rate of generation per unit volume is g, $[g] = H/tL^3$, then the rate at which heat is generated in the slice is $A\Delta x g$. (Note that g may depend on x, t, and even u.)

We have now quantified the law of conservation of energy for the slice of rod in the form

$$Aq(x,t) + A\Delta x g = Aq(x+\Delta x, t) + A\Delta x \rho c \frac{\partial u}{\partial t}.$$

After some algebraic manipulation, we have

$$\frac{q(x,t) - q(x+\Delta x, t)}{\Delta x} + g = \rho c \frac{\partial u}{\partial t}.$$

The ratio

$$\frac{q(x+\Delta x,t)-q(x,t)}{\Delta x}$$

should be recognized as a difference quotient. If we allow Δx to decrease, this quotient becomes, in the limit,

$$\lim_{\Delta x\to 0}\frac{q(x+\Delta x,t)-q(x,t)}{\Delta x}=\frac{\partial q}{\partial x}.$$

The limit process thus leaves the law of conservation of energy in the form

$$-\frac{\partial q}{\partial x}+g=\rho c\frac{\partial u}{\partial t}. \tag{2}$$

We are not finished, since there are two dependent variables, q and u, in this equation. We need another equation relating q and u. This relation is Fourier's law of heat conduction, which in one dimension may be written:

$$q=-\kappa\frac{\partial u}{\partial x}.$$

In words, heat flows downhill (q is positive when $\partial u/\partial x$ is negative) at a rate proportional to the gradient of the temperature. The proportionality factor κ, called the *thermal conductivity*, may depend on x if the rod is not uniform, and also may depend on temperature. However, we will usually assume it to be a constant.

Substituting Fourier's law in the heat balance equation yields

$$\frac{\partial}{\partial x}\left(\kappa\frac{\partial u}{\partial x}\right)+g=\rho c\frac{\partial u}{\partial t}. \tag{3}$$

Note that κ, ρ, and c may all be functions. If, however, they are independent of x, t, and u, we may write

$$\frac{\partial^2 u}{\partial x^2}+\frac{g}{\kappa}=\frac{\rho c}{\kappa}\frac{\partial u}{\partial t}. \tag{4}$$

The equation is applicable where the rod is located and after the experiment starts: for $0<x<a$ and for $t>0$. The quantity $\kappa/\rho c$ is often written as k, and is called the *diffusivity*. Table 1 shows approximate values of these constants for several materials.

For some time we will be working with the heat equation without generation:

$$\frac{\partial^2 u}{\partial x^2}=\frac{1}{k}\frac{\partial u}{\partial t},\qquad 0<x<a,\qquad 0<t, \tag{5}$$

Table 1: Typical values of constants

Material	c	ρ	κ	$k = \dfrac{\kappa}{\rho c}$
	$\dfrac{\text{cal}}{\text{g }^\circ\text{C}}$	$\dfrac{\text{g}}{\text{cm}^3}$	$\dfrac{\text{cal}}{\text{s cm }^\circ\text{C}}$	$\dfrac{\text{cm}^2}{\text{s}}$
Aluminum	0.21	2.7	0.48	0.83
Copper	0.094	8.9	0.92	1.1
Steel	0.11	7.8	0.11	0.13
Glass	0.15	2.6	0.0014	0.0036
Concrete	0.16	2.3	0.0041	0.011
Ice	0.48	0.92	0.004	0.009

which, to review, is supposed to describe the temperature u in a rod of length a with uniform properties and cross section, in which no heat is generated, and whose cylindrical surface is insulated.

This equation alone is not enough information to completely specify the temperature, however. Each of the functions

$$u(x,t) = x^2 + 2kt$$

$$u(x,t) = e^{-kt}\sin(x)$$

satisfies the partial differential equation and so do their sum and difference.

Clearly this is not a satisfactory situation either from the mathematical or physical viewpoint; we would like the temperature to be uniquely determined. More conditions must be placed on the function u. The appropriate additional conditions are those that describe:

1. the initial temperature distribution in the rod, and

2. what is happening at the ends of the rod.

The *initial condition* is described mathematically as

$$u(x,0) = f(x), \qquad 0 < x < a$$

where $f(x)$ is a given function of x alone.

The *boundary conditions* may take a variety of forms. First, the temperature at either end may be held constant, for instance by exposing the end to an ice-water bath or to condensing steam. We can describe these conditions by the equations

$$u(0,t) = T_0, \qquad u(a,t) = T_1, \qquad t > 0$$

where T_0 and T_1 may be the same or different. More generally, the temperature at the boundary may be controlled in some way, without being held constant. If x_0 symbolizes an endpoint, the condition is

$$u(x_0,t) = \alpha(t), \tag{6}$$

where α is a function of time. Of course, the case of a constant function is included here. This type of boundary condition is called a *Dirichlet condition* or *condition of the first kind.*

Another possibility is that the heat flow rate is controlled. Since Fourier's law associates the heat flow rate and the gradient of the temperature, we can write

$$\frac{\partial u}{\partial x}(x_0,t) = \beta(t) \tag{7}$$

where β is a function of time. This is called a *Neumann condition* or *condition of the second kind.* We most frequently take $\beta(t)$ to be identically zero. Then the condition

$$\frac{\partial u}{\partial x}(x_0,t) = 0$$

corresponds to an *insulated* surface, for this equation says that the heat flow is zero.

Still another possible boundary condition is

$$c_1 u(x_0,t) + c_2 \frac{\partial u}{\partial x}(x_0,t) = \gamma(t), \tag{8}$$

called *third kind* or a *Robin condition.* This kind of condition can also be realized physically. If the surface at $x = a$ is exposed to air or other fluid, then the heat conducted up to that surface from inside the rod is carried away by convection. Newton's law of cooling says that the rate at which heat is transferred from the body to the fluid is proportional to the difference in temperature between the body and the fluid. In symbols, we have

$$q(a,t) = h(u(a,t) - T(t)) \tag{9}$$

where $T(t)$ is the air temperature. After application of Fourier's law, this becomes

$$-\kappa\frac{\partial u}{\partial x}(a,t) = hu(a,t) - hT(t). \tag{10}$$

This equation can be put into the form of Eq. (8). (Note: h is called the convection coefficient or heat transfer coefficient; $[h] = H/L^2tT$.)

All of the boundary conditions given in Eqs. (6), (7) and (8) involve the function u and/or its derivative at one point. If more than one point is involved, the boundary condition is called *mixed.* For example, if a uniform rod is bent into a ring and the ends $x = 0$ and $x = a$ are joined, appropriate boundary conditions would be

$$u(0,t) = u(a,t), \quad t > 0 \tag{11}$$

$$\frac{\partial u}{\partial x}(0,t) = \frac{\partial u}{\partial x}(a,t), \quad t > 0 \tag{12}$$

both of mixed type.

Many other kinds of boundary conditions exist and are even realizable, but the four kinds already mentioned here are the most commonly encountered. An important feature common to all four types is that they involve a *linear* operation on the function u.

The heat equation, an initial condition, and a boundary condition for each end form what is called an *initial value-boundary value problem.* For instance, one possible problem would be

$$\frac{\partial^2 u}{\partial x^2} = \frac{1}{k}\frac{\partial u}{\partial t}, \qquad 0 < x < a, \qquad 0 < t \tag{13}$$

$$u(0,t) = T_0, \qquad 0 < t \tag{14}$$

$$-\kappa\frac{\partial u}{\partial x}(a,t) = h(u(a,t) - T_1), \quad 0 < t \tag{15}$$

$$u(x,0) = f(x), \qquad 0 < x < a. \tag{16}$$

Notice that the boundary conditions may be of different kinds at different ends.

Although we shall not prove it, it is true that there is one, and only one, solution to a complete initial value-boundary value problem.

We have derived the heat equation (4) as a mathematical model for the temperature in a "rod," suggesting an object that is much longer than it is wide. The equation applies equally well to a "slab," an object that is much wider than it is thick. The important feature is that we may assume in either case that the temperature varies in only one space direction (along the length of the rod or the thickness of the slab). In Chapter 5, we derive a multidimensional heat equation.

It may come as a surprise that the partial differential equations of this section have another completely different but equally important physical interpretation. Suppose that a static medium occupies a region of space between $x = 0$ and $x = a$ (a slab!), and that we wish to study the concentration u, measured in units of mass per unit volume, of another substance whose molecules or atoms can move, or diffuse, through the medium. We assume that the concentration is a function of x and t only and designate $q(x,t)$ to be the mass flux ($[q] = m/tL^2$). Then the principle of conservation of mass may be applied to a layer of the medium between x and $x + \Delta x$ to obtain the equation

$$q(x,t) + \Delta x g = q(x + \Delta x, t) + \Delta x \frac{\partial u}{\partial t}(x,t). \tag{17}$$

When we rearrange Eq. (17) and take the limit as Δx approaches 0, it becomes

$$-\frac{\partial q}{\partial x} + g = \frac{\partial u}{\partial t}. \tag{18}$$

In these equations, g is a "generation rate" ($[g] = m/tL^3$), a function that accounts for any gain or loss of the substance from the layer by means other than movement in the x-direction. For example, the substance may participate in a chemical reaction with the medium at a rate proportional to its concentration (a first-order reaction) so that in this case the generation rate is

$$g = -ku(x,t). \tag{19}$$

The concentration and the mass flux are linked by a phenomenological relation called *Fick's first law*, written in one dimension as

$$q = -D\frac{\partial u}{\partial x}. \tag{20}$$

In words, the diffusing substance moves toward regions of lower concentration at a rate proportional to the gradient of the concentration. The coefficient of proportionality D, usually constant, is called the *diffusivity*. By combining Fick's law with Eq. (18) arising from the conservation of mass, we obtain the diffusion equation

$$\frac{\partial^2 u}{\partial x^2} + \frac{g}{D} = \frac{1}{D}\frac{\partial u}{\partial t}. \tag{21}$$

At a boundary of the medium, the concentration of the diffusing substance may be controlled, leading to a condition like Eq. (6), or the flux of the substance may be controlled, leading via Fick's law to a condition like Eq. (7); an impermeable surface corresponds to zero flux. If a boundary is covered with a permeable film, then the flux through the film is usually taken to be proportional to the difference in concentrations on the two sides of the film. Suppose that the surface in question is at $x = a$. Then these statements may be expressed symbolically as

$$q(a,t) = h(u(a,t) - C(t)), \tag{22}$$

where the proportionality constant h is called the film coefficient and C is the concentration outside the medium. Using Fick's law here leads to the equation

$$-D\frac{\partial u}{\partial x}(a,t) = hu(a,t) - hC(t). \tag{23}$$

This equation is analogous to Eq. (10) and can be put into the form of Eq. (8).

Exercises

1. Give a physical interpretation for the problem in Eqs. (13)-(16).

2. Verify that the following functions are solutions of the heat equation (5):

$$u(x,t) = \exp(-\lambda^2 kt)\cos(\lambda x)$$
$$u(x,t) = \exp(-\lambda^2 kt)\sin(\lambda x)$$

3. Suppose that the rod gains heat through the cylindrical surface by convection from a surrounding fluid at temperature U (constant). Newton's law of cooling says that the rate of heat transfer is proportional to exposed area and temperature difference. What is g in Eq. (1)? What form does Eq. (4) take?

4. Suppose that the end of the rod at $x = 0$ is immersed in an insulated container of water or other fluid; that the temperature of the fluid is the same as the temperature of the end of the rod; that the heat capacity of the fluid is C units of heat per degree. Show that this situation is represented mathematically by the equation

$$C\frac{\partial u}{\partial t}(0,t) = \kappa A\frac{\partial u}{\partial x}(0,t)$$

where A is the cross-sectional area of the rod.

5. Put Eq. (10) into Eq. (8) form. Notice that the signs still indicate that heat flows in the direction of lower temperature. That is, if $u(a,t) > T(t)$, then $q(a,t)$ is positive and the gradient of u is negative. Show that, if the surface at $x = 0$ (left end) is exposed to convection, the boundary condition would read

$$\kappa \frac{\partial u}{\partial x}(0,t) = hu(0,t) - hT(t).$$

Explain the signs.

6. Suppose the surface at $x = a$ is exposed to radiation. The Stefan-Boltzmann law of radiation says that the rate of radiation heat transfer is proportional to the difference of the fourth powers of the *absolute* temperatures of the bodies:

$$q(a,t) = \sigma\left(u^4(a,t) - T^4\right).$$

Use this equation and Fourier's law to obtain a boundary condition for radiation at $x = a$ to a body at temperature T.

7. The difference cited in Exercise 6 may be written

$$u^4 - T^4 = (u - T)\left(u^3 + u^2T + uT^2 + T^3\right).$$

Under what conditions might the second factor on the right be taken approximately constant? If the factor were constant the boundary condition would be linear.

8. Interpret this problem in terms of diffusion. Be sure to explain how the boundary conditions could arise physically.

$$D\frac{\partial^2 u}{\partial x^2} - k = \frac{\partial u}{\partial t}, \quad 0 < x < a, \quad 0 < t$$

$$u(0,t) = C, \quad \frac{\partial u}{\partial x}(a,t) = 0, \quad 0 < t$$

$$u(x,0) = 0, \quad 0 < x < a.$$

2.2 STEADY-STATE TEMPERATURES

Before tackling a complete heat conduction problem, we shall solve a simplified version called the steady-state or equilibrium problem. We begin with this example:

$$\frac{\partial^2 u}{\partial x^2} = \frac{1}{k}\frac{\partial u}{\partial t}, \quad 0 < x < a, \qquad 0 < t \tag{1}$$

$$u(0,t) = T_0, \qquad 0 < t \tag{2}$$

$$u(a,t) = T_1, \qquad 0 < t \tag{3}$$

$$u(x,0) = f(x), \quad 0 < x < a. \tag{4}$$

We may think of $u(x,t)$ as the temperature in a cylindrical rod, with insulated lateral surface, whose ends are held at constant temperatures T_0 and T_1.

Experience indicates that after a long time under the same conditions, the variation of temperature with time dies away. In terms of the function $u(x,t)$ that represents temperature, we thus expect that the limit of $u(x,t)$, as t tends to infinity, exists and depends only on x:

$$\lim_{t\to\infty} u(x,t) = v(x)$$

and also that

$$\lim_{t\to\infty} \frac{\partial u}{\partial t} = 0.$$

The function $v(x)$, called the *steady-state temperature distribution*, must still satisfy the boundary conditions and the heat equation, which are valid for all $t > 0$. Therefore $v(x)$ should be the solution to the problem

$$\frac{d^2 v}{dx^2} = 0, \qquad 0 < x < a \tag{5}$$

$$v(0) = T_0, \qquad v(a) = T_1. \tag{6}$$

On integrating the differential equation twice, we find

$$\frac{dv}{dx} = A, \qquad v(x) = Ax + B.$$

The constants A and B are to be chosen so that $v(x)$ satisfies the boundary conditions:

$$v(0) = B = T_0, \qquad v(a) = Aa + B = T_1.$$

When the two equations are solved for A and B, the steady-state distribution becomes

$$v(x) = T_0 + (T_1 - T_0)\frac{x}{a}. \tag{7}$$

Of course, Eqs. (5) and (6), which together form the steady-state problem corresponding to Eqs. (1)-(4), could have been derived from scratch as was done in Chapter 0, Section 3. Here, however, we see it as part of a more comprehensive problem.

We can establish this rule for setting up the steady-state problem corresponding to a given heat conduction problem: Take limits in all equations valid for large t (the partial differential equation and the boundary conditions), replacing u and its derivatives with respect to x by v and its derivatives, and replacing $\partial u/\partial t$ by 0.

Now we take as an example Eqs. (13)-(16) of Section 1, which were

$$\frac{\partial^2 u}{\partial x^2} = \frac{1}{k}\frac{\partial u}{\partial t}, \qquad 0 < x < a, \qquad 0 < t \tag{8}$$

$$u(0,t) = T_0, \qquad 0 < t \tag{9}$$

$$-\kappa\frac{\partial u}{\partial x}(a,t) = h(u(a,t) - T_1), \quad 0 < t \tag{10}$$

$$u(x,0) = f(x), \qquad 0 < x < a. \tag{11}$$

When the rule given here is applied to this problem, we are led to the following equations:

$$\frac{d^2 v}{dx^2} = 0, \qquad 0 < x < a$$

$$v(0) = T_0, \qquad -\kappa v'(a) = h(v(a) - T_1).$$

The solution of the differential equation is $v(x) = A + Bx$. The boundary conditions require that A and B satisfy

$$v(0) = T_0 : A = T_0$$

$$-\kappa v'(a) = h(v(a) - T_1) : -\kappa B = h(A + Ba - T_1).$$

Solving simultaneously, we find

$$A = T_0, \qquad B = \frac{h(T_1 - T_0)}{\kappa + ha}.$$

Thus the steady-state solution of Eqs. (8)-(11) is (see Fig. 6)

$$v(x) = T_0 + \frac{xh(T_1 - T_0)}{\kappa + ha}. \tag{12}$$

In both of these examples, the steady-state temperature distribution has been uniquely determined by the differential equation and boundary conditions. This is usually the case, but not always. For the problem

$$\frac{\partial^2 u}{\partial x^2} = \frac{1}{k}\frac{\partial u}{\partial t}, \quad 0 < x < a, \qquad 0 < t \tag{13}$$

$$\frac{\partial u}{\partial x}(0,t) = 0, \qquad 0 < t \tag{14}$$

$$\frac{\partial u}{\partial x}(a,t) = 0, \qquad 0 < t \tag{15}$$

$$u(x,0) = f(x), \quad 0 < x < a, \tag{16}$$

which describes the temperature in an insulated rod that also has insulated ends, the corresponding steady-state problem for $v(x) = \lim_{t\to\infty} u(x,t)$ is

$$\frac{d^2v}{dx^2} = 0, \quad 0 < x < a$$

$$\frac{dv}{dx}(0) = 0, \quad \frac{dv}{dx}(a) = 0.$$

It is easy to see that $v(x) = T$ (any constant) is a solution to this problem. The solution is not unique. However, there are physical considerations by which T can be determined. As the lateral surface of the rod is insulated, and because the ends are also insulated in this problem, the rod does not exchange heat with the rest of the universe. Thus the heat energy present at $t = 0$ is still present for any other time.

If ρ, c and A have the same meaning as before, we may say that the heat energy in a slice of the rod between x and $x + \Delta x$ is given approximately by $\rho c A \Delta x u(x,t)$, and so the total heat energy in the rod at time t is

$$\int_0^a \rho c A u(x,t) dx.$$

At time $t = 0$, $u(x,0) = f(x)$, while in the limit $u(x,t) \to v(x)$. The heat content at $t = 0$ and in the limit must be the same:

$$\int_0^a \rho c A f(x) dx = \int_0^a \rho c A v(x) dx = \int_0^a \rho c A T dx = \rho c A a T.$$

The coefficients ρ, c, and A are independent of x and may be cancelled. Thus we find

$$T = \frac{1}{a}\int_0^a f(x) dx,$$

which says that T, the final temperature in the rod, is the average of the temperature at $t = 0$.

It should not be supposed that every steady-state temperature distribution has a straight-line graph. This is certainly not the case in the problem of Exercise 1.

While the steady-state solution gives us some valuable information about the solution of an initial value-boundary value problem, it also is important as the first step in finding the complete solution. We now isolate the "rest" of the unknown temperature $u(x,t)$ by defining the *transient temperature distribution*,

$$w(x,t) = u(x,t) - v(x).$$

The name *transient* is appropriate because, according to our assumptions about the behavior of u for large values of t, we expect $w(x,t)$ to tend to zero as t tends to infinity.

In general, the transient also satisfies an initial value-boundary value problem that is similar to the original one but is distinguished by having a homogeneous partial differential equation and boundary conditions. To illustrate this point, we shall treat the problem stated in Eqs. (1)-(4) whose steady-state solution is given by Eq. (7).

By using the equality $u(x,t) = w(x,t) + v(x)$, and what we know about v — that is, Eqs. (5) and (6) — we make the original problem into a new problem for $w(x,t)$. We have the following relations:

$$\frac{\partial^2 u}{\partial x^2} = \frac{\partial^2 w}{\partial x^2} + \frac{d^2 v}{dx^2} = \frac{\partial^2 w}{\partial x^2}$$

$$\frac{\partial u}{\partial t} = \frac{\partial w}{\partial t} + \frac{dv}{dt} = \frac{\partial w}{\partial t}$$

$$u(0,t) = T_0 = w(0,t) + v(0) = w(0,t) + T_0$$

$$u(a,t) = T_1 = w(a,t) + v(a) = w(a,t) + T_1$$

$$u(x,0) = w(x,0) + v(x).$$

By substituting into Eqs. (1)-(4), we get the following initial value-boundary value problem for w:

$$\frac{\partial^2 w}{\partial x^2} = \frac{1}{k}\frac{\partial w}{\partial t}, \qquad 0 < x < a, \quad 0 < t \tag{17}$$

$$w(0,t) = 0, \qquad 0 < t \tag{18}$$

$$w(a,t) = 0, \qquad 0 < t \tag{19}$$

$$w(x,0) = f(x) - \left[T_0 + (T_1 - T_0)\frac{x}{a}\right] \tag{20}$$

$$\equiv g(x), \qquad 0 < x < a. \tag{21}$$

It is evident from Eqs. (17)-(19) that the homogeneous partial differential equation and boundary conditions predicted for w have been established. We shall see in the next section how the transient temperature can be found.

Exercises

1. State and solve the steady-state problem corresponding to

$$\frac{\partial^2 u}{\partial x^2} - \gamma^2(u - U) = \frac{1}{k}\frac{\partial u}{\partial t}, \quad 0 < x < a, \qquad 0 < t$$
$$u(0,t) = T_0, \qquad u(a,t) = T_1, \qquad 0 < x < a$$
$$u(x,0) = 0, \qquad 0 < x < a.$$

Also find a physical interpretation of this problem. (See Exercise 3, Section 1.)

2. State the problem satisfied by the transient temperature distribution corresponding to the problem in Exercise 1.

3. Obtain the steady-state solution of the problem

$$\frac{\partial^2 u}{\partial x^2} - \gamma^2(u - T) = \frac{1}{k}\frac{\partial u}{\partial t}, \quad 0 < x < a, \qquad 0 < t$$
$$u(0,t) = T, \qquad u(a,t) = T, \qquad 0 < t$$
$$u(x,0) = T_1\frac{x}{a}, \qquad 0 < x < a.$$

Can you think of a physical interpretation of this problem? Note the difference between the partial differential equation in this exercise and in Exercise 1. What happens if $\gamma = \pi/a$?

4. State the initial value-boundary value problem satisfied by the transient temperature distribution corresponding to Eqs. (8)-(11).

5. Find the steady-state solution of the problem

$$\frac{\partial}{\partial x}\left(\kappa\frac{\partial u}{\partial x}\right) = c\rho\frac{\partial u}{\partial t}, \quad 0 < x < a, \qquad 0 < t$$
$$u(0,t) = T_0, \qquad u(a,t) = T_1, \qquad 0 < t$$

if the conductivity varies in a linear fashion with x: $\kappa(x) = \kappa_0 + \beta x$, where κ_0 and β are constants.

6. Find and sketch the steady-state solution of

$$\frac{\partial^2 u}{\partial x^2} = \frac{1}{k}\frac{\partial u}{\partial t}, \qquad 0 < x < a, \qquad 0 < t$$

together with boundary conditions

a. $\frac{\partial u}{\partial x}(0,t) = 0, \qquad u(a,t) = T_0$

b. $u(0,t) - \frac{\partial u}{\partial x}(0,t) = T_0, \qquad \frac{\partial u}{\partial x}(a,t) = 0$

c. $u(0,t) - \frac{\partial u}{\partial x}(0,t) = T_0, \qquad u(a,t) + \frac{\partial u}{\partial x}(a,t) = T_1.$

7. Find the steady-state solution of this problem, where r is a constant that represents heat generation.

$$\frac{\partial^2 u}{\partial x^2} + r = \frac{1}{k}\frac{\partial u}{\partial t}, \qquad 0 < x < a, \qquad 0 < t$$
$$u(0,t) = T_0, \qquad \frac{\partial u}{\partial x}(a,t) = 0, \quad 0 < t.$$

8. Find the steady-state solution of

$$\frac{\partial^2 u}{\partial x^2} + \gamma^2(U(x) - u) = \frac{1}{k}\frac{\partial u}{\partial t}, \qquad 0 < x < a, \qquad 0 < t$$
$$u(0,t) = U_0, \qquad \frac{\partial u}{\partial x}(a,t) = 0, \quad 0 < t$$

where $U(x) = U_0 + Sx$ (U_0, S are constants).

9. This problem describes the diffusion of a substance in a medium that is moving with speed S to the right. The unknown function $u(x,t)$ is the concentration of the diffusing substance. Write out the steady-state problem and solve it. (D, U, and S are constants.)

$$D\frac{\partial^2 u}{\partial x^2} = \frac{\partial u}{\partial t} + S\frac{\partial u}{\partial x}, \qquad 0 < x < a, \quad 0 < t,$$
$$u(0,t) = U, \quad u(a,t) = 0, \quad 0 < t,$$
$$u(x,0) = 0, \qquad 0 < x < a.$$

2.3 EXAMPLE: FIXED END TEMPERATURES

In Section 1 we saw that the temperature $u(x, t)$ in a uniform rod with insulated material surface would be determined by the problem

$$\frac{\partial^2 u}{\partial x^2} = \frac{1}{k}\frac{\partial u}{\partial t}, \quad 0 < x < a, \qquad 0 < t \tag{1}$$

$$u(0, t) = T_0, \quad 0 < t \tag{2}$$

$$u(a, t) = T_1, \quad 0 < t \tag{3}$$

$$u(x, 0) = f(x), \quad 0 < x < a \tag{4}$$

if the ends of the rod are held at fixed temperatures and if the initial temperature distribution is $f(x)$. In Section 2 we found that the steady-state temperature distribution,

$$v(x) = \lim_{t \to \infty} u(x, t),$$

satisfied the boundary value problem

$$\frac{d^2 v}{dx^2} = 0, \quad 0 < x < a \tag{5}$$

$$v(0) = t_0, \quad v(a) = T_1. \tag{6}$$

In fact, we were able to find $v(x)$ explicitly:

$$v(x) = T_0 + (T_1 - T_0)\frac{x}{a}. \tag{7}$$

We also defined the transient temperature distribution as

$$w(x, t) = u(x, t) - v(x)$$

and determined that w satisfies the boundary value-initial value problem

$$\frac{\partial^2 w}{\partial x^2} = \frac{1}{k}\frac{\partial w}{\partial t}, \qquad 0 < x < a, \qquad 0 < t \tag{8}$$

$$w(0, t) = 0, \qquad 0 < t \tag{9}$$

$$w(a, t) = 0, \qquad 0 < t \tag{10}$$

$$w(x, 0) = f(x) - v(x) \equiv g(x), \quad 0 < x < a. \tag{11}$$

Our objective is to determine the transient temperature distribution, $w(x, t)$, and — since $v(x)$ is already known — the unknown temperature is

$$u(x, t) = v(x) + w(x, t). \tag{12}$$

The problem in w can be attacked by a method called *product method, separation of variables*, or *Fourier's method.* For this method to work, it is essential to have homogeneous boundary conditions. Thus, the method may be applied to the transient distribution w but *not* to the original function u. Of course, because both the partial differential equation and the boundary conditions satisfied by $w(x,t)$ are homogeneous, the function $w \equiv 0$ satisfies them. Because this solution is obvious and is of no help in satisfying the initial condition, it is called the *trivial solution.* We are seeking the unobvious, nontrivial solutions, so we shall avoid the trivial solution at every turn.

The general idea of the method is to assume that the solution of the partial differential equation has the form of a product: $w(x,t) = \phi(x)T(t)$. Since each of the factors depends on only one variable, we have

$$\frac{\partial^2 w}{\partial x^2} = \phi''(x)T(t), \qquad \frac{\partial w}{\partial t} = \phi(x)T'(t).$$

The partial differential equation becomes

$$\phi''(x)T(t) = \frac{1}{k}\phi(x)T'(t)$$

and on dividing through by ϕT, we find

$$\frac{\phi''(x)}{\phi(x)} = \frac{T'(t)}{kT(t)}, \qquad 0 < x < a, \qquad 0 < t.$$

Here is the key argument: The ratio on the left contains functions of x alone and cannot vary with t. On the other hand, the ratio on the right contains functions of t alone and cannot vary with x. Since this equality must hold for all x in the interval $0 < x < a$ and for all $t > 0$, the common value of the two sides must be a constant, varying neither with x nor t:

$$\frac{\phi''(x)}{\phi(x)} = p, \qquad \frac{T'(t)}{kT(t)} = p.$$

Now we have two ordinary differential equations for the two factor functions:

$$\phi'' - p\phi = 0, \qquad T' - pkT = 0. \tag{13}$$

The two boundary conditions on w may also be stated in the product form:

$$w(0,t) = \phi(0)T(t) = 0, \qquad w(a,t) = \phi(a)T(t) = 0.$$

There are two ways that these equations can be satisfied for all $t > 0$. Either the function $T(t) \equiv 0$ for all t, or the other factors must be zero.

However, in the first case, $w = \phi(x)T(t)$ is also identically zero, leading us back to the trivial solution. Therefore we take the other alternative and choose

$$\phi(0) = 0, \qquad \phi(a) = 0. \tag{14}$$

Our job now is to solve Eqs. (13) and satisfy the boundary conditions (14) while avoiding the trivial solution. The solutions of Eq. (13) are (assuming $p > 0$)

$$\phi(x) = c_1 \cosh(\sqrt{p}x) + c_2 \sinh(\sqrt{p}x), \qquad T(t) = ce^{pkt}.$$

Applying the boundary conditions (14) to $\phi(x)$ leads us to conclude first that $c_1 = 0$ and then that $c_2 = 0$, so that $\phi(x) \equiv 0$. However, this gives the trivial solution, $w(x, t) \equiv 0$, which is unsatisfactory. The same conclusion is reached if we try $p = 0$.

We now try a negative constant. Replacing p by $-\lambda^2$ in Eq. (13) gives us the two equations

$$\phi'' + \lambda^2\phi = 0, \qquad T' + \lambda^2 kT = 0$$

whose solutions are

$$\phi(x) = c_1 \cos(\lambda x) + c_2 \sin(\lambda x), \qquad T(t) = c\exp(-\lambda^2 kt).$$

If ϕ has the form given in the preceding, the boundary conditions require that $\phi(0) = c_1 = 0$, leaving $\phi(x) = c_2 \sin(\lambda x)$. Then $\phi(a) = c_2 \sin(\lambda a) = 0$.

We now have two choices: either $c_2 = 0$, making $\phi(x) \equiv 0$ for all values of x; or $\sin(\lambda a) = 0$. We reject the first possibility, for it leads to the trivial solution $w(x, t) \equiv 0$. In order for the second possibility to hold, we must have $\lambda = n\pi/a$, where $n = \pm 1, \pm 2, \pm 3, \cdots$. The negative values of n do not give any new functions, because $\sin(-\theta) = -\sin(\theta)$. Hence we allow $n = 1, 2, 3, \cdots$ only. We shall set $\lambda_n = n\pi/a$.

Incidentally, as the differential equation (13) and the boundary conditions (14) for $\phi(x)$ are homogeneous, any constant multiple of a solution is still a solution. We shall therefore remember this fact and drop the constant c_2 in $\phi(x)$. Likewise, we delete the c in $T(t)$.

To review our position, we have for each $n = 1, 2, 3, \cdots$, a function $\phi_n(x) = \sin(\lambda_n x)$, and an associated function $T_n(t) = \exp(-\lambda_n^2 kt)$. The product $w_n(x, t) = \sin(\lambda_n x)\exp(-\lambda_n^2 kt)$ has the properties:

1. $w_n(x, t)$ satisfies the heat equation;
2. $w_n(0, t) = 0$; and

3. $w_n(a,t) = 0$.

We now apply the principle of superposition and form a linear combination of all available solutions. Because we have infinitely many solutions, we need an infinite series to combine them all:

$$w(x,t) = \sum_{n=1}^{\infty} b_n \sin(\lambda_n x) \exp(-\lambda_n^2 kt). \tag{15}$$

As each term satisfies the heat equation and it is a linear, homogeneous equation, the sum of the series should satisfy the heat equation. (There is a mathematical question about convergence that we shall ignore.) Also, as each term is 0 at $x = 0$ and at $x = a$, the sum of the series must also be 0 at those two points, for any choice of constants b_n. Hence, the function $w(x,t)$ satisfies the differential equation and the two boundary conditions of the boundary value-initial value problem.

Of the four parts of the original problem, only the initial condition has not yet been satisfied. At $t = 0$, the exponentials in Eq. (15) are all unity. Thus the initial condition takes the form

$$w(x,0) = \sum_{n=1}^{\infty} b_n \sin\left(\frac{n\pi x}{a}\right) = g(x), \qquad 0 < x < a.$$

We immediately recognize a problem in Fourier series, which is solved by choosing the constants b_n according to the formula

$$b_n = \frac{2}{a}\int_0^a g(x) \sin\left(\frac{n\pi x}{a}\right) dx.$$

If the function g is continuous and sectionally smooth, we know that the Fourier series actually converges to $g(x)$ in the interval $0 < x < a$, so the solution which we have found for $w(x,t)$ actually satisfies all requirements set on w. Even if g does not satisfy these conditions, it can be shown that the solution we have arrived at is the best we can do.

Once the transient temperature has been determined, we find the original unknown $u(x,t)$ as the sum of the transient and the steady-state solutions,

$$u(x,t) = v(x) + w(x,t).$$

Example.
Suppose the original problem to be

$$\begin{aligned} \frac{\partial^2 u}{\partial x^2} &= \frac{1}{k}\frac{\partial u}{\partial t}, \quad 0 < x < a, \qquad 0 < t \\ u(0,t) &= T_0, \qquad 0 < t \\ u(a,t) &= T_1, \qquad 0 < t \\ u(x,0) &= 0, \qquad 0 < x < a. \end{aligned}$$

The steady-state solution is

$$v(x) = T_0 + (T_1 - T_0)\frac{x}{a}.$$

The transient temperature, $w(x,t) = u(x,t) - v(x)$, satisfies

$$\begin{aligned} \frac{\partial^2 w}{\partial x^2} &= \frac{1}{k}\frac{\partial w}{\partial t}, & 0 < x < a, \qquad 0 < t \\ w(0,t) &= 0, & 0 < t \\ w(a,t) &= 0, & 0 < t \\ w(x,0) &= -T_0 - (T_1 - T_0)\frac{x}{a} \equiv g(x), & 0 < x < a. \end{aligned}$$

According to the preceding calculations, w has the form

$$w(x,t) = \sum_{n=1}^{\infty} b_n \sin(\lambda_n x)\exp(-\lambda_n^2 kt) \tag{16}$$

and the initial condition is

$$w(x,0) = \sum_{n=1}^{\infty} b_n \sin\left(\frac{n\pi x}{a}\right) = g(x), \quad 0 < x < a.$$

The coefficients b_n are given by

$$\begin{aligned} b_n &= \frac{2}{a}\int_0^a \left[-T_0 - (T_1 - T_0)\frac{x}{a}\right]\sin\left(\frac{n\pi x}{a}\right)dx \\ &= \frac{2T_0}{a}\frac{\cos(n\pi x/a)}{(n\pi/a)}\Bigg|_0^a \\ &\quad -\frac{2}{a^2}(T_1 - T_0)\frac{\sin(n\pi x/a) - (n\pi x/a)\cos(n\pi x/a)}{(n\pi/a)^2}\Bigg|_0^a \\ &= -\frac{2T_0}{n\pi}\left(1 - (-1)^n\right) + \frac{2(T_1 - T_0)}{n\pi}(-1)^n \\ b_n &= \frac{-2}{n\pi}\left(T_0 - T_1(-1)^n\right). \end{aligned}$$

Now the complete solution (see Fig. 3) is

$$
\begin{aligned}
u(x,t) &= w(x,t) + T_0 + (T_1 - T_0)\frac{x}{a} \\
w(x,t) &= -\frac{2}{\pi}\sum_{n=1}^{\infty} \frac{T_0 - T_1(-1)^n}{n} \sin(\lambda_n x)\exp(-\lambda_n^2 kt).
\end{aligned} \tag{17}
$$

We can discover certain features of $u(x,t)$ by examining the solution. First, $u(x,0)$ really is zero $(0 < x < a)$ because the Fourier series converges to $-v(x)$ at $t = 0$. Second, when t is positive but very small, the series for $w(x,t)$ will almost equal $-T_0 - (T_1 - T_0)x/a$. But at $x = 0$ and $x = a$, the series adds up to zero (and $w(x,t)$ is a continuous function of x); thus $u(x,t)$ satisfies the boundary conditions. Third, when t is large, $\exp(-\lambda_1^2 kt)$ is small, and the other exponentials are still smaller. Then $w(x,t)$ may be well approximated by the first term (or first few terms) of the series. Finally, as $t \to \infty$, $w(x,t)$ disappears completely.

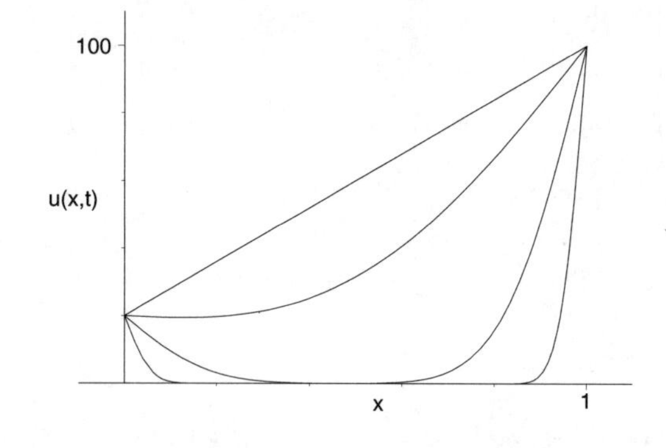

Figure 3: The solution of the example with $T_1 = 100$ and $T_0 = 20$. The function $u(x,t)$ is graphed as a function of x for four values of t, chosen so that the dimensionless time kt/a^2 has the values 0.001, 0.01, 0.1 and 1. For $kt/a^2 = 1$, the steady state is practically achieved.

Exercises

1. Write out the first few terms of the series for $w(x,t)$ in Eq. (17).

2. If $k = 1\ \text{cm}^2/\text{s}$, $a = 1$ cm, show that after $t = 0.5$ s the other terms of the series for w are negligible compared with the first term. Sketch $u(x,t)$ for $t = 0$, $t = 0.5$, $t = 1.0$, and $t = \infty$. Take $T_0 = 100$, $T_1 = 300$.

3. We can see from Eq. (17) that the dimensionless combinations x/a and kt/a^2 appear in the sine and exponential functions. Reformulate the partial differential equation (8) in terms of the dimensionless variables. $\xi = x/a$, $\tau = kt/a^2$. Set $u(x,t) = U(\xi, \tau)$.

4. Sketch the functions ϕ_1, ϕ_2, and ϕ_3, and verify that they satisfy the boundary conditions $\phi(0) = 0$, $\phi(a) = 0$.

In Exercises 5-8, solve the problem

$$\frac{\partial^2 w}{\partial x^2} = \frac{1}{k}\frac{\partial w}{\partial t}, \quad 0 < x < a, \qquad 0 < t$$
$$w(0,t) = 0, \qquad w(a,t) = 0, \qquad 0 < t$$
$$w(x,0) = g(x), \quad 0 < x < a$$

for the given function $g(x)$.

5. $g(x) = T_0$ (constant) 5.

6. $g(x) = \beta x$ (β is constant)

7. $g(x) = \beta(a - x)$ (β is constant)

8. $g(x) = \begin{cases} \dfrac{2T_0 x}{a}, & 0 < x < \dfrac{a}{2} \\ \dfrac{2T_0(a-x)}{a}, & \dfrac{a}{2} < x < a. \end{cases}$

9. A.N. Virkar, T.B. Jackson and R.A. Cutler (Thermodynamic and kinetic effects of oxygen removal on the thermal conductivity of aluminum nitride, Journal of the American Ceramic Society, 72 (1989) 2031-2042) use the following boundary value problem to study the kinetics of oxygen removal from a grain of aluminum nitride by diffusion:

$$\frac{\partial C}{\partial t} = D\frac{\partial^2 C}{\partial x^2}, \quad 0 < x < a, \qquad 0 < t$$
$$C(0,t) = C_1, \qquad C(a,t) = C_1, \quad 0 < t$$
$$C(x,0) = C_0, \qquad 0 < x < a.$$

In these equations, C is the oxygen concentration, D is the diffusion constant, a is the thickness of a grain, C_0 and C_1 are known concentrations.

a. Find the steady-state solution, $v(x)$.

b. State the problem (partial differential equation, boundary conditions and initial condition) for the transient, $w(x,t) = C(x,t) - v(x)$.

c. Solve the problem for $w(x,t)$ and write out the complete solution $C(x,t)$.

d. The concentration in the center of the grain, $C(a/2,t)$ varies from C_0 at time $t = 0$ toward C_1 as t increases. Suppose that we want to find out how long it takes for this concentration to complete 90% of the change it will make from C_0 to C_1; that is, we want to solve this equation for t:

$$C\left(\frac{a}{2}, t\right) - C_0 = 0.9(C_1 - C_0).$$

Show that this equation is equivalent to the equation

$$w\left(\frac{a}{2}, t\right) = -0.1(C_1 - C_0).$$

Find an approximate formula for the solution by using just the first term of the series for $w(x,t)$.

e. Use the formula in **d** to find t explicitly for $a = 5 \times 10^{-6}$m, $D = 10^{-11}\text{cm}^2/\text{s}$. Be careful to check dimensions.

2.4 EXAMPLE: INSULATED BAR

We shall consider again the uniform bar that was discussed in Section 1. Let us suppose now that the ends of the bar at $x = 0$ and $x = a$ are insulated instead of being held at constant temperatures. The boundary value-initial value problem which describes the temperature in this rod is:

$$\frac{\partial^2 u}{\partial x^2} = \frac{1}{k}\frac{\partial u}{\partial t}, \quad 0 < x < a, \qquad 0 < t \qquad (1)$$

$$\frac{\partial u}{\partial x}(0,t) = 0, \qquad \frac{\partial u}{\partial x}(a,t) = 0, \qquad 0 < t \qquad (2)$$

$$u(x,0) = f(x), \quad 0 < x < a, \qquad (3)$$

where $f(x)$ is supposed to be a given function.

We saw in Section 2 that the solution of the steady-state problem is not unique; by invoking a physical argument we determined that the steady-state temperature should be

$$\lim_{t\to\infty} u(x,t) = \frac{1}{a}\int_0^a f(x)dx.$$

The mathematical purpose behind finding the steady-state solution is to pave the way for a homogeneous problem (partial differential equation and boundary conditions) for the transient; however, the partial differential equation and boundary conditions in this example are already homogeneous. Thus, we do not need the steady-state solution or the transient problem. We may look for $u(x,t)$ directly.

Assume that u has the product form $u(x,t) = \phi(x)T(t)$. The heat equation becomes

$$\phi''(x)T(t) = \frac{1}{k}\phi(x)T'(t)$$

and the variables are separated by dividing through by ϕT, leaving

$$\frac{\phi''(x)}{\phi(x)} = \frac{T'(x)}{kT(t)}, \quad 0 < x < a, \quad 0 < t.$$

In order that a function of x equal a function of t, their mutual value must be a constant. If that constant were positive, T would be an increasing exponential function of time, which would be unacceptable. It is also easy to show that if the constant were positive, ϕ could not satisfy the boundary conditions without being identically zero.

Assuming then a negative constant, we can write

$$\frac{\phi''(x)}{\phi(x)} = -\lambda^2 = \frac{T'(t)}{kT(t)}$$

and separate these equalities into two ordinary differential equations linked by the common parameter λ^2:

$$\phi'' + \lambda^2\phi = 0, \quad 0 < x < a \tag{4}$$

$$T' + \lambda^2 kT = 0, \quad 0 < t. \tag{5}$$

The boundary conditions on u can be translated into conditions on ϕ, because they are homogeneous conditions. The boundary conditions in product form are

$$\frac{\partial u}{\partial x}(0,t) = \phi'(0)T(t) = 0, \quad 0 < t$$

$$\frac{\partial u}{\partial x}(a,t) = \phi'(a)T(t) = 0, \quad 0 < t.$$

To satisfy these equations, we must have the function $T(t)$ always zero (which would make $u(x,t) \equiv 0$), or else

$$\phi'(0) = 0, \quad \phi'(a) = 0.$$

The second alternative avoids the trivial solution.

We now have a homogeneous differential equation for ϕ together with homogeneous boundary conditions:

$$\phi'' + \lambda^2\phi = 0, \quad 0 < x < a \tag{6}$$

$$\phi'(0) = 0, \quad \phi'(a) = 0. \tag{7}$$

A problem of this kind is called an *eigenvalue problem.* We are looking for those values of the parameter λ^2 for which nonzero solutions of Eqs. (6) and (7) may exist. Those values are called *eigenvalues*, and the corresponding solutions are called *eigenfunctions.* Note that the significant parameter is λ^2, not λ. The square is used only for convenience. It is worth mentioning that we already saw an eigenvalue problem in Section 3 and in the Euler buckling problem of Chapter 0.

The general solution of the differential equation in Eq. (6) is

$$\phi(x) = c_1\cos(\lambda x) + c_2\sin(\lambda x).$$

Applying the boundary condition at $x = 0$, we see that $\phi'(0) = c_2\lambda = 0$, giving $c_2 = 0$ or $\lambda = 0$. We put aside the case $\lambda = 0$ and assume $c_2 = 0$, so $\phi(x) = c_1\cos(\lambda x)$. Then the second boundary condition requires that $\phi'(a) = -c_1\lambda\sin(\lambda a) = 0$. Once again, we may have $c_1 = 0$ or $\sin(\lambda a) = 0$. But $c_1 = 0$ makes $\phi(x) \equiv 0$, and therefore $u(x,t) \equiv 0$; this is a trivial solution. We choose therefore to make $\sin(\lambda a) = 0$ by restricting λ to the values π/a, $2\pi/a$, $3\pi/a, \cdots$. We label the eigenvalues with a subscript:

$$\lambda_n^2 = \left(\frac{n\pi}{a}\right)^2, \quad \phi_n(x) = \cos(\lambda_n x), \quad n = 1, 2, \cdots.$$

Notice that any constant multiple of an eigenfunction is still an eigenfunction; thus, we may take $c_1 = 1$ for simplicity.

Returning to the case $\lambda = 0$, we see that Eqs. (6) and (7) become

$$\phi'' = 0, \quad 0 < x < a,$$

$$\phi'(0) = 0, \quad \phi'(a) = 0.$$

Any constant function certainly satisfies these conditions. Therefore, we designate

$$\lambda_0^2 = 0, \quad \phi_0(x) = 1.$$

Let us summarize our findings by saying that the eigenvalue problem, Eqs. (6) and (7), has the solution

$$\begin{cases} \lambda_0^2 = 0, & \phi_0(x) = 1, \\ \lambda_n^2 = \left(\dfrac{n\pi}{a}\right)^2, & \phi_n(x) = \cos(\lambda_n x), \quad n = 1, 2, \cdots. \end{cases}$$

Now that the numbers λ_n^2 are known, we can solve Eq. (5) for $T(t)$, finding

$$T_0(t) = 1, \quad T_n(t) = \exp(-\lambda_n^2 kt).$$

The products $\phi_n(x)T_n(t)$ give solutions of the partial differential equation (1), which satisfy the boundary conditions, Eq. (2):

$$u_0(x,t) = 1, \quad u_n(x,t) = \cos(\lambda_n x)\exp(-\lambda_n^2 kt). \tag{8}$$

Because the partial differential equation and the boundary conditions are all linear and homogeneous, the principle of superposition applies, and any linear combination of solutions is also a solution. The solution $u(x,t)$ of the whole system may therefore have the form

$$u(x,t) = a_0 + \sum_1^\infty a_n \cos(\lambda_n x)\exp(-\lambda_n^2 kt). \tag{9}$$

There is only one condition of the original set remaining to be satisfied, the initial condition Eq. (3). For $u(x,t)$ in the form above, the initial condition is

$$u(x,0) = a_0 + \sum_1^\infty a_n \cos(\lambda_n x) = f(x), \quad 0 < x < a.$$

Because $\lambda_n = n\pi/a$, we recognize a problem in Fourier series and can immediately cite formulas for the coefficients:

$$a_0 = \frac{1}{a}\int_0^a f(x)dx, \quad a_n = \frac{2}{a}\int_0^a f(x)\cos\left(\frac{n\pi x}{a}\right)dx. \tag{10}$$

When these coefficients are computed and substituted in the formulas for $u(x,t)$, that function becomes the solution to the initial value-boundary value problems, Eqs. (1)-(3). Notice that the first term of the series solution a_0 is exactly the function we found for a steady-state solution before; and, in fact, when $t \to \infty$ all other terms in $u(x,t)$ do disappear leaving

$$\lim_{t\to\infty} u(x,t) = a_0 = \frac{1}{a}\int_0^a f(x)dx.$$

As an example, let us work out the complete solution for the initial temperature distribution $f(x) = T_0 + (T_1 - T_0)x/a$. It requires no integration to find that $a_0 = (T_1 + T_0)/2$. The remaining coefficients are

$$\begin{aligned} a_n &= \frac{2}{a}\int_0^a (T_0 + (T_1 - T_0)x/a)\cos\left(\frac{n\pi x}{a}\right)dx \\ &= 2(T_1 - T_0)\frac{\cos(n\pi) - 1}{n^2\pi^2} \end{aligned}$$

Thus the solution is given by Eq. (9) with these coefficients for a_0 and a_n. A graph of $u(x,t)$ as a function of x is shown in Fig. 4.

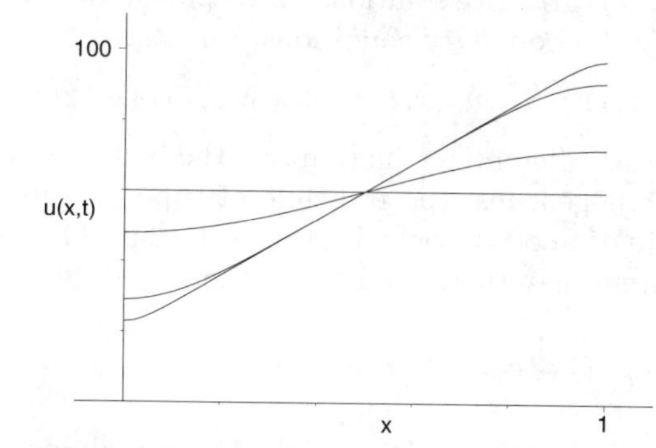

Figure 4: The solution of the example, $u(x,t)$, as a function of x for several times. The initial temperature distribution is $f(x) = T_0 + (T_1 - T_0)x/a$. For this illustration, $T_0 = 20, T_1 = 100$, and the times are chosen so that the dimensionless time kt/a^2 takes the values 0.001, 0.01, 0.1 and 1. The last case is indistinguishable from the steady state.

Exercises

1. Using the initial condition

$$u(x,0) = T_1 \frac{x}{a}, \quad 0 < x < a$$

find the solution $u(x,t)$ of this example problem. Sketch $u(x,0)$, $u(x,t)$ for some $t > 0$ (using the first three terms of the series), and the steady-state solution.

2. Repeat Exercise 1 using the initial condition

$$u(x,0) = T_0 + T_1 \left(\frac{x}{a}\right)^2, \quad 0 < x < a.$$

3. Same as Exercise 1, but with initial condition

$$u(x,0) = \begin{cases} \dfrac{2T_0 x}{a}, & 0 < x < \dfrac{a}{2} \\ \dfrac{2T_0(a-x)}{a}, & \dfrac{a}{2} < x < a \end{cases}$$

4. Solve Eqs. (1)-(3) using the initial condition $u(x, 0) = f(x)$, where

$$f(x) = \begin{cases} T_1, & 0 < x < \dfrac{a}{2} \\ T_2, & \dfrac{a}{2} < x < a. \end{cases}$$

5. Consider this heat problem, which is related to Eqs. (1)-(3):

$$\frac{\partial^2 u}{\partial x^2} = \frac{1}{k}\frac{\partial u}{\partial t}, \quad 0 < x < a, \qquad 0 < t$$
$$\frac{\partial u}{\partial x}(0, t) = S_0, \qquad \frac{\partial u}{\partial x}(a, t) = S_1, \quad 0 < t$$
$$u(x, 0) = f(x), \quad 0 < x < a.$$

a. Show that the steady-state problem has a solution if and only if $S_0 = S_1$ and give a physical reason why this should be true. (Recall that the heat flux q is proportional to the derivative of u with respect to x.) Find the steady-state solution if this condition is met.

b. If the steady-state solution $v(x)$ exists, show that the "transient," $w(x, t) = u(x, t) - v(x)$ has the boundary conditions

$$\frac{\partial w}{\partial x}(0, t) = 0, \quad \frac{\partial w}{\partial x}(a, t) = 0, \quad 0 < t.$$

c. Show that the function $u(x, t) = A(kt + x^2/2) + Bx$ satisfies the heat equation for arbitrary A and B, and that A and B can be chosen to satisfy the boundary conditions

$$\frac{\partial u}{\partial x}(0, t) = S_0, \quad \frac{\partial u}{\partial x}(a, t) = S_1, \quad 0 < t.$$

What happens to $u(x, t)$ as t increases if $S_0 \neq S_1$?

6. Verify that $u_n(x, t)$ in Eq. (8) satisfies the partial differential equation (1) and the boundary conditions, Eq. (2).

7. State the eigenvalue problem associated with the solution of the heat problem in Section 3. Also state its solution.

8. Suppose that the function $\phi(x)$ satisfies the relation

$$\frac{\phi''(x)}{\phi(x)} = p^2 > 0.$$

Show that the boundary conditions $\phi'(0) = 0$, $\phi'(a) = 0$, then force $\phi(x)$ to be identically 0. Thus, a positive "separation constant" can only lead to the trivial solution.

9. Refer to Eqs. (9) and (10), which give the solution of the problem stated in Eqs. (1), (2) and (3). If f is sectionally continuous, the coefficients $a_n \to 0$ as $n \to \infty$. For $t = t_1 > 0$, fixed, the solution is

$$u(x, t_1) = a_0 + \sum_1^\infty a_n \exp(-\lambda_n^2 k t_1) \cos(\lambda_n x)$$

and the coefficients of this cosine series are

$$A_n(t_1) = a_n \exp(-\lambda_n^2 k t_1).$$

Show that $A_n(t_1) \to 0$ so rapidly as $n \to \infty$ that the series given in the preceding converges uniformly $0 \le x \le a$. (See Chapter 1, Section 4, Theorem 1.) Show the same for the series which represents

$$\frac{\partial^2 u}{\partial x^2}(x, t_1).$$

10. Sketch the functions ϕ_1, ϕ_2, and ϕ_3 and verify graphically that they satisfy the boundary conditions of Eq.(7).

2.5 EXAMPLE: DIFFERENT BOUNDARY CONDITIONS

In many cases, boundary conditions at the two endpoints will be different kinds. In this section we shall solve the problem of finding the temperature in a rod having one end insulated and the other held at a constant temperature. The boundary value–initial value problem satisfied by the temperature in the rod is

$$\frac{\partial^2 u}{\partial x^2} = \frac{1}{k}\frac{\partial u}{\partial t}, \quad 0 < x < a, \quad 0 < t \tag{1}$$

$$u(0, t) = T_0, \quad 0 < t \tag{2}$$

$$\frac{\partial u}{\partial x}(a, t) = 0, \quad 0 < t \tag{3}$$

$$u(x, 0) = f(x), \quad 0 < x < a. \tag{4}$$

It is easy to verify that the steady-state solution of this problem is $v(x) = T_0$. Using this information, we can find the boundary value-initial value problem satisfied by the transient temperature $w(x,t) = u(x,t) - T_0$:

$$\frac{\partial^2 w}{\partial x^2} = \frac{1}{k}\frac{\partial w}{\partial t}, \qquad 0 < x < a, \quad 0 < t \tag{5}$$

$$w(0,t) = 0, \qquad 0 < t \tag{6}$$

$$\frac{\partial w}{\partial x}(a,t) = 0, \qquad 0 < t \tag{7}$$

$$w(x,0) = f(x) - T_0 = g(x), \quad 0 < x < a. \tag{8}$$

Since this problem is homogeneous, we can attack it by the method of separation of variables. The assumption that $w(x,t)$ has the form of a product, $\phi(x)T(t)$, and insertion of w in that form into the partial differential equation (5) lead to the separated equations

$$\phi'' + \lambda^2\phi = 0, \quad 0 < x < a \tag{9}$$

$$T' + \lambda^2 kT = 0, \quad 0 < t. \tag{10}$$

Moreover, the boundary conditions take the form

$$\phi(0)T(t) = 0, \quad 0 < t \tag{11}$$

$$\phi'(a)T(t) = 0, \quad 0 < t. \tag{12}$$

As before, we conclude that $\phi(0)$ and $\phi'(a)$ should both be zero:

$$\phi(0) = 0, \quad \phi'(a) = 0. \tag{13}$$

Now, the general solution of the differential equation (9) is

$$\phi(x) = c_1\cos(\lambda x) + c_2\sin(\lambda x).$$

The boundary condition, $\phi(0) = 0$, requires that $c_1 = 0$, leaving

$$\phi(x) = c_2\sin(\lambda x).$$

The boundary condition at $x = a$ now takes the form

$$\phi'(a) = c_2\lambda\cos(\lambda a) = 0.$$

The three choices are: $c_2 = 0$, which gives the trivial solution; $\lambda = 0$, which should be investigated separately (Exercise 2), and $\cos(\lambda a) = 0$. The third alternative — the only acceptable one — requires that λa be an odd multiple of $\pi/2$, which we may express as

$$\lambda_n = \frac{(2n-1)\pi}{2a}, \quad n = 1, 2, \cdots. \tag{14}$$

Thus, we have found that the eigenvalue problem consisting of Eqs. (9) and (13) has the solution

$$\lambda_n = \frac{(2n-1)\pi}{2a}, \quad \phi_n(x) = \sin(\lambda_n x), \quad n = 1, 2, 3, \cdots. \tag{15}$$

With the eigenfunctions and eigenvalues now in hand, we return to the differential equation (10), whose solution is

$$T_n(t) = \exp(-\lambda_n^2 kt).$$

As in previous cases, we assemble the general solution of the homogeneous problem expressed in Eqs. (5), (6), and (7) by forming a general linear combination of our product solutions,

$$w(x,t) = \sum_{n=1}^{\infty} b_n \sin(\lambda_n x) \exp(-\lambda_n^2 kt). \tag{16}$$

The choice of the coefficients, b_n, must be made so as to satisfy the initial condition, Eq. (8). Using the form of w given by Eq. (16), we find that the initial condition is

$$w(x,0) = \sum_{n=1}^{\infty} b_n \sin\left(\frac{(2n-1)\pi x}{2a}\right) = g(x), \quad 0 < x < a. \tag{17}$$

A routine Fourier sine series for the interval $0 < x < a$ would involve the functions $\sin(n\pi x/a)$, rather than the functions we have. By one of several means (Exercises 3-8), it may be shown that the series in Eq. (17) represents the function $g(x)$ provided that g is sectionally smooth and that we choose the coefficients by the formula

$$b_n = \frac{2}{a} \int_0^a g(x) \sin\left(\frac{(2n-1)\pi x}{2a}\right) dx. \tag{18}$$

Now the original problem is completely solved. The solution is

$$u(x,t) = T_0 + \sum_{n=1}^{\infty} b_n \sin(\lambda_n x) \exp(-\lambda_n^2 kt). \tag{19}$$

It should be noted carefully that the T_0 term in Eq. (19) is the steady-state solution in this case; it is not part of the separation-of-variables solution.

By way of example, suppose that the initial condition Eq. (4) is

$$u(x,0) = T_1, \quad 0 < x < a.$$

Then $g(x) = T_1 - T_0$, $0 < x < a$, and the coefficients as determined by Eq. (18) are

$$b_n = (T_1 - T_0)\frac{4}{\pi(2n-1)}.$$

Therefore, the complete solution of the boundary value-initial value problem with initial condition $u(x,0) = T_1$ would be

$$u(x,t) = T_0 + (T_1 - T_0)\frac{4}{\pi}\sum_{n=1}^{\infty}\frac{1}{2n-1}\sin(\lambda_n x)\exp(-\lambda_n^2 kt). \qquad (20)$$

(See Fig. 5 for graphs.)

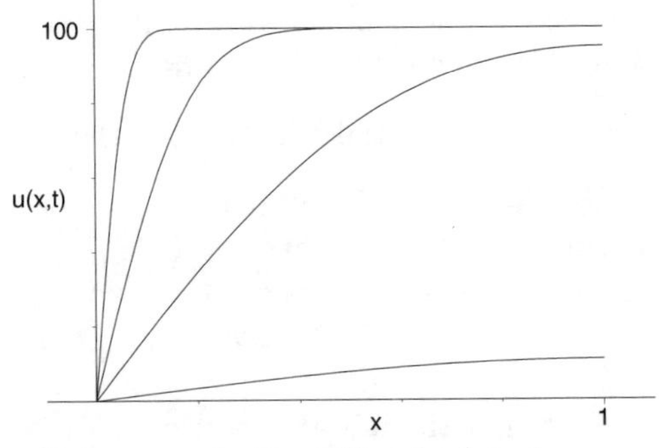

Figure 5: Solution of the example, Eq. (20): $u(x,t)$ is shown as a function of x for various times, which are chosen so that the dimensionless time kt/a^2 takes the values 0.001, 0.01, 0.1, 1.0. For the illustration, T_0 has been chosen equal to 0 and $T_1 = 100$.

Now that we have been through three examples, we can outline the method we have been using to solve linear boundary value-initial value problems. Up to this moment we have seen only homogeneous partial differential equations, but a nonhomogeneity that is independent of t can be treated by the same technique.

Summary.

1. If the partial differential equation or a boundary condition or both are not homogeneous, first find a function $v(x)$, independent of t, which satisfies the partial differential equation and the boundary

conditions. Since $v(x)$ does not depend on t, the partial differential equation applied to $v(x)$ becomes an ordinary differential equation. Finding $v(x)$ is just a matter of solving a two-point boundary value problem.

2. Determine the initial value-boundary value problem satisfied by the "transient solution" $w(x,t) = u(x,t) - v(x)$. This must be a *homogeneous problem.* That is, the partial differential equation and the boundary conditions (but not usually the initial condition) are satisfied by the constant function 0.

3. Assuming that $w(x,t) = \phi(x)T(t)$, separate the partial differential equation into two ordinary differential equations, one for $\phi(x)$ and one for $T(t)$, linked by the separation constant, $-\lambda^2$. Also reduce the boundary conditions to conditions on ϕ alone.

4. Solve the eigenvalue problem for ϕ. That is, find the values of λ^2 for which the eigenvalue problem has nonzero solutions. Label the eigenfunctions and eigenvalues $\phi_n(x)$ and λ_n^2.

5. Solve the ordinary differential equation for the time factors, $T_n(t)$.

6. Form the general solution of the homogeneous problem as a sum of constant multiples of the product solutions:

$$w(x,t) = \sum c_n \phi_n(x) T_n(t).$$

7. Choose the c_n so that the initial condition is satisfied. This may or may not be a routine Fourier series problem. If not, an orthogonality principle must be used to determine the coefficients. (We shall see the theory in Sections 7 and 8.)

8. Form the solution of the original problem

$$u(x,t) = v(x) + w(x,t)$$

and check that all conditions are satisfied.

Exercises

1. Find the steady-state solution of the problem stated in Eqs. (1)-(4).

2. Determine whether 0 is an eigenvalue of the eigenvalue problem stated in Eqs. (9) and (13).

3. To justify the expansion of Eq. (17), for an arbitrary sectionally smooth $g(x)$,

$$\sum_{n=1}^{\infty} b_n \sin\left(\frac{(2n-1)\pi x}{2a}\right) = g(x), \quad 0 < x < a,$$

construct the function $G(x)$ with these properties:

$$G(x) = g(x), \qquad 0 < x < a$$
$$G(x) = g(2a - x), \quad a < x < 2a.$$

Show that $G(x)$ corresponds to the series

$$G(x) \sim \sum_{N=1}^{\infty} B_N \sin\left(\frac{N\pi x}{2a}\right), \quad 0 < x < 2a.$$

4. Show that the B_N of the series in the preceding equation satisfy

$$B_N = 0 \ (N \text{ even}), \quad B_N = \frac{2}{a}\int_0^a g(x)\sin\left(\frac{N\pi x}{2a}\right) dx \ (N \text{ odd}).$$

5. Sketch $g(x)$ and $G(x)$, and find the corresponding series, if $g(x)$ is given by

a. $g(x) = x, \quad 0 < x < a$

b. $g(x) = T, \quad 0 < x < a.$

6. Show that the eigenfunctions found in this section are orthogonal. That is, prove that

$$\int_0^a \sin(\lambda_n x)\sin(\lambda_m x)dx = \begin{cases} 0 & (m \neq n) \\ \dfrac{a}{2} & (m = n) \end{cases}$$

when $\lambda_n = \dfrac{(2n-1)\pi}{2a}$.

7. Solve the problem stated in Eqs. (1)-(4), taking $f(x) = Tx/a$.

8. Use the orthogonality relation in Exercise 6 to justify the formula in Eq. (18).

9. Solve the nonhomogeneous problem

$$\frac{\partial^2 u}{\partial x^2} = \frac{1}{k}\frac{\partial u}{\partial t} - \frac{T}{a^2}, \qquad 0 < x < a, \quad 0 < t$$
$$u(0,t) = T_0, \quad \frac{\partial u}{\partial x}(a,t) = 0, \quad 0 < t$$
$$u(x,0) = T_0, \qquad 0 < x < a.$$

10. Solve this problem for the temperature in a rod in contact along the lateral surface with a medium at temperature 0.

$$\frac{\partial^2 u}{\partial x^2} = \frac{1}{k}\frac{\partial u}{\partial t} + \gamma^2 u, \qquad 0 < x < a, \quad 0 < t$$
$$u(0,t) = 0, \quad \frac{\partial u}{\partial x}(a,t) = 0, \quad 0 < t$$
$$u(x,0) = T_0, \qquad 0 < x < a.$$

11. Solve the problem

$$\frac{\partial^2 u}{\partial x^2} = \frac{1}{k}\frac{\partial u}{\partial t} \qquad 0 < x < a, \quad 0 < t$$
$$\frac{\partial u}{\partial x}(0,t) = 0, \quad u(a,t) = T_0, \quad 0 < t$$
$$u(x,0) = T_1, \qquad 0 < x < a.$$

12. Compare the solution of Exercise 11 with Eq. (20). Can one be turned into the other?

13. a. Solve this problem over the interval $0 < x < 2a$.

$$\frac{\partial^2 u}{\partial x^2} = \frac{1}{k}\frac{\partial u}{\partial t}, \qquad 0 < x < 2a, \quad 0 < t$$
$$u(0,t) = T_0, \quad u(2a,t) = T_0, \quad 0 < t$$
$$u(x,0) = g(x), \qquad 0 < x < 2a.$$

The function f is given over the interval $0 < x < a$, and g is an extension of f defined by

$$g(x) = \begin{cases} f(x), & 0 < x < a \\ f(2a - x), & a < x < 2a. \end{cases}$$

b. Explain why the solution of the problem comprised of Eqs. (1)-(4) is exactly the same as the solution of the problem in part **a.**

14. A flat enzyme electrode can be visualized by imagining it seen from the side. The electrode itself lies to the left of $x = 0$ (its thickness is unimportant); a gel-containing enzyme lies in a layer between $x = 0$ and $x = L$; and the test solution lies to the right of $x = L$. When the substance to be detected is introduced into the solution, it diffuses into the gel and reacts with the enzyme, yielding a product. The electrode responds to the product with a measurable electric potential.

P.W. Carr (Fourier analysis of the transient response of potentiometric enzyme electrodes, Analytical Chemistry 49 (1977) 799-802) studied the transient response of such an electrode via two partial differential equations that describe the concentrations, S and P, of the substance being detected and the enzyme-reaction product as they diffuse and react in the gel:

$$\frac{\partial S}{\partial t} = D\frac{\partial^2 S}{\partial x^2} - \frac{VS}{K+S}, \quad 0 < x < L, \quad 0 < t \tag{1*}$$

$$\frac{\partial P}{\partial t} = D\frac{\partial^2 P}{\partial x^2} + \frac{VS}{K+S}, \quad 0 < x < L, \quad 0 < t. \tag{2*}$$

In these equations, V is the specific enzyme activity (mol/ml s), K is a constant related to reaction rate, and D is the diffusion constant (cm^2/s), assumed to be the same for both substance and product. Reasonable boundary conditions are

$$\frac{\partial S}{\partial x}(0,t) = 0, \quad \frac{\partial P}{\partial x}(0,t) = 0, \quad 0 < t, \tag{3*}$$

representing no reaction or penetration at the electrode surface, and

$$S(L,t) = S_0, \quad P(L,t) = 0, \quad 0 < t, \tag{4*}$$

where the gel meets the test solution. Initially, we assume

$$S(x,0) = 0, \quad P(x,0) = 0, \quad 0 < x < L. \tag{5*}$$

Eq. (1*) is nonlinear because the unknown function S appears in the denominator of the last term. However, if S is much smaller than K, we may replace $K + S$ by K, and Eq. (1*) becomes

$$\frac{\partial S}{\partial t} = D\frac{\partial^2 S}{\partial x^2} - \frac{V}{K}S, \quad 0 < x < L, \quad 0 < t \tag{6*}$$

a. State and solve the steady-state problem for this equation, subject to the boundary conditions on S in Eqs. (3*) and (4*).

b. Find the transient solution and then the complete solution $S(x, t)$.

15. Refer to Exercise 14 above. Eq. (2*), although linear, is not easy to solve. However, if Eqs. (1*) and (2*) are added together, the nonlinear terms cancel, leaving this homogeneous linear equation for the sum of the concentrations:

$$\frac{\partial(S+P)}{\partial t} = D\frac{\partial^2(S+P)}{\partial x^2}, \quad 0 < x < L, \quad 0 < t.$$

Defining $u = S + P$, find the boundary and initial conditions for u, and solve completely. Then find $P(x, t)$ as $u(x, t) - S(x, t)$.

16. Refer to Exercises 14 and 15 here. In order to determine the response time of the enzyme electrode, one wants to know the function $P(0, t)$. Approximate this using only steady-state terms and the first term of each infinite series in your solution. Sketch. Find the "time constants," the multipliers of t in the exponential functions.

2.6 EXAMPLE: CONVECTION

We have seen three examples in which boundary conditions specified either u or $\partial u/\partial x$. Now we shall study a case where a condition of the third kind is involved. The physical model is conduction of heat in a rod with insulated lateral surface whose left end is held at constant temperature and whose right end is exposed to convective heat transfer. The boundary value-initial value problem satisfied by the temperature in the rod is

$$\frac{\partial^2 u}{\partial x^2} = \frac{1}{k}\frac{\partial u}{\partial t}, \qquad 0 < x < a, \qquad 0 < t \tag{1}$$

$$u(0, t) = T_0, \qquad 0 < t \tag{2}$$

$$-\kappa\frac{\partial u}{\partial x}(a, t) = h(u(a, t) - T_1), \quad 0 < t \tag{3}$$

$$u(x, 0) = f(x), \qquad 0 < x < a. \tag{4}$$

We found in Section 2 that the steady-state solution of this problem is

$$v(x) = T_0 + \frac{xh(T_1 - T_0)}{\kappa + ha}. \tag{5}$$

Now, since the original boundary conditions were nonhomogeneous, we form the problem for the transient solution $w(x,t) = u(x,t) - v(x)$. By direct substitution it is found that

$$\frac{\partial^2 w}{\partial x^2} = \frac{1}{k}\frac{\partial w}{\partial t}, \qquad 0 < x < a, \quad 0 < t \tag{6}$$

$$w(0,t) = 0, hw(a,t) + \kappa\frac{\partial w}{\partial x}(a,t) = 0, \quad 0 < t \tag{7}$$

$$w(x,0) = f(x) - v(x) \equiv g(x), \qquad 0 < x < a. \tag{8}$$

The solution for $w(x,t)$ can now be found by the product method. On the assumption that w has the form of a product $\phi(x)T(t)$, the variables can be separated exactly as before, giving two ordinary differential equations linked by a common parameter λ^2:

$$\phi'' + \lambda^2\phi = 0, \qquad 0 < x < a$$

$$T' + \lambda^2 kT = 0, \quad 0 < t.$$

Also, since the boundary conditions are linear and homogeneous, they can be translated directly into conditions on ϕ:

$$w(0,t) = \phi(0)T(t) = 0$$

$$\kappa\frac{\partial w}{\partial x}(a,t) + hw(a,t) = [\kappa\phi'(a) + h\phi(a)]T(t) = 0.$$

Either $T(t)$ is identically zero (which would make $w(x,t)$ identically zero), or

$$\phi(0) = 0, \quad \kappa\phi'(a) + h\phi(a) = 0.$$

Combining the differential equation and boundary conditions on ϕ, we get the eigenvalue problem

$$\phi'' + \lambda^2\phi = 0, \qquad 0 < x < a \tag{9}$$

$$\phi(0) = 0, \qquad \kappa\phi'(a) + h\phi(a) = 0. \tag{10}$$

The general solution of the differential equation is

$$\phi(x) = c_1\cos(\lambda x) + c_2\sin(\lambda x).$$

The boundary condition at $x = 0$ requires that $\phi(0) = c_1 = 0$, leaving $\phi(x) = c_2\sin(\lambda x)$. Now, at the other boundary,

$$\kappa\phi'(a) + h\phi(a) = c_2(\kappa\lambda\cos(\lambda a) + h\sin(\lambda a)) = 0.$$

Discarding the possibilities $c_2 = 0$ and $\lambda = 0$, which both lead to the trivial solution, we are left with the equation

$$\kappa\lambda\cos(\lambda a) + h\sin(\lambda a) = 0, \quad \text{or} \quad \tan(\lambda a) = -\frac{\kappa}{h}\lambda. \tag{11}$$

From sketches of the graphs of $\tan(\lambda a)$ and $-\kappa\lambda/h$ (Fig. 6), we see that there is an infinite number of solutions, λ_1, λ_2, $\lambda_3, \cdots$, and that, for very large n, λ_n is given *approximately* by

$$\lambda_n \cong \frac{2n-1}{2}\frac{\pi}{a}.$$

Table 2 shows the first five values of the product λa for several different values of the dimensionless parameter κ/ha. (More solutions are tabulated in *Handbook of Mathematical Functions* by Abramowitz and Stegun.)

Table 2: First five positive solutions of the equation $\tan(x) = -Ax$.

n	A				
	0.2500	0.5000	1.0000	2.0000	4.0000
1	2.5704	2.2889	2.0288	1.8366	1.7155
2	5.3540	5.0870	4.9132	4.8158	4.7648
3	8.3029	8.0962	7.9787	7.9171	7.8857
4	11.3348	11.1727	11.0855	11.0408	11.0183
5	14.4080	14.2764	14.2074	14.1724	14.1548

Thus we have for each $n = 1, 2, \cdots$, an eigenvalue λ_n^2 and an eigenfunction $\phi_n(x)$, which satisfies the eigenvalue problem Eqs. (9) and (10). Accompanying $\phi_n(x)$ is the function

$$T_n(t) = \exp(-\lambda_n^2 kt)$$

that makes $w_n(x,t) = \phi_n(x)T_n(t)$ a solution of the partial differential equation (6) and the boundary conditions Eq. (7). Since Eqs. (6) and (7) are linear and homogeneous, any linear combination of solutions is also a solution. Therefore, the transient solution will have the form

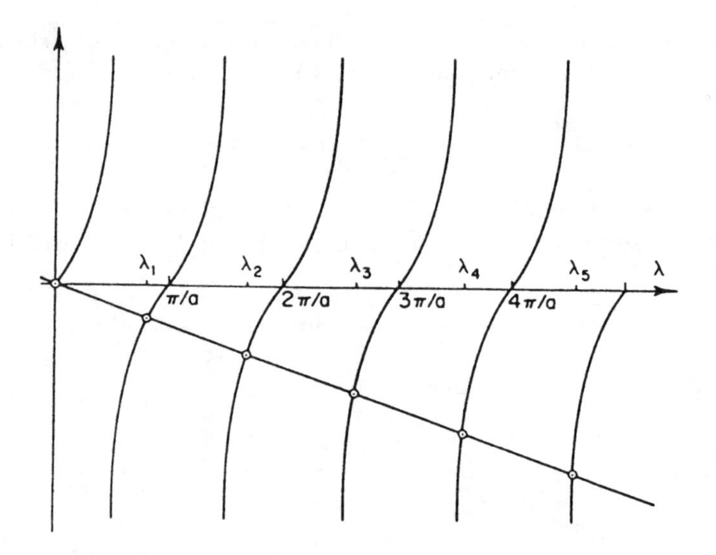

Figure 6: Graphs of $\tan(\lambda a)$ and $-\lambda\kappa/h$. The points of intersection are solutions of $\tan(\lambda a) = -\lambda\kappa/h$, eigenvalues of the problem Eqs. (6)-(7). The intersection at $\lambda = 0$ corresponds to the trivial solution.

$$w(x,t) = \sum_{n=1}^{\infty} b_n \sin(\lambda_n x) \exp(-\lambda_n^2 kt),$$

and the remaining condition to be satisfied, the initial condition Eq. (8), is

$$w(x,0) = \sum_{n=1}^{\infty} b_n \sin(\lambda_n x) = g(x), \quad 0 < x < a. \tag{12}$$

Thus the constants b_n are to be chosen so as to make the infinite series equal $g(x)$.

Although Eq. (12) looks like a Fourier series problem, it is not, because λ_2, λ_3, and so forth, are not all integer multiples of λ_1. If we attempt to use the idea of orthogonality, we can still find a way to select the b_n, for it may be shown by direct computation that

$$\int_0^a \sin(\lambda_n x) \sin(\lambda_m x) dx = 0, \quad \text{if} \quad n \neq m. \tag{13}$$

Then if we multiply both sides of the proposed Eq. (12) by $\sin(\lambda_m x)$ (where m is fixed) and integrate from 0 to a, we have

$$\int_0^a g(x)\sin(\lambda_m x)dx = \sum_{n=1}^{\infty} b_n \int_0^a \sin(\lambda_n x)\sin(\lambda_m x)dx$$

where we have integrated term by term. According to Eq. (13), all the terms of the series disappear, except the one in which $n = m$, yielding an equation for b_m:

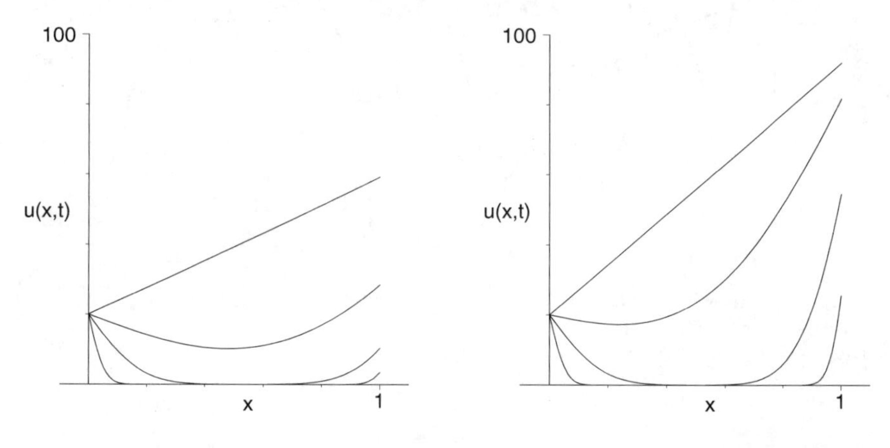

Figure 7: Solution of Eqs. (1)-(4) with $T_0 = 20$, $T_1 = 100$, and $f(x) = 0$. Graphs (a) and (b) correspond to $\kappa/ha = 0.1$ and $\kappa/ha = 1.0$ respectively. In each case, $u(x,t)$ is graphed as a function of x for times chosen so that the dimensionless time kt/a^2 takes on the values 0.001, 0.01, 0.1, 1. Note that both the temperature and its slope at the right end ($x = a$) change with time, so that the boundary condition Eq. (3) is satisfied.

$$b_m = \frac{\displaystyle\int_0^a g(x)\sin(\lambda_m x)dx}{\displaystyle\int_0^a \sin^2(\lambda_m x)dx}. \tag{14}$$

By this formula, the b_m may be calculated and inserted into the formula for $w(x,t)$. Then we may put together the solution $u(x,t)$ of the original problem Eqs. (1)-(4):

$$\begin{aligned} u(x,t) &= v(x) + w(x,t) \\ &= T_0 + \frac{xh(T_1 - T_0)}{(\kappa + ha)} + \sum_{n=1}^{\infty} b_n \sin(\lambda_n x)\exp(-\lambda_n^2 kt). \end{aligned}$$

In Fig. 7 are graphs of $u(x,t)$ for two different values of the parameter κ/ha; both have initial conditions $u(x,0)=0$.

Exercises

1. Sketch $v(x)$ as given in Eq. (5) assuming

a. $T_1 > T_0$ **b.** $T_1 = T_0$ **c.** $T_1 < T_0$.

2. If $T_1 > T_0$, as in Fig. 7, what is the maximum value of the temperature $u(x,t)$ on the interval $0 \le x \le a$ at any fixed time t? The solution will be a function of T_0, T_1 and $z = \kappa/ha$.

3. Why have we ignored the negative solutions of the equation

$$\tan(\lambda a) = \frac{-\kappa\lambda}{h}?$$

4. Derive the formula Eq. (12) for the coefficients b_m.

5. Sketch the first two eigenfunctions of this example taking $\kappa/h = 0.5$. ($\lambda_1 = 2.29/a$, $\lambda_2 = 5.09/a$).

6. Verify that

$$\int_0^a \sin^2(\lambda_m x)dx = \frac{a}{2} + \frac{\kappa}{h}\frac{\cos^2(\lambda_m a)}{2}.$$

7. Find the coefficients b_m corresponding to

$$g(x) = 1, \quad 0 < x < a.$$

8. Using the solution of Exercise 7, write out the first few terms of the solution of Eqs. (6)-(8), where $g(x) = T$, $0 < x < a$.

9. Same as Exercise 7 for $g(x) = x$, $0 < x < a$.

10. Verify the orthogonality integral by direct integration. It will be necessary to use the equation that defines the λ_n:

$$\kappa\lambda_n \cos(\lambda_n a) + h\sin(\lambda_n a) = 0.$$

2.7 STURM-LIOUVILLE PROBLEMS

At the end of the preceding section, we saw that ordinary Fourier series are not quite adequate for all the problems we can solve. We can make some generalizations, however, which do cover most cases that arise from separation of variables. In simple problems, we often find eigenvalue problems of the form

$$\phi'' + \lambda^2\phi = 0, \quad l < x < r \tag{1}$$

$$\alpha_1\phi(l) - \alpha_2\phi'(l) = 0, \tag{2}$$

$$\beta_1\phi(r) + \beta_2\phi'(r) = 0. \tag{3}$$

It is not difficult to determine the eigenvalues of this problem and to show the eigenfunctions orthogonal by direct calculation, but an indirect calculation is still easier.

Suppose that ϕ_n and ϕ_m are eigenfunctions corresponding to different eigenvalues λ_n^2 and λ_m^2. That is,

$$\phi_n'' + \lambda_n^2\phi_n = 0, \quad \phi_m'' + \lambda_m^2\phi_m = 0,$$

and both functions satisfy the boundary conditions. Let us multiply the first differential equation by ϕ_m, the second by ϕ_n, subtract the two, and move the terms containing $\phi_n\phi_m$ to the other side:

$$\phi_n''\phi_m - \phi_m''\phi_n = (\lambda_m^2 - \lambda_n^2)\phi_n\phi_m.$$

The right-hand side is a constant (nonzero) multiple of the integrand in the orthogonality relation

$$\int_l^r \phi_n(x)\phi_m(x)dx = 0, \quad n \neq m,$$

which is proved true if the left-hand side is zero:

$$\int_l^r (\phi_n''\phi_m - \phi_m''\phi_n)\,dx = 0.$$

This integral is integrable by parts:

$$\int_l^r (\phi_n''\phi_m - \phi_m''\phi_n)\,dx = [\phi_n'(x)\phi_m(x) - \phi_m'(x)\phi_n(x)]\Big|_l^r$$

$$-\int_l^r (\phi_n'\phi_m' - \phi_m'\phi_n')\,dx.$$

The last integral is obviously zero, so we have

$$(\lambda_m^2 - \lambda_n^2) \int_l^r \phi_n(x)\phi_m(x)dx = [\phi_n'(x)\phi_m(x) - \phi_m'(x)\phi_n(x)]\Big|_l^r .$$

Both ϕ_n and ϕ_m satisfy the boundary condition at $x = r$,

$$\beta_1\phi_m(r) + \beta_2\phi_m'(r) = 0$$

$$\beta_1\phi_n(r) + \beta_2\phi_n'(r) = 0.$$

These two equations may be considered simultaneous equations in β_1 and β_2. At least one of the numbers β_1 and β_2 is different from zero; otherwise, there would be no boundary condition. Hence the determinant of the equations must be zero:

$$\phi_m(r)\phi_n'(r) - \phi_n(r)\phi_m'(r) = 0.$$

A similar result holds at $x = l$. Thus

$$[\phi_n'(x)\phi_m(x) - \phi_m'(x)\phi_n(x)]|_l^r = 0$$

and, therefore, we have proved the orthogonality relation

$$\int_l^r \phi_n(x)\phi_m(x)dx = 0, \quad n \neq m$$

for the eigenfunctions of Eqs. (1)-(3).

We may make a much broader generalization about orthogonality of eigenfunctions with very little trouble. Consider the model eigenvalue problem below, which might arise from separation of variables in a heat conduction problem (see Section 9):

$$[s(x)\phi'(x)]' - q(x)\phi(x) + \lambda^2 p(x)\phi(x) = 0, \quad l < x < r$$

$$\alpha_1\phi(l) - \alpha_2\phi'(l) = 0$$

$$\beta_1\phi(r) + \beta_2\phi'(r) = 0.$$

Let us carry out the procedure used in the preceding with this problem. The eigenfunctions satisfy the differential equations

$$(s\phi_n')' - q\phi_n + \lambda_n^2 p\phi_n = 0$$

$$(s\phi_m')' - q\phi_m + \lambda_m^2 p\phi_m = 0.$$

Multiply the first by ϕ_m, the second by ϕ_n, subtract (the terms containing $q(x)$ cancel), and move the term containing $p\phi_n\phi_m$ to the other side:

$$(s\phi_n')'\phi_m - (s\phi_m')'\phi_n = (\lambda_m^2 - \lambda_n^2)p\phi_n\phi_m. \tag{4}$$

Integrate both sides from l to r, and apply integration by parts to the left-hand side:

$$\int_l^r [(s\phi_n')'\phi_m - (s\phi_m')'\phi_n]dx = [s\phi_n'\phi_m - s\phi_m'\phi_n]\Big|_l^r - \int_l^r (s\phi_n'\phi_m' - s\phi_n'\phi_m')\,dx.$$

The second integral is zero. From the boundary conditions we find that

$$\phi_n'(r)\phi_m(r) - \phi_m'(r)\phi_n(r) = 0$$
$$\phi_n'(l)\phi_m(l) - \phi_m'(l)\phi_n(l) = 0$$

by the same reasoning as before. Hence, we discover the orthogonality relation

$$\int_l^r p(x)\phi_n(x)\phi_m(x)dx = 0, \quad \lambda_n^2 \neq \lambda_m^2$$

for the eigenfunctions of the problem stated.

During these operations, we have made some tacit assumptions about integrability of functions after Eq. (4). In individual cases, where the coefficient functions s, q, and p and the eigenfunctions themselves are known, one can easily check the validity of the steps taken. In general, however, we would like to guarantee the existence of eigenfunctions and the legitimacy of computations after Eq. (4). To do so, we need the following:

Definition.
The problem

$$(s\phi')' - q\phi + \lambda^2 p\phi = 0, \quad l < x < r \tag{5}$$
$$\alpha_1\phi(l) - \alpha_2\phi'(l) = 0 \tag{6}$$
$$\beta_1\phi(r) + \beta_2\phi'(r) = 0 \tag{7}$$

is called a *regular Sturm-Liouville problem* if the following conditions are fulfilled:

a. $s(x)$, $s'(x)$, $q(x)$, and $p(x)$ are continuous for $l \le x \le r$;

b. $s(x) > 0$ and $p(x) > 0$ for $l \le x \le r$;

c. $\alpha_1^2 + \alpha_2^2 > 0$, $\beta_1^2 + \beta_2^2 > 0$.

d. The parameter λ occurs only where shown.

Condition a and the first condition b guarantee that the differential equation has solutions with continuous first and second derivatives. Notice that $s(l)$ and $s(r)$ must both be positive (not zero). Condition c just says that there are two boundary conditions: $\alpha_1^2 + \alpha_2^2 = 0$ only if $\alpha_1 = \alpha_2 = 0$, which would be no condition. The other requirements contribute to the desired properties in ways that are not obvious.

We are now ready to state the theorems that contain necessary information about eigenfunctions.

Theorem 1

The regular Sturm-Liouville problem has an infinite number of eigenfunctions $\phi_1, \phi_2, \cdots$, each corresponding to a different eigenvalue $\lambda_1^2, \lambda_2^2, \cdots$. If $n \neq m$, the eigenfunctions ϕ_n and ϕ_m are orthogonal with *weight function* $p(x)$:

$$\int_l^r \phi_n(x)\phi_m(x)p(x)dx = 0, \quad n \neq m.$$

The theorem is already proved, for the continuity of coefficients and eigenfunctions makes our previous calculations legitimate. It should be noted that any constant multiple of an eigenfunction is also an eigenfunction, but aside from a constant multiplier, the eigenfunctions of a Sturm-Liouville problem are unique.

A number of other properties of the Sturm-Liouville problem are known. We summarize a few here.

Theorem 2

(a) The regular Sturm-Liouville problem has an infinite number of eigenvalues, and $\lambda_n^2 \to \infty$ as $n \to \infty$.

(b) If the eigenvalues are numbered in order, $\lambda_1^2 < \lambda_2^2 < \cdots$, then the eigenfunction corresponding to λ_n^2 has exactly $n-1$ zeros in the interval $l < x < r$ (endpoints excluded).

(c) If $q(x) \ge 0$, and $\alpha_1, \alpha_2, \beta_1, \beta_2$, are all greater than or equal to zero, then all the eigenvalues are nonnegative.

Examples.

1. We note that the eigenvalue problems in Sections 3 to 6 of this chapter are all regular Sturm-Liouville problems, as is the problem in Eqs. (1)-(3) of this section. In particular, the problem

$$\phi'' + \lambda^2\phi = 0, \quad 0 < x < a$$

$$\phi(0) = 0, \quad h\phi(a) + \kappa\phi'(a) = 0$$

is a regular Sturm-Liouville problem, in which

$$s(x) = p(x) = 1, \;\; q(x) = 0, \;\; \alpha_1 = 1, \;\; \alpha_2 = 0, \;\; \beta_1 = h, \;\; \beta_2 = \kappa.$$

All conditions are met.

2. A less trivial example is

$$(x\phi')' + \lambda^2\left(\frac{1}{x}\right)\phi = 0, \quad 1 < x < 2$$

$$\phi(1) = 0, \quad \phi(2) = 0.$$

We identify $s(x) = x$, $p(x) = 1/x$, $q(x) = 0$. This is a regular Sturm-Liouville problem. The orthogonality relation is

$$\int_1^2 \phi_n(x)\phi_m(x)\frac{1}{x}dx = 0, \quad n \neq m.$$

Exercises

1. The general solution of the differential equation in Example 2 is

$$\phi(x) = c_1\cos(\lambda\ln(x)) + c_2\sin(\lambda\ln(x)).$$

Find the eigenvalues and eigenfunctions and verify the orthogonality relation directly by integration.

2. Check the results of Theorem 2 for the problem consisting of

$$\phi'' + \lambda^2\phi = 0, \quad 0 < x < a,$$

with boundary conditions

a. $\phi(0) = 0, \phi(a) = 0$ **b.** $\phi'(0) = 0, \;\; \phi'(a) = 0.$

In case **b**, $\lambda_1^2 = 0$.

3. Find the eigenvalues and eigenfunctions and sketch the first few eigenfunctions of the problem

$$\phi'' + \lambda^2 \phi = 0, \quad 0 < x < a$$

with boundary conditions

a. $\phi(0) = 0, \quad \phi'(a) = 0$

b. $\phi'(0) = 0, \quad \phi(a) = 0$

c. $\phi(0) = 0, \quad \phi(a) + \phi'(a) = 0$

d. $\phi(0) - \phi'(0) = 0, \quad \phi'(a) = 0$

e. $\phi(0) - \phi'(0) = 0, \quad \phi(a) + \phi'(a) = 0.$

4. In Eqs. (1)-(3), take $l = 0$, $r = a$, and show that

a. The eigenfunctions are $\phi_n(x) = \alpha_2 \lambda_n \cos(\lambda_n x) + \alpha_1 \sin(\lambda_n x)$.

b. The eigenvalues must be solutions of the equation

$$-\tan(\lambda a) = \frac{\lambda(\alpha_1 \beta_2 + \alpha_2 \beta_1)}{\alpha_1 \beta_1 - \alpha_2 \beta_2 \lambda^2}.$$

5. Show by applying Theorem 1 that the eigenfunctions of each of the following problems are orthogonal, and state the orthogonality relation.

a. $\phi'' + \lambda^2 (1 + x)\phi = 0, \quad \phi(0) = 0, \quad \phi'(a) = 0$

b. $(e^x \phi')' + \lambda^2 e^x \phi = 0, \quad \phi(0) - \phi'(0) = 0, \quad \phi(a) = 0$

c. $\phi'' + \left(\dfrac{\lambda^2}{x^2}\right)\phi = 0, \quad \phi(1) = 0, \quad \phi'(2) = 0$

d. $\phi'' - \sin(x)\phi + e^x \lambda^2 \phi = 0, \quad \phi'(0) = 0, \quad \phi'(a) = 0$

6. Consider the problem

$$(s\phi')' - q\phi + \lambda^2 p\phi = 0, \quad l < x < r$$

$$\phi(r) = 0$$

in which $s(l) = 0$, $s(x) > 0$ for $l < x \le r$, but p and q satisfy the conditions of a regular Sturm-Liouville problem. Require also that both $\phi(x)$ and $\phi'(x)$ have finite limits as $x \to l+$. Show that the eigenfunctions (if they exist) are orthogonal.

7. The problem below is not a regular Sturm-Liouville problem. Why? Show that the eigenfunctions are not orthogonal.

$$\phi'' + \lambda^2 \phi = 0, \quad 0 < x < a$$
$$\phi(0) = 0, \quad \phi'(a) - \lambda^2 \phi(a) = 0$$

8. Show that 0 is an eigenvalue of the problem

$$(s\phi')' + \lambda^2 p \phi = 0, \quad l < x < r$$
$$\phi'(l) = 0, \quad \phi'(r) = 0$$

where s and p satisfy the conditions of a regular Sturm-Liouville problem.

9. Find all values of the parameter μ for that there is a nonzero solution of this problem:

$$\phi'' + \mu \phi = 0$$
$$\phi(0) + \phi'(0) = 0, \quad \phi(a) + \phi'(a) = 0.$$

One solution is negative. Does this contradict Theorem 2?

2.8 EXPANSION IN SERIES OF EIGENFUNCTIONS

We have seen that the eigenfunctions that arise from a regular Sturm-Liouville problem

$$(s\phi')' - q\phi + \lambda^2 p \phi = 0, \quad l < x < r \tag{1}$$
$$\alpha_1 \phi(l) - a_2 \phi'(l) = 0 \tag{2}$$
$$\beta_1 \phi(r) + \beta_2 \phi'(r) = 0 \tag{3}$$

are orthogonal with weight function $p(x)$:

$$\int_l^r p(x)\phi_n(x)\phi_m(x)dx = 0, \quad n \neq m, \tag{4}$$

and it should be clear, from the way in which the question of orthogonality arose, that we are interested in expressing functions in terms of eigenfunction series.

Suppose that a function $f(x)$ is given in the interval $l < x < r$ and that we wish to express $f(x)$ in terms of the eigenfunctions $\phi_n(x)$ of Eqs. (1)-(3). That is, we wish to have

$$f(x) = \sum_{n=1}^{\infty} c_n \phi_n(x), \quad l < x < r. \tag{5}$$

The orthogonality relation Eq. (4) clearly tells us how to compute the coefficients. Multiplying both sides of the proposed Eq. (5) by $\phi_m(x)p(x)$ (where m is a fixed integer) and integrating from l to r yields

$$\int_l^r f(x)\phi_m(x)p(x)dx = \sum_{n=1}^{\infty} c_n \int_l^r \phi_n(x)\phi_m(x)p(x)dx.$$

The orthogonality relation says that all the terms in the series, except that one in which $n = m$, must disappear. Thus

$$\int_l^r f(x)\phi_m(x)p(x)dx = c_m \int_l^r \phi_m^2(x)p(x)dx$$

gives a formula for choosing c_m.

We can now cite a convergence theorem for expansion in terms of eigenfunctions. Notice the similarity to the Fourier series convergence theorem. Of course, the Fourier sine or cosine series are series of eigenfunctions on a regular Sturm-Liouville problem in which the weight function $p(x)$ is 1.

Theorem

Let $\phi_1, \phi_2, \cdots$, be eigenfunctions of a regular Sturm-Liouville problem Eqs. (1)-(3), in which the α's and β's are not negative.

If $f(x)$ is sectionally smooth on the interval $l < x < r$, then

$$\sum_{n=1}^{\infty} c_n \phi_n(x) = \frac{f(x+) + f(x-)}{2}, \quad l < x < r \tag{6}$$

where

$$c_n = \frac{\displaystyle\int_l^r f(x)\phi_n(x)p(x)dx}{\displaystyle\int_l^r \phi_n^2(x)p(x)dx}.$$

Furthermore, if the series

$$\sum_{n=1}^{\infty} |c_n| \left[\int_l^r \phi_n^2(x)p(x)dx\right]^{1/2}$$

converges, then the series Eq. (6) converges uniformly, $l \le x \le r$.

Exercises

1. Verify that

$$\lambda_n^2 = \left(\frac{n\pi}{\ln(b)}\right)^2, \quad \phi_n = \sin(\lambda_n \ln(x))$$

are the eigenvalues and eigenfunctions of

$$(x\phi')' + \lambda^2 \left(\frac{1}{x}\right)\phi = 0, \quad 1 < x < b$$

$$\phi(1) = 0, \quad \phi(b) = 0.$$

Find the expansion of the function $f(x) = x$ in terms of these eigenfunctions. To what values does the series converge at $x = 1$ and $x = b$?

2. If $\phi_1, \phi_2, \cdots,$ are the eigenfunctions of a regular Sturm-Liouville problem and are orthogonal with weight function $p(x)$ on $l < x < r$, and if $f(x)$ is sectionally smooth, then

$$\int_l^r f^2(x)p(x)dx = \sum_{n=1}^{\infty} a_n c_n^2$$

where

$$a_n = \int_l^r \phi_n^2(x)p(x)dx$$

and c_n is the coefficient of f as given in the theorem. Show why this should be true and conclude that $c_n\sqrt{a_n} \to 0$ as $n \to \infty$.

3. Verify that the eigenvalues and eigenfunctions of the problem

$$(e^x\phi')' + e^x\gamma^2\phi = 0, \quad 0 < x < a$$

$$\phi(0) = 0, \quad \phi(a) = 0$$

are

$$\gamma_n^2 = \left(\frac{n\pi}{a}\right)^2 + \frac{1}{4}, \quad \phi_n(x) = \exp\left(-\frac{x}{2}\right)\sin\left(\frac{n\pi x}{a}\right).$$

Find the coefficients for the expansion of the function $f(x) = 1$, $0 < x < a$, in terms of the ϕ_n.

4. If $\phi_1, \phi_2, \cdots$ are eigenfunctions of a regular Sturm-Liouville problem, the numbers $\sqrt{a_n}$ are called *normalizing constants*, and the functions $\psi_n = \phi_n/\sqrt{a_n}$ are called *normalized eigenfunctions.* Show that

$$\int_l^r \psi_n^2(x)p(x)dx = 1, \quad \int_l^r \psi_n(x)\psi_m(x)p(x)dx = 0, \quad n \neq m.$$

5. Find the formula for the coefficients of a sectionally smooth function $f(x)$ in the series

$$f(x) = \sum_{n=1}^{\infty} b_n \psi_n(x), \qquad l < x < r$$

where the ψ_n are normalized eigenfunctions.

6. Show that, for the function in Exercise 5

$$\int_l^r f^2(x)p(x)dx = \sum_{n=1}^{\infty} b_n^2.$$

7. What are the normalized eigenfunctions of the following problem?

$$\phi'' + \lambda^2\phi = 0, \quad 0 < x < 1$$

$$\phi'(0) = 0, \quad \phi'(1) = 0$$

2.9 GENERALITIES ON THE HEAT CONDUCTION PROBLEM

On the basis of the information we have about the Sturm-Liouville problem, we can make some observations on a fairly general heat conduction problem. We take as a physical model a rod whose lateral surface is insulated. In order to simplify slightly, we will assume that no heat is generated inside the rod.

Since material properties may vary with position, the partial differential equation that governs the temperature $u(x,t)$ in the rod will be

$$\frac{\partial}{\partial x}\left(\kappa(x)\frac{\partial u}{\partial x}\right) = \rho(x)c(x)\frac{\partial u}{\partial t}, \qquad l < x < r, \quad 0 < t. \tag{1}$$

Any of the three types of boundary conditions may be imposed at either boundary, so we use as boundary conditions

$$\alpha_1 u(l,t) - \alpha_2 \frac{\partial u}{\partial x}(l,t) = c_1, \quad t > 0 \tag{2}$$

$$\beta_1 u(r,t) + \beta_2 \frac{\partial u}{\partial x}(r,t) = c_2, \quad t > 0. \tag{3}$$

If the temperature is fixed, the coefficient of $\partial u/\partial x$ is zero. If the boundary is insulated, the coefficient of u is zero, and the right-hand side is also zero. If there is convection at a boundary, both coefficients will be positive, and the signs will be as shown.

We already know that in the case of two insulated boundaries, the steady-state solution has some peculiarities, so we set this aside as a special case. Assume, then, that either α_1 or β_1 or both are positive. Finally we need an initial condition in the form

$$u(x,0) = f(x), \quad l < x < r. \tag{4}$$

Equations (1)-(4) make up an initial value-boundary value problem.

Assuming that c_1 and c_2 are constants, we must first find the steady-state solution

$$v(x) = \lim_{t\to\infty} u(x,t).$$

The function $v(x)$ satisfies the boundary value problem

$$\frac{d}{dx}\left(\kappa(x)\frac{dv}{dx}\right) = 0, \qquad l < x < r \tag{5}$$

$$\alpha_1 v(l) - \alpha_2 v'(l) = c_1 \tag{6}$$

$$\beta_1 v(r) + \beta_2 v'(r) = c_2. \tag{7}$$

Since we have assumed that at least one of α_1 or β_1 is positive, this problem can be solved. In fact, it is possible to give a formula for $v(x)$ in terms of the function (see Exercise 1)

$$\int_l^x \frac{d\xi}{\kappa(\xi)} = I(x). \tag{8}$$

Before proceeding further, it is convenient to introduce some new functions. Let $\bar{\kappa}$, $\bar{\rho}$, and $\bar{c}$ indicate average values of the functions $\kappa(x)$, $\rho(x)$, and $c(x)$. We shall define dimensionless functions $s(x)$ and $p(x)$ by

$$\kappa(x) = \bar{\kappa}s(x), \quad \rho(x)c(x) = \bar{\rho}\bar{c}p(x).$$

Also we define the transient temperature to be

$$w(x,t) = u(x,t) - v(x).$$

By direct computation, using the fact that $v(x)$ is a solution of Eqs. (5)-(7), we can show that $w(x,t)$ satisfies the initial value-boundary value problem

$$\frac{\partial}{\partial x}\left(s(x)\frac{\partial w}{\partial x}\right) = \frac{1}{k}p(x)\frac{\partial w}{\partial t}, \quad l < x < r, \quad 0 < t \tag{9}$$

$$\alpha_1 w(l,t) - \alpha_2 \frac{\partial w}{\partial x}(l,t) = 0, \qquad 0 < t \tag{10}$$

$$\beta_1 w(r,t) + \beta_2 \frac{\partial w}{\partial x}(r,t) = 0, \qquad 0 < t \tag{11}$$

$$w(x,0) = f(x) - v(x) = g(x), \qquad l < x < r \tag{12}$$

which has homogeneous boundary conditions. The constant k is defined to be $\bar{\kappa}/\bar{\rho}\bar{c}$.

Now we use our method of separation of variables to find w. If w has the form $w(x,t) = \phi(x)T(t)$, the differential equation becomes

$$T(t)\left(s(x)\phi'(x)\right)' = \frac{1}{k}p(x)\phi(x)T'(t)$$

and, on dividing through by $p\phi T$, we find the separated equation

$$\frac{(s\phi')'}{p\phi} = \frac{T'}{kT}, \quad l < x < r, \quad 0 < t.$$

As before, the equality between a function of x and a function of t can hold only if their common value is constant. Furthermore, we expect the constant to be negative, so we put

$$\frac{(s\phi')'}{p\phi} = \frac{T'}{kT} = -\lambda^2$$

and separate two ordinary equations

$$T' + \lambda^2 kT = 0, \quad 0 < t$$

$$(s\phi')' + \lambda^2 p\phi = 0, \quad l < x < r.$$

The boundary conditions, being linear and homogeneous, can also be changed into conditions of ϕ. For instance, Eq. (10) becomes

$$\left[\alpha_1\phi(l) - \alpha_2\phi'(l)\right]T(t) = 0, \quad 0 < t$$

and, because $T(t) \equiv 0$ makes $w(x,t) \equiv 0$, we take the other factor to be zero. We have, then, the eigenvalue problem

$$(s\phi')' + \lambda^2 p\phi = 0, \quad l < x < r \tag{13}$$

$$\alpha_1\phi(l) - \alpha_2\phi'(l) = 0 \tag{14}$$

$$\beta_1\phi(r) + \beta_2\phi'(r) = 0. \tag{15}$$

Since s and p are related to the physical properties of the rod, they should be positive. We suppose also that s, s', and p are continuous. Then Eqs. (13)-(15) is a regular Sturm-Liouville problem, and we know that

1. There is an infinite number of eigenvalues

$$0 < \lambda_1^2 < \lambda_2^2 < \cdots.$$

2. To each eigenvalue corresponds just one eigenfunction (give or take a constant multiplier).

3. The eigenfunctions are orthogonal with weight $p(x)$:

$$\int_l^r \phi_n(x)\phi_m(x)p(x)dx = 0, \quad n \neq m.$$

The function $T_n(t)$ that accompanies $\phi_n(x)$ is given by

$$T_n(t) = \exp(-\lambda_n^2 kt).$$

We now begin to assemble the solution. For each $n = 1, 2, 3, \cdots, w_n(x,t) = \phi_n(x)T_n(t)$ satisfies Eqs. (9)-(11). As these are all linear homogeneous equations, any linear combination of solutions is again a solution. Thus the transient temperature has the form

$$w(x,t) = \sum_{n=1}^{\infty} a_n\phi_n(x)\exp(-\lambda_n^2 kt).$$

The initial condition Eq. (12) will be satisfied if we choose the a_n so that

$$w(x,0) = \sum_{n=1}^{\infty} a_n\phi_n(x) = g(x), \quad l < x < r.$$

The convergence theorem tells us that the equality will hold, except possibly at a finite number of points, if $f(x)$—and therefore $g(x)$—is sectionally smooth. Thus $w(x, t)$ is the solution of its problem, if we choose

$$a_n = \frac{\int_l^r g(x)\phi_n(x)p(x)dx}{\int_l^r \phi_n^2(x)p(x)dx}.$$

Finally, we can write the complete solution of Eqs. (1)-(4) in the form

$$u(x,t) = v(x) + \sum_{n=1}^{\infty} a_n\phi_n(x)\exp(-\lambda_n^2 kt). \qquad (16)$$

Working from the representation Eq. (16) we can draw some conclusions about the solution of Eqs. (1)-(4):

1. Since all the λ_n^2 are positive, $u(x,t)$ does tend to $v(x)$ as $t \to \infty$.

2. For any $t_1 > 0$, the series for $u(x, t_1)$ converges uniformly in $l \le x \le r$ because of the exponential factors; therefore $u(x, t_1)$ is a continuous function of x. Any discontinuity in the initial condition is immediately eliminated.

3. For large enough values of t, we can approximate $u(x,t)$ by

$$v(x) + a_1\phi_1(x)\exp(-\lambda_1^2 kt).$$

(To judge how large t might be, we need to know something about the a_n and the λ_n.) Because $\phi_1(x)$ is of one sign on the interval $l < x < r$ (that is, $\phi_1(x) > 0$ or $\phi_1(x) < 0$ for all x between l and r), the graph of the approximation above will lie either above or below the graph of $v(x)$, but will not cross it (provided that $a_1 \neq 0$).

Exercises

1. Find the explicit form for $v(x)$ in terms of the function in Eq. (8) assuming

a. $\alpha_1 = \beta_1 = 0, \qquad c_1 = c_2 = 0$

b. $\alpha_1 > 0$ or $\beta_1 > 0$, and no coefficient negative.

Why are these two cases separate?

2. Justify each of the conclusions.

3. Derive the general form of $u(x,t)$ if the boundary conditions are $\partial u/\partial x = 0$ at both ends. In this case, $\lambda^2 = 0$ is an eigenvalue.

2.10 SEMI-INFINITE ROD

Up to this point we have seen only problems over finite intervals. Frequently, however, it is justifiable and useful to assume that an object is infinite in length. (Sometimes this assumption is used to disguise ignorance of a boundary condition or to suppress the influence of a complicated condition.)

If the rod we have been studying is very long, we may treat it as *semi-infinite* — that is, as extending from 0 to ∞. The partial differential equation governing the temperature $u(x,t)$ remains

$$\frac{\partial^2 u}{\partial x^2} = \frac{1}{k}\frac{\partial u}{\partial t}, \quad 0 < x, \quad 0 < t \tag{1}$$

if the properties are uniform.

Suppose that at $x = 0$ the temperature is held constant, say

$$u(0,t) = 0, \quad t > 0 \tag{2}$$

in some temperature scale. In the absence of another boundary, there is no other boundary condition. However, it is desirable that $u(x,t)$ remain finite — less than some fixed bound — as $x \to \infty$. As usual, an initial condition is necessary

$$u(x,0) = f(x), \quad 0 < x. \tag{3}$$

Attacking Eqs. (1)-(3) by separation of variables, we assume that $u(x,t) = \phi(x)T(t)$, so the partial differential equation can be separated into two ordinary equations as usual

$$T' \pm \lambda^2 kT = 0, \quad 0 < t \tag{4}$$

$$\phi'' \pm \lambda^2 \phi = 0, \quad 0 < x. \tag{5}$$

There is just one boundary condition on u, which requires that $\phi(0) = 0$. The boundedness condition also requires that $\phi(x)$ remain finite as $x \to \infty$. The differential equation (5) has the solution

$$\phi(x) = c_1 \cosh(\lambda x) + c_2 \sinh(\lambda x)$$

if the negative signs are chosen in Eqs. (5) and (4). But $\phi(0) = 0$ requires that $c_1 = 0$, leaving $\phi(x) = c_2 \sinh(\lambda x)$, a function that increases without bound as $x \to \infty$. Thus we must choose the positive signs. The differential equation (5) together with the boundary and boundedness conditions forms a *singular* eigenvalue problem (singular because of the semi-infinite interval),

$$\phi'' + \lambda^2 \phi = 0, \qquad 0 < x \tag{6}$$

$$\phi(0) = 0, \qquad \phi(x) \text{ bounded as } x \to \infty. \tag{7}$$

The general solution of the differential equation is

$$\phi(x) = c_1 \cos(\lambda x) + c_2 \sin(\lambda x),$$

which is bounded for any choice of the constants and for any value of λ. The boundedness condition told us to use the plus sign in Eq. (5) and now contributes nothing further.

Applying the boundary condition at $x = 0$ shows that $c_1 = 0$, leaving $\phi(x) = c_2 \sin(\lambda x)$. In this singular eigenvalue problem, there are no "special" values of λ: *any* value produces a nonzero solution of the differential equation that also satisfies the boundary and boundedness conditions. (But negative values of λ produce no new solutions.) Recalling that any constant multiple of a solution of a homogeneous problem is still a solution, we choose $c_2 = 1$ and summarize the solution of the singular eigenvalue problem as

$$\phi(x; \lambda) = \sin(\lambda x), \quad \lambda > 0. \tag{8}$$

The solution of Eq. (4), also with the + sign, is

$$T(t) = \exp(-\lambda^2 kt).$$

For any value of λ^2, the function

$$u(x, t; \lambda) = \sin(\lambda x) \exp(-\lambda^2 kt)$$

satisfies Eqs. (1) and (2) and the boundedness condition. Equation (1) and the boundary condition Eq. (2) are homogeneous; therefore any linear combination of solutions is a solution. Since the parameter λ may take on any value, the appropriate linear combination is an integral. So u should have the form

$$u(x, t) = \int_0^\infty B(\lambda) \sin(\lambda x) \exp(-\lambda^2 kt) d\lambda. \tag{9}$$

(We need not include negative values of λ. They give no new solutions.) The initial condition will be satisfied if $B(\lambda)$ is chosen to make

$$u(x,0) = \int_0^\infty B(\lambda)\sin(\lambda x)d\lambda = f(x), \quad 0 < x.$$

We recognize this as a Fourier integral; $B(\lambda)$ is to be chosen as

$$B(\lambda) = \frac{2}{\pi}\int_0^\infty f(x)\sin(\lambda x)dx. \tag{10}$$

If $B(\lambda)$ exists, then Eq. (9) is the solution of the problem. Notice that when $t > 0$, the exponential function makes the improper integral in Eq. (9) converge very rapidly.

Some care must be taken in the interpretation of our solution. If the rod really is finite (say length L) the expression in Eq. (9) is, of course, meaningless for x greater than L. The presence of a boundary condition at $x = L$ would influence temperatures nearby, so Eq. (9) can be considered a valid approximation only for $x \ll L$.

Example.

We shall solve the problem in Eqs. (1)-(3) using the initial temperature distribution

$$f(x) = \begin{cases} T_0, & 0 < x < b \\ 0, & b < x. \end{cases}$$

This means that a section of length b at the left end of the rod starts out at temperature T_0, different from the temperature of the long right end, which is at the same temperature as the left boundary. (We assume $T_0 > 0$.) The solution is given by Eq. (9), with $B(\lambda)$ calculated from Eq. (10):

$$\begin{aligned} B(\lambda) &= \frac{2}{\pi}\int_0^\infty f(x)\sin(\lambda x)dx \\ &= \frac{2}{\pi}\int_0^b T_0\sin(\lambda x)dx \\ &= \frac{2T_0}{\lambda\pi}(1-\cos(\lambda b)). \end{aligned}$$

Therefore, the complete solution is

$$u(x,t) = \frac{2}{\pi}T_0\int_0^\infty \frac{1-\cos(\lambda b)}{\lambda}\sin(\lambda x)\exp(-\lambda^2 kt)d\lambda.$$

In Fig. 8 are graphs of $u(x,t)$ as a function of x for various values of t.

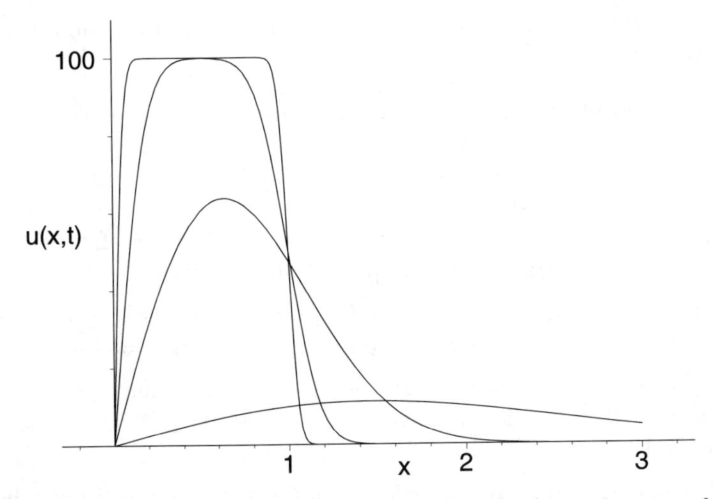

Figure 8: Graphs of the solution of the example, $u(x,t)$ as a function of x over the interval $0 < x < 3b$ where $b = 1$ and $T_0 = 100$ for convenience. The times have been chosen so that the dimensionless time kt/b^2 takes the values 0.001, 0.01, 0.1 and 1. When $kt/b^2 = 0.01$, the temperature near $x = b/2$ has not changed noticeably from its initial value.

Exercises

1. Find the solution of Eqs. (1)-(3) if the initial temperature distribution is given by

$$f(x) = \begin{cases} 0, & 0 < x < a \\ T, & a < x < b \\ 0, & b < x. \end{cases}$$

2. Verify that $u(x,t)$ as given by Eq. (6) is a solution of Eqs. (1)-(3). What is the steady-state temperature distribution?

3. Find the solution of Eqs. (1)-(3) if $f(x) = T_0 e^{-\alpha x}$, $x > 0$.

4. Find a formula for the solution of the problem

$$\begin{aligned} \frac{\partial^2 u}{\partial x^2} &= \frac{1}{k}\frac{\partial u}{\partial t}, \quad 0 < x, \quad 0 < t \\ \frac{\partial u}{\partial x}(0,t) &= 0, \qquad 0 < t \\ u(x,0) &= f(x), \quad 0 < x. \end{aligned}$$

5. Determine the solution of Exercise 4 if $f(x)$ is the function given in Exercise 1.

6. Penetration of heat into the earth. Assume that the earth is flat, occupying the region $0 < x$ (so that x measures distance down from the surface). At the surface, the temperature fluctuates according to season, time of day, etc. We cover several cases by taking the boundary condition to be $u(0,t) = \sin(\omega t)$, where the frequency ω can be chosen according to the period of interest.

a. Show that $u(x,t) = e^{-px}\sin(\omega t - px)$ satisfies the boundary condition and is a solution of the heat equation if $p = \sqrt{\omega/2k}$.

b. Sketch $u(x,t)$ as a function of t for $x = 0, 1$ and 2 m, taking $\omega = 2 \times 10^{-7}$ rad/s (approximately one cycle per year) and $k = 0.5 \times 10^{-6}\mathrm{m}^2/\mathrm{s}$.

c. With ω as in part **b**, find the depth (as a function of k) at which seasons are reversed.

7. Consider the problem

$$\frac{\partial^2 u}{\partial x^2} = \frac{1}{k}\frac{\partial u}{\partial t}, \quad 0 < x, \quad 0 < t$$
$$u(0,t) = T_0, \quad 0 < t$$
$$u(x,0) = f(x), \quad 0 < x.$$

Show that, for our method of solution to work, it is necessary to have $T_0 = \lim_{x\to\infty} f(x)$. Find a formula for $u(x,t)$ if this is the case.

8. Show that the function $u(x,t)$ given in Eq. (6) is an odd function of x.

9. R.C. Bales, M.P. Valdez and G.A. Dawson (Gaseous deposition to snow, 2: Physical-chemical model for SO_2 deposition, Journal of Geophysical Research 92 (1987), pp. 9789-9799) develop a mathematical model for the transport of SO_2 gas into snow by molecular diffusion. The governing partial differential equation is

$$\frac{\partial C}{\partial t} = D\left(\frac{\partial^2 C}{\partial x^2} - a^2 C\right)$$

where C is the concentration of SO_2 as a function of x (depth into the snow) and time; D is a diffusion constant. The term containing

C appears because the SO_2 takes part in a chemical reaction with water in the snow, forming sulphuric acid, H_2SO_4. The coefficient a^2 depends on pH, temperature and other circumstances; we treat it as a constant. The problem is to be solved for a wide range of values for the parameters.

If the snow is deep, the authors believe that it is reasonable to use a semi-infinite interval for x and to add the condition $C(x,t) \to 0$ as $x \to \infty$. In addition, a natural boundary condition at the snow surface is that concentration in the snow match that in the air: $C(0,t) = C_0$. Furthermore, if the snow is fresh, we can assume that the concentration throughout is initially 0,

$$C(x,0) = 0, \quad 0 < x.$$

a. Find a steady-state solution $v(x)$ that satisfies the partial differential equation and the boundary conditions.

b. State the problem (partial differential equation, boundary condition at $x = 0$, condition as $x \to \infty$, and initial condition) to be satisfied by the transient $w(x,t) = C(x,t) - v(x)$.

c. Solve the problem for the transient. Note that the condition as $x \to \infty$ must be relaxed to: $w(x,t)$ bounded as $x \to \infty$. Individual product solutions do not approach 0 as x increases.

2.11 INFINITE ROD

If we wish to study heat conduction in the center of a very long rod, we may assume that it extends from $-\infty$ to ∞. Then there are no boundary conditions, and the problem to be solved is

$$\frac{\partial^2 u}{\partial x^2} = \frac{1}{k}\frac{\partial u}{\partial t}, \qquad -\infty < x < \infty, \qquad 0 < t \tag{1}$$

$$u(x,0) = f(x), \qquad -\infty < x < \infty \tag{2}$$

$$|u(x,t)| \text{bounded} \quad \text{as } x \to \pm\infty. \tag{3}$$

Using the same techniques as before, we look for solutions in the form $u(x,t) = \phi(x)T(t)$, so that the heat equation (1) becomes

$$\frac{\phi''(x)}{\phi(x)} = \frac{T'(t)}{T(t)} = \text{constant}.$$

As in the previous section, the constant must be nonpositive (say $-\lambda^2$) in order for the solutions to be bounded. But then *every* solution of ϕ''/ϕ

$= -\lambda^2$ is bounded. Thus, our factors $\phi(x)$ and $T(t)$ are

$$\phi(x) = A\cos(\lambda x) + B\sin(\lambda x)$$
$$T(t) = \exp(-\lambda^2 kt).$$

We combine the solutions $\phi(x)T(t)$ in the form of an integral to obtain

$$u(x,t) = \int_0^\infty (A(\lambda)\cos(\lambda x) + B(\lambda)\sin(\lambda x))\exp(-\lambda^2 kt)d\lambda. \quad (4)$$

At time $t = 0$, the exponential factor becomes 1, and the initial condition is

$$\int_0^\infty (A(\lambda)\cos(\lambda x) + B(\lambda)\sin(\lambda x))\,d\lambda = f(x), \quad -\infty < x < \infty.$$

As this is clearly a Fourier integral problem, we must choose $A(\lambda)$ and $B(\lambda)$ to be the Fourier integral coefficient functions,

$$A(\lambda) = \frac{1}{\pi}\int_{-\infty}^\infty f(x)\cos(\lambda x)dx, \quad B(\lambda) = \frac{1}{\pi}\int_{-\infty}^\infty f(x)\sin(\lambda x)dx. \quad (5)$$

Then the function $u(x,t)$ in Eq. (4) satisfies the partial differential equation (1) and the initial condition (2), provided that f is sectionally smooth and $|f(x)|$ has a finite integral. It can be proved that the boundedness condition (3) is also satisfied, provided that the initial value $f(x)$ is bounded as $x \to \pm\infty$.

For an example, consider the problem posed in Eqs. (1)-(3) with

$$f(x) = \begin{cases} 0, & x < -a \\ T_0, & -a < x < a, \\ 0, & -a < x. \end{cases}$$

In words, the rod has a center section of length $2a$ whose temperature is different from that of the long sections to the left and right. We must compute the coefficient functions $A(\lambda)$ and $B(\lambda)$. The latter is identically 0 because $f(x)$ is an even function; and

$$\begin{aligned} A(\lambda) &= \frac{1}{\pi}\int_{-\infty}^\infty f(x)\cos(\lambda x)dx \\ &= \frac{1}{\pi}\int_{-a}^a T_0\cos(\lambda x)dx \\ &= \frac{2T_0}{\lambda\pi}\sin(\lambda a). \end{aligned}$$

Thus, the solution of the problem is

$$u(x,t) = \frac{2T_0}{\pi} \int_0^\infty \frac{\sin(\lambda a)}{\lambda} \cos(\lambda x) \exp\left(-\lambda^2 kt\right) d\lambda. \tag{6}$$

This function is graphed as a function of x for several values of t in Fig. 9.

The figure suggests that $u(x,t)$ is *positive* for all x when $t > 0$. This is indeed true and illustrates an interesting property of the solutions of the heat equation: the instantaneous transmission of information. The "hot" section in the interval $-a < x < a$ instantly raises the temperature everywhere else from the initial value of 0 to a positive value.

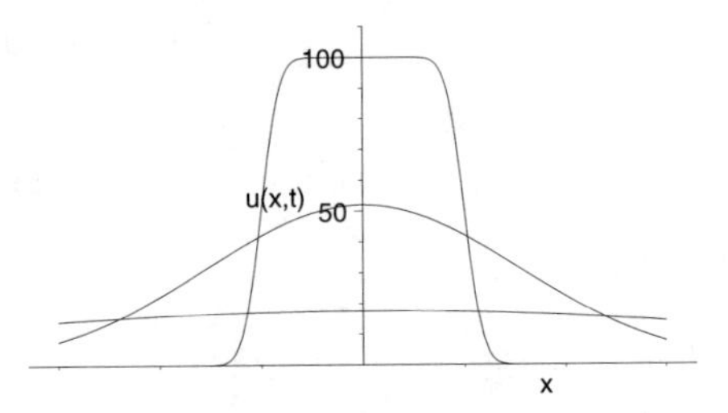

Figure 9: Solution of example problem. At $t = 0$, the temperature is $T_0 > 0$ for $-a < x < a$ and is 0 in the rest of the rod; $u(x,t)$ is shown as a function of x on the interval $-3a < x < 3a$ for three times. The times are chosen so that the dimensionless time kt/a^2 takes the values 0.01, 1 and 10 (to get a clear picture of the changes in u). Note that $u(x,t)$ is positive everywhere for any $t > 0$. The values $T_0 = 100$ and $a = 1$ have been used for convenience.

Starting from the general form of a solution in Eq. (4), we can derive some very interesting results. Change the variable of integration in Eq. (5) to x' and substitute the formulas for $A(\lambda)$ and $B(\lambda)$ into Eq. (4):

$$\begin{aligned} u(x,t) &= \frac{1}{\pi} \int_0^\infty \left[\int_{-\infty}^\infty f(x') \cos(\lambda x') dx' \cos(\lambda x) \right. \\ &\quad \left. + \int_{-\infty}^\infty f(x') \sin(\lambda x') dx' \sin(\lambda x) \right] \exp(-\lambda^2 kt) d\lambda. \end{aligned}$$

Combining terms we find

$$\begin{aligned} u(x,t) &= \frac{1}{\pi} \int_0^\infty \int_{-\infty}^\infty f(x') \left[\cos(\lambda x') \cos(\lambda x) + \right. \\ &\qquad \left. \sin(\lambda x') \sin(\lambda x) \right] dx' \exp(-\lambda^2 kt) d\lambda \\ &= \frac{1}{\pi} \int_0^\infty \int_{-\infty}^\infty f(x') \cos(\lambda(x' - x)) dx' \exp(-\lambda^2 kt) d\lambda. \end{aligned}$$

If the order of integration may be reversed, we may write

$$u(x,t) = \frac{1}{\pi}\int_{-\infty}^{\infty} f(x')\int_{0}^{\infty} \cos(\lambda(x'-x))\exp(-\lambda^2 kt)d\lambda dx'.$$

The inner integral can be computed by complex methods of integration. It is known to be (Miscellaneous Exercises 32, Chapter 1)

$$\int_{0}^{\infty} \cos(\lambda(x'-x))\exp(-\lambda^2 kt)d\lambda = \sqrt{\frac{\pi}{4kt}}\exp\left[\frac{-(x'-x)^2}{4kt}\right], t > 0.$$

This gives us, finally, a new form for the temperature distribution:

$$u(x,t) = \frac{1}{\sqrt{4k\pi t}}\int_{-\infty}^{\infty} f(x')\exp\left[\frac{-(x'-x)^2}{4kt}\right]dx'. \tag{7}$$

Using this form, we find the solution of the example problem solved above (see Eq. (6)) to be

$$u(x,t) = \frac{T_0}{\sqrt{4\pi kt}}\int_{-a}^{a} \exp\left[\frac{-(x'-x)^2}{4kt}\right]dx'. \tag{8}$$

Of the two formulas, Eqs. (4) and (7), for the solution $u(x,t)$, each has its advantages. For simple problems we may be able to evaluate the coefficients $A(\lambda)$ and $B(\lambda)$ in Eq. (5). However, it is a rare case indeed when the integral in Eq. (4) can be evaluated analytically. The same is true for the integral in Eq. (7). Thus, if the value of u at a specific x and t is needed, either integral would be calculated numerically. For large values of kt, the exponential factor in the integrand of Eq. (4) will be nearly zero, except for small λ. Thus, Eq. (4) is approximately

$$u(x,t) \cong \int_{0}^{\Lambda} (A(\lambda)\cos(\lambda x) + B(\lambda)\sin(\lambda x))\exp(-\lambda^2 kt)d\lambda$$

for Λ not large, and the right-hand side may be found to a high degree of accuracy with little effort.

On the other hand, if kt is small, the exponential in the integrand of Eq. (7) will be nearly zero, except for x' near x. The approximation

$$u(x,t) \cong \frac{1}{\sqrt{4k\pi t}}\int_{x-h}^{x+h} f(x')\exp\left[\frac{-(x'-x)^2}{4kt}\right]dx'$$

is satisfactory for h not large, and again numerical techniques are easily applied to the right-hand side.

The expression in Eq. (7) also has a number of other advantages. It requires no intermediate integrations (compare Eq. (5)). It shows directly the influence of initial conditions on the solution. Moreover, the function $f(x)$ need not satisfy the restriction

$$\int_{-\infty}^{\infty} |f(x)| dx < \infty$$

in order for Eq. (7) to satisfy the original problem.

Exercises

1. Find the solution of Eqs. (1)-(3) using the form given in Eq. (7) if the initial temperature distribution is

$$f(x) = \begin{cases} T_0, & x < 0, \\ T_1, & 0 < x. \end{cases}$$

2. Find the solution of Eqs. (1)-(3) using the form given in Eq. (4) if

$$f(x) = \begin{cases} T_0(a - |x|), & -a < x < a, \\ 0, & \text{otherwise}. \end{cases}$$

3. Same task as in Exercise 2, with $f(x) = T_0 e^{-|x/a|}$ for all x.

4. Show that the solution of the problem studied in Section 10,

$$\frac{\partial^2 u}{\partial x^2} = \frac{1}{k}\frac{\partial u}{\partial t}, \quad 0 < x, \quad 0 < t$$

$$u(0, t) = 0, \qquad 0 < t$$

$$u(x, 0) = f(x), \quad 0 < x,$$

can be expressed as

$$u(x, t) = \frac{1}{\sqrt{4\pi kt}} \int_0^{\infty} f(x') \left[\exp\left(\frac{-(x' - x)^2}{4kt} \right) - \exp\left(\frac{-(x' + x)^2}{4kt} \right) \right] dx'.$$

Hint: Start from the problem of this section with initial condition

$$u(x, 0) = f_o(x), \quad -\infty < x < \infty,$$

where f_o is the odd extension of f. Then use Eq. (7), and split the interval of integration at 0.

5. Verify that the function

$$u(x,t) = \frac{1}{\sqrt{4k\pi t}} \exp\left[-\frac{x^2}{4kt}\right]$$

is a solution of the heat equation

$$\frac{\partial^2 u}{\partial x^2} = \frac{1}{k}\frac{\partial u}{\partial t}, \quad 0 < t, \quad -\infty < x < \infty.$$

What can be said about u at $x = 0$? at $t = 0+$? What is $\lim_{t\to 0+} u(0,t)$? Sketch $u(x,t)$ for various fixed values of t.

6. Suppose that $f(x)$ is an odd periodic function with period $2a$. Show that $u(x,t)$ defined by Eq. (7) also has these properties.

7. If $f(x) = 1$ for all x, the solution of our heat conduction problem is $u(x,t) = 1$. Use this fact together with Eq. (7) to show that

$$1 = \frac{1}{\sqrt{4\pi kt}} \int_{-\infty}^{\infty} \exp\left[\frac{-(x'-x)^2}{4kt}\right] dx'.$$

8. Solve the problem that follows using Eq. (7).

$$\frac{\partial^2 u}{\partial x^2} = \frac{1}{k}\frac{\partial u}{\partial t}, \qquad -\infty < x < \infty, \quad 0 < t$$

$$u(x,0) = \begin{cases} 1, & x > 0 \\ -1, & x < 0 \end{cases}$$

9. Can Exercise 8 be solved in the form of Eq. (4)? Note that

$$\frac{2}{\pi}\int_0^{\infty} \frac{\sin(\lambda x)}{\lambda} d\lambda = \begin{cases} 1, & 0 < x \\ -1, & x < 0. \end{cases}$$

(See Chapter 1, Section 8.)

2.12 THE ERROR FUNCTION

In Section 11 we made transformations of a Fourier integral to obtain the solution of the heat problem

$$\frac{\partial^2 u}{\partial x^2} = \frac{1}{k}\frac{\partial u}{\partial t}, \qquad -\infty < x < \infty, \quad 0 < t \tag{1}$$

$$u(x,0) = f(x), \qquad -\infty < x < \infty, \tag{2}$$

in the form of a single integral,

$$u(x,t) = \frac{1}{\sqrt{4\pi kt}} \int_{-\infty}^{\infty} f(x')e^{-(x-x')^2/4kt}dx'. \tag{3}$$

Even for the simplest functions f, this integration cannot be carried out in closed form, mainly because the indefinite integral $\int e^{-x^2}dx$ is not an elementary function. We can improve our understanding of the solution Eq. (3) if we introduce the *error function*, defined as

$$\operatorname{erf}(z) = \frac{2}{\sqrt{\pi}} \int_0^z e^{-y^2} dy. \tag{4}$$

A graph of erf(z) is shown in Fig. 10. Convenient tables, together with approximations to the error function, will be found in *Handbook of Mathematical Functions*, by Abramowitz and Stegun.

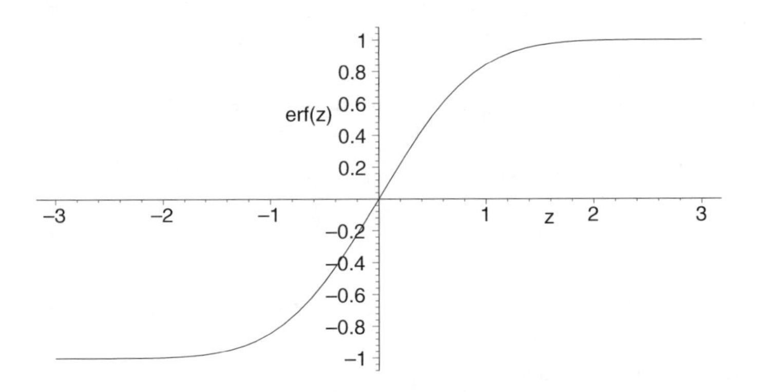

Figure 10: Graph of the error function erf(z) for $-3 < z < 3$.

Several important properties of the error function follow immediately from the definition. First, it is clear that erf(0) = 0, and it is easy to show that erf is an odd function (Exercise 1). Second, by the fundamental theorem of calculus, the derivative of the error function is

$$\frac{d}{dz}\operatorname{erf}(z) = \frac{2}{\sqrt{\pi}}e^{-z^2}. \tag{5}$$

And finally, the error function supplies the integral

$$\int_a^b e^{-y^2} dy = \frac{\sqrt{\pi}}{2} \left(\operatorname{erf}(b) - \operatorname{erf}(a)\right). \tag{6}$$

The reason for the choice of the constant in front of the integral in Eq. (4) is to make

$$\lim_{z\to\infty} \operatorname{erf}(z) = 1. \tag{7}$$

To see that this is true, define

$$A = \int_0^\infty e^{-y^2} dy.$$

We are going to show that $A = \sqrt{\pi}/2$. First write A^2 as the product of two integrals,

$$A^2 = \int_0^\infty e^{-y^2} dy \int_0^\infty e^{-x^2} dx.$$

Remember that the name of the variable of integration in a definite integral is immaterial. This expression for A^2 can be interpreted as an iterated double integral over the first quadrant of the x, y-plane, equivalent to

$$A^2 = \int_0^\infty \int_0^\infty e^{-(x^2+y^2)} dx dy.$$

Now, change to polar coordinates. The first quadrant is described by the inequalities $0 < r < \infty$, $0 < \theta < \pi/2$, and the element of area in polar coordinates is $r dr d\theta$. Thus, we have

$$A^2 = \int_0^{\pi/2} \int_0^\infty e^{-r^2} r dr d\theta. \tag{8}$$

This integral, which can be evaluated by elementary means (see Exercise 2), has value $\pi/4$. Hence Eq. (7) is validated.

Many workers also use the *complementary error function*, $\operatorname{erfc}(z)$, defined as

$$\operatorname{erfc}(z) = \frac{2}{\sqrt{\pi}} \int_z^\infty e^{-y^2} dy. \tag{9}$$

By using Eq. (7) we obtain the identity

$$\operatorname{erfc}(z) = 1 - \operatorname{erf}(z). \tag{10}$$

Some properties of the complementary error function are found in Exercise 3.

We are interested in the error function because of its role in solving the heat equation. First we shall show that *the solution of the problem*

$$\frac{\partial^2 u}{\partial x^2} = \frac{1}{k}\frac{\partial u}{\partial t}, \quad -\infty < x < \infty, \qquad 0 < t \tag{11}$$

$$u(x, 0) = \operatorname{sgn}(x), \quad -\infty < x < \infty, \tag{12}$$

is $u(x,t) = \operatorname{erf}(x/\sqrt{4kt})$. (Recall that $\operatorname{sgn}(x)$ has the value -1 if x is negative or $+1$ if x is positive.) The easy way to prove this statement is to verify it directly. (See Exercises 4 and 5.) Here, we shall arrive at the same conclusion, starting from Eq. (3),

$$u(x,t) = \frac{1}{\sqrt{4\pi kt}} \int_{-\infty}^{\infty} \operatorname{sgn}(x') e^{-(x-x')^2/4kt} dx'. \tag{13}$$

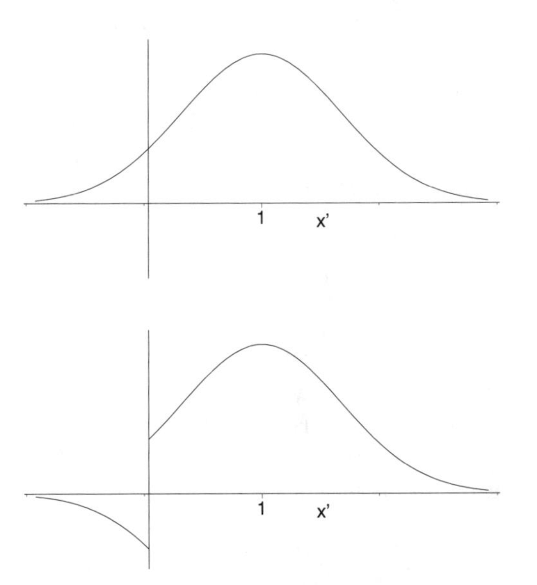

Figure 11: (a) Graph of $\exp(-(x-x')^2/4kt)$ and (b) graph of $\operatorname{sgn}(x')\exp(-(x-x')^2/4kt)$, the integrand in Eq. (13). Both graphs show x' in the range $-x$ to $3x$. The areas of the tails in (b) have the same magnitude but opposite sign.

First observe that the function $e^{-(x-x')^2/4kt}$ has even symmetry with respect to the line $x = x'$, as seen in Fig. 11a. Next, note that the integrand is related in a simple way to this function: Its sign is changed for negative values of x', as one sees in Fig. 11b. Clearly, the tail to the left of $x' = 0$ has the same area as the tail to the right of $x' = 2x$, but they are opposite in sign. Thus the interval of integration only has to run from

0 to $2x$. Now, change the variable of integration to $y = (x' - x)/\sqrt{4kt}$:

$$\begin{aligned} u(x,t) &= \frac{1}{\sqrt{4\pi kt}} \int_0^{2x} e^{-(x'-x)^2/4kt} dx' \\ &= \frac{1}{\sqrt{\pi}} \int_{-x/\sqrt{4kt}}^{x/\sqrt{4kt}} e^{-y^2} dy. \end{aligned}$$

Finally, use the symmetry of the integrand to halve the interval of integration and double the result:

$$u(x,t) = \frac{2}{\sqrt{\pi}} \int_0^{x/\sqrt{4kt}} e^{-y^2} dy = \mathrm{erf}(x/\sqrt{4kt}).$$

This is the result we wanted to arrive at. Figure 12 shows graphs of $u(x,t) = \mathrm{erf}(x/\sqrt{4kt})$ as a function of x for several values of kt.

Because the $\mathrm{erf}(0) = 0$, the function above, $u(x,t) = \mathrm{erf}(x/\sqrt{4kt})$, must also be the solution of the problem

$$\begin{aligned} \frac{\partial^2 u}{\partial x^2} &= \frac{1}{k}\frac{\partial u}{\partial t}, \quad && 0 < x, \quad 0 < t \\ u(0,t) &= 0, && 0 < t \\ u(x,0) &= 1, && 0 < x. \end{aligned}$$

A simple modification leads to the conclusion that the complementary error function, $u(x,t) = \mathrm{erfc}(x/\sqrt{4kt})$ is the solution of this problem with zero initial condition and constant boundary condition,

$$\begin{aligned} \frac{\partial^2 u}{\partial x^2} &= \frac{1}{k}\frac{\partial u}{\partial t}, \quad && 0 < x, \quad 0 < t \\ u(0,t) &= 1, && 0 < t \\ u(x,0) &= 0, && 0 < x. \end{aligned}$$

Exercises

1. Show that $\mathrm{erf}(-z) = -\mathrm{erf}(z)$, that is, that erf is an odd function.

2. Carry out the integration indicated in Eq. (8).

3. Verify these properties of the complementary error function:

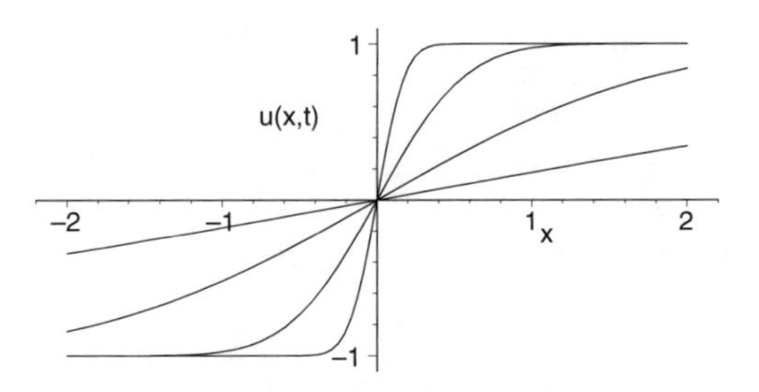

Figure 12: Graphs of the solution of the problem in Eqs. (11) and (12), $u(x,t) = \operatorname{erf}(x/\sqrt{4kt})$, for x in the range -2 to 2 and for $kt = 0.01$, 0.1, 1 and 10. As kt increases, the graph of $u(x,t)$ collapses toward the x-axis.

a. $\dfrac{d}{dz}\operatorname{erfc}(z) = -e^{-z^2}$.

b. $\operatorname{erfc}(0) = 1$.

c. $\lim\limits_{z\to\infty} \operatorname{erfc}(z) = 0$.

d. $\lim\limits_{z\to-\infty} \operatorname{erfc}(z) = 2$.

e. $\operatorname{erfc}(z)$ is neither even nor odd.

4. Verify by differentiating that $u(x,t) = \operatorname{erf}(x/\sqrt{4kt})$ satisfies this heat problem.

$$\frac{\partial^2 u}{\partial x^2} = \frac{1}{k}\frac{\partial u}{\partial t}, \qquad -\infty < x < \infty, \quad 0 < t$$

$$u(x,0) = \begin{cases} 1, & x > 0 \\ -1, & x < 0 \end{cases}$$

6. In probability and statistics, the *normal* or *Gaussian* probability density function is defined as

$$f(z) = \frac{1}{\sqrt{2\pi}} e^{-z^2/2}, \quad -\infty < z < \infty,$$

and the cumulative distribution function is

$$\Phi(x) = \int_{-\infty}^{x} f(z)dz.$$

Show that the cumulative distribution function and the error function are related by $\Phi(x) = [1 + \operatorname{erf}(x/\sqrt{2})]/2$.

7. Express the integral below in terms of the error function:

$$I(x) = \int \frac{e^{-x}}{\sqrt{x}} dx.$$

8. Use error functions to solve the problem

$$\frac{\partial^2 u}{\partial x^2} = \frac{1}{k}\frac{\partial u}{\partial t}, \quad 0 < x, \quad 0 < t,$$
$$u(0,t) = U_b, \quad 0 < t$$
$$u(x,0) = U_i, \quad 0 < x.$$

(Hint: What conditions does $u(x,t) - U_i$ satisfy?)

9. Assuming that $U_b < 0$ and $U_i > 0$, the problem in Exercise 8 might be interpreted as representing the temperature in a freezing lake. (Think of x as measuring depth from the surface.) Define $x(t)$ as the depth of the ice-water interface; then $u(x(t),t) = 0$. Now find $x(t)$ explicitly.

10. Use error functions to solve the problem

$$\frac{\partial^2 u}{\partial x^2} = \frac{1}{k}\frac{\partial u}{\partial t}, \quad -\infty < x < \infty, \quad 0 < t$$
$$u(x,0) = f(x), \quad -\infty < x < \infty,$$

where $f(x) = U_0$ for $x < 0$ and $f(x) = U_1$ for $x > 0$.

2.13 COMMENTS AND REFERENCES

In about 1810, Fourier made an intensive study of heat conduction problems, in which he used the product method of solution and developed the idea of Fourier series. Sturm and Liouville made their clear and simple generalization of Fourier series in the 1830s. Among modern works, *Conduction of Heat in Solids* by Carslaw and Jaeger, is the standard reference. *The Mathematics of Diffusion*, by Crank, and *The Heat Equation* by D.V. Widder, are newer references. (See the Bibliography.)

Although we have motivated our study in terms of heat conduction and, to a lesser extent, by diffusion, many other physical phenomena of interest in engineering are described by the heat/diffusion equation: for example, voltage and current in an inductance-free cable (see Miscellaneous Exercises 26 and 27 in Chapter 3) and vorticity transport in fluid flow. The heat/diffusion equation and allied equations are being employed in biology to model cell physiology, chemical reactions, nerve impulses, the spread of populations, and many other phenomena. Two good references are *Differential Equations and Mathematical Biology* by D.S. Jones and B.D. Sleeman, and *Mathematical Biology* by J.D. Murray.

The diffusion equation also turns up in some classical problems of probability theory, especially the description of Brownian motion. Suppose a particle moves exactly one step of length Δx in each time interval Δt. The step may be either to the left or right, each equally likely. Let $u_i(m)$ denote the probability that, at time $m\Delta t$, the particle is at point $i\Delta x(m = 0, 1, 2, \cdots, i = 0, \pm 1, \pm 2, \cdots)$. In order to arrive at point $i\Delta x$ at time $(m+1)\Delta t$, the particle must have been at one of the adjacent points $(i \pm 1)\Delta x$ at the preceding time $m\Delta t$, and must have moved toward $i\Delta x$. From this, we see that the probabilities are related by the equation

$$u_i(m+1) = \frac{1}{2}u_{i-1}(m) + \frac{1}{2}u_{i+1}(m).$$

The u's are completely determined once an initial probability distribution is given. For instance, if the particle is initially at point zero ($u_0(0) = 1$, $u_i(0) = 0$, for $i \neq 0$), the $u_i(m)$ are formed by successive applications of the difference equation, as shown in Table 3.

The equation may be transformed into a close relative of the heat equation. First, subtract $u_i(m)$ from both sides:

$$u_i(m+1) - u_(m) = \frac{1}{2}\left(u_{i+1}(m) - 2u_i(m) + u_{i-1}(m)\right).$$

Next divide by $\Delta x^2/2$ on the right and by $\Delta t \cdot (\Delta x^2/2\Delta t)$ on the left to obtain

$$\frac{u_i(m+1) - u_i(m)}{\Delta t}\frac{2\Delta t}{(\Delta x)^2} = \frac{u_{i+1}(m) - 2u_i(m) + u_{i-1}(m)}{(\Delta x)^2}.$$

If both the time interval Δt and the step length Δx are small, we may think of $u_i(m)$ as being the value of a continuous function $u(x,t)$ at $x = i\Delta x$, $t = m\Delta t$. In the limit, the difference quotient on the left approaches $\partial u/\partial t$. The right-hand side, being a difference of differences, approaches $\partial^2 u/\partial x^2$. The heat equation thus results if, in the simultaneous limit as Δx and Δt tend to zero, the quantity $2\Delta t/(\Delta x)^2$ approaches a finite, nonzero limit. In this context, the heat equation is called the *Fokker-Planck equation.* More details and references may be found in Feller, *Introduction to Probability Theory and Its Applications.*

Table 3: Random-walk probabilities

i / m	-3	-2	-1	0	1	2	3
0	0	0	0	1	0	0	
1	0	0	0.5	0	0.5	0	0
2	0	0.25	0	0.5	0	0.25	0
3	0.125	0	0.375	0	0.375	0	0.125

We have used the phrase "linear partial differential equation" several times. The most general such equation, of second order in two independent variables, can be put in the form

$$A\frac{\partial^2 u}{\partial x^2} + B\frac{\partial^2 u}{\partial x \partial t} + C\frac{\partial^2 u}{\partial t^2} + D\frac{\partial u}{\partial x} + E\frac{\partial u}{\partial t} + Fu + G = 0$$

where $A, B, \cdots, G$ are known—perhaps functions of x and t but not of u or its derivatives. If G is identically zero, the equation is homogeneous. Of course, the ordinary heat equation has this form if we take $A = 1$, $E = -1/k$, and all other coefficients equal to zero.

Some astute students will have wondered why we should seek solutions in product form. The simplest answer is in many cases that it works. A more subtle rationale is that of seeking solutions that are geometrically similar functions of x at different times. The idea of similarity—related to dimensional analysis—has been most fruitful in the mechanics of fluids.

MISCELLANEOUS EXERCISES

1-16. Find the steady-state solution, the associated eigenvalue problem, and the complete solution for each problem.

1. $\dfrac{\partial^2 u}{\partial x^2} = \dfrac{1}{k}\dfrac{\partial u}{\partial t}, \qquad 0 < x < a, \quad 0 < t$

$u(0,t) = T_0, \qquad u(a,t) = T_0, \quad 0 < t$

$u(x,0) = T_1, \qquad 0 < x < a$

2. $\dfrac{\partial^2 u}{\partial x^2} - \gamma^2 u = \dfrac{1}{k}\dfrac{\partial u}{\partial t}, \qquad 0 < x < a, \quad 0 < t$

$u(0,t) = T_0, \qquad u(a,t) = T_0, \quad 0 < t$

$u(x,0) = T_1, \qquad 0 < x < a$

3. $\dfrac{\partial^2 u}{\partial x^2} - r = \dfrac{1}{k}\dfrac{\partial u}{\partial t}, \qquad 0 < x < a, \quad 0 < t$

$u(0,t) = T_0, \qquad u(a,t) = T_0, \quad 0 < t$

$u(x,0) = T_1, \qquad 0 < x < a, \quad (r \text{ is constant})$

4. $\dfrac{\partial^2 u}{\partial x^2} = \dfrac{1}{k}\dfrac{\partial u}{\partial t}, \qquad 0 < x < a, \quad 0 < t$

$u(0,t) = T_0, \qquad \dfrac{\partial u}{\partial x}(a,t) = 0, \quad 0 < t$

$u(x,0) = \dfrac{T_1 x}{a}, \qquad 0 < x < a$

5. $\dfrac{\partial^2 u}{\partial x^2} - \gamma^2 u = \dfrac{1}{k}\dfrac{\partial u}{\partial t}, \qquad 0 < x < a, \quad 0 < t$

$\dfrac{\partial u}{\partial x}(0,t) = 0, \qquad \dfrac{\partial u}{\partial x}(a,t) = 0, \quad 0 < t$

$u(x,0) = \dfrac{T_1 x}{a}, \qquad 0 < x < a$

6. $\dfrac{\partial^2 u}{\partial x^2} = \dfrac{1}{k}\dfrac{\partial u}{\partial t}, \qquad 0 < x < a, \quad 0 < t$

$u(0,t) = 0, \qquad u(a,t) = T_0, \quad 0 < t$

$u(x,0) = 0, \qquad 0 < x < a$

7. $\dfrac{\partial^2 u}{\partial x^2} = \dfrac{1}{k}\dfrac{\partial u}{\partial t}, \qquad 0 < x < a, \quad 0 < t$

$u(0,t) = T_0, \qquad u(a,t) = T_0,$

$u(x,0) = T_0, \qquad 0 < x < a$

8. $\dfrac{\partial^2 u}{\partial x^2} = \dfrac{1}{k}\dfrac{\partial u}{\partial t}, \qquad 0 < x < a, \quad 0 < t$

$\dfrac{\partial u}{\partial x}(0,t) = \dfrac{\Delta T}{a}, \qquad \dfrac{\partial u}{\partial x}(a,t) = \dfrac{\Delta T}{a}, \quad 0 < t$

$u(x,0) = T_0, \qquad 0 < x < a$

9. $\dfrac{\partial^2 u}{\partial x^2} = \dfrac{1}{k}\dfrac{\partial u}{\partial t}, \qquad 0 < x < a, \quad 0 < t$

$u(0,t) = T_0, \qquad \dfrac{\partial u}{\partial x}(a,t) = 0, \quad 0 < t$

$u(x,0) = T_1, \qquad 0 < x < a$

10. $$\frac{\partial^2 u}{\partial x^2} = \frac{1}{k}\frac{\partial u}{\partial t}, \qquad 0 < x < a, \quad 0 < t$$
$$\frac{\partial u}{\partial x}(0,t) = 0, \qquad \frac{\partial u}{\partial x}(a,t) = 0, \quad 0 < t$$
$$u(x,0) = \begin{cases} T_0, & 0 < x < \frac{a}{2} \\ T_1, & \frac{a}{2} < x < a \end{cases}$$

11. $$\frac{\partial^2 u}{\partial x^2} = \frac{1}{k}\frac{\partial u}{\partial t}, \qquad 0 < x < \infty, \quad 0 < t$$
$$u(0,t) = T_0, \qquad 0 < t$$
$$u(x,0) = T_0(1 - e^{-\alpha x}), \quad 0 < x$$

12. $$\frac{\partial^2 u}{\partial x^2} = \frac{1}{k}\frac{\partial u}{\partial t}, \qquad 0 < x < \infty, \quad 0 < t$$
$$u(0,t) = T_0, \qquad 0 < t$$
$$u(x,0) = \begin{cases} T_0, & 0 < x < a \\ 0, & a < x \end{cases}$$

13. $$\frac{\partial^2 u}{\partial x^2} = \frac{1}{k}\frac{\partial u}{\partial t}, \qquad 0 < x < \infty, \quad 0 < t$$
$$\frac{\partial u}{\partial x}(0,t) = 0, \qquad 0 < t$$
$$u(x,0) = \begin{cases} T_0, & 0 < x < a \\ 0, & a < x \end{cases}$$

14. $$\frac{\partial^2 u}{\partial x^2} = \frac{1}{k}\frac{\partial u}{\partial t}, \qquad -\infty < x < \infty, \quad 0 < t$$
$$u(x,0) = \exp(-\alpha|x|), \quad -\infty < x < \infty$$

15. $$\frac{\partial^2 u}{\partial x^2} = \frac{1}{k}\frac{\partial u}{\partial t}, \qquad -\infty < x < \infty, \quad 0 < t$$
$$u(x,0) = \begin{cases} 0, & -\infty < x < 0 \\ T_0, & 0 < x < a \\ 0, & a < x < \infty \end{cases}$$

16. $\dfrac{\partial^2 u}{\partial x^2} = \dfrac{1}{k}\dfrac{\partial u}{\partial t}, \qquad 0 < x < a, \quad 0 < t$

$\dfrac{\partial u}{\partial x}(0,t) = 0, \quad u(a,t) = T_0, \quad 0 < t$

$u(x,0) = T_0 + S(a - x), \quad 0 < x < a$

17. Give a physical interpretation for this problem and thus explain why $u(x,t)$ should increase steadily as t increases. (Assume that S is a positive constant.)

$$\frac{\partial^2 u}{\partial x^2} = \frac{1}{k}\frac{\partial u}{\partial t}, \qquad 0 < x < a, \qquad 0 < t$$
$$\frac{\partial u}{\partial x}(0,t) = 0, \qquad \frac{\partial u}{\partial x}(a,t) = S, \quad 0 < t$$
$$u(x,0) = 0, \qquad 0 < x < a$$

18. Show that $v(x,t) = (S/2a)(x^2 + 2kt)$ satisfies the heat equation and the boundary conditions of the problem in Exercise 17. Also find $w(x,t)$, defined by $u(x,t) = v(x,t) + w(x,t)$.

19. Show that the four functions

$$u_0 = 1, \quad u_1 = x, \quad u_2 = x^2 + 2kt, \quad u_3 = x^3 + 6kxt$$

are solutions of the heat equation. (These are sometimes called heat polynomials.) Find a linear combination of them that satisfies the boundary conditions $u(0,x) = 0$, $u(a,t) = t$.

20. Suppose that $u(x,t)$ is a positive function that satisfies

$$\frac{\partial^2 u}{\partial x^2} = \frac{\partial u}{\partial t}.$$

Show that the function

$$w(x,t) = -\frac{2}{u}\frac{\partial u}{\partial x}$$

satisfies the nonlinear partial differential equation called *Burgers' equation:*

$$\frac{\partial w}{\partial t} + w\frac{\partial w}{\partial x} = \frac{\partial^2 w}{\partial x^2}.$$

21. Find a solution of the Burgers' equation that satisfies the conditions

$$w(0,t) = 0, \quad w(1,t) = 0, \quad 0 < t$$

$$w(x,0) = 1, \quad 0 < x < 1.$$

22. Taking the function $u(x,t)$ given below as a solution of the heat equation (with $k = 1$), find a solution w of Burgers' equation. Verify that w satisfies Burgers' equation.

$$u(x,t) = \frac{1}{\sqrt{4\pi t}} \exp\left(\frac{-x^2}{4t}\right).$$

23. Consider a solid metal bar surrounded by a finite quantity of water confined in a water jacket. If the bar and the water are at different temperatures, they will exchange heat. Let u_1 and u_2 be the temperatures in the bar and in the water. Heat balances for the water and the bar give these two equations:

$$c_1 \frac{du_1}{dt} = h(u_2 - u_1)$$

$$c_2 \frac{du_2}{dt} = h(u_1 - u_2).$$

Here, c_1 and c_2 are the heat capacities of the bar and the water, and h is the product of the convection coefficient with the area of the bar-water interface. Find temperatures u_1 and u_2 assuming initial conditions $u_1(0) = T_0$, $u_2(0) = 0$.

24. Solve the eigenvalue problem by setting $\phi(\rho) = \psi(\rho)/\rho$:

$$\frac{1}{\rho^2}(\rho^2\phi')' + \lambda^2\phi = 0, \quad 0 < \rho < a$$

$$\phi(0) \text{ bounded}, \quad \phi(a) = 0.$$

Is this a regular Sturm-Liouville problem? Are the eigenfunctions orthogonal?

25. Solve this problem for heat conduction in a sphere. (Hint: let $u(\rho,t) = v(\rho,t)/\rho$.)

$$\frac{1}{\rho^2}\frac{\partial}{\partial\rho}\left(\rho^2\frac{\partial u}{\partial\rho}\right) = \frac{1}{k}\frac{\partial u}{\partial t}, \quad 0 < \rho < a, \quad 0 < t$$

$$u(0,t) \text{ bounded}, \quad u(a,t) = 0, \quad 0 < t$$

$$u(\rho,0) = T_0, \quad 0 < \rho < a$$

26. State and solve the eigenvalue problem associated with

$$e^{-x}\frac{\partial}{\partial x}\left(e^{x}\frac{\partial u}{\partial x}\right) = \frac{1}{k}\frac{\partial u}{\partial t}, \quad 0 < x < a, \quad 0 < t$$

$$u(0,t) = 0, \quad \frac{\partial u}{\partial x}(a,t) = 0.$$

27. Find the steady-state solution of the problem

$$\frac{\partial^2 u}{\partial x^2} + \gamma^2(T(x) - u) = \frac{1}{k}\frac{\partial u}{\partial t}, \quad 0 < x < a, \quad 0 < t$$

$$u(0,t) = T_0, \quad \frac{\partial u}{\partial x}(a,t) = 0, \quad 0 < t$$

where $T(x) = T_0 + Sx$.

28. Determine whether or not $\lambda = 0$ is an eigenvalue of the problem

$$\phi'' + \lambda^2 x\phi = 0, \quad 0 < x < a$$

$$\phi'(0) = 0, \quad \phi(a) = 0.$$

29. Same question as Exercise 28, but with boundary conditions:

$$\phi'(0) = 0, \phi'(a) = 0.$$

30. Prove the following identity:

$$\frac{1}{\sqrt{4\pi kt}}\int_b^a \exp\left[-\frac{(\xi - x)^2}{4kt}\right]d\xi = \frac{1}{2}\left[\operatorname{erf}\left(\frac{b-x}{\sqrt{4kt}}\right) - \operatorname{erf}\left(\frac{a-x}{\sqrt{4kt}}\right)\right].$$

31. In Exercise 6 of Section 10, it was shown that the function

$$w(x,t;\omega) = e^{-px}\sin(\omega t - px)$$

where $p = \sqrt{\omega/2k}$, satisfies the heat equation and also the boundary condition

$$w(0,t;\omega) = \sin(\omega t).$$

Show how to choose the coefficient $B(\omega)$ so that the function

$$u(x,t) = \int_0^\infty B(\omega)e^{-px}\sin(\omega t - px)d\omega$$

satisfies the boundary condition

$$u(0,t) = f(t), \qquad 0 < t$$

for a suitable function t.

32. Use the idea of Exercise 31 to find a solution of

$$\frac{\partial^2 u}{\partial x^2} = \frac{1}{k}\frac{\partial u}{\partial t}, \quad 0 < x, \quad 0 < t$$
$$u(0,t) = h(t), \quad 0 < t,$$

where

$$h(t) = \begin{cases} 1, & 0 < t < T, \\ 0, & T < t. \end{cases}$$

33. S.E. Serrano and T.E. Unny develop probabilistic mathematical models for ground water flow under uncertain conditions (Predicting ground water flow in a phreatic aquifer, Journal of Hydrology, 95 (1987) 241-268) and compare the results to measurements. One of the models uses this nonlinear Boussinesq equation,

$$S\frac{\partial y}{\partial t} - \frac{\partial}{\partial x}\left(Kh\frac{\partial y}{\partial x}\right) = I + \phi, \quad 0 < x < L, \quad 0 < t,$$

together with the conditions

$$y(0,t) = y_1(t), \quad y(L,t) = y_2(t), \quad 0 < t,$$
$$y(x,0) = y_0(x), \quad 0 < x < L.$$

In these equations, $y(x,t)$ is the water table elevation above sea level, $h(x,t)$ is water table elevation above bedrock, K is hydraulic conductivity (in meters per day, m/da), S is the aquifer specific yield, I is a function representing input by percolation from the aquifer, and $\phi(x,t)$ is a random function that accounts for uncertainty in input.

The partial differential equation is nonlinear because h and y represent the same thing, relative to two different references. We assume that the bedrock elevation has constant slope a, so that $y = h + ax$. Then the equation can be written in terms of h alone as

$$S\frac{\partial h}{\partial t} - \frac{\partial}{\partial x}\left(Kh\frac{\partial h}{\partial x}\right) - a\frac{\partial}{\partial x}(Kh) = I + \phi.$$

Next, this equation is linearized. Assume that K is constant, and that h can be broken down as $h = \bar{h} + h'$, where $\bar{h}$ is a constant mean value of h, h' is a fluctuation much smaller than $\bar{h}$. (In this case, $\bar{h}$ is about 150 m and h' is less than 1m.) Then the product Kh is approximately equal to $K\bar{h} = T$ (transmissivity) and, as a coefficient in the second term, can be treated as a constant. The equation is now linear in h' (we drop the prime for convenience):

$$S\frac{\partial h}{\partial t} - T\frac{\partial^2 h}{\partial x^2} - aK\frac{\partial h}{\partial x} = I + \phi, \quad 0 < x < L, \quad 0 < t.$$

a. Treating I as a constant, find a steady-state solution $v(x)$ for the statistical mean value of h, which is obtained by replacing $\phi(x,t)$ with 0. The boundary and initial conditions are

$$h(0,t) = h_1, \qquad h(L,t) = h_2, \quad 0 < t,$$
$$h(x,0) = h_0(x), \quad 0 < x < L.$$

b. State the problem (partial differential equation, boundary conditions, and initial condition) to be satisfied by the mean transient, $w(x,t) = h(x,t) - v(x)$. (Again, the statistical mean corresponds to $\phi \equiv 0$.)

c. Solve the problem in **b.**

d. Values for the parameters are: $a = 0.0292$m/m, $K = 17.28$m/da, $T = 218.4\text{m}^2/\text{da}$, $S = 0.15$, $L = 116.25$m. Find the eigenvalues.

34. Consider a steel plate that is much larger in length and width (x- and z-directions) than in thickness (y-direction), and suppose the plate is free to expand or contract under the effects of heating. Assume that the temperature T in the plate is a function of y and t only. Timoshenko and Goodier (*Theory of Elasticity*, pp. 399-403) derive the following expression for the stresses due to thermal effects:

$$\sigma_x = \sigma_z = -\frac{\alpha T E}{1-\nu} + \frac{1}{2c(1-\nu)}\int_{-c}^{+c} \alpha T E dy + \frac{3y}{2c^3(1-\nu)}\int_{-c}^{+c} \alpha T E y dy.$$

The parameters, and their values for steel are as follows: α is the coefficient of expansion, 6.5×10^{-6} per degree F; E is Young's modulus, $28 \times 10^6 \text{lb/in}^2$; ν is Poisson's ratio, 0.7; and $2c$ is the thickness of the plate. Note that the origin is located so that the plate lies between $y = c$ and $y = -c$.

a. Show that if $T(y) = T_0 + Sy$, where T_0 and S are constants, then the thermal stress is 0. (This is a typical steady-state temperature distribution.)

b. Suppose that the plate is initially at temperature 500 °F throughout, and then the temperature on the face $y = c$ is suddenly changed to 200° while the temperature at $y = -c$ remains at 500°. Find $T(y, t)$.

c. Assume the initial and boundary conditions given in **b**. Use your understanding of the function $T(y, t)$ to explain why the thermal stress near the face $y = c$ is large just after time 0.

Chapter 3

The Wave Equation

3.1 THE VIBRATING STRING

A simple and historically important example of a problem that includes the wave equation is provided by the study of the vibration of a string, like a violin or guitar string. We set up a coordinate system as shown in Fig. 1. The unknown is the transverse displacement, $u(x,t)$, measured up from the x-axis. The situation is similar to that of the hanging cable discussed in Chapter 0, but here the string is taut, and of course motion is allowed. In order to find the equation of motion of the string, we consider a short piece whose ends are at x and $x + \Delta x$ and apply Newton's second law of motion to it.

First, we must analyze the nature of the forces on the string. We assume that the only external force is the attraction of gravity, acting perpendicular to the x-direction. Internal forces are exerted on the segment by the rest of the string. We will *assume that the string is perfectly flexible* and offers no resistance to bending. Then the only force that can be transmitted by the string is a pull or tension, which acts in a direction tangential to the center line of the string. Its magnitude we denote by $T(x,t)$.

The forces on the small segment of string are shown in Fig. 2. We shall further *assume that each point on the string moves only in the vertical direction.* Thus, the horizontal component of acceleration is zero. Application of Newton's second law for the horizontal direction to the segment leads to the equation

$$-T(x,t)\cos(\phi(x,t)) + T(x+\Delta x,t)\cos(\phi(x+\Delta x,t)) = 0,$$

or

$$T(x,t)\cos(\phi(x,t) = T(x+\Delta x,t))\cos(\phi(x+\Delta x,t)). \tag{1}$$

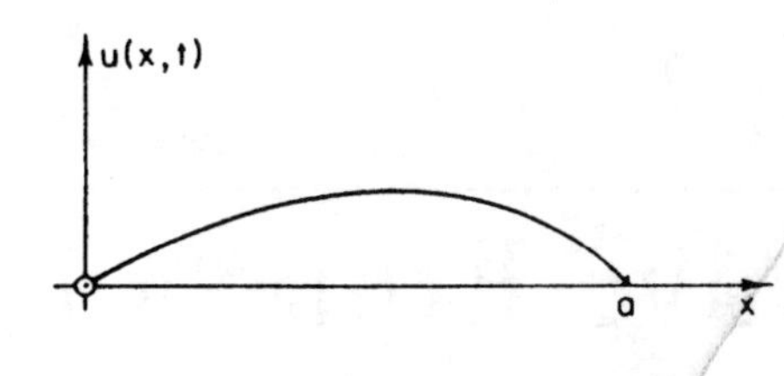

Figure 1: String fixed at ends.

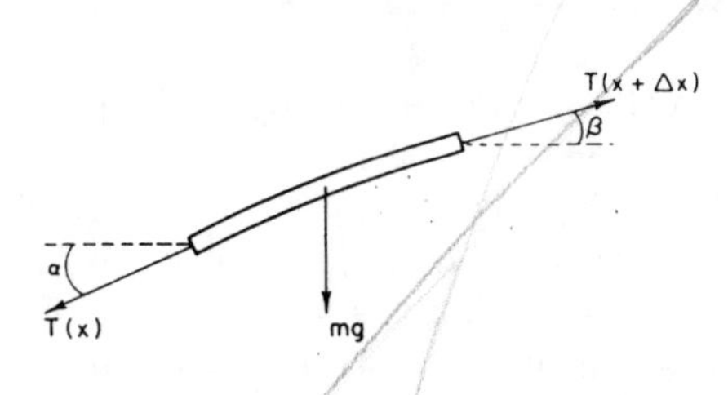

Figure 2: Section of string showing forces exerted on it. The angles are $\alpha = \phi(x,t)$ and $\beta = \phi(x+\Delta x,t)$.

This says that the horizontal component of tension in the string is the same at every point: $T(x,t)\cos(\phi(x,t)) = T(x+\Delta x,t)\cos(\phi(x+\Delta x,t)) = T$, independent of x. If the string is taut, T can vary only slightly with t, so we will assume that T is constant.

In the absence of external forces other than gravity, Newton's second law for the vertical direction yields

$$-T(x,t)\sin(\phi(x,t)) + T(x+\Delta x,t))\sin(\phi(x+\Delta x,t)) - mg = m\frac{\partial^2 u}{\partial t^2}(x,t). \tag{2}$$

(Because $u(x,t)$ measures displacement in the vertical direction, its second partial derivative with respect to t is the vertical acceleration.) The mass of the short piece of string we are examining is proportional to its length, $m = \rho\Delta x$, where ρ is the linear density, measured in units of mass per unit length.

Now we use Eq. (1) to solve for the tensions at the ends of the segment of string as

$$T(x,t) = \frac{T}{\cos(\phi(x,t))}, \quad T(x+\Delta x,t) = \frac{T}{\cos(\phi(x+\Delta x,t))}.$$

When these expressions are substituted into Eq. (2), we have

$$-T\tan(\phi(x,t)) + T\tan(\phi(x+\Delta x,t)) - \rho\Delta x g = \rho\Delta x\frac{\partial^2 u}{\partial t^2}. \tag{3}$$

Recall from elementary calculus that $\tan(\phi(x,t))$ is the slope of the string at (x,t) and hence may be expressed in terms of the (partial) derivative with respect to x:

$$\tan(\phi(x,t)) = \frac{\partial u}{\partial x}(x,t), \quad \tan(\phi(x+\Delta x,t)) = \frac{\partial u}{\partial x}(x+\Delta x,t).$$

Substituting these into Eq. (3), we have

$$T\left(\frac{\partial u}{\partial x}(x+\Delta x,t) - \frac{\partial u}{\partial x}(x,t)\right) = \rho\Delta x\left(\frac{\partial^2 u}{\partial t^2} + g\right).$$

On dividing through by Δx, we see a difference quotient on the left:

$$\frac{T}{\Delta x}\left(\frac{\partial u}{\partial x}(x+\Delta x,t) - \frac{\partial u}{\partial x}(x,t)\right) = \rho\left(\frac{\partial^2 u}{\partial t^2} + g\right).$$

In the limit as $\Delta x \to 0$, the difference quotient becomes a partial derivative with respect to x, leaving Newton's second law in the form

$$T\frac{\partial^2 u}{\partial x^2} = \rho\frac{\partial^2 u}{\partial t^2} + \rho g, \quad \text{or} \tag{4}$$

$$\frac{\partial^2 u}{\partial x^2} = \frac{1}{c^2}\frac{\partial^2 u}{\partial t^2} + \frac{1}{c^2}g \tag{5}$$

where $c^2 = T/\rho$. If c^2 is very large (usually on the order of $10^5\mathrm{m}^2/\mathrm{s}^2$), we neglect the last term, giving the equation of the vibrating string

$$\frac{\partial^2 u}{\partial x^2} = \frac{1}{c^2}\frac{\partial^2 u}{\partial t^2}, \quad 0 < x < a, \quad 0 < t. \tag{6}$$

This equation is called the *wave equation* in one dimension. Two- and three-dimensional versions will be treated in Chapter 5.

In describing the motion of an object, one must specify not only the equation of motion, but also both the initial position and velocity of the object. The initial conditions for the string, then, must state the initial displacement of every particle — that is $u(x, 0)$ — and the initial velocity of every particle, $\partial u/\partial t(x, 0)$.

For the vibrating string as we have described it, the boundary conditions are zero displacement at the ends, so the boundary value-initial value problem for the string is

$$\frac{\partial^2 u}{\partial x^2} = \frac{1}{c^2}\frac{\partial^2 u}{\partial t^2}, \quad 0 < x < a, \quad 0 < t \tag{7}$$

$$u(0,t) = 0, \quad u(a,t) = 0, \quad 0 < t \tag{8}$$

$$u(x,0) = f(x), \quad 0 < x < a \tag{9}$$

$$\frac{\partial u}{\partial t}(x,0) = g(x), \quad 0 < x < a \tag{10}$$

under the assumptions noted plus the assumption that gravity is negligible.

Exercises

1. Find the dimensions of each of the following quantities, using the facts that force is equivalent to mL/t^2, and that the dimension of tension is F (force): u, $\partial^2 u/\partial x^2$, $\partial^2 u/\partial t^2$, c, g/c^2. Check the dimension of each term in Eq. (5).

2. Suppose a distributed vertical force $F(x, t)$ (positive upwards) acts on the string. Derive the equation of motion:
$$\frac{\partial^2 u}{\partial x^2} = \frac{1}{c^2}\frac{\partial^2 u}{\partial t^2} - \frac{1}{T}F(x,t).$$
The dimension of a distributed force is F/L. (If the weight of the string is considered as a distributed force, and is the only one, then we would have $F(x, t) = -\rho g$. Check dimensions and signs.)

3. Find a solution $v(x)$ of Eq. (5) with boundary conditions Eq. (8) that is independent of time. (This corresponds to a "steady-state solution," but the words "steady-state" are no longer appropriate. "Equilibrium solution" is more accurate.)

4. Suppose that the string is located in a medium that resists its movement, such as air. The resistance is expressed as a force opposite in

direction and proportional in magnitude to velocity. Thus it affects only Eq. (2). Proceed to derive the equation that replaces Eq. (7) for this case.

3.2 SOLUTION OF THE VIBRATING STRING PROBLEM

The initial value-boundary value problem that describes the displacement of the vibrating string,

$$\frac{\partial^2 u}{\partial x^2} = \frac{1}{c^2}\frac{\partial^2 u}{\partial t^2}, \quad 0 < x < a, \quad 0 < t \tag{1}$$

$$u(0,t) = 0, \quad u(a,t) = 0, \quad 0 < t \tag{2}$$

$$u(x,0) = f(x), \quad 0 < x < a \tag{3}$$

$$\frac{\partial u}{\partial t}(x,0) = g(x), \quad 0 < x < a \tag{4}$$

contains a linear, homogeneous partial differential equation and linear, homogeneous boundary conditions. Thus we may apply the method of separation of variables with hope of success. If we assume that[1] $u(x,t) = \phi(x)T(t)$, Eq. (1) becomes

$$\phi''(x)T(t) = \frac{1}{c^2}\phi(x)T''(t).$$

Dividing through by ϕT, we obtain

$$\frac{\phi''(x)}{\phi(x)} = \frac{T''(t)}{c^2 T(t)}, \quad 0 < x < a, \quad 0 < t.$$

For the equality to hold, both members of this equation must be constant. We write the constant as $-\lambda^2$ and separate the preceding equation into two ordinary differential equations linked by the common parameter λ^2.

$$T'' + \lambda^2 c^2 T = 0, \quad 0 < t \tag{5}$$

$$\phi'' + \lambda^2 \phi = 0, \quad 0 < x < a. \tag{6}$$

The boundary conditions become

$$\phi(0)T(t) = 0, \quad \phi(a)T(t) = 0, \quad 0 < t$$

and, since $T(t) \equiv 0$ gives a trivial solution for $u(x,t)$, we must have

$$\phi(0) = 0, \quad \phi(a) = 0. \tag{7}$$

[1] T no longer symbolizes tension.

The eigenvalue problem Eqs. (6) and (7) is exactly the same as the one we have seen and solved before. (See Chapter 2, Section 3.) We know that the eigenvalues and eigenfunctions are

$$\lambda_n^2 = \left(\frac{n\pi}{a}\right)^2, \quad \phi_n(x) = \sin(\lambda_n x), \quad n = 1, 2, 3, \cdots.$$

Equation (5) is also of a familiar type, and its solution is known to be

$$T_n(t) = a_n \cos(\lambda_n ct) + b_n \sin(\lambda_n ct)$$

where a_n and b_n are arbitrary. (In other words, there are two independent solutions.) Note, however, that there is a substantial difference between the T that arises here and the T that we found in the heat conduction problem. The most important difference is the behavior as t tends to infinity. In the heat conduction problem $T(t)$ tends to 0; whereas here $T(t)$ has no limit but oscillates periodically in agreement with our intuition.

For each $n = 1, 2, 3, \cdots$, we now have a function

$$u_n(x,t) = \sin(\lambda_n x)\left[a_n \cos(\lambda_n ct) + b_n \sin(\lambda_n ct)\right]$$

which, for any choice of a_n and b_n, is a solution of the partial differential equation (1) and also satisfies the boundary conditions Eq. (2). Therefore, linear combinations of the $u_n(x,t)$ also satisfy both Eqs. (1) and (2). In making our linear combinations, we need no new constants because the a_n and b_n are arbitrary. We have, then,

$$u(x,t) = \sum_{n=1}^{\infty} \sin(\lambda_n x)\left[a_n \cos(\lambda_n ct) + b_n \sin(\lambda_n ct)\right]. \tag{8}$$

The initial conditions, which remain to be satisfied, have the form

$$u(x,0) = \sum_{n=1}^{\infty} a_n \sin\left(\frac{n\pi x}{a}\right) = f(x), \qquad 0 < x < a$$

$$\frac{\partial u}{\partial t}(x,0) = \sum_{n=1}^{\infty} b_n \frac{n\pi}{a} c \sin\left(\frac{n\pi x}{a}\right) = g(x), \quad 0 < x < a.$$

(Here we have assumed that

$$\frac{\partial u}{\partial t}(x,t) = \sum_{n=1}^{\infty} \sin(\lambda_n x)\left[-a_n \lambda_n c \sin(\lambda_n ct) + b_n \lambda_n c \cos(\lambda_n ct)\right].$$

In other words, we assume that the series for u may be differentiated term-by-term.) Both initial conditions take the form of Fourier series problems: a given function is to be expanded in a series of sines. In each case, then, the constant multiplying $\sin(n\pi x/a)$ must be the Fourier sine coefficient for the given function. Thus we determine that

$$a_n = \frac{2}{a}\int_0^a f(x)\sin\left(\frac{n\pi x}{a}\right)dx, \tag{9}$$

and

$$b_n\frac{n\pi}{a}c = \frac{2}{a}\int_0^a g(x)\sin\left(\frac{n\pi x}{a}\right)dx$$

or

$$b_n = \frac{2}{n\pi c}\int_0^a g(x)\sin\left(\frac{n\pi x}{a}\right)dx. \tag{10}$$

If the functions $f(x)$ and $g(x)$ are sectionally smooth on the interval $0 < x < a$, then we know that the initial conditions really are satisfied, except possibly at points of discontinuity of f or g. By the nature of the problem, however, one would expect that f, at least, would be continuous and would satisfy $f(0) = f(a) = 0$. Thus we expect the series for f to converge uniformly.

Let us take specific initial conditions to see an example solution. If the string is lifted in the middle and then released, appropriate initial conditions are

$$u(x,0) = f(x) = \begin{cases} h\cdot\dfrac{2x}{a}, & 0 < x < \dfrac{a}{2} \\ h\left(2-\dfrac{2x}{a}\right), & \dfrac{a}{2} < x < a \end{cases} \tag{11}$$

$$\frac{\partial u}{\partial t}(x,0) = 0, \qquad 0 < x < a.$$

Then $b_n = 0$ for $n = 1, 2, 3, \cdots$, and

$$\begin{aligned} a_n &= \frac{2}{a}\left[\int_0^{a/2} h\cdot\frac{2x}{a}\sin\left(\frac{n\pi x}{a}\right)dx + \int_{a/2}^a h\left(2-\frac{2x}{a}\right)\sin\left(\frac{n\pi x}{a}\right)dx\right] \\ &= \frac{8h}{\pi^2}\frac{\sin(n\pi/2)}{n^2}. \end{aligned}$$

Therefore the complete solution is

$$u(x,t) = \frac{8h}{\pi^2}\sum_{n=1}^{\infty}\frac{\sin(n\pi/2)}{n^2}\sin\left(\frac{n\pi x}{a}\right)\cos\left(\frac{n\pi ct}{a}\right). \tag{12}$$

Although the solution above may be considered valid, it is difficult to see, in the present form, what shape the string will take at various times.

However, because of the simplicity of the sines and cosines, it is possible to rewrite the solution in such a way that $u(x,t)$ may be determined without summing a series.

By applying the trigonometric identity

$$\sin(A)\cos(B) = \frac{1}{2}\left[\sin(A-B) + \sin(A+B)\right]$$

we can express $u(x,t)$ as

$$u(x,t) = \frac{1}{2}\left[\frac{8h}{\pi^2}\sum_{n=1}^{\infty}\frac{\sin(n\pi/2)}{n^2}\sin\left(\frac{n\pi(x-ct)}{a}\right) + \frac{8h}{\pi^2}\sum_{n=1}^{\infty}\frac{\sin(n\pi/2)}{n^2}\sin\left(\frac{n\pi(x+ct)}{a}\right)\right].$$

We know that the series

$$\frac{8h}{\pi^2}\sum_{n=1}^{\infty}\frac{\sin(n\pi/2)}{n^2}\sin\left(\frac{n\pi x}{a}\right)$$

actually converges to the odd periodic extension, with period $2a$, of the function $f(x)$. Let us designate this extension by $\bar{f}_o(x)$ and note that it is defined for all values of its argument. Using this observation, we can express $u(x,t)$ more simply as

$$u(x,t) = \frac{1}{2}\left[\bar{f}_o(x-ct) + \bar{f}_o(x+ct)\right]. \tag{13}$$

In this form, the solution $u(x,t)$ can easily be sketched for various values of t. The graph of $\bar{f}_o(x+ct)$ has the same shape as that of $\bar{f}_o(x)$ but is shifted ct units to the left. Similarly, the graph of $\bar{f}_o(x-ct)$ is the graph of $\bar{f}_o(x)$ shifted ct units to the right. When the graphs of $\bar{f}_o(x+ct)$ and $\bar{f}_o(x-ct)$ are drawn on the same axes, they may be averaged graphically to get the graph of $u(x,t)$.

In Fig. 3 are graphs of $\bar{f}_o(x+ct)$, $\bar{f}_o(x-ct)$, and

$$u(x,t) = \frac{1}{2}\left[\bar{f}_o(x+ct) + \bar{f}_o(x-ct)\right]$$

for the particular example discussed here and for various values of t. The displacement $u(x,t)$ is periodic in time with period $2a/c$. During the second half period (not shown) the string returns to its initial position through the positions shown. The horizontal portions of the string have

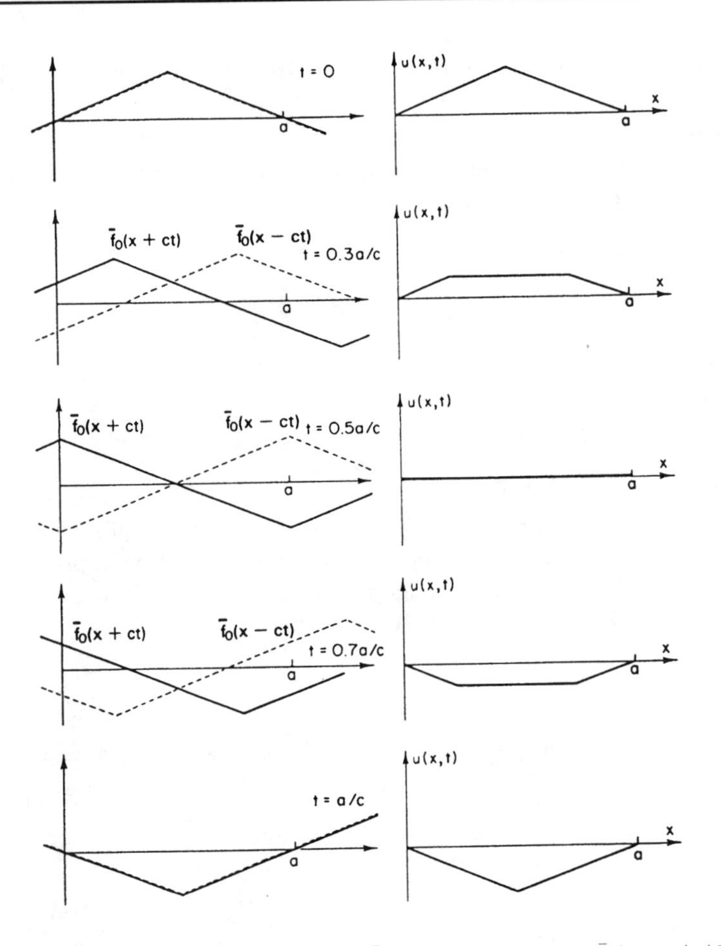

Figure 3: On the left are the graphs of $\bar{f}_o(x+ct)$ (solid) and $\bar{f}_o(x-ct)$ (dashed) for the given value of ct. On the right is the graph of $u(x,t)$ for $0 < x < a$, made by averaging the graphs on the left.

a nonzero velocity. Equation (12) can also be used to find $u(x,t)$ for any given x and t. For instance, if we take $x = 0.2a$ and $t = 0.9a/c$, we find that

$$\begin{aligned} u\left(0.2a, 0.9\frac{a}{c}\right) &= \frac{1}{2}\left[\bar{f}_o(-0.7a) + \bar{f}_o(1.1a)\right] \\ &= \frac{1}{2}\left[(-0.6h) + (-0.2h)\right] \\ &= -0.4h. \end{aligned}$$

The function values can be read directly from a graph of $f(x)$.

Exercises

In Exercises 1-3, solve the vibrating string problem, Eqs. (1)-(4), with the initial conditions given.

1. $f(x) = 0, \quad g(x) = 1, \quad 0 < x < a.$

2. $f(x) = \sin\left(\frac{\pi x}{a}\right), \quad g(x) = 0, \quad 0 < x < a.$

3. $f(x) = \begin{cases} U_0, & 0 < x < a/2, \\ 0, & a/2 < x < a \end{cases}$

$g(x) = 0, \qquad 0 < x < a.$

(This initial condition is difficult to justify for a vibrating string, but may be reasonable where the unknown function is pressure in a pipe with a membrane at the midpoint. See Miscellaneous Exercise 18 of this chapter for some derivations.)

4. If

$$\sum_{n=1}^{\infty} a_n \sin\left(\frac{n\pi x}{a}\right) = \bar{f}_o(x), \quad \sum_{n=1}^{\infty} b_n \cos\left(\frac{n\pi x}{a}\right) = \bar{G}_e(x)$$

show that $u(x,t)$ as given in Eq. (8) may be written

$$u(x,t) = \frac{1}{2}\left(\bar{f}_o(x-ct) + \bar{f}_o(x+ct)\right) + \frac{1}{2}\left(\bar{G}_e(x+ct) - \bar{G}_e(x-ct)\right).$$

Here, $\bar{f}_o(x)$ and $\bar{G}_e(x)$ are periodic with period $2a$.

5. The pressure of the air in an organ pipe satisfies the equation

$$\frac{\partial^2 p}{\partial x^2} = \frac{1}{c^2}\frac{\partial^2 p}{\partial t^2}, \quad 0 < x < a, \quad 0 < t$$

with the boundary conditions (p_0 is atmospheric pressure)

a. $p(0,t) = p_0$, $p(a,t) = p_0$ if the pipe is open, or

b. $p(0,t) = p_0$, $\dfrac{\partial p}{\partial x}(a,t) = 0$ if the pipe is closed at $x = a$.

Find the eigenvalues and eigenfunctions associated with the wave equation for each of these sets of boundary conditions.

6. Find the lowest frequency of vibration of the air in the organ pipes referred to in Exercise 5 **a** and **b**.

7. If a string vibrates in a medium that resists the motion, the problem for the displacement of the string is

$$\frac{\partial^2 u}{\partial x^2} = \frac{1}{c^2}\frac{\partial^2 u}{\partial t^2} + k\frac{\partial u}{\partial t}, \quad 0 < x < a, \quad 0 < t$$

$$u(0,t) = 0, \quad u(a,t) = 0, \quad 0 < t$$

plus initial conditions. Find eigenfunctions, eigenvalues, and product solutions for this problem. (Assume that k is small and positive.)

8. In the text, we assumed that the ratio $\phi''\phi$ had to be a negative constant. Show that, if $\phi''\phi = p^2 > 0$ (or equivalently, if $\phi'' - p^2\phi = 0$), then the only function that also satisfies the boundary conditions, Eq. (7), is $\phi(x) \equiv 0$.

9. The displacements $u(x,t)$ of a uniform thin beam satisfy

$$\frac{\partial^4 u}{\partial x^4} = -\frac{1}{c^2}\frac{\partial^2 u}{\partial t^2}, \quad 0 < x < a, \quad 0 < t.$$

If the beam is simply supported at the ends, the boundary conditions are

$$u(0,t) = 0, \quad \frac{\partial^2 u}{\partial x^2}(0,t) = 0, \quad u(a,t) = 0, \quad \frac{\partial^2 u}{\partial x^2}(a,t) = 0.$$

Find product solutions to this problem. What are the frequencies of vibration?

10. Write out formulas for the first four frequencies of vibration for a thin beam (Exercise 9) and for a string (text). Then find their values, assuming that parameters c and a have values that make the lowest frequency of each equal to 256 cycles per second. The difference in the set of frequencies accounts for some of the difference between the sound of a stringed instrument and that of a xylophone or glockenspiel.

11. My car's antenna vibrates in the wind under various conditions in one of the two shapes shown in Fig. 4. If the antenna is modeled as a uniform thin beam with centerline displacement $u(x,t)$, then u satisfies the equation

$$\frac{\partial^4 u}{\partial x^4} = -\frac{1}{c^2}\frac{\partial^2 u}{\partial t^2} + f(x,t), \quad 0 < x < a, \quad 0 < t,$$

where f is a "forcing function" that represents the effect of wind or other distributed forces. Because the base of the antenna is built into the car, the boundary conditions at the base are zero displacement and slope:

$$u(0,t) = 0, \quad \frac{\partial u}{\partial x}(0,t) = 0, \quad 0 < t.$$

The top of the antenna is free to move. There, the internal moment and shear are both zero, leading to the conditions

$$\frac{\partial^2 u}{\partial x^2}(a,t) = 0, \quad \frac{\partial^3 u}{\partial x^3}(a,t) = 0, \quad 0 < t.$$

(These four boundary conditions are standard for a cantilevered beam.)

It can be shown that the solution of the foregoing problem, together with initial conditions on u and u_t, can be represented as a series of products of the form

$$\left(a_n \cos(\lambda_n^2 ct) + b_n \sin(\lambda_n^2 ct) + F_n(t)\right) \phi_n(x)$$

where $F_n(t)$ comes from the forcing function, and $\phi_n(x)$ and λ_n are related through the eigenvalue problem

$$\phi'''' - \lambda^4 \phi = 0, \quad 0 < x < a$$

$$\phi(0) = 0, \quad \phi'(0) = 0, \quad \phi''(a) = 0, \quad \phi'''(a) = 0.$$

This arises in the obvious way from the boundary conditions and the homogeneous partial differential equation.

Solve the eigenvalue problem, sketch the first two eigenfunctions, and compare them to the Figure.

In Exercises 12-14, find a solution by separation of variables.

12. $\dfrac{\partial^2 u}{\partial x^2} = \dfrac{1}{c^2}\left(\dfrac{\partial^2 u}{\partial t^2} + 2k\dfrac{\partial u}{\partial t}\right), \quad 0 < x < a, \quad 0 < t$

$u(0,t) = 0, \quad u(a,t) = 0, \quad 0 < t$

$u(x,0) = f(x), \quad \dfrac{\partial u}{\partial t}(x,0) = 0, \quad 0 < x < a,$

where $f(x)$ is as in Eq. (11). (Assume that k is small and positive.)

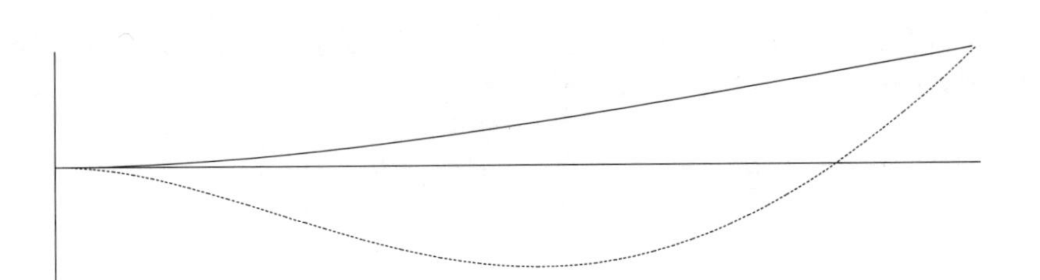

Figure 4: Shapes of car antenna.

13. $\dfrac{\partial^2 u}{\partial x^2} = \dfrac{1}{c^2}\dfrac{\partial^2 u}{\partial t^2} + \gamma^2 u, \quad 0 < x < a, \quad 0 < t$

$u(0,t) = 0, \quad u(a,t) = 0, \quad 0 < t$

$u(x,0) = h, \quad \dfrac{\partial u}{\partial t}(x,0) = 0, \quad 0 < x < a,$

where h and γ^2 are constants.

14. $\dfrac{\partial^2 u}{\partial x^2} = \dfrac{1}{c^2}\dfrac{\partial^2 u}{\partial t^2}, \quad 0 < x < a, \quad 0 < t$

$u(0,t) = 0, \quad \dfrac{\partial u}{\partial x}(a,t) = 0, \quad 0 < t$

$u(x,0) = 0, \quad \dfrac{\partial u}{\partial t}(x,0) = 1, \quad 0 < x < a$

15. Does the series in Eq. (12) converge uniformly?

16. Verify that the product solution

$$u_n(x,t) = \sin(\lambda_n x)\left[a_n \cos(\lambda_n ct) + b_n \sin(\lambda_n ct)\right]$$

satisfies the wave equation (1) and the boundary conditions, Eq. (2).

17. Sketch $u_1(x,t)$ and $u_2(x,t)$ as functions of x for several values of t. Assume a_1 and $a_2 = 1$, b_1 and $b_2 = 0$. (The solutions $u_n(x,t)$ are called *standing waves.*)

3.3 D'ALEMBERT'S SOLUTION

In Section 2 we saw that, in one particular case at least, we could express the solution of the wave equation directly in terms of the initial data. From this evidence we might suspect that there is something special about $x + ct$ and $x - ct$. To test this idea, we change variables and see what the wave equation looks like. Let $w = x + ct$, $z = x - ct$, and $u(x,t) = v(w,z)$. By the chain rule we calculate

$$\begin{aligned}\frac{\partial u}{\partial x} &= \frac{\partial v}{\partial w}\frac{\partial w}{\partial x} + \frac{\partial v}{\partial z}\frac{\partial z}{\partial x} = \frac{\partial v}{\partial w} + \frac{\partial v}{\partial z}\\ \frac{\partial^2 u}{\partial x^2} &= \frac{\partial}{\partial w}\left(\frac{\partial v}{\partial w} + \frac{\partial v}{\partial z}\right)\frac{\partial w}{\partial x} + \frac{\partial}{\partial z}\left(\frac{\partial v}{\partial w} + \frac{\partial v}{\partial z}\right)\frac{\partial z}{\partial x}\\ &= \frac{\partial^2 v}{\partial w^2} + 2\frac{\partial^2 v}{\partial w \partial z} + \frac{\partial^2 v}{\partial z^2}\end{aligned}$$

and similarly

$$\frac{\partial^2 u}{\partial t^2} = c^2\left(\frac{\partial^2 v}{\partial w^2} - 2\frac{\partial^2 v}{\partial z \partial w} + \frac{\partial^2 v}{\partial z^2}\right).$$

(We have assumed that the two mixed partials $\partial^2 v/\partial z \partial w$ and $\partial^2 v/\partial w \partial z$ are equal.) If $u(x,t)$ satisfies the wave equation, then

$$\frac{\partial^2 u}{\partial t^2} = \frac{1}{c^2}\frac{\partial^2 u}{\partial t^2}.$$

In terms of the function v and the new independent variables this equation becomes

$$\frac{\partial^2 v}{\partial w^2} + 2\frac{\partial^2 v}{\partial z \partial w} + \frac{\partial^2 v}{\partial z^2} = \frac{\partial^2 v}{\partial w^2} - 2\frac{\partial^2 v}{\partial z \partial w} + \frac{\partial^2 v}{\partial z^2}$$

or, simply,

$$\frac{\partial^2 v}{\partial z \partial w} = 0.$$

It is actually possible to find the general solution of this last equation. Put in another form it says

$$\frac{\partial}{\partial z}\left(\frac{\partial v}{\partial w}\right) = 0,$$

which means that $\partial v/\partial w$ is independent of z, or

$$\frac{\partial v}{\partial w} = \theta(w).$$

Integrating this equation, we find that

$$v = \int \theta(w) dw + \phi(z).$$

Here, $\phi(z)$ plays the role of an integration "constant." Since the integral of $\theta(w)$ is also a function of w, we may write the *general* solution of the partial differential equation above as

$$v(w, z) = \psi(w) + \phi(z),$$

where ψ and ϕ are *arbitrary* functions with continuous derivatives. Transforming back to our original variables, we obtain

$$u(x, t) = \phi(x + ct) + \phi(x - ct) \tag{1}$$

as a form for the general solution of the one-dimensional wave equation. This is known as *d'Alembert's solution.* It represents the solution as the superposition of two waves, one moving to the left and the other to the right, with propagation speed c.

Now let us look at the vibrating string problem:

$$\frac{\partial^2 u}{\partial x^2} = \frac{1}{c^2}\frac{\partial^2 u}{\partial t^2}, \quad 0 < x < a, \quad 0 < t \tag{2}$$

$$u(0, t) = 0, \qquad u(a, t) = 0, \quad 0 < t \tag{3}$$

$$u(x, 0) = f(x), \qquad 0 < x < a \tag{4}$$

$$\frac{\partial u}{\partial t}(x, 0) = g(x), \qquad 0 < x < a. \tag{5}$$

We already know a form for u. The problem is to choose ψ and ϕ in such a way that the initial and boundary conditions are satisfied. We assume then that

$$u(x, t) = \psi(x + ct) + \phi(x - ct).$$

The initial conditions are

$$\begin{aligned} \psi(x) + \phi(x) &= f(x), \quad 0 < x < a \\ c\psi'(x) - c\phi'(x) &= g(x), \quad 0 < x < a. \end{aligned} \tag{6}$$

If we divide through the second equation by c and integrate, it becomes

$$\psi(x) - \phi(x) = G(x) + A, \quad 0 < x < a \tag{7}$$

where $G(x)$ stands for

$$G(x) = \frac{1}{c}\int_0^x g(y)dy \tag{8}$$

and A is an arbitrary constant. Equations (6) and (7) can now be solved simultaneously to determine

$$\psi(x) = \frac{1}{2}\left(f(x) + G(x) + A\right), \quad 0 < x < a$$

$$\phi(x) = \frac{1}{2}\left(f(x) - G(x) - A\right), \quad 0 < x < a.$$

These equations give ψ and ϕ only for values of the argument between 0 and a. But $x \pm ct$ may take on any value whatsoever, so we must extend these functions to define them for arbitrary values of their arguments. Let us designate them as

$$\psi(x) = \frac{1}{2}\left(\tilde{f}(x) + \tilde{G}(x) + A\right)$$

$$\phi(x) = \frac{1}{2}\left(\tilde{f}(x) - \tilde{G}(x) - A\right)$$

where $\tilde{f}$ and $\tilde{G}$ are some extensions of f and G. (That is $\tilde{f}(x) = f(x)$ and $\tilde{G}(x) = G(x)$ for $0 < x < a$.) However we choose these extensions, the wave equation and the initial conditions are satisfied. Thus, they must be determined by the boundary conditions,

$$u(0,t) = \phi(ct) + \phi(-ct) = 0, \quad t > 0 \tag{9}$$

$$u(a,t) = \phi(a+ct) + \phi(a-ct) = 0, \quad t > 0. \tag{10}$$

The first of these equations says that

$$\tilde{f}(ct) + \tilde{G}(ct) + A + \tilde{f}(-ct) - \tilde{G}(-ct) - A = 0$$

or

$$\tilde{f}(ct) + \tilde{f}(-ct) + \tilde{G}(ct) - \tilde{G}(-ct) = 0.$$

As these equations must be true for arbitrary functions f and G (because the two functions are not interdependent), we must have individually

$$\tilde{f}(ct) = -\tilde{f}(-ct), \quad \tilde{G}(ct) = \tilde{G}(-ct). \tag{11}$$

That is, $\tilde{f}$ is odd and $\tilde{G}$ is even.

At the second endpoint, a similar calculation shows that

$$\tilde{f}(a+ct) + \tilde{f}(a-ct) + \tilde{G}(a+ct) - \tilde{G}(a-ct) = 0.$$

Once again, the independence of f and G implies that

$$\tilde{f}(a+ct) = -\tilde{f}(a-ct), \quad \tilde{G}(a+ct) = \tilde{G}(a-ct). \tag{12}$$

The oddness of $\tilde{f}$ and evenness of $\tilde{G}$ can be used to transform the right-hand sides. Then

$$\tilde{f}(a+ct) = \tilde{f}(-a+ct), \quad \tilde{G}(a+ct) = \tilde{G}(-a+ct).$$

These equations say that $\tilde{f}$ and $\tilde{G}$ are both periodic with period $2a$, because changing their arguments by $2a$ does not change the functional value. Thus we want $\tilde{f}$ to be the odd periodic extension of f and $\tilde{G}$ to be the even periodic extension of G. In the notation we used in Chapter 1, the explicit expressions for ϕ and ψ are

$$\psi(x+ct) = \frac{1}{2}\left(\bar{f}_o(x+ct) + \bar{G}_e(x+ct) + A\right)$$

$$\phi(x-ct) = \frac{1}{2}\left(\bar{f}_o(x-ct) - \bar{G}_e(x-ct) - A\right).$$

Finally, we arrive at an expression for the solution $u(x,t)$:

$$u(x,t) = \frac{1}{2}\left[\bar{f}_o(x+ct) + \bar{f}_o(x-ct)\right] + \frac{1}{2}\left[\bar{G}_e(x+ct) - \bar{G}_e(x-ct)\right]. \tag{13}$$

This form of the solution of Eqs. (2)-(5) allows us to see directly how the initial data influence the solution at later times. From a practical point of view, it also permits us to calculate $u(x,t)$ at any x and t and even to sketch u as a function of one independent variable for a fixed value of the other. The following procedure is helpful in sketching $u(x,t)$ as a function of x for a fixed $t = t^*$, when the initial condition (5) has $g(x) \equiv 0$. It is easily adapted to other cases.

1. Sketch the odd periodic extension of f; call this $\bar{f}_o(x)$.

2. Sketch $\bar{f}_o(x+ct^*)$ against x; this is just the graph of $\bar{f}_o$ shifted ct^* units to the left.

3. Sketch $\bar{f}_o(x-ct^*)$ against x on the same axes. This graph is the same as that of $\bar{f}_o$ but shifted ct^* units to the right.

4. Average graphically the graphs made in the two preceding steps. Check that the boundary conditions are satisfied.

Similarly, if $f(x) \equiv 0$, sketch $G(x)$ and its even periodic extension $\bar{G}_e(x)$. Then sketch the graphs of $\bar{G}_e(x+ct^*)$ (same shape as $\bar{G}_e(x)$ but shifted ct^* units to the left) and $-\bar{G}_e(x-ct^*)$ (graph of $\bar{G}_e(x)$ shifted ct^* units to the right and reflected in the horizontal axis). These two are then averaged graphically to obtain the graph of $u(x,t^*)$. Check that the boundary conditions are satisfied.

Exercises

1. Let $u(x,t)$ be a solution of Eqs. (2)-(5) with $g(x) \equiv 0$ and $f(x)$ a function whose graph is an isosceles triangle of width a and height h. Find $u(x,t)$ for $x = 0.25a$ and $0.5a$, and for $t = 0$, $0.2a/c$, $0.4a/c$, $0.8a/c$, $1.4a/c$.

2. Sketch $u(x,t)$ of Exercise 1 as a function of x for the times given. Compare your results with Fig. 3.

3. Let $u(x,t)$ be a solution of Eqs. (2)-(5) with $f(x) \equiv 0$ and $g(x) = \alpha c$, $0 < x < a$. Find $u(x,t)$ at: $x = 0$, $t = 0.5a/c$; $x = 0.2a$, $t = 0.6a/c$; $x = 0.5a$, $t = 1.2a/c$.

4. Sketch $u(x,t)$ of Exercise 3 as a function of x for times $t = 0$, $0.25a/c$, $0.5a/c$, a/c.

5. Find the function $G(x)$ corresponding to (see Eq. (8))

$$g(x) = \begin{cases} 0, & 0 < x < 0.4a \\ 5c, & 0.4a < x < 0.6a \\ 0, & 0.6a < x < a. \end{cases}$$

6. Justify this alternate description of the function $G(x)$ that is specified in Eq. (8): G is the solution of the initial value problem

$$\frac{dG}{dx} = \frac{1}{c}g(x), \quad 0 < x$$
$$G(0) = 0.$$

7. Using Eq. (8) or Exercise 6, sketch the function $G(x)$ of Exercise 5.

8. Let $u(x,t)$ be the solution of the vibrating string problem, Eqs. (2)-(5) with $f(x) \equiv 0$ and $g(x)$ as in Exercise 5. Sketch $u(x,t)$ as a function of x for times $ct = 0$, $0.2a$, $0.4a$, $0.5a$, a, $1.2a$. Hint: sketch $\bar{G}_e(x+ct)$ and $-\bar{G}_e(x-ct)$; then average them graphically.

9. Sketch the solution of the vibrating string problem, Eqs. (2)-(5) at times $ct = 0,\ 0.1a,\ 0.3a,\ 0.4a,\ 0.5a,\ 0.6a$, if $g(x) = 0$ and

$$f(x) = \begin{cases} 0, & 0 < x < 0.4a \\ 10h(x - 0.4), & 0.4a < x < 0.5a \\ 10h(0.6 - x), & 0.5a < x < 0.6a \\ 0, & 0.6a < x < a. \end{cases}$$

10. The equation for the forced vibrations of a string is

$$\frac{\partial^2 u}{\partial x^2} - \frac{1}{c^2}\frac{\partial^2 u}{\partial t^2} = -\frac{1}{T}F(x,t) \qquad (*)$$

(see Section 1, Exercise 2). Changing variables to

$$w = x + ct, \quad z = x - ct, \quad u(x,t) = v(w,z), \quad f(w,z) = F(x,t),$$

this equation becomes

$$\frac{\partial^2 v}{\partial w \partial z} = -\frac{1}{4T} f(w,z).$$

Show that the general solution of this equation is

$$v(w,z) = -\frac{1}{4T}\int\int f(w,z)\,dw\,dz + \psi(w) + \phi(z).$$

11. Find the general solution of Eq. (*) in Exercise 10 in terms of x and t, if $F(x,t) = T\cos(t)$.

12. Verify directly that $u(x,t)$ as given by Eq. (1) is a solution of the wave equation (2) if ϕ and ψ have at least two derivatives.

3.4 ONE-DIMENSIONAL WAVE EQUATION: GENERALITIES

As for the one-dimensional heat equation, we can make some comments for a generalized one-dimensional wave equation. For the sake of generality, we assume that some nonuniform properties are present. For the sake of simplicity, we assume that the equation is homogeneous, and free of u.

Our initial value-boundary value problem will be

$$\frac{\partial}{\partial x}\left(s(x)\frac{\partial u}{\partial x}\right) = \frac{p(x)}{c^2}\frac{\partial^2 u}{\partial t^2}, \quad l < x < r, \quad 0 < t \tag{1}$$

$$\alpha_1 u(l,t) - \alpha_2 \frac{\partial u}{\partial x}(l,t) = c_1, \quad 0 < t \tag{2}$$

$$\beta_1 u(r,t) + \beta_2 \frac{\partial u}{\partial x}(r,t) = c_2, \quad 0 < t \tag{3}$$

$$u(x,0) = f(x), \quad l < x < r \tag{4}$$

$$\frac{\partial u}{\partial t}(x,0) = g(x), \quad l < x < r. \tag{5}$$

We assume that the functions $s(x)$ and $p(x)$ are positive for $l \le x \le r$, because they represent physical properties; that s, s' and p are all continuous, and that s and p have no dimensions. Also suppose that none of the coefficients α_1, α_2, β_1, β_2 are negative.

To obtain homogeneous boundary conditions we can write

$$u(x,t) = v(x) + w(x,t)$$

just as before. In the wave equation, however, neither of the names "steady-state solution" nor "transient solution" is appropriate; for, as we shall see, there is no steady state, or limiting case, nor is there a part of the solution that tends to zero as t tends to infinity. Nevertheless, v represents an equilibrium solution and, more important, it is a useful mathematical device to consider u in the form provided in the preceding equation.

The function $v(x)$ is required to satisfy the conditions

$$(sv')' = 0, \quad l < x < r$$

$$\alpha_1 v(l) - \alpha_2 v'(l) = c_1$$

$$\beta_1 v(r) + \beta_2 v'(r) = c_2.$$

Thus $v(x)$ is exactly equivalent to the "steady-state solution" discussed for the heat equation.

The function $w(x,t)$, being the difference between $u(x,t)$ and $v(x)$, satisfies the initial value-boundary value problem

$$\frac{\partial}{\partial x}\left(s(x)\frac{\partial w}{\partial x}\right) = \frac{p(x)}{c^2}\frac{\partial^2 w}{\partial t^2}, \quad l < x < r, \quad 0 < t \tag{6}$$

$$\alpha_1 w(l,t) - \alpha_2 \frac{\partial w}{\partial x}(l,t) = 0, \quad 0 < t \tag{7}$$

$$\beta_1 w(r,t) + \beta_2 \frac{\partial w}{\partial x}(r,t) = 0, \qquad 0 < t \tag{8}$$

$$w(x,0) = f(x) - v(x), \quad l < x < r \tag{9}$$

$$\frac{\partial w}{\partial t}(x,0) = g(x), \qquad l < x < r. \tag{10}$$

Since the equation and the boundary conditions are homogeneous and linear, we attempt a solution by separation of variables. If $w(x,t)$ $=\phi(x)T(t)$, we find in the usual way that the factor functions ϕ and T must satisfy

$$T'' + c^2\lambda^2 T = 0, \quad 0 < t \tag{11}$$

$$(s(x)\phi')' + \lambda^2 p(x)\phi = 0, \quad l < x < r \tag{12}$$

$$\alpha_1\phi(l) - \alpha_2\phi'(l) = 0, \tag{13}$$

$$\beta_1\phi(r) + \beta_2\phi'(r) = 0. \tag{14}$$

The eigenvalue problem represented in the last three lines is a regular Sturm-Liouville problem, because of the assumptions we have made about s, p, and the coefficients. We know that there are an infinite number of nonnegative eigenvalues λ_1^2, $\lambda_2^2, \cdots$, and corresponding eigenfunctions ϕ_1, $\phi_2, \cdots$ which have the orthogonality property

$$\int_l^r \phi_n(x)\phi_m(x)p(x)dx = 0, \quad n \neq m.$$

The solution of the equation for T is

$$T_n(t) = a_n \cos(\lambda_n ct) + b_n \sin(\lambda_n ct).$$

Having solved the subsidiary problems that arose after separation of variables, we can begin to assemble the solution. The function w will have the form

$$w(x,t) = \sum_{n=1}^{\infty} \phi_n(x)\left(a_n \cos(\lambda_n ct) + b_n \sin(\lambda_n ct)\right) \tag{15}$$

and its two initial conditions, yet to be satisfied, are

$$w(x,0) = \sum_{n=1}^{\infty} a_n\phi_n(x) = f(x) - v(x), \quad l < x < r$$

$$\frac{\partial w}{\partial t}(x,t) = \sum_{n=1}^{\infty} b_n\lambda_n c\phi_n(x) = g(x), \qquad l < x < r.$$

By employing the orthogonality of the ϕ_n, we determine that the coefficients a_n and b_n are given by

$$a_n = \frac{1}{I_n}\int_l^r [f(x)v(x)]\,\phi_n(x)p(x)dx/I_n$$
$$b_n = \frac{1}{I_n\lambda_n c}\int_l^r g(x)\phi_n(x)p(x)dx$$

where

$$I_n = \int_l^r \phi_n^2(x)p(x)dx.$$

Finally, $u(x,t) = v(x) + w(x,t)$ is the solution of the original problem, and each of its parts is completely specified. From the form of $w(x,t)$, we can make certain observations about u:

1. $u(x,t)$ does not have a limit as $t \to \infty$. Each term of the series form of w is periodic in time and thus does not die away.

2. Except in very special cases, the eigenvalues λ_n^2 are not closely related to each other. So in general, if u causes acoustic vibrations, the result will not be musical to the ear. (A sound would be musical if, for instance, $\lambda_n = n\lambda_1$, as in the case of the uniform string.)

3. In general, $u(x,t)$ is not even periodic in time. Although each term in the series for w is periodic, the terms do not have a *common* period (except in special cases), and so the sum is not periodic.

Exercises

1. Verify the formulas for the a_n and b_n. Under what conditions on f and g can we say that the initial conditions are satisfied?

2. Check the statement that $v(x)$ is the same for the heat conduction problem and for the problem considered here.

3. Identify the period of $T_n(t)$ and the associated frequency.

4. Although $u(x,t)$ has no limit as $t \to \infty$, show that the following generalized limit is valid:

$$v(x) = \lim_{T\to\infty}\frac{1}{T}\int_0^T u(x,t)dt.$$

(Hint: do the integration and limiting term by term.)

5. Formally solve the problem

$$\frac{\partial}{\partial x}\left(s(x)\frac{\partial u}{\partial x}\right) = \frac{p(x)}{c^2}\left(\frac{\partial^2 u}{\partial t^2} + \gamma\frac{\partial u}{\partial t}\right) + q(x)u, \quad l < x < r, \quad 0 < t$$

with the boundary conditions Eqs. (2) and (3) and initial conditions Eqs. (4) and (5), taking γ to be constant.

6. Verify that $w(x,t)$ as given in Eq. (15) satisfies the differential equation and the boundary conditions Eqs. (6)-(8).

7. In reference to the observations at the end of the section, prove the following statement: The product solutions of the problem in Eqs. (6)-(8) all have a common period in time if the eigenvalues of the problem in Eqs. (12)-(14) satisfy the relation

$$\lambda_n = \alpha(n + \beta)$$

where β is a rational number.

8. Find a separation-of-variables solution of the problem

$$\frac{\partial u^2}{\partial x^2} = \frac{1}{c^2}\left(\frac{\partial^2 u}{\partial t^2} + \gamma^2 u\right), \qquad 0 < x < a, \quad 0 < t$$

$$u(0,t) = 0, \quad u(a,t) = 0, \qquad 0 < t$$

$$u(x,t) = f(x), \quad \frac{\partial u}{\partial t}(x,0) = g(x), \quad 0 < x < a.$$

Is this an instance of the problem in this section? Which of the observations at the end of the section are valid for the solution of this problem?

3.5 ESTIMATION OF EIGENVALUES

In many instances, one is interested not in the full solution to the wave equation, but only in the possible frequencies of vibration that may occur. For example, it is of great importance that bridges, airplane wings, and other structures should not vibrate; so it is important to know the frequencies at which a structure can vibrate, in order to avoid them. By inspecting the solution of the generalized wave equation, which we investigate in the preceding section, we can see that the frequencies of vibration are $\lambda_n c/2\pi$, $n = 1, 2, 3, \cdots$. Thus we must find the eigenvalues λ_n^2 in order to identify the frequencies of vibration.

Consider the following Sturm-Liouville problem:

$$(s(x)\phi')' - q(x)\phi + \lambda^2 p(x)\phi = 0, \quad l < x < r \tag{1}$$

$$\phi(l) = 0, \quad \phi(r) = 0 \tag{2}$$

where s, s', q, and p are continuous, and s and p are positive for $l \le x \le r$. (Note that we have a rather general differential equation, but very special boundary conditions.)

If ϕ_1 is the eigenfunction corresponding to the smallest eigenvalue λ_1^2, then ϕ_1 satisfies Eq. (1) for $\lambda = \lambda_1$. Alternatively, we can write

$$-(s\phi_1')' + q\phi_1 = \lambda_1^2 p\phi_1, \qquad l < x < r.$$

Multiplying through this equation by ϕ_1 and integrating from l to r, we obtain

$$\int_l^r -(s\phi_1')'\phi_1 dx + \int_l^r q\phi_1^2 dx = \lambda_1^2 \int_l^r p\phi_1^2 dx.$$

If the first integral is integrated by parts, it becomes

$$-\left. s\phi_1'\phi_1\right|_l^r + \int_l^r s\phi_1'\phi_1' dx.$$

But $\phi_1(l) = \phi_1(r) = 0$, so the first term vanishes and we are left with the equality

$$\int_l^r s[\phi_1']^2 dx + \int_l^r q\phi_1^2 dx = \lambda_1^2 \int_l^r p\phi_1^2 dx.$$

Because $p(x)$ is positive for $l \le x \le r$, the integral on the right is positive and we may define λ_1^2 as

$$\lambda_1^2 = \frac{\displaystyle\int_l^r s[\phi_1']^2 dx + \int_l^r q\phi_1^2 dx}{\displaystyle\int_l^r p\phi_1^2 dx} = \frac{N(\phi_1)}{D(\phi_1)}. \tag{3}$$

It can be shown in the calculus of variations that, if $y(x)$ is any function with two continuous derivatives ($l \le x \le r$) that satisfies $y(l) = y(r) = 0$, then

$$\lambda_1^2 \le \frac{N(y)}{D(y)}. \tag{4}$$

By choosing any convenient function y that satisfies the boundary conditions, we obtain from the ratio $N(y)/D(y)$ an upper bound on λ_1^2. Usually this bound is quite a good estimate. One should keep in mind that the graph of the eigenfunction $\phi_1(x)$ does not cross the x-axis between l and r, so the graph of $y(x)$ should not cross the axis either.

Example.
Estimate the first eigenvalue of

$$\phi'' + \lambda^2\phi = 0, \quad 0 < x < 1$$
$$\phi(0) = \phi(1) = 0.$$

Let us try $y(x) = x(1-x)$, which satisfies the boundary conditions. Then $y'(x) = 1 - 2s$ and

$$N(y) = \int_0^1 [y'(x)]^2 dx = \int_0^1 (1-2x)^2 dx = \frac{1}{3},$$
$$D(y) = \int_0^1 y^2(x) dx = \int_0^1 x^2(1-x)^2 dx = \frac{1}{30}.$$

Therefore $N(y)/D(y) = 10$. We know, of course, that $\phi_1(x) = \sin(\pi x)$, and

$$N(\phi_1) = \int_0^1 \pi^2 \cos^2(\pi x) dx = \frac{\pi^2}{2}$$
$$D(\phi_1) = \int_0^1 \sin^2(\pi x) dx = \frac{1}{2}$$

so $N(\phi_1)/D(\phi_1) = \lambda_1^2 = \pi^2 < 10$, confirming Eq. (4).

Example.
Estimate the first eigenvalue of

$$(x\phi')' + \lambda^2 \frac{1}{x}\phi = 0, \quad 1 < x < 2$$
$$\phi(1) = \phi(2) = 0.$$

The integrals to be calculated are

$$N(y) = \int_1^2 x(y')^2 dx, \quad D(y) = \int_1^2 \frac{1}{x} y^2 dx.$$

The tabulation gives results for several trial functions. It is known that the first eigenvalue and eigenfunction are

$$\lambda_1^2 = \left(\frac{\pi}{\ln 2}\right)^2 = 20.5423$$
$$\phi_1(x) = \sin\left(\frac{\pi \ln x}{\ln 2}\right).$$

The error for the best of the trial functions is about 1.44

$y(x)$	$\sqrt{x}(2-x)(x-1)$	$(2-x)(x-1)$	$\dfrac{(2-x)(x-1)}{x}$
$\dfrac{N(y)}{D(y)}$	23.7500	22.1349	20.8379

This method of estimating the first eigenvalue is called *Rayleigh's method*, and the ratio $N(y)/D(y)$ is called the *Rayleigh quotient.* In some mechanical systems, the Rayleigh quotient may be interpreted as the ratio between potential and kinetic energy. There are many other methods for estimating eigenvalues and for systematically improving the estimates.

Exercises

1. Using Eq. (3), show that if $q \geq 0$, then $\lambda_1^2 \geq 0$ also.

2. Verify the results for at least one of the trial functions used in the second example.

3. Estimate the first eigenvalue of the problem

$$\phi'' + \lambda^2(1+x)\phi = 0, \quad 0 < x < 1$$
$$\phi(0) = \phi(1) = 0.$$

4. Verify that the general solution of the differential equation below is $ax\cos(\lambda/x) + bx\sin(\lambda/x)$, and then solve the eigenvalue problem

$$\phi'' + \frac{\lambda^2}{x^4}\phi = 0, \quad 1 < x < 2$$
$$\phi(1) = 0, \quad \phi(2) = 0.$$

5. Estimate the lowest eigenvalue of the problem in Exercise 4 using $y = (x-1)(2-x)$.

6. Estimate the lowest eigenvalue of the problem

$$\phi'' + \lambda^2 x\phi = 0, \quad 0 < x < 1$$
$$\phi(0) = 0, \quad \phi(1) = 0.$$

Use the trial function $x^b(1-x)$, and minimize the Rayleigh quotient with respect to b.

3.6 WAVE EQUATION IN UNBOUNDED REGIONS

When the wave equation is to be solved for $0 < x < \infty$ or for $-\infty < x < \infty$, we can proceed as we did for the solution of the heat equation in these unbounded regions. That is to say, we separate variables and use a Fourier integral to combine the product solutions.

Consider the problem

$$\frac{\partial^2 u}{\partial x^2} = \frac{1}{c^2}\frac{\partial^2 u}{\partial t^2}, \qquad 0 < t, \quad 0 < x \tag{1}$$

$$u(x,0) = f(x), \qquad 0 < x \tag{2}$$

$$\frac{\partial u}{\partial t}(x,0) = g(x), \qquad 0 < x \tag{3}$$

$$u(0,t) = 0, \qquad 0 < t. \tag{4}$$

We require in addition that the solution $u(x,t)$ be bounded as $x \to \infty$.

On separating variables, we make $u(x,t) = \phi(x)T(t)$ and find that the factors satisfy

$$T'' + \lambda^2 c^2 T = 0, \quad 0 < t$$

$$\phi'' + \lambda^2 \phi = 0, \quad 0 < x$$

$$\phi(0) = 0, \quad |\phi(x)| \text{ bounded.}$$

The solutions are easily found to be

$$\phi(x;\lambda) = \sin(\lambda x), \quad T(t) = A\cos(\lambda ct) + B\sin(\lambda ct)$$

and we combine the products $\phi(x;\lambda)T(t)$ in a Fourier integral

$$u(x,t) = \int_0^\infty (A(\lambda)\cos(\lambda ct) + B(\lambda)\sin(\lambda ct))\sin(\lambda x)d\lambda. \tag{5}$$

The initial conditions become

$$u(x,0) = f(x) = \int_0^\infty A(\lambda)\sin(\lambda x)d\lambda, \qquad 0 < x$$

$$\frac{\partial u}{\partial t}(x,0) = g(x) = \int_0^\infty \lambda c B(\lambda)\sin(\lambda x)d\lambda, \quad 0 < x.$$

Both of these equations are Fourier integrals. Thus the coefficient functions are given by

$$A(\lambda) = \frac{2}{\pi}\int_0^\infty f(x)\sin(\lambda x)dx, \quad B(\lambda) = \frac{2}{\pi\lambda c}\int_0^\infty g(x)\sin(\lambda x)dx.$$

It is sufficient to demand that $\int_0^\infty |f(x)|dx$ and $\int_0^\infty |g(x)|dx$ both be finite in order to guarantee the existence of A and B.

The deficiency of the Fourier integral form of the solution given in Eq. (5) is that the formula gives no idea of what $u(x,t)$ looks like. The d'Alembert solution of the wave equation can come to our aid again here. We know that the solution of Eq. (1) has the form

$$u(x,t) = \psi(x+ct) + \phi(x-ct).$$

The two initial conditions boil down to

$$\psi(x) + \psi(x) = f(x), \qquad 0 < x$$

$$\psi(x) - \psi(x) = G(x) + A, \quad 0 < x.$$

As in the finite case, we have defined

$$G(x) = \frac{1}{c}\int_0^x g(y)dy$$

and A is any constant.

From the two initial conditions we obtain

$$\psi(x) = \frac{1}{2}\left(f(x) + G(x) + A\right), \quad x > 0$$

$$\phi(x) = \frac{1}{2}\left(f(x) - G(x) - A\right), \quad x > 0.$$

Both f and G are known for $x > 0$. Thus

$$\psi(x+ct) = \frac{1}{2}\left(f(x+ct) + G(x+ct) + A\right)$$

is defined for all $x > 0$ and $t \geq 0$. But $\phi(x-ct)$ is not yet defined for $x - ct < 0$. That means that we must extend the functions f and G in such a way as to define ϕ for negative arguments and also satisfy the boundary condition. The sole boundary condition is Eq. (4), which becomes

$$u(0,t) = 0 = \psi(ct) + \phi(-ct).$$

In terms of $\bar{f}$ and $\bar{G}$, extensions of f and G, this is

$$0 = f(ct) + G(ct) + A + \bar{f}(-ct) - \bar{G}(-ct) - A.$$

Since f and G are not dependent on each other, we must have individually

$$f(ct) + \bar{f}(-ct) = 0, \qquad G(ct) - \bar{G}(-ct) = 0.$$

That is, $\bar{f}$ is f_o, the odd extension of f, and $\bar{G}$ is G_e the even extension of G.

Finally, we arrive at a formula for the solution

$$u(x,t) = \frac{1}{2}\left[f_o(x+ct) + G_e(x+ct)\right] + \frac{1}{2}\left[f_o(x-ct) - G_e(x-ct)\right]. \quad (6)$$

Now given the functions $f(x)$ and $g(x)$, it is a simple matter to construct f_o and G_e and thus to graph $u(x,t)$ as a function of either variable or to evaluate it for specific values of x and t. By way of illustration, Fig. 5 shows the solution of Eqs. (1)-(4) as a function of x at various times, for $f(x)$ as shown and $g(x) \equiv 0$.

Another interesting problem that can be treated by the d'Alembert method is one in which the boundary condition is a function of time. For simplicity, we take zero initial conditions. Our problem becomes

$$\frac{\partial^2 u}{\partial x^2} = \frac{1}{c^2}\frac{\partial^2 u}{\partial t^2}, \qquad 0 < t, \quad 0 < x \quad (7)$$

$$u(x,0) = 0, \qquad 0 < x \quad (8)$$

$$\frac{\partial u}{\partial t}(x,0) = 0, \qquad 0 < x \quad (9)$$

$$u(0,t) = h(t), \qquad 0 < t. \quad (10)$$

As u is to be a solution of the wave equation, it must have the form

$$u(x,t) = \psi(x+ct) + \phi(x-ct). \quad (11)$$

The two initial conditions, Eqs. (8) and (9), can be treated exactly as in the first problem. Of course, $G(x) \equiv 0$, and the constant A, being arbitrary, may be taken as 0. The conclusion is that

$$\psi(x) = 0, \quad \phi(x) = 0, \quad 0 < x.$$

Because both x and t are positive in this problem, we see that $\psi(x+ct) = 0$ always, so that Eq. (11) may be simplified to

$$u(x,t) = \phi(x-ct). \quad (12)$$

The boundary condition Eq. (10) will tell us how to evaluate ϕ for negative arguments. The equation is

$$u(0,t) = \phi(-ct) = h(t), \quad 0 < t. \quad (13)$$

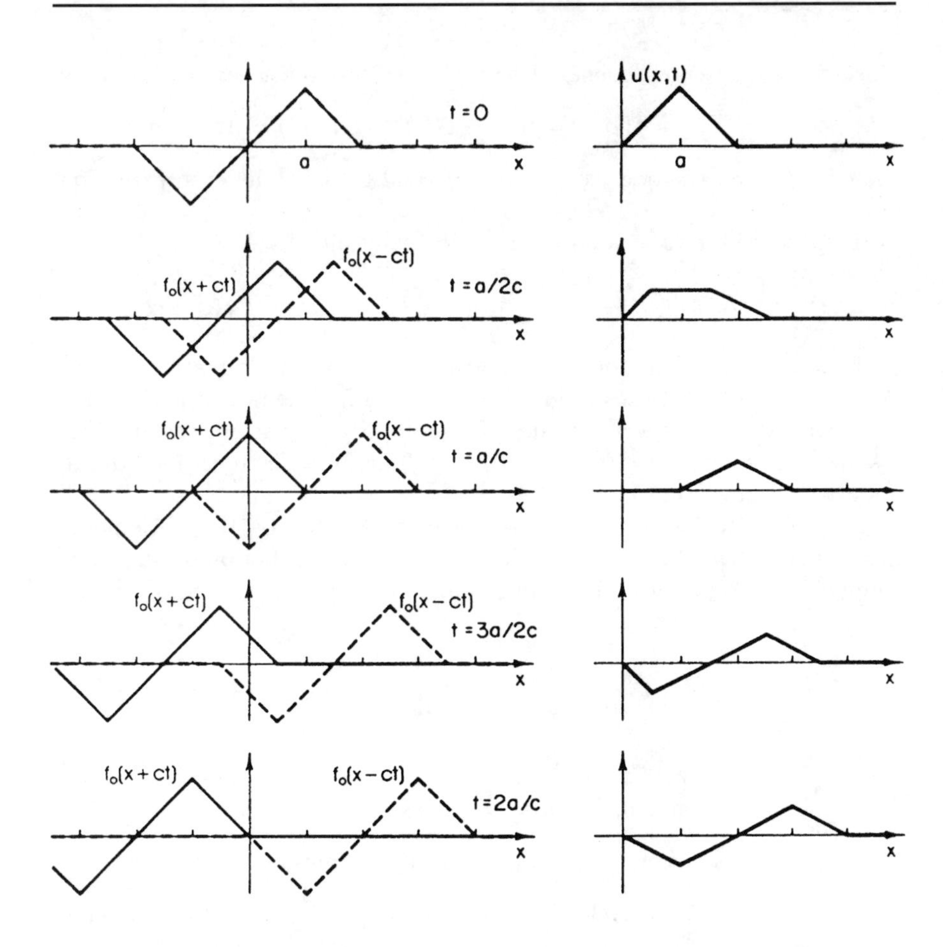

Figure 5: Solution of Eqs. (1)-(4) with $g(x) \equiv 0$. On the left are graphs of $f_o(x+ct)$ (solid) and of $f_o(x-ct)$ (dashed) at the times shown. On the right are the graphs of $u(x,t)$ for $0 < x$, made by averaging the graphs of the left.

We now put together what we know of the function ϕ:

$$\phi(q) = \begin{cases} 0, & q > 0 \\ h\left(-\frac{q}{c}\right), & q < 0. \end{cases} \tag{14}$$

The argument q is a dummy, used to avoid association with either x or t. Equations (12) and (14) now specify the solution $u(x,t)$ completely.

For a specific example, we take $h(t)$ as shown in Fig. 6. There also is a graph of $\phi(q)$. Note that the graph of ϕ for negative argument is that of h, reflected.

The graphs in Fig. 7 show $u(x,t)$ as a function of x for various values of t. It is clear from both the graphs and the formula that the disturbance caused by the variable boundary condition arrives at a fixed point x at time x/c. Thus the disturbance travels with the velocity c, the wave speed.

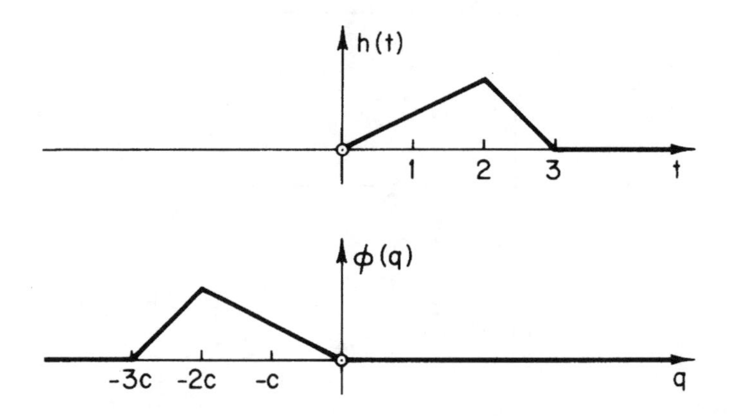

Figure 6: Graphs of $h(t)$ and $\phi(q)$ for semi-infinite string with time-varying boundary condition.

A wave equation accompanied by nonzero initial conditions and time-varying boundary conditions can be solved by breaking it into two problems, one like Eqs. (1)-(4) with zero boundary condition, and the other like Eqs. (7)-(10) with zero initial conditions.

Exercises

1. Derive a formula similar to Eq. (6) for the case in which the boundary condition Eq. (4) is replaced by

$$\frac{\partial u}{\partial x}(0,t) = 0, \quad 0 < t.$$

2. Derive Eq. (6) from Eq. (5) by using trigonometric identities for the product $\sin(\lambda x)\cdot\cos(\lambda ct)$, and so forth, and recognizing certain Fourier integrals.

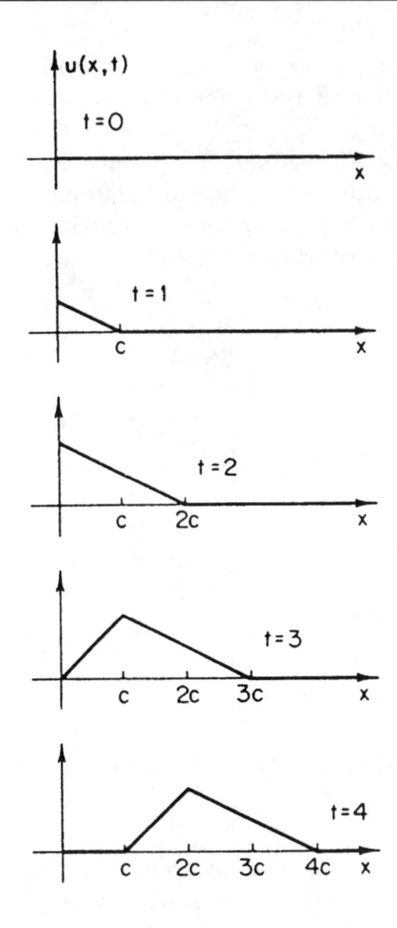

Figure 7: Graphs of $u(x, t)$ versus x for the semi-infinite string under the time-varying boundary condition $u(0, t) = h(t)$, with $h(t)$ as shown in Fig. 6. The shape seen in the last drawing will continue to travel to the right.

3. Sketch the solution of Eqs. (1)-(4) as a function of x at times $t = 0,\ 1/2c,\ 1/c,\ 2/c,\ 3/c$, if $g(x) = 0$ everywhere and $f(x)$ is the rectangular pulse

$$f(x) = \begin{cases} 0, & 0 < x < a \\ 1, & 1 < x < 2 \\ 0, & 2 < x. \end{cases}$$

4. Same as Exercise 3, but $f(x) = 0$ and $g(x)$ is the rectangular pulse

$$g(x) = \begin{cases} 0, & 0 < x < a \\ c, & 1 < x < 2 \\ 0, & 2 < x. \end{cases}$$

5. Sketch the solution of Eqs. (7)-(10) at times $t = 0,\ \pi/2,\ \pi,\ 3\pi/2,\ 2\pi,\ 5\pi/2$, if $h(t) = \sin(t)$. Take $c = 1$.

6. Sketch the solution of Eqs. (7)-(10) at times $t = 0$, $1/2$, $3/2$, and $5/2$, if $c = 1$ and

$$h(t) = \begin{cases} 0, & 0 < t < a \\ 1, & 1 < t < 2 \\ 0, & 2 < t. \end{cases}$$

7. Use the d'Alembert solution of the wave equation to solve the problem

$$\begin{aligned} \frac{\partial^2 u}{\partial x^2} &= \frac{1}{c^2}\frac{\partial^2 u}{\partial t^2}, && -\infty < x < \infty, \quad 0 < t \\ u(x, 0) &= f(x), && -\infty < x < \infty \\ \frac{\partial u}{\partial t}(x, 0) &= g(x), && -\infty < x < \infty. \end{aligned}$$

8. The solution of the problem stated in Exercise 7 is sometimes written

$$u(x,t) = \frac{1}{2}\left(f(x+ct) + f(x-ct)\right) + \frac{1}{2c}\int_{x-ct}^{x+ct} g(z)dz.$$

Show that this is correct. You will need Leibniz's rule (see the Appendix) to differentiate the integral.

3.7 COMMENTS AND REFERENCES

The wave equation is one of the oldest equations of mathematical physics. Euler, Bernoulli and d'Alembert all solved the problem of the vibrating string about 1750, using either separation of variables or what we called d'Alembert's method. This latter is, in fact, a very special case of the method of characteristics, in essence a way of identifying new independent variables having special significance. Street's *Analysis and Solution of Partial Differential Equations* has a chapter on characteristics, including their use in numerical solutions. Wan's *Mathematical Models and their Analysis* gives applications to traffic flow and also discusses other wave phenomena. (See the Bibliography.)

Because many physical phenomena described by the wave equation are part of our everyday experience — the sounds of musical instruments, for instance — featured in popular expositions of mathematical physics. The book of Davis, Hersch and Marchisotto explains standing waves (product solutions) and superposition in an elementary way. Of course, many other phenomena are described by the wave equation. Among the most important for modern life are electrical and magnetic waves that are solutions of special cases of the Maxwell field equations. These and other kinds of waves (including water waves) are studied in Main's *Vibrations and Waves in Physics*; both exposition and figures are first rate.

The potential difference V between the interior and exterior of a nerve axon can be modeled approximately by the Fitzhugh-Nagumo equations,

$$\frac{\partial V}{\partial t} = \frac{\partial^2 V}{\partial x^2} + V - \frac{1}{3}V^3 - R$$

$$\frac{\partial R}{\partial t} = k(V + a - bR).$$

Here R represents a restoring effect and a, b, and k are constants. At first glance, one would expect V to behave like the solution of a heat equation. But a traveling wave solution

$$V(x,t) = F(x - ct), \quad R(x,t) = G(x - ct),$$

of these equations can be found that shows many important features of nerve impulses. This system and many other exciting biological applications of mathematics reported by Murray's excellent book, *Mathematical Biology*.

More information about the Rayleigh quotient and estimation of eigenvalues is in *Boundary and Eigenvalue Problems in Mathematical Physics*, by Sagan. The classic reference for eigenvalues, and indeed for the partial differential equations of mathematical physics in general, is the work by Courant and Hilbert, *Methods of Mathematical Physics*.

MISCELLANEOUS EXERCISES

Exercises 1-5 refer to the problem

$$\frac{\partial^2 u}{\partial x^2} = \frac{1}{c^2}\frac{\partial^2 u}{\partial t^2}, \qquad 0 < x < a, \quad 0 < t$$
$$u(0,t) = 0, \quad u(x,t) = 0, \qquad 0 < t$$
$$u(x,0) = f(x), \quad \frac{\partial u}{\partial t}(x,0) = g(x), \quad 0 < x < a.$$

1. Take $f(x) = 1$, $0 < x < a$ and $g(x) \equiv 0$. (This is rather unrealistic if $u(x,t)$ is the displacement of a vibrating string.) Find a series (separation of variables) solution.

2. Sketch $u(x,t)$ of Exercise 1 as a function of x at various times throughout one period.

3. The solution $u(x,t)$ of Exercise 1 takes on only the three values 1, 0, or -1. Make a sketch of the region $0 < x < a$, $0 < t$ and locate the places where u takes on each of the values.

4. Take $g(x) = 0$ and $f(x)$ to be this function:

$$f(x) = \begin{cases} \dfrac{3hx}{2a}, & 0 < x < \dfrac{2a}{3} \\ \dfrac{3h(a-x)}{a}, & \dfrac{2a}{3} < x < a. \end{cases}$$

The graph of f is triangular, with peak at $x = 2a/3$. Find a series solution for $u(x,t)$.

5. Sketch $u(x,t)$ of Exercise 4 as a function of x at times 0 to a/c in steps of $a/6c$.

6. Find an analytic (integral) solution of this wave problem

$$\frac{\partial^2 u}{\partial x^2} = \frac{1}{c^2}\frac{\partial^2 u}{\partial t^2}, \qquad -\infty < x < \infty, \quad 0 < t$$
$$u(x,0) = f(x), \quad \frac{\partial u}{\partial t}(x,0) = g(x), \quad -\infty < x < \infty,$$

with $g(x) = 0$ and $f(x) = \begin{cases} h, & |x| < \epsilon \\ 0, & |x| > \epsilon. \end{cases}$

7. Sketch the solution of the problem in Exercise 6 at times $t = 0$, ϵ/c, $2\epsilon/c$, $3\epsilon/c$.

8. Same as Exercise 7, but $f(x) = 0$ and

$$g(x) = \begin{cases} c, & |x| < \epsilon \\ 0, & |x| > \epsilon. \end{cases}$$

9. Let $u(x,t)$ be the solution of the problem

$$\begin{aligned} &\frac{\partial^2 u}{\partial x^2} = \frac{1}{c^2}\frac{\partial^2 u}{\partial t^2}, && 0 < x, \quad 0 < t \\ &u(0,t) = 0, && 0 < t \\ &u(x,0) = f(x), \quad \frac{\partial u}{\partial t}(x,0) = g(x), && 0 < x. \end{aligned}$$

Sketch the solution $u(x,t)$ as a function of x at times $t = 0$, $a/6c$, $a/2c$, $5a/6c$, $7a/6c$. Use $g(x) \equiv 0$ and

$$f(x) = \begin{cases} \dfrac{3hx}{2a}, & 0 < x < \dfrac{2a}{3} \\ \dfrac{3h(a-x)}{a}, & \dfrac{2a}{3} < x < a \\ 0, & a < x. \end{cases}$$

10. Same task as in Exercise 9 but

$$f(x) = \begin{cases} \sin(x), & 0 < x < \pi \\ 0, & \pi < x \end{cases}$$

and $g(x) = 0$. Sketch at times $t = 0$, $\pi/4c$, $\pi/2c$, $3\pi/4c$, π/c, $2\pi/c$.

11. Let $u(x,t)$ be the solution of the wave equation on the semi-infinite interval $0 < x < \infty$, with both initial conditions equal to zero, but with the time-varying boundary condition

$$u(0,t) = \begin{cases} \sin\left(\dfrac{ct}{a}\right), & 0 < t < \dfrac{\pi a}{c} \\ 0, & \dfrac{\pi a}{c} < t. \end{cases}$$

Sketch $u(x,t)$ as a function of x at various times.

12. Same as Exercise 11, but the boundary condition is $u(0,t) = h$, for all $t > 0$.

13. Same as Exercise 11, but the boundary condition is

$$u(0,t) = \begin{cases} \dfrac{hct}{a}, & 0 < t < \dfrac{a}{c} \\ \dfrac{h(2a - ct)}{a}, & \dfrac{a}{c} < t < \dfrac{2a}{c} \\ 0, & \dfrac{2a}{c} < t. \end{cases}$$

14. Estimate the lowest eigenvalue of the problem

$$(e^{\alpha x}\phi')' + \lambda^2 e^{\alpha x}\phi = 0, \quad 0 < x < 1$$

$$\phi(0) = 0, \quad \phi(1) = 0.$$

(This problem can be solved exactly.)

15. Estimate the lowest eigenvalue of the problem

$$\phi'' - \frac{3}{4x^2}\phi + \lambda^2\phi = 0, \quad \frac{1}{4} < x < \frac{5}{4},$$

$$\phi\left(\frac{1}{4}\right) = 0, \quad \phi\left(\frac{5}{4}\right) = 0.$$

16. Show that the nonlinear wave equation

$$\frac{\partial u}{\partial t} + u\frac{\partial u}{\partial x} + \frac{\partial^3 u}{\partial x^3} = 0$$

(the Korteweg-deVries equation) has, as one solution,

$$u(x,t) = 12a^2 \mathrm{sech}^2(ax - 4a^3 t).$$

A wave of this form is called a soliton or solitary wave.

17. The solution in Exercise 16 is of the form $u(x,t) = f(x - ct)$. What is the function f, and what is the wave speed c?

18. For $t < 0$, water flows steadily through a long pipe connected at $x = 0$ to a large reservoir and open at $x = a$ to the air. At time

$t = 0$, a valve at $x = a$ is suddenly closed. Reasonable expressions for the conservation of momentum and of mass are

$$\frac{\partial u}{\partial t} = -\frac{\partial p}{\partial x}, \quad 0 < x < a, \quad 0 < t \tag{A}$$

$$\frac{\partial p}{\partial t} = -c^2\frac{\partial u}{\partial x}, \quad 0 < x < a, \quad \text{all } t \tag{B}$$

where p is gauge pressure and u is mass flow rate. If the pipe is rigid, $c^2 = K/\rho$, the ratio of bulk modulus of water to its density. Show that both p and u satisfy the wave equation. The phenomenon described here is called *water hammer*.

19. Introduce a function v with the definition $u = \partial v/\partial x$, $p = -\partial v/\partial t$. Show that (A) becomes an identity and that (B) becomes the wave equation for v.

20. Reasonable boundary and initial conditions for u and p are

$$\begin{aligned} &u(x,0) = U_0(\text{constant}), && 0 < x < a \\ &p(x,0) = -kx, && 0 < x < a \\ &p(0,t) = 0, && \text{all } t \\ &u(a,t) = 0, && t > 0. \end{aligned}$$

Restate these as conditions on v. Show that the first and third equations may be replaced by

$$\begin{aligned} &v(x,0) = U_0 x, && 0 < x < a \\ &v(0,t) = 0, && \text{all } t. \end{aligned}$$

21. Solve the problem in Exercise 20; find a series form for $v(x,t)$.

22. In many problems involving fluid flow, the combination

$$\frac{\partial u}{\partial t} + V\frac{\partial u}{\partial x}$$

(called the *Stokes derivative*) appears. Here V is the speed of the fluid in the x direction. If V equals u or otherwise depends on u, this operator is nonlinear and difficult to work with. Let us assume that V is a constant, so that the operator is linear, and define new variables

$$\xi = x + Vt, \quad \tau = x - Vt, \quad u(x,t) = v(\xi, \tau).$$

Show that

$$\frac{\partial u}{\partial t} + V\frac{\partial u}{\partial x} = 2V\frac{\partial v}{\partial \xi}.$$

23. Assume that $u(x, y, t)$ has the product form shown in what follows. Separate the variables in the given partial differential equation.

$$u(x,y,t) = \psi(x+Vt)\phi(x-Vt)Y(y),$$

$$\frac{\partial^2 u}{\partial y^2} = \frac{1}{k}\left(\frac{\partial u}{\partial t} + V\frac{\partial u}{\partial x}\right).$$

24. A fluid flows between two parallel plates held at temperature 0. At the inlet, fluid temperature is T_0 and initially the fluid is at temperature T_1. If V is the speed of the fluid in the x direction, a problem describing the temperature $u(x, y, t)$ is

$$\frac{\partial^2 u}{\partial y^2} = \frac{1}{k}\left(\frac{\partial u}{\partial t} + V\frac{\partial u}{\partial x}\right), \quad 0 < y < b, \quad 0 < x, \quad 0 < t$$

$$u(x,0,t) = 0, \quad u(x,b,t) = 0, \quad 0 < x, \quad 0 < t$$

$$u(0,y,t) = T_0, \quad 0 < y < b, \quad 0 < t$$

$$u(x,y,0) = T_1, \quad 0 < x, \quad 0 < y < b.$$

Make a separation of variables as in Exercise 23. State and solve the eigenvalue problem for Y. Show that

$$u_n(x,y,t) = \phi_n(x-Vt)\exp(-\lambda_n^2 k(x+Vt)/2V)\sin(\lambda_n y)$$

satisfies partial differential equation and boundary conditions at $y = 0$ and $y = b$, without restriction of ϕ_n (except differentiability).

25. Show how to satisfy the initial and inlet conditions in the problem of Exercise 24, by forming a sum of product solutions and correctly choosing the ϕ_n.

26. In a uniform electric transmission line, extending along the x-axis from $x = 0$ to $x = a$, the current and voltage to ground are represented by $i(x, t)$ and $e(x, t)$. The line is characterized by these parameters, all per unit length: resistance R, inductance L, leakage conductance G, and capacitance C. A section of the line lying between x and $x + \Delta x$ has the equivalent circuit shown in Fig. 8. Kirchhoff's laws applied to this circuit yield the equations

$$i(x,t) = i(x + \Delta x, t) + G\Delta x e(x,t) + C\Delta x \frac{\partial e}{\partial t}(x,t)$$

$$e(x,t) = e(x + \Delta x, t) + R\Delta x i(x,t) + L\Delta x \frac{\partial i}{\partial t}(x,t).$$

By rearranging terms and taking a limit as $\Delta x \to 0$, derive the simultaneous partial differential equations

$$\frac{\partial i}{\partial x} + Ge + C\frac{\partial e}{\partial t} = 0 \qquad \text{(C)}$$

$$\frac{\partial e}{\partial x} + Re + L\frac{\partial i}{\partial t} = 0. \qquad \text{(D)}$$

These are called the *transmission-line* equations or *telegraph* equations.

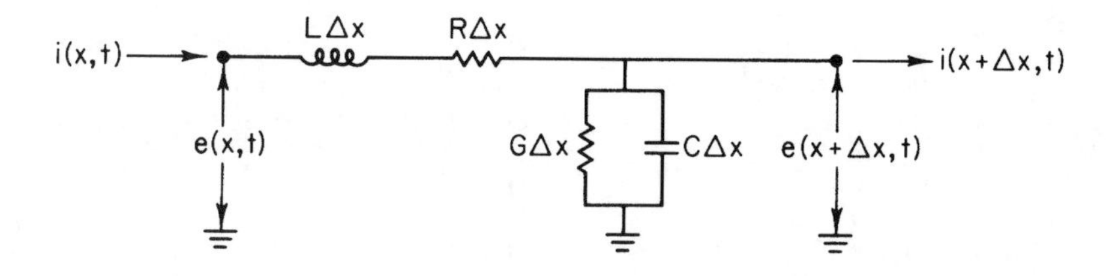

Figure 8: Equivalent circuit for a section of transmission line.

27. In telegraph lines, the leakage conductance and the inductance are negligible. Taking $G = 0$ and $L = 0$, combine equations (C) and (D) so as to get a single second-order partial differential equation for $i(x,t)$. Do the same for $e(x,t)$.

28. In other cases the resistance and leakage conductance are negligible. Taking $G = 0$ and $R = 0$, show that both e and i satisfy the wave equation ("high frequency equations") with

$$c = \sqrt{\frac{1}{LC}}.$$

29. Find solutions (in series form) of the equations (C) and (D) with $G = 0$ and $R = 0$ and with the auxiliary conditions

$$e(x,0) = V, \quad i(x,0) = 0, \quad 0 < x < a$$
$$e(0,t) = 0, \quad e(a,t) = 0, \quad 0 > t.$$

(The boundary conditions are referred to as short-circuit conditions.)

30. Obtain a solution by d'Alembert's method for the problem in Exercise 29.

31. Find series solutions of equations (C) and (D) for the case where $G = 0$ and $L = 0$, with the initial and boundary conditions

$$e(0,t) = V_0, \quad e(a,t) = 0, \quad 0 < t$$
$$e(x,0) = V_0, \quad i(x,0) = 0, \quad 0 < x < a.$$

32. Solve the equations (C) and (D) with $R = 0$ and $G = 0$, if initially both voltage and current are zero, and if at time $t = 0$ a constant voltage V is applied at $x = 0$, while the end at $x = a$ is short-circuited.

33. Solve equations (C) and (D) with $R = 0$ and $G = 0$, on the assumption that the line extends from $x = 0$ to infinity. Take the initial and boundary conditions

$$e(x,0) = Ve^{-\alpha x}, \quad i(x,0) = 0, \quad 0 < x$$
$$e(0,t) = V, \qquad 0 < t.$$

34. Find all functions ϕ such that $u(x,t) = \phi(x - ct)$ is a solution of the heat equation,

$$\frac{\partial^2 u}{\partial x^2} = \frac{1}{k}\frac{\partial u}{\partial t}.$$

35. Take the constant $c = (1+i)\sqrt{\omega k/2}$ in Exercise 34 and show that the functions

$$e^{-px}\sin(\omega t - px), \quad e^{-px}\cos(\omega t - px)$$

can be obtained from $\phi(x - ct)$ and $\phi(x - \bar{c}t)$. (Here $p = \sqrt{\omega/2k}$ and $\bar{c}$ is the complex conjugate of c. Refer to Exercise 6 in Chapter 2, Section 10.)

36. Some nonlinear equations can also result in "traveling wave solutions," $u(x,t) = \phi(x-ct)$. Show that *Fisher's equation*,

$$\frac{\partial^2 u}{\partial x^2} = \frac{\partial u}{\partial t} - u(1-u),$$

has a solution of this form if ϕ satisfies the nonlinear differential equation

$$\phi'' + c\phi' + \phi(1-\phi) = 0.$$

37. Show that the function u is a solution of the problem:

$$\frac{\partial^2 u}{\partial x^2} = \frac{1}{c^2}\frac{\partial^2 u}{\partial t^2}, \qquad 0 < x < a, \quad 0 < t$$

$$u(0,t) = 0, \quad u(a,t) = \sin(\omega t), \quad 0 < t$$

provided that the parameters are such that $\sin(\omega a/c) \neq 0$.

$$u(x,t) = \frac{\sin(\omega x/c)\sin(\omega t)}{\sin(\omega a/c)}.$$

38. If $\omega = \pi c/a$, the denominator of the function in Exercise 37 is 0. Show that, for this value of ω, a function satisfying the wave equation and the given boundary condition is

$$u(x,t) = -\frac{ct}{a}\sin\left(\frac{\pi x}{a}\right)\cos\left(\frac{\pi ct}{a}\right) - \frac{x}{a}\cos\left(\frac{\pi x}{a}\right)\sin\left(\frac{\pi ct}{a}\right).$$

Chapter 4

The Potential Equation

4.1 POTENTIAL EQUATION

The equation for the steady-state temperature distribution in two dimensions (see Chapter 5) is

$$\frac{\partial^2 u}{\partial y^2} + \frac{\partial^2 u}{\partial x^2} = 0.$$

The same equation describes the equilibrium (time-independent) displacements of a two-dimensional membrane, and so is an important common part of both the heat and wave equations in two dimensions. Many other physical phenomena — gravitational and electrostatic potentials, certain fluid flows — and an important class of functions are described by this equation, thus making it one of the most important of mathematics, physics, and engineering. The analogous equation in three dimensions is

$$\frac{\partial^2 u}{\partial x^2} + \frac{\partial^2 u}{\partial y^2} + \frac{\partial^2 u}{\partial z^2} = 0.$$

Either equation may be written $\nabla^2 u = 0$ and is commonly called the *potential equation* or Laplace's equation.

The solutions of the potential equation (called *harmonic functions*) have many interesting properties. An important one, which can be understood intuitively, is the maximum principle: If $\nabla^2 u = 0$ in a region, then u cannot have a relative maximum or minimum inside the region unless u is constant. (Thus, if $\partial u/\partial x$ and $\partial u/\partial y$ are both zero at some point, it is a saddle point.) If u is thought of as the steady-state temperature distribution in a metal plate, it is clear that the temperature cannot be greater at one point than at all other nearby points. For, if such were the case, heat would flow away from the hot point to cooler points nearby, thus reducing the temperature at the hot point. But then the temperature would not be unchanging with time. We return to this matter in Section 4.

A complete boundary value problem consists of the potential equation in a region plus boundary conditions. These may be of any of the three types

$$u \text{ given}, \quad \frac{\partial u}{\partial n} \text{ given}, \quad \text{or } \alpha u + \beta \frac{\partial u}{\partial n} \text{ given}$$

along any section of the boundary. (By $\partial u/\partial n$, we mean the directional derivative in the direction normal or perpendicular to the boundary.) When u is specified along the whole boundary, the problem is called *Dirichlet's problem*; if $\partial u/\partial n$ is specified along the whole boundary, it is *Neumann's problem*. The solutions of Neumann's problem are not unique, for if u is a solution, so is u plus a constant.

It is often useful to consider the potential equation in other coordinate systems. One of the most important is the polar coordinate system, in which the variables are

$$r = \sqrt{x^2 + y^2}, \quad \theta = \tan^{-1}\left(\frac{y}{x}\right),$$

$$x = r\cos(\theta), \quad y = r\sin(\theta).$$

By convention we require $r \geq 0$. We shall define

$$u(x, y) = u(r\cos(\theta), r\sin(\theta)) = v(r, \theta)$$

and find an expression for the Laplacian of u,

$$\frac{\partial^2 u}{\partial x^2} + \frac{\partial^2 u}{\partial y^2},$$

in terms of v and its derivatives by using the chain rule. The calculations are elementary, but tedious. (See Exercise 7.) The results are

$$\frac{\partial^2 u}{\partial x^2} = \cos^2(\theta)\frac{\partial^2 v}{\partial r^2} - \frac{2\sin(\theta)\cos(\theta)}{r}\frac{\partial^2 v}{\partial \theta \partial r} + \frac{\sin^2(\theta)}{r^2}\frac{\partial^2 v}{\partial \theta^2}$$

$$+ \frac{\sin^2(\theta)}{r}\frac{\partial v}{\partial r} + \frac{2\sin(\theta)\sin(\theta)}{r^2}\frac{\partial v}{\partial \theta}$$

$$\frac{\partial^2 u}{\partial y^2} = \sin^2(\theta)\frac{\partial^2 v}{\partial r^2} + \frac{2\sin(\theta)\cos(\theta)}{r}\frac{\partial^2 v}{\partial \theta \partial r} + \frac{\cos^2(\theta)}{r^2}\frac{\partial^2 v}{\partial \theta^2}$$

$$+\frac{\cos^2(\theta)}{r}\frac{\partial v}{\partial r} - \frac{2\sin(\theta)\sin(\theta)}{r^2}\frac{\partial v}{\partial \theta}.$$

From these equations we easily find that the potential equation in polar coordinates is

$$\frac{\partial^2 v}{\partial r^2} + \frac{1}{r}\frac{\partial v}{\partial r} + \frac{1}{r^2}\frac{\partial^2 v}{\partial \theta^2} = \frac{1}{r}\frac{\partial}{\partial r}\left(r\frac{\partial v}{\partial r}\right) + \frac{1}{r^2}\frac{\partial^2 v}{\partial \theta^2} = 0.$$

In cylindrical (r, θ, z) coordinates, the potential equation is

$$\frac{1}{r}\frac{\partial}{\partial r}\left(r\frac{\partial v}{\partial r}\right) + \frac{1}{r^2}\frac{\partial^2 v}{\partial \theta^2} + \frac{\partial^2 v}{\partial z^2} = 0.$$

Exercises

1. Find a relation among the coefficients of the polynomial

$$p(x, y) = a + bx + cy + dx^2 + exy + fy^2$$

that makes it satisfy the potential equation. Choose a specific polynomial that satisfies the equation and show that, if $\partial p/\partial x$ and $\partial p/\partial y$ are both zero at some point, the surface there is saddle shaped.

2. Show that $u(x, y) = x^2 - y^2$ and $u(x, y) = xy$ are solutions of Laplace's equation. Sketch the surfaces $z = u(x, y)$. What boundary conditions do these functions fulfill on the lines $x = 0$, $x = a$, $y = 0$, $y = b$?

3. If a solution of the potential equation in the square $0 < x < 1$, $0 < y < 1$ has the form $u(x, y) = Y(y)\sin(\pi x)$, of what form is the function Y? Find a function Y that makes $u(x, y)$ satisfy the boundary conditions $u(x, 0) = 0$, $u(x, 1) = \sin(\pi x)$.

4. Find a function $u(x)$, independent of y, that satisfies the potential equation.

5. What functions $v(r)$, independent of θ, satisfy the potential equation in polar coordinates?

6. Show that $r^n\sin(n\theta)$ and $r^n\cos(n\theta)$ both satisfy the potential equation in polar coordinates $(n = 0, 1, 2, \cdots)$.

7. Find expressions for the partial derivatives of u with respect to x and y in terms of derivatives of v with respect to r and θ.

8. If u and v are the x- and y-components of the velocity in a fluid, it can be shown (under certain assumptions) that these functions satisfy the equations

$$\frac{\partial u}{\partial x} + \frac{\partial v}{\partial y} = 0 \tag{A}$$

$$\frac{\partial u}{\partial x} - \frac{\partial v}{\partial x} = 0 \tag{B}$$

Show that the definition of a velocity potential function ϕ by the equations

$$u = \frac{\partial \phi}{\partial x}, \quad v = \frac{\partial \phi}{\partial y}$$

causes (B) to be identically satisfied and turns (A) into the potential equation. (See the Comments and References at the end of this chapter.)

4.2 POTENTIAL IN A RECTANGLE

One of the simplest and most important problems in mathematical physics is Dirichlet's problem in a rectangle. To take an easy case, we consider a problem with just two nonzero boundary conditions:

$$\frac{\partial^2 u}{\partial x^2} + \frac{\partial^2 u}{\partial y^2} = 0, \qquad 0 < x < a, \quad 0 < y < b \tag{1}$$

$$u(x, 0) = f_1(x), \quad 0 < x < a \tag{2}$$

$$u(x, b) = f_2(x), \quad 0 < x < a \tag{3}$$

$$u(0, y) = 0, \qquad 0 < y < b \tag{4}$$

$$u(a, y) = 0, \qquad 0 < y < b. \tag{5}$$

If we assume that $u(x, y)$ has a product form $u = X(x)Y(y)$, then Eq. (1) becomes

$$X''(x)Y(y) + X(x)Y''(y) = 0.$$

This equation can be separated by dividing through by XY to yield:

$$\frac{X''(x)}{X(x)} = -\frac{Y''(y)}{Y(y)}. \tag{6}$$

The nonhomogeneous conditions Eqs. (2) and (3) will not, in general, become conditions on X or Y, but the homogeneous conditions Eqs. (4) and (5), as usual, require that

$$X(0) = 0, \quad X(a) = 0. \tag{7}$$

Now, both sides of Eq. (6) must be constant, but the sign of the constant is not obvious. If we try a positive constant (say μ^2), Eq. (6) represents two ordinary equations

$$X'' - \mu^2 X = 0, \quad Y'' + \mu^2 Y = 0.$$

The solutions of these equations are

$$X(x) = A\cosh(\mu x) + B\sinh(\mu x), \quad Y(y) = C\cos(\mu y) + D\sin(\mu y).$$

In order to make X satisfy the boundary conditions Eq. (7), both A and B must be zero, leading to a solution $u(x, y) \equiv 0$. Thus we try the other possibility for sign, taking both members in Eq. (6) to equal $-\lambda^2$.

Under the new assumption, Eq. (6) separates into

$$X'' + \lambda^2 X = 0, \quad Y'' - \lambda^2 Y = 0. \tag{8}$$

The first of these equations, along with the boundary conditions, is recognizable as an eigenvalue problem, whose solutions are

$$X_n(x) = \sin(\lambda_n x), \quad \lambda_n^2 = \left(\frac{n\pi}{a}\right)^2.$$

The functions Y that accompany the X's are

$$Y_n(y) = a_n \cosh(\lambda_n y) + b_n \sinh(\lambda_n y).$$

The a's and b's are for the moment unknown.

We see that $X_n(x)Y_n(y)$ is a solution of the potential Eq. (1), which satisfies the homogeneous conditions Eqs. (4) and (5). A sum of these functions should satisfy the same conditions and equation, so u may have the form

$$u(x, y) = \sum_{n=1}^{\infty} (a_n \cosh(\lambda_n y) + b_n \sinh(\lambda_n y)) \sin(\lambda_n x). \tag{9}$$

The nonhomogeneous boundary conditions Eqs. (2) and (3) are yet to be satisfied. If u is to be of the form above, the boundary condition Eq. (2) becomes

$$u(x, 0) = \sum_{n=1}^{\infty} a_n \sin\left(\frac{n\pi x}{a}\right) = f_1(x), \quad 0 < x < a. \tag{10}$$

We recognize a problem in Fourier series immediately. The a_n must be the Fourier sine coefficients of $f_1(x)$

$$a_n = \frac{2}{a}\int_0^a f_1(x)\sin\left(\frac{n\pi x}{a}\right)dx.$$

The second boundary condition reads

$$\begin{aligned} u(x,b) &= \sum_{n=1}^{\infty}\left(a_n\cosh(\lambda_n b) + b_n\sinh(\lambda_n b)\right)\sin\left(\frac{n\pi x}{a}\right) \\ &= f_2(x), \quad 0 < x < a. \end{aligned}$$

This also is a problem in Fourier series, but not as neat. The constant

$$a_n\cosh(\lambda_n b) + b_n\sinh(\lambda_n b)$$

must be the nth Fourier sine coefficient of f_2. Since a_n is known, b_n can be determined from the following computations:

$$\begin{aligned} a_n\cosh(\lambda_n b) + b_n\sinh(\lambda_n b) &= \frac{2}{a}\int_0^a f_2(x)\sin(\lambda_n x)dx = c_n \\ b_n &= \frac{c_n - a_n\cosh(\lambda_n b)}{\sinh(\lambda_n b)}. \end{aligned}$$

If we use this last expression for b_n and substitute into Eq. (9), we find the solution

$$u(x,y) = \\ \sum_{n=1}^{\infty}\left\{c_n\frac{\sinh(\lambda_n y)}{\sinh(\lambda_n b)} + a_n\left[\cosh(\lambda_n y) - \frac{\cosh(\lambda_n b)}{\sinh(\lambda_n b)}\sinh(\lambda_n y)\right]\right\}\sin(\lambda_n x). \tag{11}$$

Notice that the function multiplying c_n is 0 at $y = 0$ and is 1 at $y = b$. Similarly, the function multiplying a_n is 1 at $y = 0$ and 0 at $y = b$. An easier way to write this latter function is

$$\frac{\sinh(\lambda_n(b-y))}{\sinh(\lambda_n b)},$$

as can readily be found from hyperbolic identities.

Let us take a specific example in order to see more clearly how the solution looks. Suppose f_1 and f_2 are both given by

$$f_1(x) = f_2(x) = \begin{cases} \dfrac{2x}{a}, & 0 < x < \dfrac{a}{2} \\ 2\left(\dfrac{a-x}{a}\right), & \dfrac{a}{2} < x < a. \end{cases}$$

Then

$$c_n = a_n = \frac{8}{\pi^2}\frac{\sin(n\pi/2)}{n^2}.$$

The solution of the potential equation for these boundary conditions is

$$u(x,y) = \frac{8}{\pi^2}\sum_{n=1}^{\infty}\frac{\sin(\frac{n\pi}{2})}{n^2}\frac{\sinh(\frac{n\pi}{a}y)+\sinh(\frac{n\pi}{a}(b-y))}{\sinh(\frac{n\pi b}{a})}\sin\left(\frac{n\pi x}{a}\right). \tag{12}$$

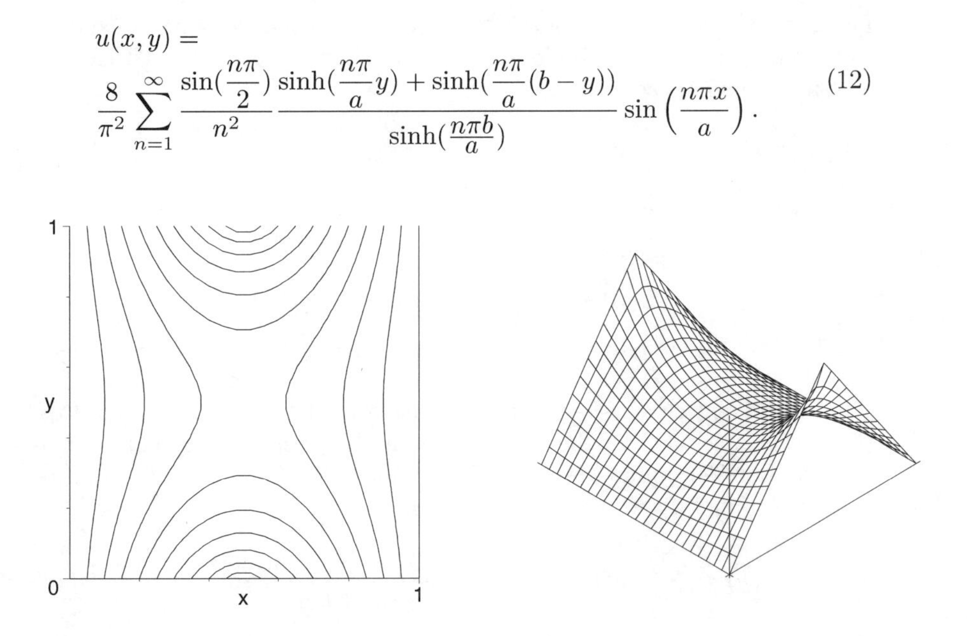

Figure 1: (a) Level curves of the solution $u(x,y)$ of the example problem (see Eq. (12)) for the case $b = a$. Each curve is part of the locus of points that satisfy $u(x,y)$ = constant for constants 0 to 0.9 in steps of 0.1. For some constants, the locus consists of more than one connected curve. (b) Perspective view of the surface $z = u(x,y)$.

In Fig. 1 is a graph of some level curves, $u(x,y)$ = constant, for the case where $a = b$, and also a view of the surface $z = u(x,y)$.

Now we have seen a solution of Dirichlet's problem in a rectangle with homogeneous conditions on two parallel sides. In general, of course, the boundary conditions will be nonhomogeneous on all four sides of the rectangle. But this more general problem can be broken down into two problems like the one we have solved.

Consider the problem

$$\nabla^2 u = 0, \qquad 0 < x < a, \quad 0 < y < b \tag{13}$$

$$u(x,0) = f_1(x), \qquad 0 < x < a \tag{14}$$

$$u(x,b) = f_2(x), \qquad 0 < x < a \tag{15}$$

$$u(0,y) = g_1(y), \qquad 0 < y < b \tag{16}$$

$$u(a,y) = g_2(y), \qquad 0 < y < b. \tag{17}$$

Let $u(x,y) = u_1(x,y) + u_2(x,y)$. We will put conditions on u_1 and u_2 so that they can readily be found, and from them u can be put together. The most obvious conditions are the following:

$$\begin{aligned} \nabla^2 u_1 &= 0, & \nabla^2 u_2 &= 0 \\ u_1(x,0) &= f_1(x) & u_2(x,0) &= 0 \\ u_1(x,b) &= f_2(x) & u_2(x,b) &= 0 \\ u_1(0,y) &= 0 & u_2(0,y) &= g_1(y) \\ u_1(a,y) &= 0 & u_2(a,y) &= g_2(y). \end{aligned}$$

It is evident that $u_1 + u_2$ is the solution of the original problem Eqs. (13)-(17). Also, each of the functions u_1 and u_2 has homogeneous conditions on parallel boundaries. We already have determined the form of u_1. The other function would be of the form

$$u_2(x,y) = \sum_{n=1}^{\infty} \sin(\mu_n y) \frac{A_n \sinh(\mu_n x) + B_n \sinh(\mu_n (a-x))}{\sinh(\mu_n a)} \tag{18}$$

where $\mu_n = n\pi/b$, and

$$A_n = \frac{2}{b} \int_0^b g_2(y) \sin(\mu_n y) dy$$

$$B_n = \frac{2}{b} \int_0^b g_1(y) \sin(\mu_n y) dy.$$

In the individual problems for u_1 and u_2, the technique of separation of variable works, because the homogeneous conditions on parallel sides of the rectangle can be translated into conditions on one of the factor functions.

When the boundary conditions are not complicated functions, it may be possible to satisfy some of them with a polynomial function. (See Exercises 1 and 2 of Section 4.1.) Then the difference between u and the polynomial is a solution of the potential equation that satisfies some homogeneous boundary conditions.

Exercises

1. Show that $\sinh(\lambda y)$ and $\sinh(\lambda(b-y))$ are independent solutions of $Y'' - \lambda^2 Y = 0$ with $\lambda \neq 0$. Thus a combination of these two functions may replace a combination of sinh and cosh as the general solution of this differential equation.

2. Show that the solution of the example problem may be written

$$u(x,y) = \frac{8}{\pi^2}\sum_{n=1}^{\infty} \frac{\sin(\frac{n\pi}{2})}{n^2} \frac{\cosh(\frac{n\pi}{a}(y - \frac{1}{2}b))}{\cosh(\frac{n\pi b}{2a})} \sin\left(\frac{n\pi x}{a}\right).$$

3. Use the form in Exercise 2 to compute u in the center of the rectangle in the three cases: $b = a$, $b = 2a$, $b = a/2$. (Hint: Check the magnitude of the terms.)

4. Verify that each term of Eq. (9) satisfies Eqs. (1), (4), and (5).

5. Solve the problem:

$$\begin{aligned}
&\nabla^2 u = 0, && 0 < x < a, \quad 0 < y < b\\
&u(0,y) = 0, \quad u(a,y) = 0, && 0 < y < b\\
&u(x,0) = 0, \quad u(x,b) = f(x), && 0 < x < a
\end{aligned}$$

where f is the same as in the example. Sketch some level curves of $u(x,y)$.

6. Solve the potential problem on the rectangle $0 < x < a$, $0 < y < b$, subject to the boundary conditions: $u(a,y) = 1$, $0 < y < b$, and $u = 0$ on the rest of the boundary.

7. Solve the problem of the potential equation in the rectangle $0 < x < a$, $0 < y < b$, for each of the following sets of boundary conditions:

a. $\dfrac{\partial u}{\partial x}(0, y) = 0$; $u = 1$ on the remainder of the boundary;

b. $\dfrac{\partial u}{\partial x}(0, y) = 0, \quad \dfrac{\partial u}{\partial x}(a, y) = 0; \quad u(x, 0) = 0, \quad u(x, b) = 1$;

c. $\dfrac{\partial u}{\partial y}(x, 0) = 0, \quad u(x, b) = 0; \quad u(0, y) = 1, \quad u(a, y) = 0$.

8. Solve the problems for u_2. (That is, derive Eq. (17).)

4.3 POTENTIAL IN A SLOT

The potential equation, as well as the heat and wave equations, can be solved in unbounded regions. Consider the following problem in which the region involved is half a vertical strip, or a slot:

$$\frac{\partial^2 u}{\partial x^2} + \frac{\partial^2 u}{\partial y^2} = 0, \qquad 0 < x < a, \qquad 0 < y \tag{1}$$

$$u(x, 0) = f(x), \quad 0 < x < a \tag{2}$$

$$u(0, y) = g_1(y), \quad 0 < y \tag{3}$$

$$u(a, y) = g_2(y), \quad 0 < y. \tag{4}$$

As usual, we required that $u(x, y)$ remain bounded as $y \to \infty$.

In order to make the separation of variables work, we must break this up into two problems. Following the model of the preceding section, we set $u(x, y) = u_1(x, y) + u_2(x, y)$ and require that the parts satisfy the two solvable problems:

$$\begin{array}{lll}
\nabla^2 u_1 = 0, & \nabla^2 u_2 = 0, & 0 < x < a, \quad 0 < y \\
u_1(x, 0) = f(x), & u_2(x, 0) = 0, & 0 < x < a \\
u_1(0, y) = 0, & u_2(0, y) = g_1(y), & 0 < y \\
u_1(a, y) = 0, & u_2(a, y) = g_2(y), & 0 < y.
\end{array}$$

We attack the problem for u_1 by assuming the product form and separating variables:

$$u_1(x, y) = X(x)Y(y), \qquad \frac{X''(x)}{X(x)} = -\frac{Y''(y)}{Y(y)} = -\lambda^2.$$

The sign of the constant $-\lambda^2$ is determined by the boundary conditions at $x = 0$ and $x = a$, which become homogeneous conditions on the factor $X(x)$:

$$X(0) = 0, \quad X(a) = 0. \tag{5}$$

(We also can see that the condition to be satisfied along $y = 0$ demands functions of x that permit a representation of an arbitrary function.)

The boundary conditions, Eq. (5), together with the differential equation

$$X'' + \lambda^2 X = 0 \tag{6}$$

that comes from the separation of variables, constitute a familiar eigenvalue problem whose solution is

$$X_n(x) = \sin\left(\frac{n\pi x}{a}\right), \quad \lambda_n^2 = \left(\frac{n\pi}{a}\right)^2, \quad n = 1, 2, 3, \cdots.$$

The equation for Y is

$$Y'' - \lambda^2 Y = 0, \quad 0 < y.$$

In addition to satisfying this differential equation, Y must remain bounded as $y \to \infty$. The solutions of the equation are $e^{\lambda y}$ and $e^{-\lambda y}$. Of these, the first is unbounded, so

$$Y_n(y) = \exp(-\lambda_n y).$$

Finally, we can write the solution of the first problem as

$$u_1(x, y) = \sum_{n=1}^{\infty} a_n \sin\left(\frac{n\pi x}{a}\right) \exp\left(\frac{-n\pi y}{a}\right). \tag{7}$$

The constants a_n are to be determined from the condition at $y = 0$.

The solution of the second problem is somewhat different. Again we seek solutions in the product form $u_2(x, y) = X(x)Y(y)$. The homogeneous boundary condition at $y = 0$ and the boundedness condition become conditions on $Y(y)$:

$$Y(0) = 0, \quad Y(y) \text{ bounded as } y \to \infty.$$

Then the potential equation becomes

$$\frac{X''(x)}{X(x)} + \frac{Y''(y)}{Y(y)} = 0, \tag{8}$$

and both ratios must be constant. If Y''/Y is positive, the auxiliary conditions force Y to be identically 0. Thus, we take $Y''/Y = -\mu^2$ or $Y'' + \mu^2 Y = 0$ and find that the solution that satisfies the auxiliary conditions is

$$Y(y) = \sin(\mu y),$$

for any $\mu > 0$. Then, the general solution of the equation $X''/X = \mu^2$ is

$$X(x) = A\frac{\sinh(\mu x)}{\sinh(\mu a)} + B\frac{\sinh(\mu(a-x))}{\sinh(\mu a)}.$$

We have chosen this special form on the basis of our experience in solving the potential equation in the rectangle.

Since μ is a continuous parameter, we combine our product solutions by means of an integral, finding

$$u_2(x,y) = \int_0^\infty \left[A(\mu)\frac{\sinh(\mu x)}{\sinh(\mu a)} + B(\mu)\frac{\sinh(\mu(a-x))}{\sinh(\mu a)}\right]\sin(\mu y)d\mu. \quad (9)$$

The nonhomogeneous boundary conditions at $x = 0$ and $x = a$ are satisfied if

$$u_2(0,y) = \int_0^\infty B(\mu)\sin(\mu y)d\mu = g_1(y), \quad 0 < y$$

$$u_2(a,y) = \int_0^\infty A(\mu)\sin(\mu y)d\mu = g_2(y), \quad 0 < y.$$

Obviously these two equations are Fourier integral problems, so we know how to determine the coefficients $A(\mu)$ and $B(\mu)$.

The potential equation can also be solved in a strip ($0 < x < a$, $-\infty < y < \infty$), a quarter plane ($0 < x$, $0 < y$), or a half plane ($0 < x$, $-\infty < y < \infty$). Along each boundary line, a boundary condition is imposed, and the solution is required to remain bounded in remote portions of the region considered. In general, a Fourier integral is employed in the solution, because the separation constant is a continuous parameter, as in the second problem here.

Exercises

1. Find a formula for the constants a_n in Eq. (7).

2. Verify that $u_1(x,y)$ in the form given in Eq. (7) satisfies the potential equation and the homogeneous boundary conditions.

3. Find formulas for $A(\mu)$ and $B(\mu)$ of Eq. (9).

4. Solve the potential equation in the slot, $0 < x < a$, $0 < y$, for each of these sets of boundary conditions.

a. $u(0,y) = 0$, $u(a,y) = 0$, $0 < y$; $u(x,0) = 1$, $0 < x < a$

b. $u(0,y) = 0$, $u(a,y) = e^{-y}$, $0 < y$; $u(x,0) = 0$, $0 < x < a$

c. $u(0,y) = f(y) = \begin{cases} 1, & 0 < y < b \\ 0, & b < y \end{cases}$, $u(a,y) = 0, \quad 0 < y;$

$u(x,0) = 0, \quad 0 < x < a.$

5. Solve the potential equation in the slot, $0 < x < a$, $0 < y$, for each of these sets of boundary conditions.

a. $\dfrac{\partial u}{\partial x}(0,y) = 0$, $u(a,y) = 0$, $0 < y$; $u(x,0) = 1$, $0 < x < a$

b. $\dfrac{\partial u}{\partial x}(0,y) = 0$, $u(a,y) = e^{-y}$, $0 < y$; $u(x,0) = 0$, $0 < x < a$

c. $u(0,y) = 0, \quad u(a,y) = f(y) = \begin{cases} 1, & 0 < y < b \\ 0, & b < y \end{cases}$

$\dfrac{\partial u}{\partial y}(x,0) = 0, \quad 0 < x < a.$

6. Show that, if the separation constant had been chosen as $-\mu^2$ instead of μ^2 in solving for u_2 (leading to $Y'' - \mu^2 Y = 0$), then $Y(y) \equiv 0$ is the only function that satisfies the differential equation, satisfies the condition $Y(0) = 0$, and remains bounded as $y \to \infty$.

7. Solve the problem of potential in a slot under the boundary conditions

$$u(x,0) = 1, \quad u(0,y) = u(a,y) = e^{-y}.$$

8. Show that the function $v(x,y)$ given here satisfies the potential equation and the boundary conditions on the "long" sides in Exercise 7, provided that $\cos(a/2) \neq 0$.

$$v(x,y) = \frac{\cos(x - \frac{1}{2}a)}{\cos(\frac{1}{2}a)} e^{-y}.$$

What partial differential equation and boundary conditions are satisfied by $w(x,y) = u(x,y) - v(x,y)$, if u is the function of Exercise 7?

9. Solve the potential equation in the slot $0 < y < b$, $0 < x$ for each of the following sets of boundary conditions:

a. $u(0,y) = 0$, $u(x,0) = 0$, $u(x,b) = f(x) = \begin{cases} 1, & 0 < x < a \\ 0, & a < x \end{cases}$

b. $u(0,y) = 0$, $u(x,0) = e^{-x}$, $u(x,b) = 0$.

10. Find product solutions of the potential problem in a strip:

$$\frac{\partial^2 u}{\partial x^2} + \frac{\partial^2 u}{\partial y^2} = 0, \quad 0 < x < a, \quad -\infty < y < \infty$$

subject to the boundedness condition: $u(x,y)$ bounded as $y \to \pm\infty$.

11. Solve the potential problem consisting of the equation and boundedness conditions from Exercise 10 and the boundary conditions

$$u(0,y) = 0, \quad u(a,y) = e^{-|y|}, \quad -\infty < y < \infty.$$

12. Show how to solve the potential problem of Exercise 10 together with the boundary conditions

$$u(0,y) = g_1(y), \quad u(a,y) = g_2(y), \quad -\infty < y < \infty,$$

where g_1 and g_2 are suitable functions.

13. Find product solutions of this potential problem in the quarter plane:

$$\frac{\partial^2 u}{\partial x^2} + \frac{\partial^2 u}{\partial y^2} = 0, \quad 0 < x, \quad 0 < y$$

$$u(0,y) = 0, \quad u(x,0) = f(x).$$

Note that $u(x,y)$ must remain bounded as $x \to \infty$ and as $y \to \infty$.

14. Solve the potential equation in the quarter plane, $x > 0$, $y > 0$, subject to the boundary conditions

$$u(0,y) = e^{-y}, \quad y > 0; \quad u(x,0) = e^{-x}, \quad x > 0.$$

15. Find product solutions of the potential equation in the half-plane $y > 0$:

$$\frac{\partial^2 u}{\partial x^2} + \frac{\partial^2 u}{\partial y^2} = 0, \qquad -\infty < x < \infty, \quad 0 < y < \infty$$

$$u(x,0) = f(x), \quad -\infty < x < \infty.$$

What boundedness conditions must $u(x,y)$ satisfy?

16. Convert your solution of Exercise 15 into the following formula. (See Exercise 8 of Section 4 and Section 11 of Chapter 2.)

$$u(x,y) = \frac{1}{\pi}\int_{-\infty}^{\infty} f(x')\frac{y}{y^2+(x-x')^2}dx'$$

17. Use the formula in Exercise 16 to solve the potential problem in the upper half plane, with boundary condition

$$u(x,0) = f(x) = \begin{cases} 1, & 0 < x \\ 0, & x < 0. \end{cases}$$

18. Solve the problem stated in Exercise 15 if the boundary function is

$$f(x) = \begin{cases} 1, & |x| < a, \\ 0, & |x| > a. \end{cases}$$

19. Show that $u(x,y) = x$ is the solution of the potential equation in a slot under the boundary conditions: $f(x) = x$, $g_1(y) = 0$, $g_2(y) = a$. Can this solution be found by the method of this section?

4.4 POTENTIAL IN A DISK

If we need to solve the potential equation in a circular disk $x^2 + y^2 < c^2$, it is natural to use polar coordinates r, θ, in terms of which the disk is described by $0 < r < c$. The Dirichlet problem on the disk is

$$\frac{1}{r}\frac{\partial}{\partial r}\left(r\frac{\partial v}{\partial r}\right) + \frac{1}{r^2}\frac{\partial^2 v}{\partial \theta^2} = 0, \qquad 0 < r < c, \quad -\pi < \theta \le \pi \tag{1}$$

$$v(c,\theta) = f(\theta), \qquad -\pi < \theta \le \pi. \tag{2}$$

Because θ and $\theta + 2\pi$ refer to the same angle, really, we restrict θ to an interval from $-\pi$ to π. However, the ray $\theta = \pi$ is not a "boundary" in the sense we have been using, because the outer world does not affect v at that point. In order to ensure that the introduction of the false boundary does not cause a discontinuity in v and its derivatives, we require that

$$v(r,\pi) = v(r,-\pi), \qquad 0 < r < c \tag{3}$$

$$\frac{\partial v}{\partial \theta}(r,\pi) = \frac{\partial v}{\partial \theta}(r,-\pi), \qquad 0 < r < c. \tag{4}$$

Alternatively, we may allow θ to assume any value but require $v(r, \theta)$ and $f(\theta)$ to be periodic in θ with periodic 2π.

By assuming $v(r, \theta) = R(r)\Theta(\theta)$ we can separate variables. The potential equation becomes

$$\frac{1}{r}(rR')'\Theta + \frac{1}{r^2}R\Theta'' = 0.$$

Separation is effected by dividing through by $R\Theta/r^2$:

$$\frac{r(rR'(r))'}{R(r)} + \frac{\Theta''(\theta)}{\Theta(\theta)} = 0.$$

Evidently the boundary condition Eq. (2) will have to be satisfied with a linear combination of solutions; hence we choose $\Theta''/\Theta = -\lambda^2$. Conditions (3) and (4) become conditions on Θ. The individual problems for the functions R and Θ are

$$\Theta'' + \lambda^2\Theta = 0, \qquad -\pi < \theta \leq \pi \tag{5}$$

$$\Theta(-\pi) = \Theta(\pi) \tag{6}$$

$$\Theta'(-\pi) = \Theta'(\pi) \tag{7}$$

$$r(rR')' - \lambda^2 R = 0, \qquad 0 < r < c. \tag{8}$$

If $\lambda \neq 0$, the general solution of Eq. (5) is

$$\Theta(\theta) = A\cos(\lambda\theta) + B\sin(\lambda\theta). \tag{9}$$

If we write Eqs. (6) and (7) in terms of the function Θ of Eq. (9) and use the properties of sine and cosine, we have

$$A\cos(\lambda\pi) - B\sin(\lambda\pi) = A\cos(\lambda\pi) + B\sin(\lambda\pi)$$

$$\lambda A\sin(\lambda\pi) + \lambda B\cos(\lambda\pi) = -\lambda A\sin(\lambda\pi) + \lambda B\cos(\lambda\pi).$$

After simple manipulations, these equations reduce to

$$B\sin(\lambda\pi) = 0, \quad \lambda A\sin(\lambda\pi) = 0.$$

Not both A and B should be 0, because then $\Theta(\theta)$ would be identically 0. Thus, $\sin(\lambda\pi) = 0$ and $\lambda = 1, 2, 3, \cdots$. If $\lambda = 0$, Eq. (5) becomes $\Theta'' = 0$, with general solution $\Theta(\theta) = A + B\theta$. The conditions of Eqs. (6) and (7) require $B = 0$. The possibilities for nonzero solutions of Eqs. (5)-(7) are thus

$$\lambda = 0, \qquad \Theta = 1$$

$$\lambda = 1, 2, 3, \cdots, \quad \Theta = \cos(\lambda\theta) \text{ or } \sin(\lambda\theta).$$

Now we have found that the eigenvalue $\lambda^2 = 0$ corresponds to the eigenfunction $\Theta = 1$ (constant), and the eigenvalues $\lambda_n^2 = n^2 (n = 1, 2, 3, \cdots)$ each correspond to *two* independent eigenfunctions: $\cos(n\theta)$ and $\sin(n\theta)$.

Knowing that $\lambda_n^2 = n^2$, we can easily find $R(r)$. The equation for R becomes

$$r^2 R'' + rR' - n^2 R = 0, \quad 0 < r < c$$

when the indicated differentiations are carried out. This is a Cauchy-Euler equation, whose solutions are known to have the form $R(r) = r^\alpha$, where α is constant. Substituting $R = r^\alpha$, $R' = \alpha r^{\alpha-1}$, and $R'' = \alpha(\alpha-1)r^{\alpha-2}$ into it leaves

$$\left(\alpha(\alpha-1) + \alpha - n^2\right) r^\alpha = 0, \quad 0 < r < c.$$

As r^α is not zero, the constant factor in parentheses must be zero — that is, $\alpha = \pm n$. The general solution of the differential equation is any combination of r^n and r^{-n}. The latter, however, is unbounded as r approaches zero, so we discard that solution, retaining $R_n(r) = r^n$. In the special case $n = 0$, the two solutions are the constant function 1 and $\ln r$. The logarithm is discarded because of its behavior at $r = 0$.

Now we reassemble our solution. The functions

$$r^0 \cdot 1 = 1, \quad r^n \cos(n\theta), \quad r^n \sin(n\theta)$$

are all solutions of the potential equation, so a general linear combination of these solutions will also be a solution. Thus $v(r, \theta)$ may have the form

$$v(r, \theta) = a_0 + \sum_{n=1}^{\infty} a_n r^n \cos(n\theta) + \sum_{n=1}^{\infty} b_n r^n \sin(n\theta). \tag{10}$$

At the true boundary $r = c$, the boundary condition reads

$$v(c, \theta) = a_0 + \sum_{n=1}^{\infty} c^n \left(a_n \cos(n\theta) + b_n \sin(n\theta)\right) = f(\theta), \quad -\pi < \theta \leq \pi.$$

This is a Fourier series problem, solved by choosing

$$a_0 = \frac{1}{2\pi} \int_{-\pi}^{\pi} f(\theta) d\theta,$$

$$a_n = \frac{1}{\pi c^n} \int_{-pi}^{\pi} f(\theta) \cos(n\theta) d\theta, \quad b_n = \frac{1}{\pi c^n} \int_{-\pi}^{\pi} f(\theta) \sin(n\theta) d\theta. \tag{11}$$

As an example, consider the problem consisting of the potential equation in the disk, Eq. (1), with the boundary condition

$$v(c,\theta) = f(\theta) = \begin{cases} 0, & -\pi < \theta < \frac{-\pi}{2}, \\ 1, & \frac{-\pi}{2} < \theta < \frac{\pi}{2}, \\ 0, & \frac{\pi}{2} < \theta < \pi. \end{cases}$$

The solution is given by Eq. (10), provided that the coefficients are chosen according to Eq. (11). Since $f(\theta)$ is an even function, $b_n = 0$, and

$$a_0 = \frac{1}{\pi}\int_0^{\pi} f(\theta)d\theta = \frac{1}{2},$$

$$a_n = \frac{2}{\pi c^n}\int_0^{\pi} f(\theta)\cos(n\theta)d\theta = \frac{2\sin(n\pi/2)}{n\pi c^n}.$$

Therefore, the solution of the problem is

$$v(r,\theta) = \frac{1}{2} + \sum_{n=1}^{\infty} \frac{2\sin(n\pi/2)}{n\pi}\frac{r^n}{c^n}\cos(n\theta). \tag{12}$$

The level curves of this function are all arcs of circles that pass through the boundary points $r = c$, $\theta = \pm\pi/2$, where $f(\theta)$ jumps between 0 and 1. Along the x-axis, the function has the simple closed form $\frac{1}{2} + \frac{2}{\pi}\tan^{-1}(x)$.

Now that we have the form Eq. (10) of the solution of the potential equation, we can see some important properties of the function $v(r,\theta)$. In particular, by setting $r = 0$ we obtain

$$v(0,\theta) = a_0 = \frac{1}{2\pi}\int_{-\pi}^{\pi} f(\theta)d\theta = \frac{1}{2\pi}\int_{-\pi}^{\pi} v(c,\theta)d\theta.$$

This says that the solution of the potential equation at the center of a disk is equal to the average of its values around the edge of the disk. It is easy to show also that

$$v(0,\theta) = \frac{1}{2\pi}\int_{-\pi}^{\pi} v(r,\theta)d\theta \tag{13}$$

for any r between 0 and c! This characteristic of solutions of the potential equation is called the *mean value property.* From the mean value property, it is just a step to prove the maximum principle mentioned in Section 1, for the mean value of a function lies between the minimum and the maximum and cannot equal either unless the function is constant.

An important consequence of the maximum principle — and thus of the mean value property — is a proof of the uniqueness of the solution of the Dirichlet problem. Suppose that u and v are two solutions of the potential equation in some region R and that they have the same values on the boundary of R. Then their difference, $w = u - v$, is also a solution of the potential equation in R and has value 0 all along the boundary of R. By the maximum principle, w has maximum and minimum values 0, and therefore w is identically 0 throughout R. In other words, u and v are identical.

Exercises

1. Solve the potential equation in the disk $0 < r < c$ if the boundary condition is $v(c, \theta) = |\theta|$, $-\pi < \theta \leq \pi$.

2. Same as Exercise 1 if $v(c, \theta) = \theta$, $-\pi < \theta < \pi$. Is the boundary condition satisfied at $\theta = \pm\pi$?

3. Same as Exercise 1, with boundary condition

$$v(c, \theta) = f(\theta) = \begin{cases} \cos(\theta), & -\pi/2 < \theta < \pi/2 \\ 0, & \text{otherwise.} \end{cases}$$

4. Find the value of the solution at $r = 0$ for the problems of Exercises 1, 2, and 3.

5. If the function $f(\theta)$ in Eq. (2) is continuous, sectionally smooth, and satisfies $f(-\pi+) = f(\pi-)$, what can be said about convergence of the series for $v(c, \theta)$?

6. Show that

$$v(r, \theta) = a_0 + \sum_{n=1}^{\infty} r^{-n} \left(a_n \cos(n\theta) + b_n \sin(n\theta)\right)$$

is a solution of Laplace's equation in the region $r > c$ (exterior of a disk) and has the property that $|v(r, \theta)|$ is bounded as $r \to \infty$.

7. If the condition $v(c, \theta) = f(\theta)$ is given, what are the formulas for the a's and b's in Exercise 6?

8. The solution of Eqs. (1)-(4) can be written in a single formula by the following sequence of operations:

a. Replace θ by ϕ in Eq. (11) for the a's and b's;

b. replace the a's and b's in Eq. (10) by the integrals in part **a**;

c. use the trigonometric identity

$$\cos(n\theta)\cos(n\phi) + \sin(n\theta)\sin(n\phi) = \cos(n(\theta - \phi));$$

d. take the integral outside the series;

e. add up the series (see Chapter 1, Section 10, Exercise 1). Then $v(r, \theta)$ is given by the single integral (Poisson integral formula)

$$v(r, \theta) = \frac{1}{2\pi} \int_{-\pi}^{\pi} f(\phi) \frac{c^2 - r^2}{c^2 + r^2 - 2rc\cos(\theta - \phi)} d\phi.$$

9. Solve Laplace's equation in the quarter disk $0 < \theta < \pi/2$, $0 < r < c$, subject to the boundary conditions $v(r, 0) = 0$, $v(r, \pi/2) = 0$, $v(c, \theta) = 1$.

10. Generalize the results of Exercise 9 by solving this problem:

$$\frac{1}{r}\frac{\partial}{\partial r}\left(r\frac{\partial v}{\partial r}\right) + \frac{1}{r^2}\frac{\partial^2 v}{\partial \theta^2} = 0, \qquad 0 < \theta < \alpha\pi, \quad 0 < r < c$$

$$v(r, 0) = 0, \qquad v(r, \alpha\pi) = 0, \quad 0 < r < c,$$

$$v(c, \theta) = f(\theta), \quad 0 < \theta < \alpha\pi.$$

Here, α is a parameter between 0 and 2.

11. Suppose that $\alpha > 1$ in Exercise 10. Show that there is a product solution with the property that $\dfrac{\partial v}{\partial r}(r, \theta)$ is not bounded as $r \to 0+$.

12. Instead of restricting θ to the interval $-\pi < \theta \leq \pi$ and imposing conditions (3) and (4), we could let θ be unrestricted and require that $v(r, \theta)$ be periodic in θ with period 2π. Show that the separation-of-variables process leads to the eigenvalue problem

$$\Theta'' + \lambda^2 \Theta = 0$$

$$\Theta(\theta + 2\pi) = \Theta(\theta)$$

instead of Eqs. (5), (6), and (7). Also show that the two eigenvalue problems have exactly the same solutions.

4.5 CLASSIFICATION OF PARTIAL DIFFERENTIAL EQUATIONS AND LIMITATIONS OF THE PRODUCT METHOD

By this time, we have seen a variety of equations and solutions. We have concentrated on three different, homogeneous equations (heat, wave, and potential) and have found the qualitative features summarized in the following tabulation:

Equation	Features
Heat	Exponential behavior in time. Existence of a limiting (steady-state) solution. Smooth graph for $t > 0$.
Wave	Oscillatory (not always periodic) behavior in time. Retention of discontinuities for $t > 0$.
Potential	Smooth surface. Maximum principle. Mean value property.

These three two-variable equations are the most important representatives of the three classes of second-order linear partial differential equations in two variables. The most general equation that fits this description is

$$A\frac{\partial^2 u}{\partial \xi^2} + B\frac{\partial^2 u}{\partial \xi \partial \eta} + C\frac{\partial^2 u}{\partial \eta^2} + D\frac{\partial u}{\partial \xi} + E\frac{\partial u}{\partial \eta} + Fu + G = 0$$

where A, B, C, and so forth, are, in general, functions of ξ and η. (We use Greek letters for the independent variables to avoid implying any relations to space or time.) Such an equation can be classified according to the sign of $B^2 - 4AC$:

$$B^2 - 4AC < 0: \quad \text{elliptic}$$

$$B^2 - 4AC = 0: \quad \text{parabolic}$$

$$B^2 - 4AC > 0: \quad \text{hyperbolic.}$$

Because A, B, and C are functions of ξ and η (not of u), the classification of an equation may vary from point to point. It is easy to see that the heat equation is parabolic, the wave equation is hyperbolic, and the potential equation is elliptic. The classification of an equation determines important features of the solution and also dictates the method of attack when numerical techniques are used for solution.

The question naturally arises whether separation of variables works on all equations. The answer is no. For instance, the equation

$$(\xi + \eta^2)\frac{\partial^2 u}{\partial \xi^2} + \frac{\partial^2 u}{\partial \eta^2} = 0$$

does not admit separation of variables. In general, it is difficult to say just which equations can be solved by this method. However, it is necessary to have $B \equiv 0$.

The region in which the solution is to be found also limits the applicability of the method we have used. The region must be a *generalized rectangle.* By this we mean a region bounded by coordinate curves of the coordinate system of the partial differential equation. Put another way, the region is described by inequalities on the coordinates, whose endpoints are fixed quantities. For instance, we have worked in regions described by the following sets of inequalities:

$$\begin{aligned} &0 < x < a, && 0 < t \\ &0 < x, && 0 < t \\ -&\infty < x < \infty, && 0 < t \\ &0 < x < a, && 0 < y < b \\ &0 < r < c, && -\pi < \theta \leq \pi. \end{aligned}$$

All of these are generalized rectangles, but only one is an ordinary rectangle. An L-shaped region is not a generalized rectangle, and our methods would break down if applied to, for instance, the potential equation there.

There are, as we know, restrictions on the kinds of boundary conditions that can be handled. From the examples in this chapter it is clear that we need homogeneous or "homogeneous-like" conditions on opposite sides of a generalized rectangle. Examples of "homogeneous-like" conditions are the requirement that a function remain bounded as some variable tends to infinity, or the periodic conditions at $\theta = \pm\pi$ (see Section 4). The point is that if two or more functions satisfy the conditions, so does a sum of those functions.

In spite of the limitations of the method of separation of variables, it works well on many important problems in two or more variables and provides insight into the nature of their solutions. Moreover, it is known that in those cases where separation of variables can be carried out, it will find a solution if one exists.

Exercises

1. Classify the following equations:

a. $\dfrac{\partial^2 u}{\partial x \partial y} = 0$

b. $\dfrac{\partial^2 u}{\partial x^2} + \dfrac{\partial^2 u}{\partial x \partial y} + \dfrac{\partial^2 u}{\partial y^2} = 2x$

c. $\dfrac{\partial^2 u}{\partial x^2} - \dfrac{\partial^2 u}{\partial x \partial y} + \dfrac{\partial^2 u}{\partial y^2} = 2u$

d. $\dfrac{\partial^2 u}{\partial x^2} - 2\dfrac{\partial^2 u}{\partial x \partial y} + \dfrac{\partial^2 u}{\partial y^2} = \dfrac{\partial u}{\partial y}$

e. $\dfrac{\partial^2 u}{\partial x^2} - \dfrac{\partial^2 u}{\partial y^2} - \dfrac{\partial u}{\partial y} = 0.$

2. Show that, in polar coordinates, an annulus, a sector, and a sector of an annulus are all generalized rectangles.

3. In which of the equations in Exercise 1 can the variables be separated?

4. Sketch the regions listed in the text as generalized rectangles.

5. Solve these three problems and compare the solutions.

a. $\dfrac{\partial^2 u}{\partial x^2} + \dfrac{\partial^2 u}{\partial y^2} = 0, \quad 0 < x < 1, \quad 0 < y$

$u(x,0) = f(x), \quad 0 < x < 1$

$u(0,y) = 0, \quad u(1,y) = 0, \quad 0 < y$

b. $\dfrac{\partial^2 u}{\partial x^2} = \dfrac{\partial^2 u}{\partial y^2}, \quad 0 < x < 1, \quad 0 < y$

$u(x,0) = f(x), \quad \dfrac{\partial u}{\partial y}(x,0) = 0, \quad 0 < x < 1$

$u(0,y) = 0, \quad u(1,y) = 0, \quad 0 < y$

c. $\dfrac{\partial^2 u}{\partial x^2} = \dfrac{\partial u}{\partial y} = 0, \quad 0 < x < 1, \quad 0 < y$

$u(x,0) = f(x), \quad 0 < x < 1$

$u(0,y) = 0, \quad u(1,y) = 0, \quad 0 < y$

6. Show that if $f_1, f_2, \cdots$ all satisfy the periodic boundary conditions

$$f(-\pi) = f(\pi), \quad f'(-\pi) = f'(\pi),$$

then so does the function $c_1 f_1 + c_2 f_2 + \cdots$, where the c's are constants.

4.6 COMMENTS AND REFERENCES

While the potential equation describes many physical phenomena, there is one that makes the solution of the Dirichlet problem very easy to visualize. Suppose a piece of wire is bent into a closed curve or frame. When the frame is held over a level surface, its projection onto the surface is a plane curve $\mathcal{C}$ enclosing a region $\mathcal{R}$. If one forms a soap film on the frame, the height $u(x,y)$ of the film above the level surface is a function that satisfies the potential equation approximately, if the effects of gravity are negligible (see Chapter 5). The height of the frame above the curve $\mathcal{C}$ gives the boundary condition on u. For example, Fig. 1b shows the surface corresponding to the problem solved in Section 2. A great deal of information about soap films is in the book *The Science of Soap Films and Soap Bubbles* by C. Isenberg. (See the Bibliogrpahy.)

Maximum Principles in Differential Equations by Protter and Weinberger is an elegant study of maximum principles for the potential equation and other partial differential equations.

It turns out that the potential equation (but not all elliptic equations) is best studied through the use of complex variables. A complex variable may be written $z = x + iy$, where x and y are real and $i^2 = -1$; similarly a function of z is denoted by $f(z) = u(x,y) + iv(x,y)$, u and v being real functions of real variables. If f has a derivative with respect to z, then both u and v satisfy the potential equation. Easy examples such as polynomials and exponentials lead to familiar solutions:

$$z^2 = (x+iy)^2 = x^2 - y^2 + i2xy$$

$$e^z = e^{x+iy} = e^x e^{iy} = e^x \cos(y) + ie^x \sin(y).$$

$$\ln(z) = \frac{1}{2}\ln(x^2+y^2) + i\tan^{-1}(y/x).$$

(See Section 1, Exercises 1, 2; Section 3, Exercises 13, 14; Miscellaneous Exercise 18 in this chapter.) Knowing these elementary solutions often helps in simplifying a problem. *Advanced Engineering Mathematics* by O'Neil, includes a complete presentation of complex analysis. Applications to the potential equation and other areas of applied mathematics are found in Chapters 18 and 19 of O'Neil's book.

In certain idealized fluid flows (steady, irrotational, two-dimensional flow of an inviscid, incompressible fluid) the velocity vector is given by $\mathbf{V} = \text{grad } \phi$, where the *velocity potential* ϕ is a solution of the potential equation. The streamlines along which the fluid flows are level curves of a related function ψ, called the *stream function*, which also is a solution of the potential equation. The two functions ϕ and ψ are the real and imaginary parts of a function of the complex variable z. The level curves $\phi = \text{constant}$ and $\psi = \text{constant}$ form two families of orthogonal curves called a *flow net*. The flow net in Fig. 2, for $\phi = x^2 - y^2$ and $\psi = 2xy$ (the real and imaginary parts of the function $f(z) = z^2$), illustrates flow near a corner formed by two walls. Many other flow nets are shown in the book *Potential Flows: Computer Graphic Solutions* by R.H. Kirchhoff. Civil engineers sometimes sketch a flow net by eye to get a rough graphical solution of the potential equation for hydrodynamics problems.

Where a physical boundary is formed by an impervious wall, the velocity vector $\mathbf{V}$ must be parallel to the boundary. This fact leads to two boundary conditions. First, the wall must coincide with a streamline; thus $\psi = \text{constant}$ along a boundary. Second, the component of $\mathbf{V}$ that is normal to the wall must be zero there, as no fluid passes through it; thus the normal derivative of ϕ is zero, $\partial\phi/\partial n = 0$, at a boundary. See Miscellaneous Exercises 30-32, and the excellent book, *Viscous Flows*, by S.W. Churchill, which contains the mathematics and compares the results to experiment.

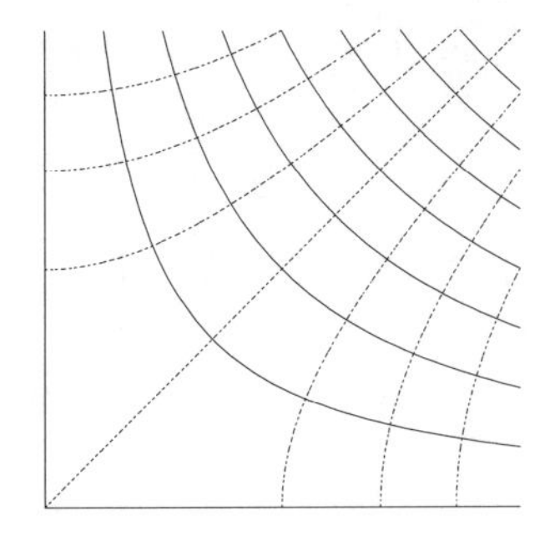

Figure 2: Streamlines (solid) and equipotential curves (dashed) for flow in a corner. The streamlines are described by the equation $2xy = \text{constant}$, with a different constant for each one. Similarly, the equipotential curves are described by the equation $x^2 - y^2 = \text{constant}$.

MISCELLANEOUS EXERCISES

1. Solve the potential equation in the rectangle $0 < x < a$, $0 < y < b$ with the boundary conditions

$$u(0,y) = 1, \quad u(a,y) = 0, \quad 0 < y < b$$
$$u(x,0) = 0, \quad u(x,b) = 0, \quad 0 < x < a.$$

2. If $a = b$ in Exercise 1, then $u(a/2, a/2) = 1/4$. Use symmetry to explain this fact.

3. Solve the potential equation on the rectangle $0 < x < a$, $0 < y < b$ with the boundary conditions

$$u(0,y) = 1, \quad u(a,y) = 1, \quad 0 < y < b$$
$$\frac{\partial u}{\partial y}(x,0) = 0, \quad \frac{\partial u}{\partial y}(x,b) = 0, \quad 0 < x < a.$$

4. Same as Exercise 3, but the boundary conditions are

$$u(0,y) = 1, \quad \frac{\partial u}{\partial x}(a,y) = 0, \quad 0 < y < b$$
$$u(x,0) = 1, \quad \frac{\partial u}{\partial y}(x,b) = 0, \quad 0 < x < a.$$

5. Same as Exercise 3, but the boundary conditions are

$$u(0,y) = 1, \quad u(a,y) = 1, \quad 0 < y < b$$
$$\frac{\partial u}{\partial y}(x,0) = 0, \quad u(x,b) = 0, \quad 0 < x < a.$$

6. Same as Exercise 3, but the boundary conditions are

$$u(0,y) = 1, \quad u(a,y) = 0, \quad 0 < y < b$$
$$u(x,0) = 1, \quad u(x,b) = 0, \quad 0 < x < a.$$

7. Same as Exercise 3, but the region is a square ($b = a$) and the boundary conditions are

$$u(0,y) = f(y), \quad u(a,y) = 0, \quad 0 < y < a$$
$$u(x,0) = f(x), \quad u(x,a) = 0, \quad 0 < x < a,$$

where f is a function whose graph is an isosceles triangle of height h and width a.

8. Solve the potential equation in the region $0 < x < a$, $0 < y$ with the boundary conditions

$$u(x,0) = 1, \qquad 0 < x < a$$
$$u(0,y) = 0, \quad u(a,y) = 0, \quad 0 < y.$$

9. Find the solution of the potential equation on the strip $0 < y < b$, $-\infty < x < \infty$, subject to the conditions that follow. Supply boundedness conditions as necessary.

$$u(x,y) = \begin{cases} 1, & -a < x < a \\ 0, & |x| > a \end{cases}$$
$$u(x,b) = 0, \quad -\infty < x < \infty$$

10. Show that the function $u(x,y) = \tan^{-1}(y/x)$ is a solution of the potential equation in the first quadrant. What conditions does u satisfy along the positive x- and y-axes?

11. Solve the potential problem in the upper half plane,

$$\frac{\partial^2 u}{\partial x^2} + \frac{\partial^2 u}{\partial y^2} = 0, \qquad -\infty < x < \infty, \quad 0 < y$$
$$u(x,0) = f(x), \quad -\infty < x < \infty,$$

taking $f(x) = \exp(-\alpha|x|)$.

12. Apply the formula below (see Section 4.3, Exercise 16) for the solution of the potential problem in the upper half plane if the boundary condition is $u(x,0) = f(x)$, where

$$f(x) = \begin{cases} 0, & x < 0, \\ 1, & x > 0. \end{cases}$$
$$u(x,y) = \frac{1}{\pi}\int_{-\infty}^{\infty} f(x')\frac{y}{y^2 + (x - x')^2}dx'.$$

13. Apply the formula in Exercise 12 to the case where $f(x) = 1$, $-\infty < x < \infty$. The solution of the problem should be $u(x, y) \equiv 1$.

14. a. Find the separation-of-variables solution of the potential problem in a disk of radius 1 if the boundary condition is $u(1, \theta) = f(\theta)$, where

$$f(\theta) = \begin{cases} -\pi - \theta, & -\pi < \theta < 0, \\ \pi - \theta, & 0 < \theta < \pi. \end{cases}$$

b. Show that the function given in polar and Cartesian coordinates by

$$\begin{aligned} u(r, \theta) &= 2\tan^{-1}\left(\frac{r\sin(\theta)}{1 - r\cos(\theta)}\right) \\ &= 2\tan^{-1}\left(\frac{y}{1-x}\right) \end{aligned}$$

satisfies the potential equation (use the Cartesian coordinates) and the boundary condition. The following identity is useful.

$$\frac{\sin(\theta)}{1 - \cos(\theta)} = \tan\left(\frac{\pi - \theta}{2}\right).$$

c. Sketch some level curves of the solution inside the circle of radius 1.

15. Solve the potential equation in a disk of radius c with boundary conditions

$$u(c, \theta) = \begin{cases} 1, & 0 < \theta < \pi \\ 0, & -\pi < \theta < 0 \end{cases}.$$

16. What is the value of u at the center of the disk, in Exercise 15?

17. Same as Exercise 15, but the boundary condition is

$$u(c, \theta) = |\sin(\theta)|.$$

18. For the potential problem on an annular ring

$$\frac{1}{r}\frac{\partial}{\partial r}\left(r\frac{\partial^2 u}{\partial r}\right) + \frac{1}{r^2}\frac{\partial^2 u}{\partial \theta^2} = 0, \quad a < r < b,$$

show that product solutions have the form

$$A_0 + B_0 \ln(r), \quad (C_0 + D_0\theta)\ln(r),$$

or

$$r^n \left(A_n \cos(n\theta) + B_n \sin(n\theta)\right) + r^{-n} \left(C_n \cos(n\theta) + D_n \sin(n\theta)\right).$$

19. Solve the potential problem on the annular ring as stated in Exercise 18 with boundary conditions

$$u(a, \theta) = 1, \quad u(b, \theta) = 0.$$

20. Find product solutions of the potential equation on a sector of a disk with zero boundary conditions on the straight edges.

$$\begin{aligned} &\nabla^2 u = 0, \quad 0 \le r < c, \quad 0 < \theta < \alpha \\ &u(r, 0) = 0, \quad u(r, \alpha) = 0 \end{aligned}$$

21. Solve the potential problem in a slit disk:

$$\begin{aligned} &\nabla^2 u = 0, && 0 \le r < c, \quad 0 < \theta < 2\pi \\ &u(r, 0) = 0, && u(r, 2\pi) = 0 \\ &u(r, \theta) = f(\theta), && 0 < \theta < 2\pi. \end{aligned}$$

22. Show that the function $u(x, y) = \sin(\pi x/a)\sinh(\pi y/a)$ satisfies the potential problem

$$\begin{aligned} &\nabla^2 u = 0, \quad 0 < x < a, \quad 0 < y \\ &u(0, y) = 0, \quad u(a, y) = 0, \quad 0 < y \\ &u(x, 0) = 0, \quad 0 < x < a. \end{aligned}$$

This solution is eliminated if it is also required that $u(x, y)$ be bounded as $y \to \infty$.

23. Solve the potential equation in the rectangle $0 < x < a$, $0 < y < b$, with the boundary conditions

$$\begin{aligned} &u(0, y) = 0, \quad \frac{\partial u}{\partial x}(a, y) = 0, \quad 0 < y < b \\ &u(x, 0) = 0, \quad u(x, b) = x, \qquad 0 < x < a. \end{aligned}$$

24. Find a polynomial of second degree in x and y,

$$v(x, y) = A + Bx + Cy + Dx^2 + Exy + Fy^2,$$

that satisfies the potential equation and these boundary conditions:

$$\begin{aligned} &v(0, y) = 0, \quad 0 < y < b \\ &v(x, 0) = 0, \quad v(x, b) = x, \quad 0 < x < a. \end{aligned}$$

25. Find the problem (partial differential equation and boundary conditions) satisfied by $w(x, y) = v(x, y) - u(x, y)$, where u and v are the solutions of the problems in Exercises 23 and 24. Solve the problem. Is this problem easier to solve than the one in Exercise 23?

26. Solve the potential equation in the quarter plane $0 < x$, $0 < y$, subject to the boundary conditions

$$u(x, 0) = f(x), \quad 0 < x$$

$$u(0, y) = f(y), \quad 0 < y.$$

The function f that appears in both boundary conditions is given by the equation

$$f(x) = \begin{cases} 1, & 0 < x < a \\ 0, & a < x. \end{cases}$$

27. (Flow past a plate) A fluid occupies the half plane $y > 0$ and flows past (left to right, approximately) a plate located near the x-axis. If the x and y components of velocity are $U_0 + u(x, y)$, and $v(x, y)$, respectively (U_0 = constant free-stream velocity), under certain assumptions, the equations of motion, continuity, and state can be reduced to

$$\frac{\partial u}{\partial y} = \frac{\partial v}{\partial x}, \quad (1 - M^2)\frac{\partial u}{\partial x} + \frac{\partial v}{\partial y} = 0,$$

valid for all x and $y > 0$. M is the free-stream Mach number. Define the velocity potential ϕ by the equations $u = \partial\phi/\partial x$ and $v = \partial\phi/\partial y$. Show that the first equation is automatically satisfied and the second is a partial differential equation that is elliptic if $M < 1$ or hyperbolic if $M > 1$.

28. If the plate is wavy — say its equation is $y = \epsilon\cos(\alpha x)$ — then the boundary condition, that the vector velocity be parallel to the wall, is

$$v(x, \epsilon\cos(\alpha x)) = -\epsilon\alpha\sin(\alpha x)\left(U_0 + u(x, \epsilon\cos(\alpha x))\right).$$

This equation is impossible to use, so it is replaced by

$$v(x, 0) = -\epsilon\alpha U_0 \sin(\alpha x)$$

on the assumption that ϵ is small and u is much smaller than U_0. Using this boundary condition, and the condition that $u(x, y) \to 0$ as $y \to \infty$, set up and solve a complete boundary value problem for ϕ, assuming $M < 1$.

29. By superposition of solutions (α ranging from 0 to ∞) find the flow past a wall whose equation is $y = f(x)$. Hint: Use the boundary condition

$$v(x,0) = U_0 f'(x) = \int_0^\infty [A(\alpha)\cos(\alpha x) + B(\alpha)\sin(\alpha x)]\, d\alpha.$$

30. In hydrodynamics, the velocity vector in a fluid is $v = \text{grad}\ u$, where u is a solution of the potential equation. The normal component of velocity, $\partial u/\partial n$, is 0 at a wall. Thus the problem

$$\begin{aligned} &\nabla^2 u = 0, && 0 < x < 1, && 0 < y < 1, \\ &\frac{\partial u}{\partial x}(0,y) = 0, && \frac{\partial u}{\partial x}(1,y) = 1, && 0 < y < 1, \\ &\frac{\partial u}{\partial y}(x,0) = 0, && \frac{\partial u}{\partial y}(x,1) = -1, && 0 < x < 1, \end{aligned}$$

represents a flow around a corner: flow inward at the top, outward at the right, with walls at left and bottom. Explain why, in a fluid flow problem, it must be true that

$$\int_C \frac{\partial u}{\partial n} ds = 0 \qquad (*)$$

if u is a solution of the potential equation in a region $\mathcal{R}$, $\partial u/\partial n$ is the outward normal derivative, $\mathcal{C}$ is the boundary of the region and s is arc length.

31. Under the conditions stated in Exercise 30, prove the validity of (*). Hint: Use Green's theorem.

32. The Neumann problem consists of the potential equation in a region $\mathcal{R}$ and conditions on $\partial u/\partial n$ along $\mathcal{C}$, the boundary of $\mathcal{R}$. Show (a) that

$$\int_C \frac{\partial u}{\partial n} ds = 0$$

is a necessary condition for a solution to exist, and (b) if u is a solution of the Neumann problem, so is $u + c$ (c is constant).

33. Show that $u(x,y) = \frac{1}{2}(x^2 - y^2)$ is a solution of the problem in Exercise 30.

34. Solve this potential problem in a half-annulus (sketch the region). At some point, it may be useful to make the substitution $s = \ln(r)$.

$$\frac{1}{r}\frac{\partial}{\partial r}\left(r\frac{\partial u}{\partial r}\right) + \frac{1}{r^2}\frac{\partial^2 u}{\partial \theta^2} = 0, \quad 1 < r < e, \qquad 0 < \theta < \pi$$

$$u(1,\theta) = 0, \quad u(e,\theta) = 0, \quad 0 < \theta < \pi$$

$$u(r,0) = 0, \quad u(r,\pi) = 1, \quad 1 < r < e$$

35. Find a relationship among the coefficients of the polynomial

$$p(x,y) = A + Bx + Cy + Dx^2 + Exy + Fy^2$$

so that it satisfies the nonhomogeneous equation

$$\frac{\partial^2 p}{\partial x^2} + \frac{\partial^2 p}{\partial y^2} = -1.$$

36. Find a polynomial $p(x,y)$ that satisfies the partial differential equation of Exercise 35 and also the boundary conditions

$$p(x,0) = 0, \quad p(x,b) = 0.$$

37. Solve the following boundary value problem

$$\frac{\partial^2 u}{\partial x^2} + \frac{\partial^2 u}{\partial y^2} = -1, \quad 0 < x < a, \qquad 0 < y < b$$

$$u(x,0) = 0, \qquad u(x,b) = 0, \quad 0 < x < a$$

$$u(0,y) = 0, \qquad u(a,y) = 0, \quad 0 < y < b.$$

38. An equation of the form $\nabla^2 u = -f$, where f is a function of the space variables, is called *Poisson's equation.* (See Exercises 35-37.) When f depends on just one variable, it is often easy to find a solution. Solve these problems in polar coordinates:

a. $\nabla^2 u = -1$

b. $\nabla^2 u = -\dfrac{1}{x^2 + y^2}$.

39. Many problems in engineering and physics require the solution of the Poisson equation,

$$\nabla^2 u = -H \text{ in a region } \mathcal{R}.$$

Most frequently the boundary condition is $u = 0$ on the boundary of $\mathcal{R}$. Here are three examples:

(1) u is the deflection of a membrane that is fastened at its edges, so that $u = 0$ on the boundary of $\mathcal{R}$; H is proportional to the pressure difference across the membrane. (See Section 5.1.)

(2) u is the steady-state temperature in a cross-section of a long cylindrical rod that is carrying an electrical current; H is proportional to the power in resistance heating. (See Section 5.2.)

(3) u is the stress function on the cross section $\mathcal{R}$ of a cylindrical bar or rod in torsion (The shear stresses are proportional to the partial derivatives of u); H is proportional to the rate of twist and to the shear modulus of the material; $u = 0$ on the boundary of $\mathcal{R}$.

In all these applications, H can be taken to be a constant, which can be made equal to 1 by appropriate scaling. We are going to solve two such problems. The first involves a region R that is a rectangle.

Problem 1.

$$\frac{\partial^2 u}{\partial x^2} + \frac{\partial^2 u}{\partial y^2} = -1, \quad 0 < x < a, \quad 0 < y < b$$

$$u(0, y) = 0, \quad u(a, y) = 0, \quad 0 < x < a$$

$$u(x, 0) = 0, \quad u(x, b) = 0, \quad 0 < y < b.$$

A related problem involves a rectangle that is "infinitely wide."

Problem 2.

$$\frac{d^2 v}{dx^2} = -1, \quad 0 < x < a$$

$$v(0) = 0, \quad v(a) = 0.$$

The solution to Problem 2 is sometimes used as a simple approximation to the solution of Problem 1. In any case, it is convenient to decompose u as $u(x, y) = v(x) - w(x, y)$.

a. Solve Problem 2.

b. Determine the partial differential equation and boundary conditions satisfied by w.

c. Solve the problem found in part **b**, and obtain the solution of Problem 1.

d. Estimate relative error incurred by using v as an approximation to u by evaluating this ratio of integrals

$$\frac{\int_0^a \int_0^b [v(x) - u(x,y)]\, dydx}{\int_0^a \int_0^b u(x,y) dydx}$$

for $b/a = 1, 2$, and 4. Use just one term in the series for $w(x,y) = v(x) - u(x,y)$.

Chapter 5

Problems in Several Dimensions

5.1 TWO-DIMENSIONAL WAVE EQUATION: DERIVATION

For an example of a two-dimensional wave equation, we consider a membrane that is stretched taut over a flat frame in the xy-plane (Fig. 1). The displacement of the membrane above the point (x, y) at time t is $u(x, y, t)$. We assume that the surface tension σ (dimensions F/L) is constant and independent of position. We also suppose that the membrane is perfectly flexible; that is, it does not resist bending. (A soap film satisfies these assumptions quite accurately.) Let us imagine that a small rectangle (of dimensions Δx by Δy aligned with the coordinate axes) is cut out of the membrane and then apply Newton's law of motion to it. On each edge of the rectangle, the rest of the membrane exerts a distributed force of magnitude σ (symbolized by the arrows in Fig. 2a); these distributed forces can be resolved into concentrated forces of magnitude $\sigma\Delta x$ or $\sigma\Delta y$, according to the length of the segment involved (see Fig. 2b).

Looking at projection on the xu- and yu-planes (Figs. 5.4a, 5.4b), we see that the sum of forces in the x-direction is $\sigma\Delta y(\cos(\beta) - \cos(\alpha))$, and the sum of forces in the y-direction is $\Delta x(\cos(\delta) - \cos(\gamma))$. It is desirable that both these sums be zero or at least negligible. Therefore we shall assume that α, β, γ, and δ are all small angles. Because we know that

$$\tan(\alpha) = \frac{\partial u}{\partial x}, \quad \tan(\gamma) = \frac{\partial u}{\partial y}$$

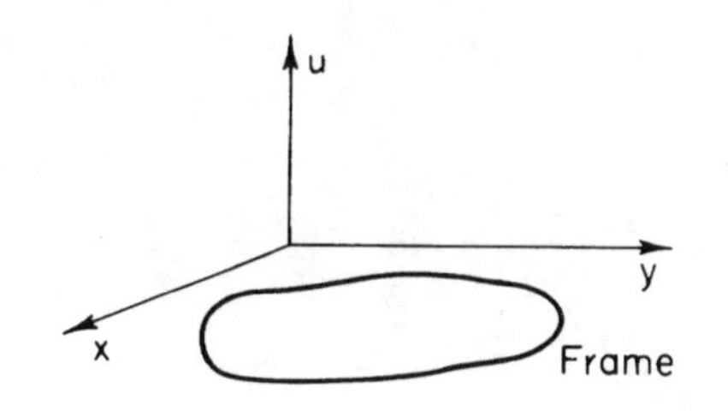

Figure 1:

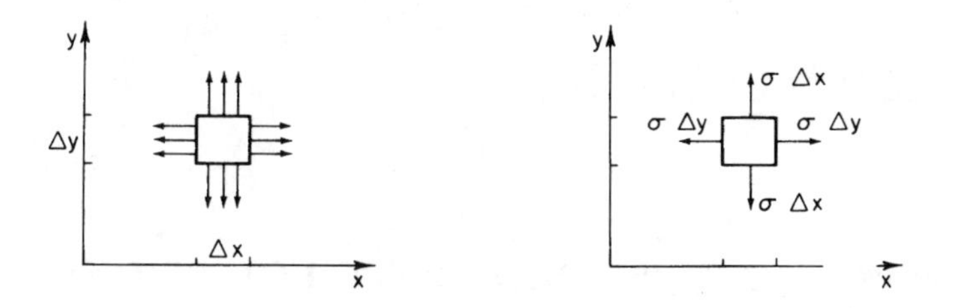

Figure 2: (a) Distributed forces. (b) Concentrated forces.

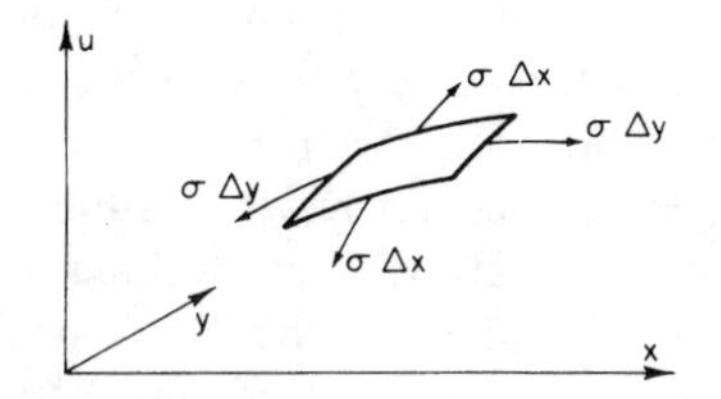

Figure 3: Forces on a piece of membrane.

and so forth, when the derivatives are evaluated at some appropriate point near (x, y), we are assuming that the slopes $\partial u/\partial x$ and $\partial u/\partial y$ of the membrane are very small.

Adding up forces in the vertical direction, and equating the sum to the mass times acceleration (in the vertical direction) we obtain

$$\sigma\Delta y(\sin(\beta) - \sin(\alpha)) + \sigma\Delta x(\sin(\delta) - \sin(\gamma)) = \rho\Delta x\Delta y\frac{\partial^2 u}{\partial t^2}$$

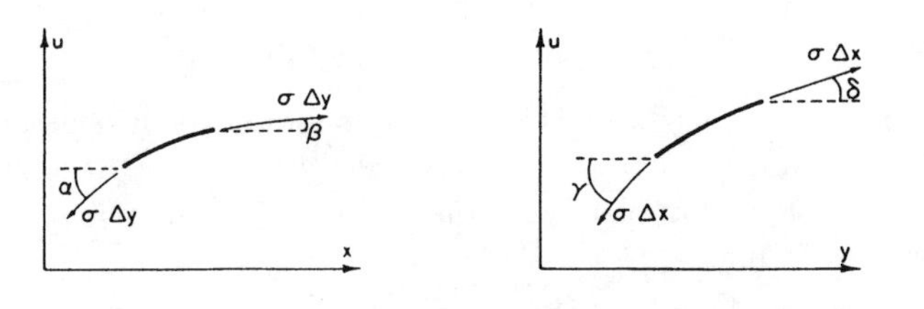

Figure 4: Forces (a) in the xu-plane; (b) in the yu-plane.

where ρ is the surface density $[m/L^2]$. Because the angles α, β, γ, and δ are small, the sine of each is approximately equal to its tangent:

$$\sin(\alpha) \cong \tan(\alpha) = \frac{\partial u}{\partial x}(x, y, t),$$

and so forth. With these approximations used throughout, the equation above becomes

$$\sigma \Delta y \left(\frac{\partial y}{\partial x}(x + \Delta x, y, t) - \frac{\partial u}{\partial x}(x, y, t) \right) + \sigma \Delta x \left(\frac{\partial u}{\partial y}(x, y + \Delta y, t) - \frac{\partial u}{\partial y}(x, y, t) \right) = \rho \Delta x \Delta y \frac{\partial^2 u}{\partial t^2}.$$

On dividing through by $\Delta x \Delta y$, we recognize two difference quotients in the left-hand member. In the limit they become partial derivatives, yielding the equation

$$\sigma \left(\frac{\partial^2 u}{\partial x^2} + \frac{\partial^2 u}{\partial y^2} \right) = \rho \frac{\partial^2 u}{\partial t^2},$$

or

$$\frac{\partial^2 u}{\partial x^2} + \frac{\partial^2 u}{\partial y^2} = \frac{1}{c^2} \frac{\partial^2 u}{\partial t^2},$$

if $c^2 = \sigma/\rho$. This is the two-dimensional wave equation.

If the membrane is fixed to the flat frame, the boundary condition would be

$$u(x, y, t) = 0 \quad for \quad (x, y) \text{ on the boundary.}$$

Naturally, it is necessary to give initial conditions describing the displacement and velocity of each point on the membrane at $t = 0$:

$$u(x, y, 0) = f(x, y)$$
$$\frac{\partial u}{\partial t}(x, y, 0) = g(x, y).$$

Exercises

1. Suppose that the frame is rectangular, bounded by segments of the lines $x = 0$, $x = a$, $y = 0$, $y = b$. Write an initial value-boundary value problem, complete with inequalities, for a membrane stretched over this frame.

2. Suppose that the frame is circular, and its equation is $x^2 + y^2 = a^2$. Write an initial value-boundary value problem for a membrane on a circular frame. (Use polar coordinates.)

3. What should the three-dimensional wave equation be?

5.2 THREE-DIMENSIONAL HEAT EQUATION: DERIVATION

To illustrate a different technique, we are going to derive the three-dimensional heat equation using vector methods. Suppose that we are investigating the temperature in a body that occupies a region $\mathcal{R}$ in space. (See Fig. 5). Let $\mathcal{V}$ be a subregion of $\mathcal{R}$ bounded by the surface $\mathcal{S}$. The law of conservation of energy, applied to $\mathcal{V}$ says:

net rate of heat in + rate of generation inside = rate of accumulation.

Our next job is to quantify this statement. The heat flow rate at any point inside $\mathcal{R}$ is a vector function, $\mathbf{q}$, measured in joules/m^2 s or similar units. The rate of heat flow through a small piece of the surface $\mathcal{S}$ with area ΔA is approximately $\hat{\mathbf{n}} \cdot \mathbf{q}\Delta A$, (see Fig. 6), where $\hat{\mathbf{n}}$ is the outward unit normal. This quantity is positive for outward flow, so the inflow is its negative. The net inflow over the entire surface $\mathcal{S}$ is a sum of quantities like this, which becomes, in the limit as ΔA shrinks, the integral

$$\iint_{\mathcal{S}} -\mathbf{q} \cdot \hat{\mathbf{n}} \ dA.$$

The term "rate of generation inside" in the energy balance is intended to include conversion of energy from other forms (chemical, electrical, nuclear) to thermal. We assume that it is specified as an intensity g measured in joules/m^3 s or similar units. Then the rate at which heat is generated in a small region of volume V centered on point P is approximately $g(P,t)\Delta V$. These contributions are summed over the whole subregion $\mathcal{V}$, and, as ΔV shrinks, their total becomes the integral

$$\iiint_{\mathcal{V}} g(P,t)dV.$$

The rate at which heat is stored in a small region of volume ΔV centered on point P is proportional to the rate at which temperature changes there. That is, the storage rate is $\rho c \Delta V u_t(P,t)$. The storage rate for the whole subregion $\mathcal{V}$ is the sum of such contributions, which passes to the integral

$$\iiint_{\mathcal{V}} \rho c \frac{\partial u}{\partial t}(P,t) dV.$$

Now the heat balance equation in mathematical terms becomes

$$\iint_{\mathcal{S}} -\mathbf{q} \cdot \hat{\mathbf{n}} \; dA + \iiint_{\mathcal{V}} g dV = \iiint_{\mathcal{V}} \rho c \frac{\partial u}{\partial t} dV. \tag{1}$$

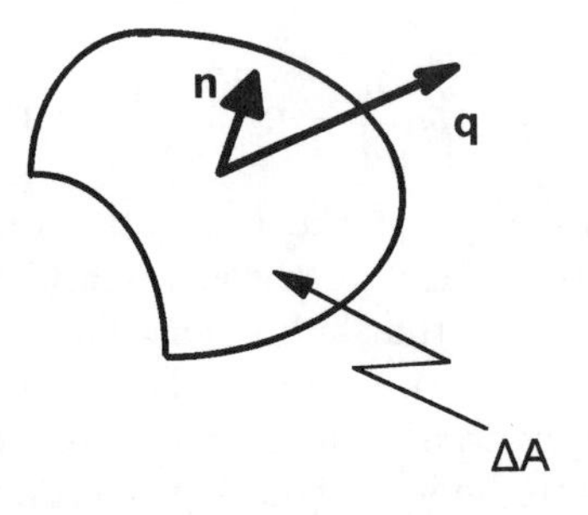

Figure 5: A solid body occupying a region $\mathcal{R}$ in space and a subregion $\mathcal{V}$ with boundary $\mathcal{S}$.

At this point, we call on the divergence theorem, which states that the integral over a surface $\mathcal{S}$ of the outward normal component of a vector function equals the integral over the volume bounded by $\mathcal{S}$ of the divergence of the function. Thus

$$\iint_{\mathcal{S}} \mathbf{q} \cdot \hat{\mathbf{n}} \; dA = \iiint_{\mathcal{V}} \nabla \cdot \mathbf{q} \; dV \tag{2}$$

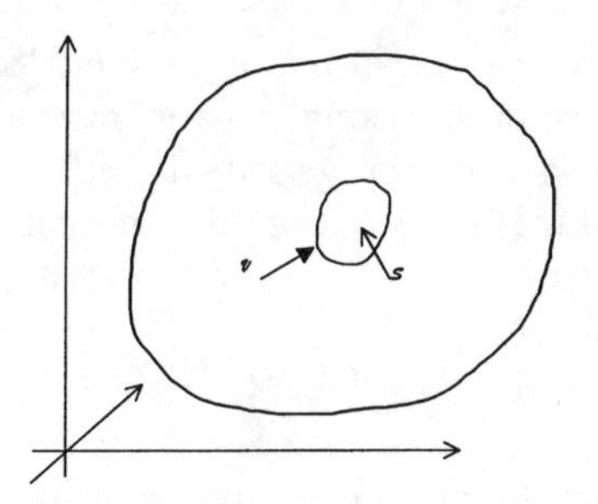

Figure 6: The heat flow rate through a small section of surface with area ΔA is $\mathbf{q} \cdot \hat{\mathbf{n}} \Delta A$.

and we make this replacement in Eq. (1). Next collect all terms on one side of the equation to find

$$\iiint_{\mathcal{V}} \left[-\nabla \cdot \mathbf{q} + g - \rho c \frac{\partial u}{\partial t} \right] dV = 0. \tag{3}$$

Because the subregion $\mathcal{V}$ was arbitrary, we conclude that the integrand must be 0 at every point:

$$-\nabla \cdot \mathbf{q} + g - \rho c \frac{\partial u}{\partial t} = 0 \quad \text{in} \quad \mathcal{R}, \quad 0 < t. \tag{4}$$

The argument goes this way. If the integrand were not identically 0, we could find some subregion of $\mathcal{R}$ throughout which it is positive (or negative). The integral over that subregion then would be positive (or negative), contradicting Eq. (3), which holds for any subregion.

The vector form of Fourier's law of heat conduction says that the heat flow rate in an isotropic solid (same properties in all directions) is negatively proportional to the temperature gradient,

$$\mathbf{q} = -\kappa \nabla u. \tag{5}$$

Again, the minus sign makes the heat flow "downhill" — from hotter to colder regions. Assuming that the conductivity κ is constant, we find on substituting Fourier's law into Eq. (4) the three-dimensional heat equation,

$$\kappa \nabla^2 u + g = \rho c \frac{\partial u}{\partial t} \quad \text{in} \quad \mathcal{R}, \quad 0 < t. \tag{6}$$

Of course, we must add an initial condition of the form

$$u(P, 0) = f(P) \quad \text{for} \quad P \quad \text{in} \quad \mathcal{R}. \tag{7}$$

In addition, at every point of the surface $\mathcal{B}$ bounding the region $\mathcal{R}$, some boundary condition must be specified. Commonly we have conditions like these that follow, any one of which may be given on $\mathcal{B}$ or some portion of it, $\mathcal{B}'$.

(1) Temperature specified, $u(P,t) = h_1(P,t)$, for P any point in $\mathcal{B}'$, where h_1 is a given function.

(2) Heat flow rate specified. The outward heat flow rate through a small portion of surface surrounding point P on $\mathcal{B}'$ is $\mathbf{q}(P,t) \cdot \hat{\mathbf{n}}$ times the area. If this is controlled, then by Fourier's law $\nabla u \cdot \hat{\mathbf{n}}$ is controlled. But this dot product is just the directional derivative of u in the outward normal direction at the point P. Thus, this type of boundary condition takes the form

$$\frac{\partial u}{\partial n}(P,t) = h_2(P,t) \quad \text{for } P \text{ on } \mathcal{B}' \tag{8}$$

where h_2 is a given function.

(3) Convection. If a part of the surface is exposed to a fluid at temperature $T(P,t)$, then an accounting of energy passing through a small piece of surface centered at P leads to the equation

$$\mathbf{q}(P,t) \cdot \hat{\mathbf{n}} = h(u(P,t) - T(P,t)) \quad \text{for } P \text{ on } \mathcal{B}'.$$

Again using Fourier's law, we obtain the boundary condition

$$\kappa \frac{\partial u}{\partial n}(P,t) + hu(P,t) = hT(P,t) \quad \text{for } P \text{ on } \mathcal{B}'. \tag{9}$$

As an example, we set up the three-dimensional problem for a solid in the form of a rectangular parallelepiped. In this case, Cartesian coordinates are appropriate, and we may describe the region $\mathcal{R}$ by the three inequalities, $0 < x < a$, $0 < y < b$, $0 < z < c$. Assuming no generation inside the object, we have the partial differential equation

$$\frac{\partial^2 u}{\partial x^2} + \frac{\partial^2 u}{\partial y^2} + \frac{\partial^2 u}{\partial z^2} = \frac{1}{k}\frac{\partial u}{\partial t}, \quad 0 < x < a, \quad 0 < y < b, \quad 0 < z < c, \quad 0 < t. \tag{10}$$

Suppose that on the faces at $x = 0$ and a, the temperature is controlled, so that the boundary condition there is

$$u(0,y,z,t) = T_0, \quad u(a,y,z,t) = T_1, \quad 0 < y < b, \quad 0 < z < c, \quad 0 < t. \tag{11}$$

Furthermore, assume that the top and bottom surfaces are insulated. Then the boundary conditions at $z = 0$ and c are

$$\frac{\partial u}{\partial z}(x,y,0,t) = 0, \quad \frac{\partial u}{\partial z}(x,y,c,t) = 0, \quad 0 < x < a, \quad 0 < y < b, \quad 0 < t. \tag{12}$$

(The outward normal directions on the top and bottom are the positive and negative z-directions, respectively.) Finally, assume that the faces at $y = 0$ and at $y = b$ are exposed to a fluid at temperature T_2, so that they transfer heat by convection there. The resulting boundary conditions are

$$-\kappa\frac{\partial u}{\partial y}(x,0,z,t)+hu(x,0,z,t) = hT_2,\ \ \kappa\frac{\partial u}{\partial y}(x,b,z,t)+hu(x,b,z,t) = hT_2,$$

$$0 < x < a, \quad 0 < z < c, \quad 0 < t. \tag{13}$$

Finally, we add an initial condition,

$$u(x, y, z, 0) = f(x, y, z), \quad 0 < x < a, \quad 0 < y < b, \quad 0 < z < c. \tag{14}$$

A full, three-dimensional problem is complicated to solve, so we often look for reasons to reduce it to two or even one dimension. In the example problem of Eqs. (10)-(14), we might eliminate z by finding the temperature averaged over the interval $0 < z < c$,

$$v(x, y, t) = \frac{1}{c}\int_0^c u(x, y, z, t)dz.$$

Because differentiation with respect to x, y or t gives the same result inside or outside the integral with respect to z, and because of the boundary condition (12), we find that

$$\frac{1}{c}\int_0^c \left(\frac{\partial^2 u}{\partial x^2} + \frac{\partial^2 u}{\partial y^2} + \frac{\partial^2 u}{\partial z^2}\right) dz = \frac{\partial^2 v}{\partial x^2} + \frac{\partial^2 v}{\partial y^2}$$

and v satisfies the two-dimensional heat equation

$$\frac{\partial^2 v}{\partial x^2} + \frac{\partial^2 v}{\partial y^2} = \frac{1}{k}\frac{\partial v}{\partial t}, \quad 0 < x < a, \quad 0 < y < b, \quad 0 < t.$$

(See the exercises for details and for boundary and initial conditions.)

If z-variation cannot be ignored, we could try to get rid of the y-variation by introducing an average in that direction,

$$w(x, z, t) = \frac{1}{b}\int_0^b u(x, y, z, t)dy.$$

From the boundary condition (13) we find that

$$\begin{aligned}\int_0^b \frac{\partial^2 u}{\partial y^2}dy &= \frac{\partial u}{\partial y}(x, b, z, t) - \frac{\partial u}{\partial y}(x, 0, z, t)\\ &= \left(\frac{h}{\kappa}\right)\left[(T_2 - u(x, b, z, t)) + (T_2 - u(x, 0, z, t))\right].\end{aligned}$$

If b is small — the parallepiped is more like a plate — we may accept the approximation $u(x,b,z,t) + u(x,0,z,t) \equiv 2w(x,z,t)$, which would make, from the expression above,

$$\frac{1}{b}\int_0^b \frac{\partial^2 u}{\partial y^2} dy \equiv \left(\frac{2h}{b\kappa}\right)(T_2 - w(x,z,t)).$$

After applying the averaging process to Eqs. (10), (11), (12) and (14) we obtain the following two-dimensional problem for w.

$$\frac{\partial^2 w}{\partial x^2} + \frac{\partial^2 w}{\partial z^2} + \frac{2h}{b\kappa}(T_2 - w) = \frac{1}{k}\frac{\partial w}{\partial t}, 0 < x < a, \quad 0 < z < c, \quad 0 < t \quad (15)$$

$$w(0,z,t) = T_0, \quad w(a,z,t) = t_1, \quad 0 < z < c, \qquad 0 < t \qquad (16)$$

$$\frac{\partial w}{\partial z}(x,0,t) = 0, \quad \frac{\partial w}{\partial z}(x,c,t) = 0, \quad 0 < x < a, \qquad 0 < t \qquad (17)$$

$$w(x,z,0) = \frac{1}{b}\int_0^b f(x,y,z)dy, \qquad 0 < x < a, \qquad 0 < z < c. \qquad (18)$$

Exercises

1. For the function $u(x,y,z,t)$ that satisfies Eqs. (10)-(14), show that

$$\int_0^c \frac{\partial^2 u}{\partial z^2} dz = 0.$$

2. Find the initial and boundary conditions satisfied by the function

$$v(x,y,t) = \frac{1}{c}\int_0^c u(x,y,z,t)dz$$

where u satisfies Eqs. (10)-(14).

3. In Eqs. (15)-(18), suppose that $w(x,z,t) \to W(x,z)$ as $t \to \infty$. State and solve the boundary value problem for W. (This problem is much easier than it appears, because there is no variation with z.)

4. Find the dimensions of ρ, c, κ, q, and g, and verify that the dimensions of the right and left members of the heat equation are the same.

5. Suppose the plate lies in the rectangle $0 < x < a$, $0 < y < b$. State a complete initial value-boundary value problem for temperature in the plate if: there is no heat generation; the temperature is held at T_0 along $x = a$ and $y = 0$; the edges at $x = 0$ and $y = b$ are insulated.

5.3 TWO-DIMENSIONAL HEAT EQUATION: SOLUTION

In order to see the technique of solution for a two-dimensional problem, we shall consider the diffusion of heat in a rectangular plate of uniform, isotropic material. The steady-state temperature distribution is a solution of the potential equation (see Exercise 6). Suppose that the initial value-boundary value problem for the transient temperature $u(x, y, t)$ is

$$\frac{\partial^2 u}{\partial x^2} + \frac{\partial^2 u}{\partial y^2} = \frac{1}{k}\frac{\partial u}{\partial t}, \qquad 0 < x < a, \qquad 0 < y < b, \quad 0 < t \qquad (1)$$

$$u(x, 0, t) = 0, \quad u(x, b, t) = 0, \quad 0 < x < a, \qquad 0 < t \qquad (2)$$

$$u(0, y, t) = 0, \quad u(a, y, t) = 0, \quad 0 < y < b, \qquad 0 < t \qquad (3)$$

$$u(x, y, 0) = f(x, y), \qquad 0 < x < a, \qquad 0 < y < b. \qquad (4)$$

This problem contains a homogeneous partial differential equation and homogeneous boundary conditions. We may thus proceed with separation of variables by seeking solutions in the form

$$u(x, y, t) = \phi(x, y)T(t).$$

On substituting u in product form into Eq. (1), we find that it becomes

$$\left(\frac{\partial^2 \phi}{\partial x^2} + \frac{\partial^2 \phi}{\partial y^2}\right) T = \frac{1}{k}\phi T'.$$

Separation can be achieved by dividing through by ϕT, which leaves

$$\left(\frac{\partial^2 \phi}{\partial x^2} + \frac{\partial^2 \phi}{\partial y^2}\right) \frac{1}{\phi} = \frac{T'}{kT}.$$

We may argue, as usual, that the common value of the members of this equation must be a constant, which we expect to be negative $(-\lambda^2)$. The equations that result are

$$T' + \lambda^2 kT = 0, \qquad 0 < t \qquad (5)$$

$$\frac{\partial^2 \phi}{\partial x^2} + \frac{\partial^2 \phi}{\partial y^2} = -\lambda^2 \phi, \qquad 0 < x < a, \quad 0 < y < b. \qquad (6)$$

In terms of the product solutions, the boundary conditions become

$$\phi(x, 0)T(t) = 0, \quad \phi(x, b)T(t) = 0$$

$$\phi(0, y)T(t) = 0, \quad \phi(a, y)T(t) = 0.$$

In order to satisfy all four equations, either $T(t) \equiv 0$ for all t, or $\phi = 0$ on the boundary. We have seen many times that the choice of $T(t) \equiv 0$ wipes out our solution completely. Therefore, we require that ϕ satisfy the conditions

$$\phi(x,0) = 0, \quad \phi(x,b) = 0, \quad 0 < x < a \tag{7}$$

$$\phi(0,y) = 0, \quad \phi(a,y) = 0, \quad 0 < y < b. \tag{8}$$

We are not yet out of difficulty, because Eqs. (6)-(8) constitute a new problem, a two-dimensional eigenvalue problem. It is evident, however, that the partial differential equation and the boundary conditions are linear and homogeneous; thus separation of variables may work again. Supposing that ϕ has the form

$$\phi(x,y) = X(x)Y(y)$$

we find that the partial differential equation (6) becomes

$$\frac{X''(x)}{X(x)} + \frac{Y''(y)}{Y(y)} = -\lambda^2, \quad 0 < x < a, \quad 0 < y < b.$$

The sum of a function of x and a function of y can be constant only if those two functions are individually constant:

$$\frac{X''}{X} = \text{constant}, \quad \frac{Y''}{Y} = \text{constant}.$$

Before naming the constants, let us look at the boundary conditions on $\phi = XY$:

$$X(x)Y(0) = 0, \quad X(x)Y(b) = 0, \quad 0 < x < a$$

$$X(0)Y(y) = 0, \quad X(a)Y(y) = 0, \quad 0 < y < b.$$

If either of the functions X or Y is zero throughout the whole interval of its variable, the conditions are certainly satisfied, but ϕ is identically zero. We therefore require each of the functions X and Y to be zero at the endpoints of its interval:

$$Y(0) = 0, \qquad Y(b) = 0 \tag{9}$$

$$X(0) = 0, \qquad X(a) = 0. \tag{10}$$

Now it is clear that each of the ratios X''/X and Y''/Y should be a negative constant, designated by $-\mu^2$ and $-\nu^2$, respectively. The separate equations for X and Y are

$$X'' + \mu^2 X = 0, \qquad 0 < x < a \tag{11}$$

$$Y'' + \nu^2 Y = 0, \qquad 0 < y < b. \tag{12}$$

Finally, the original separation constant $-\lambda^2$ is determined by

$$\lambda^2 = \mu^2 + v^2. \tag{13}$$

Now we see two independent eigenvalue problems: Eqs. (9) and (12) form one problem and Eqs. (10) and (11) the other. Each is of a very familiar form; the solutions are

$$X_m(x) = \sin\left(\frac{m\pi x}{a}\right), \quad \mu_m^2 = \left(\frac{m\pi}{a}\right)^2, \quad m = 1, 2, \cdots,$$

$$Y_n(y) = \sin\left(\frac{n\pi y}{b}\right), \quad \nu_n^2 = \left(\frac{n\pi}{b}\right)^2, \quad n = 1, 2, \cdots.$$

Notice that the indices n and m are independent. This means that ϕ will have a double index. Specifically, the solutions of the two-dimensional eigenvalue problem Eqs. (6)-(8) are

$$\phi_{mn}(x, y) = X_m(x)Y_n(y)$$

$$\lambda_{mn}^2 = \mu_m^2 + \nu_n^2$$

and the corresponding function T is

$$T_{mn} = \exp(-\lambda_{mn}^2 kt).$$

We now begin to assemble the solution. For each pair of indices m, n ($m = 1, 2, 3, \cdots, n = 1, 2, 3, \cdots$) there is a function

$$\begin{aligned} u_{mn}(x, y, t) &= \phi_{mn}(x, y)T_{mn}(t) \\ &= \sin\left(\frac{m\pi x}{a}\right)\sin\left(\frac{n\pi y}{b}\right)\exp(-\lambda_{mn}^2 kt) \end{aligned}$$

that satisfies the partial differential equatoin (1) and the boundary conditions Eqs. (2) and (3). We may form linear combinations of these solutions to get other solutions. The most general linear combination would be the double series

$$u(x, y, t) = \sum_{m=1}^{\infty}\sum_{n=1}^{\infty} a_{mn}\phi_{mn}(x, y)T_{mn}(t), \tag{14}$$

and any such combination should satisfy Eqs. (1)-(3). There remains the initial condition Eq. (4) to be satisfied. If u has the form given in Eq. (14), then the initial condition becomes

$$\sum_{m=1}^{\infty}\sum_{n=1}^{\infty} a_{mn}\phi_{mn}(x, y) = f(x, y), \quad 0 < x < a, \quad 0 < y < b. \tag{15}$$

The idea of orthogonality is once again applicable to the problem of selecting the coefficients a_{mn}. One can show by direct computation that

$$\int_0^b \int_0^a \phi_{mn}(x,y)\phi_{pq}(x,y)dxdy = \begin{cases} \dfrac{ab}{4} & \text{if } m = p \text{ and } n = q \\ 0, & \text{otherwise.} \end{cases} \tag{16}$$

Thus, the appropriate formula for the coefficients a_{mn} is

$$a_{mn} = \frac{4}{ab}\int_0^b \int_0^a f(x,y)\sin\left(\frac{m\pi x}{a}\right)\sin\left(\frac{n\pi y}{b}\right)dxdy.$$

If f is a sufficiently regular function, the series in Eq. (15) will converge and equal $f(x,y)$ in the rectangular region $0 < x < a$, $0 < y < b$. We may then say that the problem is solved. It is reassuring to notice that each term in the series of Eq. (14) contains a decaying exponential, and thus, as t increases, $u(x,y,t)$ tends to zero, as expected.

Let us take the specific initial condition

$$f(x,y) = xy, \quad 0 < x < a, \quad 0 < y < b.$$

The coefficients are easily found to be

$$a_{mn} = \frac{4ab}{\pi^2}\frac{\cos(m\pi)\cos(n\pi)}{mn} = \frac{4ab}{\pi^2}\frac{(-1)^{m+n}}{mn}$$

whence the solution to this problem is

$$u(x,y,t) = \frac{4ab}{\pi^2}\sum_{m=1}^{\infty}\sum_{n=1}^{\infty}\frac{(-1)^{m+n}}{mn}\sin\left(\frac{m\pi x}{a}\right)\sin\left(\frac{n\pi y}{b}\right)\exp(-\lambda_{mn}^2 kt). \tag{17}$$

The double series that appear here are best handled by converting them into single series. To do this, arrange the terms in order of increasing values of λ_{mn}^2. Then the first terms in the single series are the most significant, those that decay least rapidly. For example, if $a = 2b$, so that

$$\lambda_{mn}^2 = \frac{(m^2 + 4n^2)\pi^2}{a^2},$$

then the following list gives the double index (m,n) in order of increasing values of λ_{mn}^2:

$$(1,1), (2,1), (3,1), (1,2), (2,2), (4,1), (3,2), \cdots.$$

Exercises

1. Write out the "first few" terms of the series of Eq. (17). By "first few," we mean those for which λ_{mn}^2 is smallest. (Assume $a = b$ in determining relative magnitudes of the λ^2.)

2. Provide the details of the separation of variables by which Eqs. (9)-(13) are derived.

3. Find the frequencies of vibration of a rectangular membrane. See Section 1, Exercise 1.

4. Verify that $u_{mn}(x, y, t)$ satisfies Eqs. (1)-(3).

5. Show that $X_m(x) = \cos(m\pi x/a)$ $(m = 0, 1, 2, \cdots)$ if the boundary conditions Eq. (3) are replaced by

$$\frac{\partial u}{\partial x}(0, y, t) = 0, \quad \frac{\partial u}{\partial x}(a, y, t) = 0, \quad 0 < y < b, \quad 0 < t.$$

What values will the λ_{mn}^2 have, and of what form will the solution $u(x, y, t)$ be?

6. Suppose that, instead of boundary conditions Eqs. (2) and (3), we have

$$u(x, 0, t) = f_1(x), \quad u(x, b, t) = f_2(x), \quad 0 < x < a, \quad 0 < t \quad (2')$$

$$u(0, y, t) = g_1(7), \quad u(a, y, t) = g_2(y), \quad 0 < y < b, \quad 0 < t. \quad (3')$$

Show that the steady-state solution involves the potential equation, and indicate how to solve it.

7. Solve the two-dimensional heat conduction problem in a rectangle if there is insulation on all boundaries and the initial condition is

a. $u(x, y, 0) = 1$

b. $u(x, y, 0) = x + y$

c. $u(x, y, 0) = xy$.

8. Verify the orthogonality relation in Eq. (16) and the formula for a_{mn}.

9. Show that the separation constant $-\lambda^2$ must be negative by showing that $-\mu^2$ and $-\nu^2$ must both be negative.

5.4 PROBLEMS IN POLAR COORDINATES

We found that the one-dimensional wave and heat problems have a great deal in common. Namely, the steady-state or time-independent solutions and the eigenvalue problems that arise are identical in both cases. Also, in solving problems in a rectangular region, we have seen that those same features are shared by the heat and wave equations.

If we consider now the vibrations of a circular membrane or heat conduction in a circular plate, we shall see common features again. In what follows, these two problems are given side by side for the region $0 < r < a$, $-\pi < \theta \leq \pi$, $0 < t$.

Wave	**Heat**
$\nabla^2 v = \frac{1}{c^2}\frac{\partial^2 v}{\partial t^2}$	$\nabla^2 v = \frac{1}{k}\frac{\partial v}{\partial t}$
$v(a,\theta,t) = f(\theta)$	$v(a,\theta,t) = f(\theta)$
$v(r,\theta,0) = g(r,\theta)$	$v(r,\theta,0) = g(r,\theta)$
$\frac{\partial v}{\partial t}(r,\theta,0) = h(r,\theta)$	

In both problems we require

$$v(r,-\pi,t) = v(r,\pi,t), \quad 0 < r < a, \quad 0 < t$$

$$\frac{\partial v}{\partial \theta}(r,-\pi,t) = \frac{\partial v}{\partial \theta}(r,\pi,t), \quad 0 < r < a, \quad 0 < t$$

so that no discontinuity will arise at the artificial boundaries $\theta = \pm\pi$.

Although the interpretation of the function v is different in the two cases, we see that the solution of the problem

$$\nabla^2 v = 0, \quad v(a,\theta) = f(\theta)$$

is the rest-state or steady-state solution for both problems, and will be needed in both problems to make the boundary condition at $r = a$ homogeneous. Let us suppose that the time-independent solution has been found and subtracted; that is, we will replace $f(\theta)$ by zero. Then we have

$\nabla^2 v = \frac{1}{c^2}\frac{\partial^2 v}{\partial t^2}$	$\nabla^2 v = \frac{1}{k}\frac{\partial v}{\partial t}$
$v(a,\theta,t) = 0$	$v(a,\theta,t) = 0$

plus the appropriate initial conditions. If we attempt to solve by separation of variables, setting $v(r,\theta,t) = \phi(r,\theta)T(t)$, in both cases we will find that $\phi(r, \theta)$ must satisfy

$$\nabla^2\phi = -\lambda^2\phi, \qquad 0 < r < a, \quad -\pi < \theta \pm \pi \tag{1}$$

$$\phi(a, \theta) = 0, \qquad -\pi < \theta \leq \pi \tag{2}$$

$$\phi(r, -\pi) = \phi(r, \pi), \qquad 0 < r < a \tag{3}$$

$$\frac{\partial \phi}{\partial \theta}(r, -\pi) = \frac{\partial \phi}{\partial \theta}(r, \pi), \quad 0 < r < a. \tag{4}$$

Now we shall concentrate on the solution of this two-dimensional eigenvalue problem. Written out in polar coordinates, Eq. (1) becomes

$$\frac{1}{r}\frac{\partial}{\partial r}\left(r\frac{\partial \phi}{\partial r}\right) + \frac{1}{r^2}\frac{\partial^2 \phi}{\partial \theta^2} = -\lambda^2\phi.$$

We can separate variables again by assuming that $\phi(r, \theta) = R(r)\Theta(\theta)$. After some algebra, we find that

$$\frac{(rR')'}{rR} + \frac{\Theta''}{r^2\Theta} = -\lambda^2 \tag{5}$$

$$R(a) = 0 \tag{6}$$

$$\Theta(-\pi) = \Theta(\pi) \tag{7}$$

$$\Theta'(-\pi) = \Theta'(\pi). \tag{8}$$

The ratio Θ''/Θ must be constant; otherwise, λ^2 could not be constant. Choosing $\Theta''/\Theta = -\mu^2$, we get a familiar problem

$$\Theta'' + \mu^2\Theta = 0, \qquad -\pi < \theta \leq \pi \tag{9}$$

$$\Theta(-\pi) = \Theta(\pi) \tag{10}$$

$$\Theta'(-\pi) = \Theta'(\pi). \tag{11}$$

We found (in Chapter 4) that the solutions of this problem are

$$\begin{aligned} \mu_0^2 &= 0, \quad \Theta(\theta) = 1 \\ \mu_m^2 &= m^2, \quad \Theta(\theta) = \cos(m\theta) \;\text{ and }\; \sin(m\theta) \end{aligned} \tag{12}$$

where $m = 1, 2, 3, \cdots$.

There remains a problem in R:

$$(rR')' - \frac{\mu^2}{r}R + \lambda^2 rR = 0, \quad 0 < r < a \tag{13}$$

$$R(a) = 0. \tag{14}$$

Equation (13) is called *Bessel's equation*, and we shall solve it in the next section.

Exercises

1. State the full initial value-boundary value problems that result from the problems as originally given when the steady-state or time-independent solution is subtracted from v.

2. Verify the separation of variables that leads to Eqs. (1) and (2).

3. Substitution of $v(r, \theta, t)$ in the form of a product led to the problem of Eqs. (1)-(4) for the factor $\phi(r, \theta)$. What differential equation is to be satisfied by the factor $T(t)$?

4. Solve Eqs. (9)-(11) and check the solutions given.

5. Suppose the problems originally stated were to be solved in the half-disk $0 < r < a$, $0 < \theta < \pi$, with additional conditions.

$$v(r, 0, t) = 0, \quad 0 < r < a, \quad 0 < t$$

$$v(r, \pi, t) = 0, \quad 0 < r < a, \quad 0 < t.$$

What eigenvalue problem arises in place of Eqs. (9)-(11)? Solve it.

6. Suppose that the boundary condition

$$\frac{\partial v}{\partial r}(a, \theta, t) = 0, \quad -\pi < \theta \leq \pi, \quad 0 < t$$

were given instead of $v(a, \theta, t) = f(\theta)$. Carry out the steps involved in separation of variables. Show that the only change is in Eq. (14), which becomes $R'(a) = 0$.

7. One of the consequences of Green's theorem is the integral relation

$$\iint_{\mathcal{R}} (f\nabla^2 g - g\nabla^2 f)\, dA = \int_{\mathcal{C}} \left(f\frac{\partial g}{\partial n} - g\frac{\partial f}{\partial n} \right) ds$$

where $\mathcal{R}$ is a region in the plane, $\mathcal{C}$ is the closed curve that bounds $\mathcal{R}$, and $\partial f/\partial n$ is the directional derivative in the direction normal to the curve $\mathcal{C}$. Use this relation to show that eigenfunctions of the problem

$$\nabla^2 \phi = -\lambda^2 \phi, \quad \text{in} \quad \mathcal{R}$$

$$\phi = 0, \qquad \text{on} \quad \mathcal{C}$$

are orthogonal if they correspond to different eigenvalues. (Hint: use $f = \phi_k$, $g = \phi_m$, $m \neq k$.)

8. Same problem as Exercise 7, except that the boundary condition is

$$\phi + \lambda \frac{\partial \phi}{\partial n} = 0 \quad \text{on } \mathcal{C}.$$

5.5 BESSEL'S EQUATION

In order to solve the Bessel equation

$$(rR')' - \frac{\mu^2}{r} R + \lambda^2 r R = 0 \tag{1}$$

we apply the method of Frobenius. Assume that $R(r)$ has the form of a power series multiplied by an unknown power of r:

$$R(r) = r^\alpha (c_0 + c_1 r + \cdots + c_k r^k + \cdots). \tag{2}$$

When the differentiations in Eq. (1) are carried out and the equation is multiplied by r, it becomes

$$\begin{aligned}
r^2 R'' &= \alpha(\alpha-1)c_0\, r^\alpha + (\alpha+1)\alpha c_1\, r^{\alpha+1} + (\alpha+2)(\alpha+1)c_2 r^{\alpha+2} + \cdots \\
&\qquad + (\alpha+k)(\alpha+k-1)c_k\, r^{\alpha+k} + \cdots \\
rR' &= \alpha c_0\, r^\alpha + (\alpha+1)c_1\, r^{\alpha+1} + (\alpha+2)c_2 r^{\alpha+2} + \cdots \\
&\qquad + (\alpha+k)c_k\, r^{\alpha+k} + \cdots \\
-\mu^2 R &= -\mu^2 c_0 r^\alpha - \mu^2 c_1\, r^{\alpha+1} - \mu^2 c_2 r^{\alpha+2} - \cdots \\
&\qquad - \mu^2 c_k\, r^{\alpha+k} + \cdots \\
\lambda^2 r^2 R &= \lambda^2 c_0 r^{\alpha+2} + \cdots \\
&\qquad + \lambda^2 c_{k-2}\, r^{\alpha+k} + \cdots
\end{aligned}$$

The expression for $\lambda^2 r^2 R$ is jogged to the right to make like powers of r line up vertically. Note that the lowest power of r present in $\lambda^2 r^2 R$ is $r^{\alpha+2}$.

Now we add the tableau vertically. The sum of the left-hand sides is, according to the differential equation, equal to zero. Therefore

$$\begin{aligned}
0 = c_0(\alpha^2 - \mu^2) r^\alpha &+ c_1 \left[(\alpha+1)^2 - \mu^2\right] r^{\alpha+1} \\
&+ \left[c_2\left((\alpha+2)^2 - \mu^2\right) + \lambda^2 c_0\right] r^{\alpha+2} \\
&+ \cdots + \left[c_k\left((\alpha+k)^2 - \mu^2\right) + \lambda^2 c_{k-2}\right] r^{\alpha+k} + \cdots.
\end{aligned}$$

Each term in this power series must be zero in order for the equality to hold. Therefore, the coefficient of each term must be zero:

$$\begin{aligned} c_0(\alpha^2 - \mu^2) &= 0 \\ c_1\left((\alpha+1)^2 - \mu^2\right) &= 0 \\ &\vdots \\ c_k\left((\alpha+k)^2 - \mu^2\right) + \lambda^2 c_{k-2} &= 0, \quad k \geq 2. \end{aligned}$$

As a bookkeeping agreement, we take $c_0 \neq 0$. Thus $\alpha = \pm\mu$. Let us study the case $\alpha = \mu \geq 0$. The second equation becomes

$$c_1\left((\mu+1)^2 - \mu^2\right) = 0$$

and this implies $c_1 = 0$. Now in general the relation

$$c_k = -\frac{\lambda^2 c_{k-2}}{(\mu+k)^2 - \mu^2} = -\lambda^2 \frac{c_{k-2}}{k(2\mu+k)}, \quad k \geq 2 \tag{3}$$

says that c_k can be found from c_{k-2}. In particular, we find

$$\begin{aligned} c_2 &= -\frac{\lambda^2}{2(2\mu+2)} c_0 \\ c_4 &= -\frac{\lambda^2}{4(2\mu+4)} c_2 = \frac{\lambda^4}{2 \cdot 4 \cdot (2\mu+2)(2\mu+4)} c_0 \end{aligned}$$

and so forth. All c's with odd index are zero, since they are all multiples of c_1. The general formula for a coefficient with even index $k = 2m$ is

$$c_{2m} = \frac{(-1)^m}{m!(\mu+1)(\mu+2)\cdots(\mu+m)} \left(\frac{\lambda}{2}\right)^{2m} c_0. \tag{4}$$

For integral values of μ, c_0 is chosen by convention to be

$$c_0 = \left(\frac{\lambda}{2}\right)^{\mu} \cdot \frac{1}{\mu!}.$$

Then the solution of Eq. (1) that we have found is called the *Bessel function of the first kind of order* μ:

$$J_\mu(\lambda r) = \left(\frac{\lambda r}{2}\right)^{\mu} \sum_{m=0}^{\infty} \frac{(-1)^m}{m!(\mu+m)!} \left(\frac{\lambda r}{2}\right)^{2m}. \tag{5}$$

This series serves us for evaluating the function and for obtaining its properties. (See the Exercises.) However, *from now on, we consider the Bessel functions of the first kind to be as well known as sines and cosines,* although less familiar.

There must be a second independent solution of Bessel's equation, which can be found by using variation of parameters. This method yields a solution in the form

$$J_\mu(\lambda r) \cdot \int \frac{dr}{r J_\mu^2(\lambda r)}. \tag{6}$$

In its standard form, the second solution of Bessel's equation is called the *Bessel function of second kind of order* μ and is denoted by $Y_\mu(\lambda r)$.

The most important feature of the second solution is its behavior near $r = 0$. When r is very small, we can approximate $J_\mu(\lambda r)$ by the first term of its series expansion:

$$J_\mu(\lambda r) \cong \left(\frac{\lambda}{2}\right)^\mu \frac{1}{\mu!} r^\mu, \quad r \ll 1.$$

The solution Eq. (6) then can be approximated by

$$\text{constant} \times r^\mu \int \frac{dr}{r^{1+2\mu}} = \text{constant} \times \begin{cases} \ln(r), & \text{if} \quad \mu = 0 \\ r^{-\mu}, & \text{if} \quad \mu > 0. \end{cases}$$

In either case, it is easy to see that

$$|Y_\mu(\lambda r)| \to \infty, \quad \text{as} \quad r \to 0.$$

Both kinds of Bessel functions have an infinite number of zeros. That is, there are an infinite number of values of α (and β) for which

$$J_\mu(\alpha) = 0, \quad Y_\mu(\beta) = 0.$$

Also, as $r \to \infty$, both $J_\mu(\lambda r)$ and $Y_\mu(\lambda r)$ tend to zero. Figure 7 gives graphs of several Bessel functions, and Table 1 provides values of their zeros. Further information can be found in most books of tables.

The modified Bessel equation differs from the Bessel equation only in the sign of one term. It is

$$(rR')' - \frac{\mu^2}{r} R - \lambda^2 r R = 0. \tag{7}$$

Using the same method as in the preceding, an infinite series can be developed for the solutions (see Exercise 8). The solution that is bounded at $r = 0$, in standard form, is called the *modified Bessel function of the first kind of order* μ, designated $I_\mu(\lambda r)$, and its series is

$$I_\mu(\lambda r) = \left(\frac{\lambda r}{2}\right)^\mu \sum_{m=0}^{\infty} \frac{1}{m!(\mu+m)!} \left(\frac{\lambda r}{2}\right)^{2m}.$$

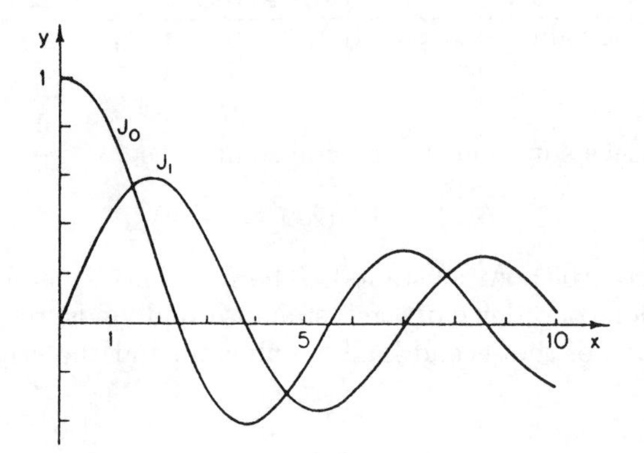

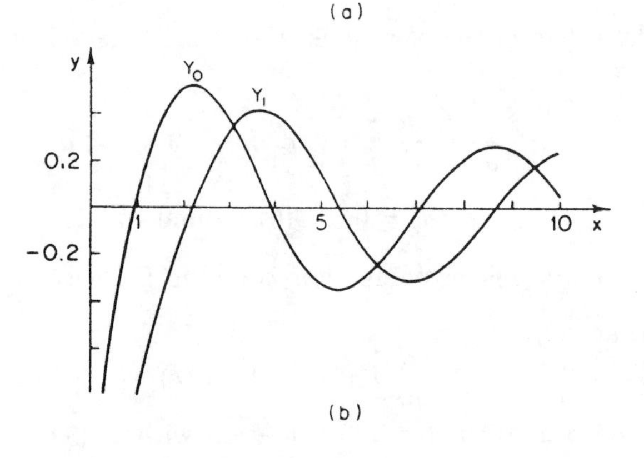

Figure 7: Graphs of Bessel functions of the first kind.

Summary.
The differential equation

$$\frac{d}{dr}\left(r\frac{dR}{dr}\right) - \frac{\mu^2}{r}R + \lambda^2 rR = 0$$

Table 1: Zeros of Bessel functions[a]

m \ n	1	2	3	4
0	2.405	5.520	8.654	11.792
1	3.832	7.016	10.173	13.324
2	5.136	8.417	11.620	14.796
3	6.380	9.761	13.015	16.223

[a]The values α_{mn} satisfy the equation $J_m(\alpha_{mn}) = 0$.

is called Bessel's equation. Its general solution is

$$R(r) = AJ_\mu(\lambda r) + BY_\mu(\lambda r)$$

(A and B are arbitrary constants). The functions J_μ and Y_μ are called Bessel functions of order μ of the first and second kinds, respectively. The Bessel function of the second kind is unbounded at the origin.

Exercises

1. Find the values of the parameter λ for which the following problem has a nonzero solution:

$$\frac{1}{r}\frac{d}{dr}\left(r\frac{d\phi}{dr}\right) + \lambda^2\phi = 0, \quad 0 < r < a$$

$$\phi(a) = 0, \quad \phi(0) \text{ bounded.}$$

2. Sketch the first few eigenfunctions found in Exercise 1.

3. Show that

$$\frac{d}{dr}J_\mu(\lambda r) = \lambda J'_\mu(\lambda r)$$

where the prime denotes differentiation with respect to the argument.

4. Show from the series that

$$\frac{d}{dr}J_0(\lambda r) = -\lambda J_1(\lambda r).$$

5. By using Exercise 4 and Rolle's theorem, and knowing that $J_0(x) = 0$ for an infinite number of values of x, show that $J_1(x) = 0$ has an infinite number of solutions.

6. Using the infinite series representations for the Bessel functions, verify the formulas

$$\frac{d}{dx}\left(x^{-\mu}J_\mu(x)\right) = -x^{-\mu}J_{\mu+1}(x)$$

$$\frac{d}{dx}\left(x^{\mu}J_\mu(x)\right) = x^{\mu}J_{\mu-1}(x).$$

7. Use the second formula in Exercise 6 to derive the integral formula

$$\int x^{\mu}J_\mu(x)x\,dx = x^{\mu+1}J_{\mu+1}(x).$$

8. Use the method of Frobenius to obtain a solution of the modified Bessel equation (7). Show that the coefficients of the power series are just the same as those for the Bessel function of the first kind, except for signs.

9. Use the modified Bessel function to solve this problem for the temperature in a circular plate, when the surface is exposed to convection.

$$\frac{1}{r}\frac{d}{dr}\left(r\frac{du}{dr}\right) - \gamma^2(u - T) = 0, \quad 0 < r < a$$

$$u(a) = T_1.$$

10. Using the result of Exercise 4, solve the eigenvalue problem

$$\frac{1}{r}\frac{d}{dr}\left(r\frac{d\phi}{dr}\right) + \lambda^2\phi = 0, \quad 0 < r < a$$

$$\frac{d\phi}{dr}(a) = 0, \quad \phi(0) \text{ bounded.}$$

5.6 TEMPERATURE IN A CYLINDER

In Section 4, we observed that both the heat and the wave equations have a great deal in common, especially the equilibrium solution and the eigenvalue problem. To reinforce that observation, we will solve a heat problem and a wave problem with analogous conditions so that their similarities may be seen. These examples illustrate another important point: Problems which would be two-dimensional in one coordinate system (rectangular) may become one-dimensional in another system (polar). In order to obtain this simplification, we will assume that the unknown function,

$v(r, \theta, t)$, is actually independent of the angular coordinate θ. (We write $v(r, t)$ then.) As a consequence of this assumption, the two-dimensional Laplacian operator becomes

$$\nabla^2 v = \frac{1}{r}\frac{\partial}{\partial r}\left(r\frac{\partial v}{\partial r}\right).$$

Suppose that the temperature $v(r, t)$ in a large cylinder (radius a) satisfies the problem

$$\frac{1}{r}\frac{\partial}{\partial r}\left(r\frac{\partial v}{\partial r}\right) = \frac{1}{k}\frac{\partial v}{\partial t}, \quad 0 < r < a, \qquad 0 < t \tag{1}$$

$$v(a, t) = 0, \qquad 0 < t \tag{2}$$

$$v(r, 0) = f(r), \quad 0 < r < a. \tag{3}$$

As the differential equation (1) and boundary condition (2) are homogeneous, we may start the separation of variables by assuming $v(r, t) = \phi(r)T(t)$. Using this form for v, we find that the partial differential equation (1) becomes

$$\frac{1}{r}(r\phi')'T = \frac{1}{k}\phi T'.$$

After dividing through this equation by ϕT, we arrive at the equality

$$\frac{(r\phi'(r))'}{r\phi(r)} = \frac{T'(t)}{kT(t)}. \tag{4}$$

The two members of this equation must both be constant; call their mutual value $-\lambda^2$. Then we have two linked ordinary differential equations,

$$T' + \lambda^2 kT = 0, \qquad 0 < t \tag{5}$$

$$(r\phi')' + \lambda^2 r\phi = 0, \qquad 0 < r < a. \tag{6}$$

The boundary condition, Eq. (2), becomes $\phi(a)T(t) = 0$, $0 < t$. It will be satisfied by requiring that

$$\phi(a) = 0. \tag{7}$$

We can recognize Eq. (6) as Bessel's equation with $\mu = 0$. (See Summary, Section 5.) The general solution, therefore, has the form

$$\phi(r) = AJ_0(\lambda r) + BY_0(\lambda r).$$

If $B \neq 0$, $\phi(r)$ must become infinite as r approaches zero. The physical implications of this possibility are unacceptable, so we require that $B = 0$. In effect we have added the boundedness condition

$$|v(r,t)| \quad \text{bounded at} \quad r = 0, \tag{8}$$

which we shall employ frequently.

The function $\phi(r) = J_0(\lambda r)$ is a solution of Eq. (6), and we wish to choose λ so that Eq. (7) is satisfied. Then we must have

$$J_0(\lambda a) = 0$$

or

$$\lambda_n = \frac{\alpha_n}{a}, \quad n = 1, 2, \cdots,$$

where α_n are the zeros of the function J_0. Thus the eigenfunctions and eigenvalues of Eqs. (6), (7), and (8) are

$$\phi_n(r) = J_0(\lambda_n r), \quad \lambda_n^2 = \left(\frac{\alpha_n}{a}\right)^2. \tag{9}$$

Returning to Eq. (5), we determine that the time factors T_n are

$$T_n(t) = \exp(-\lambda_n^2 kt).$$

We may now assemble the general solution of the partial differential equation (1), under the boundary condition (2) and boundedness condition (8), as a general linear combination of our product solutions:

$$v(r,t) = \sum_{n=1}^{\infty} a_n J_0(\lambda_n r) \exp(-\lambda_n^2 kt). \tag{10}$$

It remains to determine the coefficients a_n so as to satisfy the initial condition (3), which now takes the form

$$v(r,0) = \sum_{n=1}^{\infty} a_n J_0(\lambda_n r) = f(r), \quad 0 < r < a. \tag{11}$$

While this problem is not a routine exercise in Fourier series, nor even a regular Sturm-Liouville problem (see Section 7 of Chapter 2, and especially Exercise 6 there), it is nevertheless true that the eigenfunctions of Eqs. (6) and (7) are orthogonal, indexBessel function!orthogonality as expressed by the relation

$$\int_0^a \phi_n(r)\phi_m(r) r \, dr = 0 \quad (n \neq m)$$

or

$$\int_0^a J_0(\lambda_n r) J_0(\lambda_m r) r\, dr = 0 \quad (n \neq m).$$

More importantly, the following theorem gives us justification for Eq. (11).

Theorem

If $f(r)$ is sectionally smooth on the interval $0 < r < a$, then at every point r on that interval,

$$\sum_{n=1}^{\infty} a_n J_0(\lambda_n r) = \frac{f(r+) + f(r-)}{2}, \quad 0 < r < a, \tag{12}$$

where the λ_n are solutions of $J_0(\lambda a) = 0$, and

$$a_n = \frac{\displaystyle\int_0^a f(r) J_0(\lambda_n r) r dr}{\displaystyle\int_0^a J_0^2(\lambda_n r) r dr}. \tag{13}$$

Now we may proceed with the problem at hand. If the function $f(r)$ in the initial condition (3) is sectionally smooth, the use of Eq. (13) to chose the coefficients a_n guarantees that Eq. (11) is satisfied (as nearly as possible) and hence the function

$$v(r,t) = \sum_{n=1}^{\infty} a_n J_0(\lambda_n r) \exp(-\lambda_n^2 kt)$$

satisfies the problem expressed by Eqs. (1), (2), (3), and (8).

By way of example, let us suppose that the function $f(r) = T_0$, $0 < r < a$. It is necessary to determine the coefficients a_n by formula (13). The numerator is the integral

$$\int_0^a T_0 J_0(\lambda_n r) r dr.$$

This integral is evaluated by means of the relation (see Exercise 6 of Section 5)

$$\frac{d}{dx}(x J_1(x)) = x J_0(x). \tag{14}$$

Hence, we find

$$\int_0^a J_0(\lambda_n r) r dr = \frac{1}{\lambda_n} r J_1(\lambda_n r) \Big|_0^a$$

$$= \frac{a}{\lambda_n} J_1(\lambda_n a) = \frac{a^2}{\alpha_n} J_1(\alpha_n). \tag{15}$$

The denominator of Eq. (13) is known to have the value (Exercise 5)

$$\begin{aligned} \int_0^a J_0^2(\lambda_n r) r dr &= \frac{a^2}{2} J_1^2(\lambda_n a) \\ &= \frac{a^2}{2} J_1^2(\alpha_n). \end{aligned} \tag{16}$$

Table 2: Values for equation (18)

n	α_n	$J_1(\alpha_n)$	$\frac{2}{\alpha_n J_1(\alpha_n)}$
1	2.405	+0.5191	+1.6020
2	5.520	−0.3403	−1.0647
3	8.654	+0.2715	+0.8512
4	11.792	−0.2325	−0.7295

Putting together the numerator and denominator from Eqs. (15) and (16), we find that the coefficients we need are

$$a_n = \frac{2T_0}{\alpha_n J_1(\alpha_n)}. \tag{17}$$

Thus, the solution to the heat conduction problem is

$$v(r,t) = T_0 \sum_{n=1}^{\infty} \frac{2}{\alpha_n J_1(\alpha_n)} J_0(\lambda_n r) \exp(-\lambda_n^2 kt). \tag{18}$$

In Table 2 are listed the first few values of the ratio $2/[\alpha_n J_1(\alpha_n)]$. Figure 8 shows graphs of $v(r,t)$ as a function of r for several times. Also, see Exercise 1.

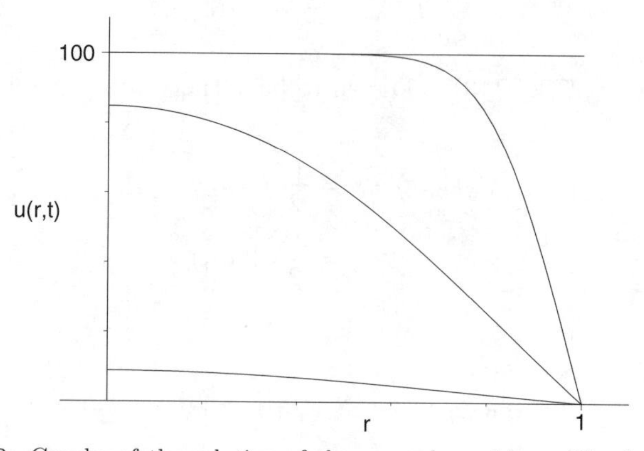

Figure 8: Graphs of the solution of the example problem. The function $v(r,t)$ as given in Eq. (18) is shown vs r for times chosen so that kt/a^2 takes the values 0, 0.01, 0.1 and 0.5. $T_0 = 100$, and $a = 1$.

Exercises

1. Use Eq. (18) to find an expression for the function $v(0,t)/T_0$. Evaluate the function for

$$\frac{kt}{a^2} = 0.1, 0.2, 0.3.$$

(The first two terms of the series are sufficient.)

2. Write out the first three terms of the series in Eq. (18).

3. Solve the heat problem consisting of Eqs. (1)-(3) if $f(r)$ is

$$f(r) = \begin{cases} T_0, & 0 < r < \dfrac{a}{2} \\ 0, & \dfrac{a}{2} < r < a. \end{cases}$$

4. Let $\phi(r) = J_0(\lambda r)$, so that $\phi(r)$ satisfies Bessel's equation of order 0. Multiply through the differential equation by $r\phi'$ and conclude that

$$\frac{d}{dr}\left[(r\phi')^2\right] + \lambda^2 r^2 \frac{d}{dr}[\phi^2] = 0.$$

5. Assuming that λ is chosen so that $\phi(a) = 0$, integrate the equation above over the interval $0 < r < a$ to find

$$\int_0^a \phi^2(r) r dr = \frac{1}{2\lambda^2}(a\phi'(a))^2.$$

6. Use Exercise 5 to validate Eq. (16).

5.7 VIBRATIONS OF A CIRCULAR MEMBRANE

We shall now attempt to solve the problem of describing the displacement of a circular membrane that is fixed at its edge. To begin with, we treat the simple case in which the initial conditions are independent of θ. Thus the displacement $v(r,t)$ satisfies the problem

$$\frac{1}{r}\frac{\partial}{\partial r}\left(r\frac{\partial v}{\partial r}\right) = \frac{1}{c^2}\frac{\partial^2 v}{\partial t^2}, \quad 0 < r < a, \quad 0 < t \tag{1}$$

$$v(a,t) = 0, \quad 0 < t \tag{2}$$

$$v(r,0) = f(r), \quad 0 < r < a \tag{3}$$

$$\frac{\partial v}{\partial t}(r,0) = g(r), \quad 0 < r < a. \tag{4}$$

We start immediately with separation of variables, assuming $v(r,t) = \phi(r)T(t)$. The differential equation (1) becomes

$$\frac{1}{r}(r\phi')'T = \frac{1}{c^2}\phi T''$$

and the variables may be separated by dividing by ϕT. Then we find

$$\frac{(r\phi'(r))'}{r\phi(r)} = \frac{T''(t)}{c^2 T(t)}.$$

The two sides must both be equal to a constant (say $-\lambda^2$) yielding two linked, ordinary differential equations

$$T'' + \lambda^2 c^2 T = 0, \quad 0 < t \tag{5}$$

$$(r\phi')' + \lambda^2 r\phi = 0, \quad 0 < r < a. \tag{6}$$

The boundary condition Eq. (2) is satisfied if

$$\phi(a) = 0. \tag{7}$$

Of course, because $r = 0$ is a singular point of the differential equation (6), we add the requirement

$$|\phi(r)| \text{ bounded at } r = 0, \tag{8}$$

which is equivalent to requiring that $|v(r,t)|$ be bounded at $r = 0$. We recognize that Eq. (6) is Bessel's equation, of which the function $\phi(r) = J_0(\lambda r)$ is the solution bounded at $r = 0$. In order to satisfy the boundary condition Eq. (7), we must have

$$J_0(\lambda a) = 0,$$

which implies that

$$\lambda_n = \frac{\alpha_n}{a}, \quad n = 1, 2, \cdots \tag{9}$$

where α_n are the zeros of the function J_0. Thus the eigenfunctions and eigenvalues of Eqs.(6)-(8) are

$$\phi_n(r) = J_0(\lambda_n r), \quad \lambda_n^2 = \left(\frac{\alpha_n}{a}\right)^2.$$

The rest of our problem can now be dispatched easily. Returning to Eq. (5), we see that

$$T_n(t) = a_n \cos(\lambda_n ct) + b_n \sin(\lambda_n ct),$$

and then for each $n = 1, 2, \cdots$ we have a solution of Eqs. (1), (2), and (8)

$$v_n(r,t) = \phi_n(r) T_n(t).$$

The most general linear combination of the v_n would be

$$v(r,t) = \sum_{n=1}^{\infty} J_0(\lambda_n r) \left[a_n \cos(\lambda_n ct) + b_n \sin(\lambda_n ct)\right]. \tag{10}$$

The initial conditions Eqs. (3) and (4) are satisfied if

$$v(r,0) = \sum_{n=1}^{\infty} a_n J_0(\lambda_n r) = f(r), \quad 0 < r < a$$

$$\frac{\partial v}{\partial t}(r,0) = \sum_{n=1}^{\infty} b_n \lambda_n c J_0(\lambda_n r) = g(r), \quad 0 < r < a.$$

As in the preceding section, the coefficients of these series are to be found through the integral formulas

$$a_n = \frac{1}{I_n}\int_0^a f(r)J_0(\lambda_n r)rdr, \quad b_n = \frac{1}{\lambda_n c I_n}\int_0^a g(r)J_0(\lambda_n r)rdr,$$

$$I_n = \int_0^a [J_0(\lambda_n r)]^2 rdr.$$

With the coefficients determined by these formulas, the function given in Eq. (10) is the solution to the vibrating membrane problem that we started with.

Having seen the simplest case of the vibrations of a circular membrane, we return to the more general case. The full problem was

$$\frac{1}{r}\frac{\partial}{\partial r}\left(r\frac{\partial u}{\partial r}\right) + \frac{1}{r^2}\frac{\partial^2 u}{\partial \theta^2} = \frac{1}{c^2}\frac{\partial^2 u}{\partial t^2}, \quad 0 < r < a, \quad 0 < t, -\pi < \theta \le \pi. \tag{11}$$

$$u(a,\theta,t) = 0, \qquad 0 < t, \qquad -\pi < \theta \le \pi \tag{12}$$

$$|u(0,\theta,t)| \text{ bounded}, \qquad 0 < t, \qquad -\pi < \theta \le \pi \tag{13}$$

$$u(r,-\pi,t) = u(r,\pi,t), \qquad 0 < r < a, \qquad 0 < t$$

$$\frac{\partial u}{\partial \theta}(r,-\pi,t) = \frac{\partial u}{\partial \theta}(r,\pi,t), \qquad 0 < r < a, \qquad 0 < t \tag{14}$$

$$u(r,\theta,0) = f(r,\theta), \qquad 0 < r < a, \qquad -\pi < \theta \le \pi \tag{15}$$

$$\frac{\partial u}{\partial t}(r,\theta,0) = g(r,\theta), \qquad 0 < r < a, \qquad -\pi < \theta \le \pi. \tag{16}$$

Condition (13) has been added because $r = 0$ is a singular point of the partial differential equation.

Following the procedure suggested in Section 4, we assume that u has the product form

$$u = \phi(r,\theta)T(t)$$

and we find that Eq. (11) separates into two linked equations

$$T'' + \lambda^2 c^2 T = 0, \qquad 0 < t \tag{17}$$

$$\frac{1}{r}\frac{\partial}{\partial r}\left(r\frac{\partial \phi}{\partial r}\right) + \frac{1}{r^2}\frac{\partial^2 \phi}{\partial \theta^2} = -\lambda^2\phi, \quad 0 < r < a, \quad -\pi < \theta \le \pi. \tag{18}$$

If we separate variables of the function ϕ by assuming $\phi(r,\theta) = R(r)\Theta(\theta)$, Eq. (18) takes the form

$$\frac{1}{r}(rR')'\Theta + \frac{1}{r^2}R\Theta'' = -\lambda^2 R\Theta.$$

The variables will separate if we multiply by r^2 and divide by $R\Theta$. Then the preceding equation may be put in the form

$$\frac{r(rR')'}{R} + \lambda^2 r^2 = -\frac{\Theta''}{\Theta} = \mu^2.$$

Finally we obtain two problems for R and Θ.

$$\begin{aligned} \Theta'' + \mu^2\Theta &= 0, \qquad -\pi < \theta \le \pi \\ \Theta(-\pi) &= \Theta(\pi) \\ \Theta'(-\pi) &= \Theta'(\pi) \end{aligned} \tag{19}$$

$$(rR')' - \frac{\mu^2}{r}R + \lambda^2 rR = 0, \qquad 0 < r < a \tag{20}$$

$$|R(0)| \text{ bounded}$$
$$R(a) = 0.$$

As we observed before, the problem (19) has the solutions

$$\mu^2 = 0, \qquad \Theta_0 = 1$$
$$\mu^2 = m^2, \quad \Theta_m = \cos(m\theta) \text{ and } \sin(m\theta), \quad m = 1, 2, 3, \cdots.$$

Also the differential equation of Eq. (20) will be recognized as Bessel's equation, the general solution of which is (using $\mu = m$)

$$R(r) = CJ_m(\lambda r) + DY_m(\lambda r).$$

In order for the boundedness condition in Eq. (20) to be fulfilled, D must be zero. Then we are left with

$$R(r) = J_m(\lambda r).$$

(Because any multiple of a solution is another solution, we can drop the constant C.) The boundary condition of Eq. (20) becomes

$$R(a) = J_m(\lambda a) = 0,$$

implying that λa must be a root of the equation

$$J_m(\alpha) = 0.$$

(See Table 1.) For each fixed integer m,

$$\alpha_{m1}, \quad \alpha_{m2}, \quad \alpha_{m3}, \cdots$$

are the first, second, third, ..., solutions of the preceding equation. The values of λ for which $J_m(\lambda r)$ solves the differential equation and satisfies the boundary condition are

$$\lambda_{mn} = \frac{\alpha_{mn}}{a}, \quad m = 0, 1, 2, \cdots, \quad n = 1, 2, 3, \cdots.$$

Now that the functions R and Θ are determined, we can construct ϕ. For $m = 1, 2, 3, \cdots$ and $n = 1, 2, 3, \cdots$, both of the functions

$$J_m(\lambda_{mn} r)\cos(m\theta), \quad J_m(\lambda_{mn} r)\sin(m\theta) \tag{21}$$

are solutions of the problem Eq. (18), both corresponding to the same eigenvalue λ_{mn}^2. For $m = 0$ and $n = 1, 2, 3, \cdots$, we have the functions

$$J_0(\lambda_{0n} r) \tag{22}$$

that correspond to the eigenvalues λ_{0n}^2. (Compare with the simple case.) The function $T(t)$ that is a solution of Eq. (17) is any combination of $\cos(\lambda_{mn} ct)$ and $\sin(\lambda_{mn} ct)$.

Now the solutions of Eqs. (11)-(14) have any of the forms

$$\begin{aligned} &J_m(\lambda_{mn} r)\cos(m\theta)\cos(\lambda_{mn} ct), \quad J_m(\lambda_{mn} r)\sin(m\theta)\cos(\lambda_{mn} ct) \\ &J_m(\lambda_{mn} r)\cos(m\theta)\sin(\lambda_{mn} ct), \quad J_m(\lambda_{mn} r)\sin(m\theta)\sin(\lambda_{mn} ct) \end{aligned} \tag{23}$$

for $m = 1, 2, 3, \cdots$ and $n = 1, 2, 3, \cdots$. In addition, there is the special case $m = 0$, for which solutions have the form

$$J_0(\lambda_{0n} r)\cos(\lambda_{0n} ct), \quad J_0(\lambda_{0n} r)\sin(\lambda_{0n} ct). \tag{24}$$

The general solution of the problem Eqs. (11)-(14) will thus have the form of a linear combination of the solutions in Eqs. (23) and (24). We shall use several series to form the combination:

$$\begin{aligned} u(r, \theta, t) &= \sum_n a_{0n} J_0(\lambda_{0n} r)\cos(\lambda_{0n} ct) \\ &\qquad + \sum_{m,n} a_{mn} J_m(\lambda_{mn} r)\cos(m\theta)\cos(\lambda_{mn} ct) \\ &\qquad + \sum_{m,n} b_{mn} J_m(\lambda_{mn} r)\sin(m\theta)\cos(\lambda_{mn} ct) \\ &\quad + \sum_n A_{0n} J_0(\lambda_{0n} r)\sin(\lambda_{0n} ct) \\ &\qquad + \sum_{m,n} A_{mn} J_m(\lambda_{mn} r)\cos(m\theta)\sin(\lambda_{mn} ct) \\ &\qquad + \sum_{m,n} B_{mn} J_m(\lambda_{mn} r)\sin(m\theta)\sin(\lambda_{mn} ct). \end{aligned} \tag{25}$$

When $t = 0$, the last three sums disappear, and the cosines of t in the first three sums are all equal to 1. Thus

$$\begin{aligned} u(r,\theta,0) &= \sum_n a_{0n} J_0(\lambda_{0n} r) + \sum_{m,n} a_{mn} J_m(\lambda_{mn} r)\cos(m\theta) \\ &\quad + \sum_{m,n} b_{mn} J_m(\lambda_{mn} r)\sin(m\theta) \\ &= f(r,\theta), \quad 0 < r < a, \quad -\pi < \theta \le \pi. \end{aligned} \tag{26}$$

We expect to fulfill this equality by choosing the a's and b's according to some orthogonality principle. Since each function present in the series is an eigenfunction of the problem

$$\begin{aligned} \nabla^2\phi &= -\lambda^2\phi, \qquad 0 < r < a, \quad -\pi < \theta \le \pi \\ \phi(a,\theta) &= 0, \qquad -\pi \le \theta \le \pi \\ \phi(r,-\pi) &= \phi(r,\pi), \quad \frac{\partial\phi}{\partial\theta}(r,-\pi) = \frac{\partial\phi}{\partial\theta}(r,\pi), \qquad 0 < r < a, \end{aligned}$$

we expect it to be orthogonal to each of the others (see Section 4, Exercise 7). This is indeed true: Any function from one series is orthogonal to all of the functions in the other series, and also to the rest of the functions in its own series. To illustrate this orthogonality, we have

$$\begin{aligned} &\iint_{\mathcal{R}} J_0(\lambda_{0n} r) J_m(\lambda_{mn} r)\cos(m\theta)\,dA \\ &\quad = \int_0^a J_0(\lambda_{0n} r) J_m(\lambda_{mn} r) \int_{-\pi}^{\pi} \cos(m\theta)\,d\theta\, r\,dr = 0, \quad m \ne 0. \end{aligned} \tag{27}$$

There are two other relations like this one involving functions from two different series.

We already know that the functions within the first series are orthogonal to each other:

$$\int_{-\pi}^{\pi}\int_0^a J_0(\lambda_{0n} r) J_0(\lambda_{0q} r)\ \ r\,dr\,d\theta = 0, \quad n \ne q.$$

Within the second series we must show that, if $m \ne p$ or $n \ne q$, then

$$0 = \int_{-\pi}^{\pi}\int_0^a J_m(\lambda_{mn} r)\cos(m\theta) J_p(\lambda_{pq} r)\cos(p\theta)\ \ r\,dr\,d\theta. \tag{28}$$

(Recall that $rdrd\theta = dA$ in polar coordinates.) Integrating with respect to θ first, we see that the integral must be zero if $m \neq p$, by the orthogonality of $\cos(m\theta)$ and $\cos(p\theta)$. If $m = p$, the preceding integral becomes

$$\pi \int_0^a J_m(\lambda_{mn}r)J_m(\lambda_{mq}r)rdr$$

after the integration with respect to θ. Finally, if $n \neq q$, this integral is zero; the demonstration follows the same lines as the usual Sturm-Liouville proof. Thus the functions within the second series are shown orthogonal to each other. For the functions of the last series, the proof of orthogonality is similar.

Equipped now with an orthogonality relation, we can determine formulas for the a's and b's. For instance,

$$a_{0n} = \frac{\displaystyle\int_{-\pi}^{\pi} \int_0^a f(r,\theta) J_0(\lambda_{0n}r) \; rdrd\theta}{\displaystyle 2\pi \int_0^a J_0^2(\lambda_{0n}r) \; rdr}. \tag{29}$$

The A's and B's are calculated from the second initial condition.

It should now be clear that, while the computation of the solution to the original problem is possible in theory, it will be very painful in practice. Worse yet, the final form of the solution Eq. (25) does not give a clear idea of what u looks like. All is not wasted, however. We can say, from an examination of the λ's , that the tone produced is not musical — that is, u is not periodic in t. Also we can sketch some of the fundamental modes of vibration of the membrane corresponding to some low eigenvalues (Fig. 9). The curves represent points for which displacement is zero in that mode (nodal curves).

Exercises

1. Verify that each of the functions in the series in Eq. (10) satisfies Eqs. (1), (2), and (8).

2. Derive the formulas for the a's and b's of Eq. (10).

3. List the five lowest frequencies of vibration of a circular membrane.

4. Sketch the function $J_0(\lambda_n r)$, for $n = 1, 2, 3$.

5. What boundary conditions must the function ϕ of Eq. (18) satisfy?

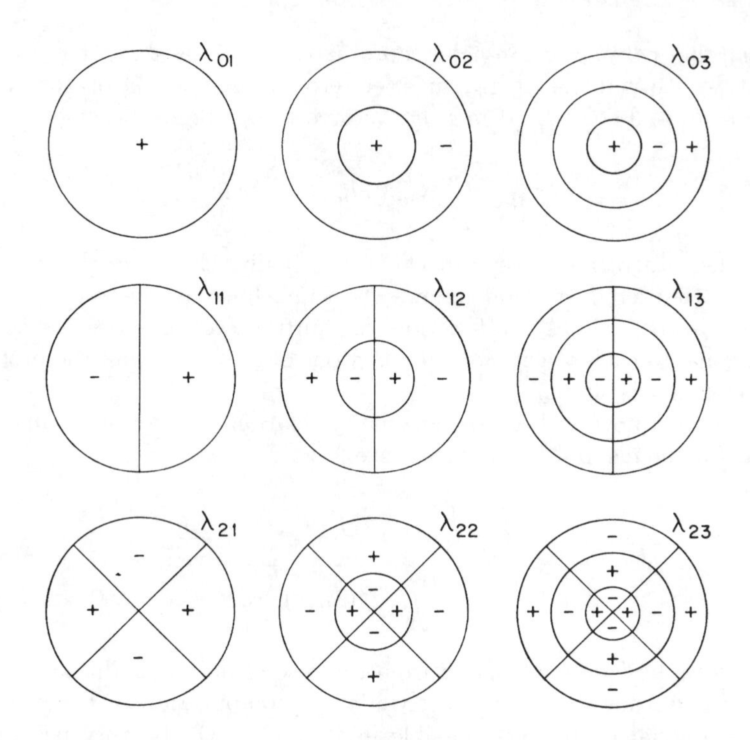

Figure 9: Nodal curves: The curves in these graphs represent solutions of $\phi_{mn}(r, \theta) = 0$. Adjacent regions bulge up or down, according to the sign. Only those ϕ's containing the factor $\cos(m\theta)$ have been used.

6. Justify the derivation of Eqs. (19) and (20) from Eqs. (12)-(14) and (18).

7. Show that

$$\int_0^a J_m(\lambda_{mn} r) J_m(\lambda_{mq} r) r dr = 0, \quad n \neq q$$

if

$$J_m(\lambda_{ms} a) = 0, \quad s = 1, 2, \cdots.$$

8. Sketch the nodal curves of the eigenfunctions Eq. (21) corresponding to λ_{31}, λ_{32}, and λ_{33}.

5.8 SOME APPLICATIONS OF BESSEL FUNCTIONS

After the elementary functions, the Bessel functions are among the most useful in engineering and physics. One reason for their usefulness is the fact that they solve a fairly general differential equation. The general solution of

$$\phi'' + \frac{1-2\alpha}{x}\phi' + \left[(\lambda\gamma x^{\gamma-1})^2 - \frac{p^2\gamma^2 - \alpha^2}{x^2} \right] \phi = 0 \tag{1}$$

is given by

$$\phi(x) = x^{\alpha} \left[AJ_p(\lambda x^{\gamma}) + BY_p(\lambda x^{\gamma}) \right].$$

Several problems in which the Bessel functions play an important role follow. The details of separation of variables, which should now be routine, are kept to a minimum.

A. Potential Equation in a Cylinder

The steady-state temperature distribution in a circular cylinder with insulated surface is determined by the problem

$$\frac{1}{r}\frac{\partial}{\partial r}\left(r\frac{\partial u}{\partial r} \right) + \frac{\partial^2 u}{\partial z^2} = 0, \qquad 0 < r < a, \qquad 0 < z < b \tag{2}$$

$$\frac{\partial u}{\partial r}(a, z) = 0, \qquad 0 < z < b \tag{3}$$

$$u(r, 0) = f(r), \qquad 0 < r < a \tag{4}$$

$$u(r, b) = g(r), \qquad 0 < r < a. \tag{5}$$

Here we are considering the boundary conditions to be independent of θ, so that u is independent of θ also.

Assuming that $u = R(r)Z(z)$ we find that

$$(rR')' + \lambda^2 rR = 0, \qquad 0 < r < a \tag{6}$$

$$R'(a) = 0 \tag{7}$$

$$|R(0)| \text{ bounded} \tag{8}$$

$$Z'' - \lambda^2 Z = 0. \tag{9}$$

Condition (8) has been added because $r = 0$ is a singular point. The solution of Eqs. (6)-(8) is

$$R_n(r) = J_0(\lambda_n r) \tag{10}$$

where the eigenvalues λ_n^2 are defined by the solutions of

$$R'(a) = \lambda J_0'(\lambda a) = 0. \tag{11}$$

Because $J_0' = -J_1$, the λ's are related to the zeros of J_1. The first three eigenvalues are: 0, $(3.832/a)^2$, and $(7.016/a)^2$. Note that $R(0) = J_0(0) = 1$.

The solution of the problem Eqs. (2)-(5) may be put in the form

$$u(r,z) = a_0 + b_0 z + \sum_{n=1}^{\infty} J_0(\lambda_n r)\left[a_n \frac{\sinh(\lambda_n z)}{\sinh(\lambda_n b)} + b_n \frac{\sinh(\lambda_n (b-z))}{\sinh(\lambda_n b)}\right]. \tag{12}$$

The a's and b's are determined from Eqs. (4) and (5) by using the orthogonality relation

$$\int_0^a J_0(\lambda_n r) J_0(\lambda_m r) r dr = 0, \quad n \neq m.$$

B. Spherical Waves

In spherical (ρ, θ, ϕ) coordinates (see Section 9), the Laplacian operator ∇^2 becomes

$$\nabla^2 u = \frac{1}{\rho^2}\frac{\partial}{\partial \rho}\left(\rho^2 \frac{\partial u}{\partial \rho}\right) + \frac{1}{\rho^2 \sin(\phi)}\frac{\partial}{\partial \phi}\left(\sin(\phi)\frac{\partial u}{\partial \phi}\right) + \frac{1}{\rho^2 \sin^2(\phi)}\frac{\partial^2 u}{\partial \theta^2}.$$

Consider a wave problem in a sphere when the initial conditions depend only on the radial coordinate ρ:

$$\frac{1}{\rho^2}\frac{\partial}{\partial \rho}\left(\rho^2 \frac{\partial u}{\partial \rho}\right) = \frac{1}{c^2}\frac{\partial^2 u}{\partial t^2}, \quad 0 < \rho < a, \quad 0 < t \tag{13}$$

$$u(a,t) = 0, \quad 0 < t \tag{14}$$

$$u(\rho, 0) = f(\rho), \quad 0 < \rho < a \tag{15}$$

$$\frac{\partial u}{\partial t}(\rho, 0) = g(\rho), \quad 0 < \rho < a. \tag{16}$$

Assuming $u(\rho, t) = R(\rho)T(t)$, we separate variables and find

$$T'' + \lambda^2 c^2 T = 0, \tag{17}$$

$$(\rho^2 R')' + \lambda^2 \rho^2 R = 0, \quad 0 < \rho < a \tag{18}$$

$$R(a) = 0, \tag{19}$$

$$|R(0)| \text{ bounded}. \tag{20}$$

Again, the condition (20) has been added because $\rho = 0$ is a singular point. Equation (18) may be put into the form

$$R'' + \frac{2}{\rho}R' + \lambda^2 R = 0$$

and comparison with Eq. (1) shows that $\alpha = -1/2$, $\gamma = 1$, and $\rho = 1/2$; thus the general solution of Eq. (18) is

$$R(\rho) = \rho^{-1/2}\left[AJ_{1/2}(\lambda\rho) + BY_{1/2}(\lambda\rho)\right].$$

We know that near $\rho = 0$,

$$J_{1/2}(\lambda\rho) \sim \text{const} \times \rho^{1/2}$$
$$Y_{1/2}(\lambda\rho) \sim \text{const} \times \rho^{-1/2}.$$

Thus in order to satisfy Eq. (20), we must have $B = 0$. It is possible to show that

$$J_{1/2}(\lambda\rho) = \frac{2}{\pi}\frac{\sin(\lambda\rho)}{\sqrt{\lambda\rho}}, \quad Y_{1/2}(\lambda\rho) = -\frac{2}{\pi}\frac{\cos(\lambda\rho)}{\sqrt{\lambda\rho}}.$$

Our solution to Eqs. (18) and (20) is, therefore,

$$R(\rho) = \frac{\sin(\lambda\rho)}{\rho} \tag{21}$$

and Eq. (19) is satisfied if $\lambda_n^2 = (n\pi/a)^2$. The solution of the problem of Eqs. (13)-(16) can be written in the form

$$u(\rho, t) = \sum_{n=1}^{\infty} \frac{\sin(\lambda_n\rho)}{\rho}\left[a_n \cos(\lambda_n ct) + b_n \sin(\lambda_n ct)\right]. \tag{22}$$

The a's and b's are, as usual, chosen so that the initial conditions Eqs. (15) and (16) are satisfied.

C. Pressure in a Bearing

The pressure in the lubricant inside a plane-pad bearing satisfies the problem

$$\frac{\partial}{\partial x}\left(x^3\frac{\partial p}{\partial x}\right) + x^3\frac{\partial^2 p}{\partial y^2} = -1, \qquad a < x < b, \quad -c < y < c \tag{23}$$

$$p(a, y) = 0, \quad p(b, y) = 0, \qquad -c < y < c \tag{24}$$

$$p(x, -c) = 0, \quad p(x, c) = 0, \qquad a < x < b. \tag{25}$$

(Here a and c are positive constants, and $b = a + 1$.) Equation (23) is elliptic and nonhomogeneous. To reduce this equation to a more familiar one, let $p(x, y) = v(x) + u(x, y)$, where $v(x)$ satisfies the problem

$$(x^3 v')' = -1, \qquad a < x < b \tag{26}$$

$$v(a) = 0, \qquad v(b) = 0. \tag{27}$$

Then, when v is found, u must be the solution of the problem

$$\frac{\partial}{\partial x}\left(x^3 \frac{\partial u}{\partial x}\right) + x^3 \frac{\partial^2 u}{\partial y^2} = 0, \qquad a < x < b, \qquad -c < y < c \tag{28}$$

$$u(a, y) = 0, \qquad u(b, y) = 0, \qquad -c < y < c \tag{29}$$

$$u(x, \pm c) = -v(x), \quad a < x < b. \tag{30}$$

If we now assume that $u(x, y) = X(x)Y(y)$, the variables can be separated:

$$(x^3 X')' + \lambda^2 x^3 X = 0, \qquad a < x < b \tag{31}$$

$$X(a) = 0, \qquad X(b) = 0 \tag{32}$$

$$Y'' - \lambda^2 Y = 0, \qquad -c < y < c. \tag{33}$$

Equation (31) may be put in the form

$$X'' + \frac{3}{x} X' + \lambda^2 X = 0, \quad a < x < b.$$

By comparing to Eq. (1) we find that $\alpha = -1$, $\gamma = 1$, and $p = 1$, and that the general solution of Eq. (31) is

$$X(x) = \frac{1}{x}\left(A J_1(\lambda x) + B Y_1(\lambda x)\right).$$

Because the point $x = 0$ is not included in the interval $a < x < b$, there is no problem with boundedness. Instead we must satisfy the boundary conditions Eq. (32), which after some algebra have the form

$$A J_1(\lambda a) + B Y_1(\lambda a) = 0,$$

$$A J_1(\lambda b) + B Y_1(\lambda b) = 0.$$

Not both A and B may be zero, so the determinant of these simultaneous equations must be zero:

$$J_1(\lambda a) Y_1(\lambda b) - J_1(\lambda b) Y_1(\lambda a) = 0.$$

Some solutions of the equation are tabulated for various values of b/a. For instance, if $b/a = 2.5$, the first three eigenvalues λ^2 are

$$\left(\frac{2.156}{a}\right)^2, \quad \left(\frac{4.223}{a}\right)^2, \quad \left(\frac{6.307}{a}\right)^2.$$

We now can take X_n to be

$$X_n(x) = \frac{1}{x}\left(Y_1(\lambda_n a)J_1(\lambda_n x) - J_1(\lambda_n a)Y_1(\lambda_n x)\right) \tag{34}$$

and the solution of Eqs. (28)-(30) has the form

$$u(x,y) = \sum_{n=1}^{\infty} a_n X_n(x) \frac{\cosh(\lambda_n y)}{\cosh(\lambda_n c)}. \tag{35}$$

The a's are chosen to satisfy the boundary conditions Eq. (30), using the orthogonality principle

$$\int_a^b X_n(x)X_m(x)x^3\,dx = 0, \quad n \neq m.$$

Notice that Eqs. (31) and (32) make up a regular Sturm-Liouville problem.

Exercises

1. Find the general solution of the differential equation

$$(x^n\phi')' + \lambda^2 x^n \phi = 0$$

where $n = 0, 1, 2, \cdots$.

2. Find the solution of the equation in Exercise 1 that is bounded at $x = 0$.

3. Find the solutions of Eq. (9), including the case $\lambda^2 = 0$, and prove that Eq. (12) is a solution of Eqs. (2)-(4).

4. Show that any function of the form

$$u(\rho,t) = \frac{1}{\rho}\left(\phi(\rho + ct) + \psi(\rho - ct)\right)$$

is a solution of Eq. (13) if ϕ and ψ have at least two derivatives.

5. Find functions ϕ and ψ, such that $u(\rho, t)$ as given in Exercise 4 satisfies Eqs. (14)-(16).

6. Give the formula for the a's and b's in Eq. (12).

7. What is the orthogonality relation for the eigenfunctions of Eqs. (18)-(20)? Use it to find the a's and b's in Eq. (22).

8. Sketch the first few eigenfunctions of Eqs. (18)-(20).

9. Find the function $v(x)$ that is the solution of Eqs. (26) and (27).

10. Use the technique of Example C to change the following problem into a potential problem:

$$\frac{\partial^2 u}{\partial x^2} + \frac{\partial^2 u}{\partial y^2} = -f(x), \quad 0 < x < a, \quad 0 < y < b,$$

$u = 0$ on all boundaries.

11. Will the same technique work if $f(x)$ is replaced by $f(x, y)$?

12. Verify that Eqs. (31) and (32) form a regular Sturm-Liouville problem. Show the eigenfunctions' orthogonality by using the orthogonality of the Bessel functions.

13. Find a formula for the a_n of Eq. (35).

14. Verify that Eq. (34) is a solution of Eqs. (28)-(30).

5.9 SPHERICAL COORDINATES; LEGENDRE POLYNOMIALS

After the Cartesian and cylindrical coordinate systems, the one most frequently encountered is the spherical system (Fig. 10), in which

$$x = \rho \sin(\phi) \cos(\theta)$$
$$y = \rho \sin(\phi) \sin(\theta)$$
$$z = \rho \cos(\phi).$$

The variables are restricted by $0 \le \rho$, $0 \le \theta < 2\pi$, $0 \le \phi \le \pi$. In this coordinate system the Laplacian operator is

$$\nabla^2 u = \frac{1}{\rho^2} \left\{ \frac{\partial}{\partial \rho} \left(\rho^2 \frac{\partial u}{\partial \rho} \right) + \frac{1}{\sin(\phi)} \frac{\partial}{\partial \phi} \left(\sin(\phi) \frac{\partial u}{\partial \phi} \right) + \frac{1}{\sin^2(\phi)} \frac{\partial^2 u}{\partial \theta^2} \right\}.$$

From what we have seen in other cases, we expect solvable problems in spherical coordinates to reduce to one of the following:

Problem 1
$\nabla^2 u = -\lambda^2 u$ in $\mathcal{R}$, plus homogeneous boundary conditions.

Problem 2
$\nabla^2 u = 0$ in $\mathcal{R}$, plus homogeneous boundary conditions on facing sides (where $\mathcal{R}$ is a generalized rectangle in spherical coordinates).

Problem 1 would come from a heat or wave equation after separating out the time variable. Problem 2 is a part of the potential problem.

The complete solution of either of these problems is very complicated, but a number of special cases are simple, important and not uncommon. We have already seen Problem 1 solved (Section 8) when u is a function of ρ only. A second important case is Problem 2, when u is independent of the variable θ. We shall state a complete boundary value problem and solve it by separation of variables.

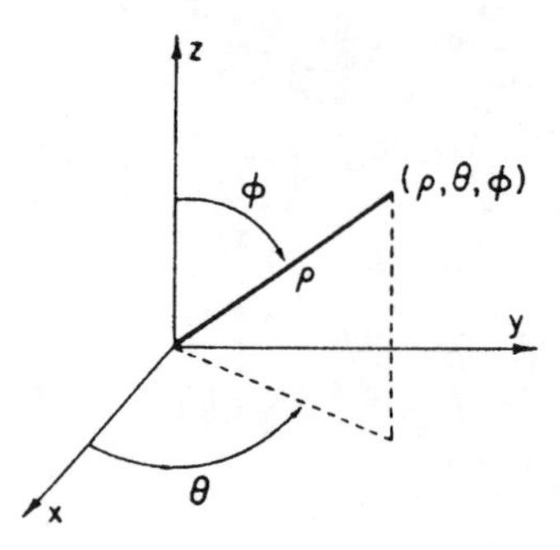

Figure 10: Spherical coordinates.

$$\frac{1}{\rho^2}\left\{\frac{\partial}{\partial \rho}\left(\rho^2 \frac{\partial u}{\partial \rho}\right) + \frac{1}{\sin(\phi)}\frac{\partial}{\partial \phi}\left(\sin(\phi)\frac{\partial u}{\partial \phi}\right)\right\} = 0, 0 < \rho < c, 0 < \phi < \pi \tag{1}$$

$$u(c, \phi) = f(\phi), \quad 0 < \phi < \pi. \tag{2}$$

From the assumption $u(\rho, \phi) = R(\rho)\Phi(\phi)$, it follows that

$$\frac{(\rho^2 R'(\rho))'}{R(\rho)} + \frac{(\sin(\phi)\Phi'(\phi))'}{\sin(\phi)\Phi(\phi)} = 0.$$

Both terms are constant, and the second is negative, $-\mu^2$, because the boundary condition at $\rho = c$ will have to be satisfied by a linear combination of functions of ϕ. The separated equations are

$$(\rho^2 R')' - \mu^2 R = 0, \quad 0 < \rho < c \tag{3}$$

$$(\sin(\phi)\Phi')' + \mu^2 \sin(\phi)\Phi = 0, \quad 0 < \phi < \pi. \tag{4}$$

Neither equation has a boundary condition. However, $\rho = 0$ is a singular point of the first equation, and both $\phi = 0$ and $\rho = \pi$ are singular points of the second equation. (At these points, the coefficient of the highest-order derivative is zero, while some other coefficient is nonzero.) At each of the singular points, we impose a boundedness condition:

$$R(0) \text{ bounded}, \quad \Phi(0) \text{ and } \Phi(\pi) \text{ bounded}.$$

Equation (4) can be simplified by the change of variables $x = \cos(\phi)$, $\Phi(\phi) = y(x)$. By the chain rule, the relevant derivatives are

$$\frac{d\Phi}{d\phi} = -\sin(\phi)\frac{dy}{dx}$$

$$\frac{d}{d\phi}\left(\sin(\phi)\frac{d\Phi}{d\phi}\right) = \sin^3(\phi)\frac{d^2y}{dx^2} - 2\sin(\phi)\cos(\phi)\frac{dy}{dx}.$$

The differential equation becomes

$$\sin^2(\phi)\frac{d^2y}{dx^2} - 2\cos(\phi)\frac{dy}{dx} + \mu^2 y = 0$$

or, in terms of x alone

$$(1 - x^2)y'' - 2xy' + \mu^2 y = 0, \quad -1 < x < 1. \tag{5}$$

In addition, we require that $y(x)$ be bounded at $x = \pm 1$.

Solutions of the differential equation are usually found by the power series method. Assume that $y(x) = a_0 + a_1 x + \cdots + a_k x^k + \cdots$. The terms of the differential equations are then

$$\begin{aligned}
y'' &= 2a_2 + 3\cdot 2a_3 x + 4\cdot 3a_4 x^2 + \cdots + (k+2)(k+1)a_{k+2}x^k + \cdots \\
-x^2 y'' &= \qquad\qquad -2a_2 x^2 - \cdots \qquad -k(k-1)a_k x^k + \cdots \\
-2xy' &= \qquad -2a_1 x \quad -2a_2 x^2 - \cdots \qquad -2ka_k x^k - \cdots \\
\mu^2 y &= \mu^2 a_0 + \mu^2 a_1 x + \mu^2 a_2 x^2 + \cdots \qquad +\mu^2 a_k x^k + \cdots .
\end{aligned}$$

When this tableau is added vertically, the left-hand side is zero, according to the differential equation. The right-hand side adds up to a power series, each of whose coefficients must be zero. We therefore obtain the following relations

$$2a_2 + \mu^2 a_0 = 0$$
$$6a_3 + (\mu^2 - 2)a_1 = 0$$
$$(k+2)(k+1)a_{k+2} + [\mu^2 - k(k+1)]a_k = 0.$$

The last equation actually includes the first two, apparently special, cases. We may write the general relation as

$$a_{k+2} = \frac{k(k+1) - \mu^2}{(k+2)(k+1)} a_k$$

valid for $k = 0, 1, 2, \cdots$.

Suppose for the moment that μ^2 is given. A short calculation gives the first few coefficients.

$$\begin{aligned} a_2 &= \frac{-\mu^2}{2} a_0, & a_3 &= \frac{2-\mu^2}{6} a_1 \\ a_4 &= \frac{6-\mu^2}{12} a_2 & a_5 &= \frac{12-\mu^2}{20} a_3 \\ &= \frac{6-\mu^2}{12} \cdot \frac{-\mu^2}{2} a_0, & &= \frac{12-\mu^2}{20} \cdot \frac{2-\mu^2}{6} a_1. \end{aligned}$$

It is clear that all the a's with even index will be multiples of a_0 and those with odd index will be multiples of a_1. Thus $y(x)$ equals a_0 times an even function plus a_1 times an odd function, with both a_0 and a_1 arbitrary.

It is not difficult to prove that odd and even series produced by this process diverge at both $x = \pm 1$, for general μ^2. However, when μ^2 has one of the special values

$$\mu^2 = \mu_n^2 = n(n+1), \quad n = 0, 1, 2, \cdots$$

one of the two series turns out to have all zero coefficients after a_n. For instance, if $\mu^2 = 3 \cdot 4$, then $a_5 = 0$, and all subsequent coefficients with odd index are also zero. Hence, one of the solutions of

$$(1 - x^2)y'' - 2xy' + 12y = 0$$

is the polynomial $a_1(x - 5x^3/3)$. The other solution is an even function unbounded at both $x = \pm 1$.

Now we see that the boundedness conditions can be satisfied only if μ^2 is one of the numbers $0, 2, 6, \cdots, n(n+1), \cdots$. In such a case, one solution of the differential equation is a polynomial (naturally bounded at $x = \pm 1$). When normalized by the condition $y(1) = 1$, these are called *Legendre polynomials*, written $P_n(x)$. Table 3 provides the first five Legendre polynomials. Figure 11 shows their graphs.

Table 3: Legendre polynomials

$P_0(x)$	$= 1$
$P_1(x)$	$= x$
$P_2(x)$	$= (3x^2 - 1)/2$
$P_3(x)$	$= (5x^3 - 3x)/2$
$P_4(x)$	$= (35x^4 - 30x^2 + 3)/8$

Since the differential equation (5) is easily put into self-adjoint form

$$((1 - x^2)y')' + \mu^2 y = 0, \quad -1 < x < 1,$$

it is routine to show that the Legendre polynomials satisfy the orthogonality relation

$$\int_{-1}^{1} P_n(x)P_m(x)dx = 0, \quad n \neq m.$$

By direct calculation, it can be shown that

$$\int_{-1}^{1} P_n^2(x)dx = \frac{2}{2n+1}. \tag{6}$$

A compact way of representing the Legendre polynomials is by means of Rodrigues' formula,

$$P_n(x) = \frac{1}{n!2^n}\frac{d^n}{dx^n}\left[(x^2 - 1)^n\right]. \tag{7}$$

Elementary algebra and calculus show that the nth derivative of $(x^2 - 1)^n$ is a polynomial of degree n. Substituting this polynomial into the differential equation (5), with $\mu^2 = n(n+1)$, shows that it is a solution—bounded, of course. Therefore, it is a multiple of the Legendre polynomial $P_n(x)$. Through Rodrigues' formula or otherwise, it is possible to prove the following two formulas, which relate three consecutive Legendre polynomials:

$$(2n+1)P_n(x) = P'_{n+1}(x) - P'_{n-1}(x) \tag{8}$$

$$(n+1)P_{n+1}(x) + nP_{n-1}(x) = (2n+1)xP_n(x). \tag{9}$$

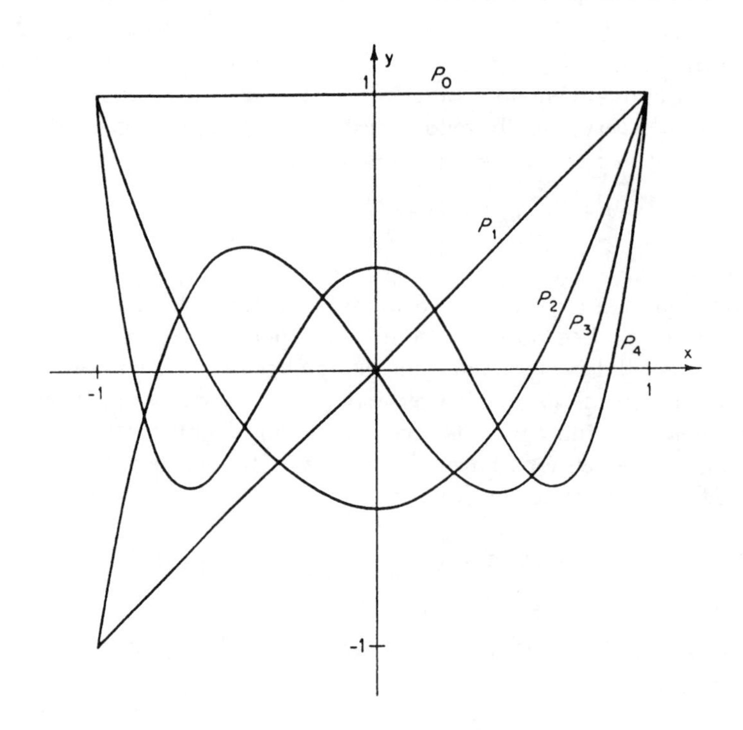

Figure 11: Graphs of the first five Legendre polynomials.

In order to use Legendre polynomials in boundary value problems, we need to be able to express a given function $f(x)$ in the form of a Legendre series,

$$f(x) = \sum_{n=0}^{\infty} b_n P_n(x), \quad -1 < x < 1.$$

From the orthogonality relation and the integral, Eq. (6), it follows that the coefficient in the series must be

$$b_n = \frac{2n+1}{2} \int_{-1}^{1} f(x) P_n(x) dx. \tag{10}$$

The convergence theorem for Legendre series is analogous to the one for Fourier series in Chapter 1.

Theorem
If $f(x)$ is sectionally smooth on the interval $-1 < x < 1$, then at every point of that interval the Legendre series of f is convergent, and

$$\sum_{n=0}^{\infty} b_n P_n(x) = \frac{f(x+) + f(x-)}{2}.$$

From equation (10) for the coefficient of a Legendre series, and from the fact that the Legendre polynomials are odd or even, we see that an odd function will have only odd-indexed coefficients that are nonzero, and an even function will have only even-indexed coefficients that are nonzero. Furthermore, if a function f is given on the interval $0 < x < 1$, then its odd and even extensions have odd and even Legendre series, and f is represented by either in that interval:

$$\begin{aligned} f(x) &= \sum_{n \text{ even}} b_n P_n(x), \quad 0 < x < 1, \\ b_n &= (2n+1) \int_0^1 f(x) P_n(x) dx \quad (n \text{ even}) \end{aligned} \tag{11}$$

$$\begin{aligned} f(x) &= \sum_{n \text{ odd}} b_n P_n(x), \qquad 0 < x < 1, \\ b_n &= (2n+1) \int_0^1 f(x) P_n(x) dx \quad (n \text{ odd}). \end{aligned} \tag{12}$$

Because the $P_n(x)$ are polynomials, the integral Eq. (10) for any specific coefficient can be done in closed form for a variety of functions $f(x)$. However, getting a_n as a function of n is not so easy. Fortunately, some elementary integrals can be done using the differential equation

$$((1-x^2)P_n')' + n(n+1)P_n = 0.$$

(1) First, separate the two terms of the differential equation and integrate:

$$\begin{aligned} n(n+1) \int P_n(x) dx &= \int -((1-x^2)P_n')' \, dx \\ &= -(1-x^2)P_n'(x). \end{aligned}$$

This equation may be solved for the integral if $n \neq 0$.

(2) Now multiply through the differential equation by x, separate terms and integrate:

$$\begin{aligned} n(n+1)\int xP_n(x)dx &= \int -x\left((1-x^2)P_n'\right)' dx \\ &= -x(1-x^2)P_n' + \int (1-x^2)P_n' dx \\ &= -x(1-x^2)P_n' + (1-x^2)P_n - \int (-2x)P_n dx. \end{aligned}$$

Next, move the last term to the left-hand member of the equation to find

$$(n+2)(n-1)\int xP_n(x)dx = (1-x^2)\left(P_n(x) - xP_n'(x)\right).$$

This equation can be solved for the integral on the left, provided that $n \neq 1$. (For $n = 1$, the integration is done directly.)

Summary.

$$\int P_n(x)dx = \frac{-(1-x^2)}{n(n+1)}P_n'(x) \tag{13}$$

$$\int xP_n(x)dx = \frac{(1-x^2)}{(n+2)(n-1)}\left(P_n(x) - xP_n'(x)\right) \tag{14}$$

These integration formulas are useful if we can evaluate $P_n(x)$ and $P_n'(x)$ easily for any x. The relations in Eqs. (8) and (9) are useful for this purpose. We illustrate by finding $P_n(0)$. First, note that $P_n(0) = 0$ for odd values of n, because the Legendre polynomials with odd index are odd functions of x. For odd n, Eq. (9) gives

$$(n+1)P_{n+1}(0) + nP_{n-1}(0) = 0,$$

or

$$P_{n+1}(0) = -\frac{n}{n+1}P_{n-1}(0).$$

Because $P_0(0) = 1$, we find successively that

$$P_2(0) = -1/2, \quad P_4(0) = 1\cdot 3/(2\cdot 4), \quad P_6(0) = -1\cdot 3\cdot 5/(2\cdot 4\cdot 6),$$

or in general

$$\begin{aligned} P_n(0) &= (-1)^{n/2}\frac{1\cdot 3\cdots(n-1)}{2\cdot 4\cdots cdotn}, \quad && n = 2, 4, 6, \cdots \\ P_n(0) &= 0, && n = 1, 3, 5, \cdots. \end{aligned} \tag{15}$$

Similarly, but not as easily, Eq. (8) can be used to find the values of $P_n'(0)$. It is simpler to use the relation

$$P_n'(0) = nP_{n-1}(0), \tag{16}$$

which can be derived from Eqs. (8) and (9).

Example.

Let $f(x) = \begin{cases} -1, & -1 < x < 0 \\ 1, & 0 < x < 1 \end{cases}$. The Legendre series will contain only odd-indexed polynomials, and their coefficients are

$$\begin{aligned} b_n &= (2n+1)\int_0^1 P_n(x)dx \quad (n \ \text{odd}) \\ &= -\frac{2n+1}{n(n+1)}\left[(1-x^2)P_n'(x)\right]_0^1 \\ &= \frac{2n+1}{n(n+1)}P_n'(0) = \frac{2n+1}{n+1}P_{n-1}(0) \\ &= (-1)^{(n-1)/2}\frac{1\cdot 3\cdot 5\cdots(n-2)}{2\cdot 4\cdot 6\cdots(n-1)}\cdot\frac{2n+1}{n+1} \quad (n = 3, 5, 7, \cdots). \end{aligned}$$

Specifically we find $b_1 = 3/2$ (by a separate calculation), $b_3 = -7/8$, $b_5 = 11/16, \cdots$. Because $f(x)$ is indeed sectionally smooth,

$$f(x) = \frac{3}{2}P_1(x) - \frac{7}{8}P_3(x) + \frac{11}{16}P_5(x) - + \cdots.$$

See Fig. 12 for graphs of the partial sums of this series.

Summary.
The solution of the eigenvalue problem

$$((1-x^2)y')' + \mu^2 y = 0, \quad -1 < x < 1$$

$$y(x) \ \text{bounded at} \ x = -1 \ \text{and at} \ x = 1,$$

is $y(x) = P_n(x)$, $\mu_n^2 = n(n+1)$, $n = 0, 1, 2, \cdots$.

The solution of the eigenvalue problem

$$(\sin(\phi)\Phi')' + \mu^2 \sin(\phi)\Phi = 0,$$

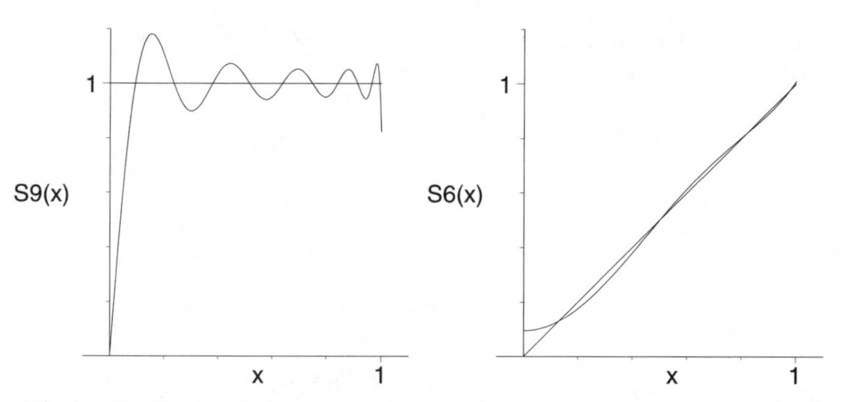

Figure 12: Graphs of a function and a partial sum of its Legendre series: (a) through $P_9(x)$ for the function $f(x)$ in the example; (b) through $P_6(x)$ for $f(x) = |x|$, $-1 < x < 1$. Compare with the partial sums of the Fourier series, Figures 9 and 10 of Chapter 1.

$$\Phi(\phi) \text{ bounded at } \phi = 0 \text{ and at } \phi = \pi,$$

is $\Phi(\phi) = P_n(\cos(\phi))$, $\mu^2 = n(n+1)$, $n = 0, 1, 2, \cdots$.

The Legendre polynomials $P_n(\cos(\phi))$ are often called *zonal harmonics* because their nodal lines (loci of solutions of $P_n(\cos(\phi)) = 0$) divide a sphere into zones, as shown in Fig. 13.

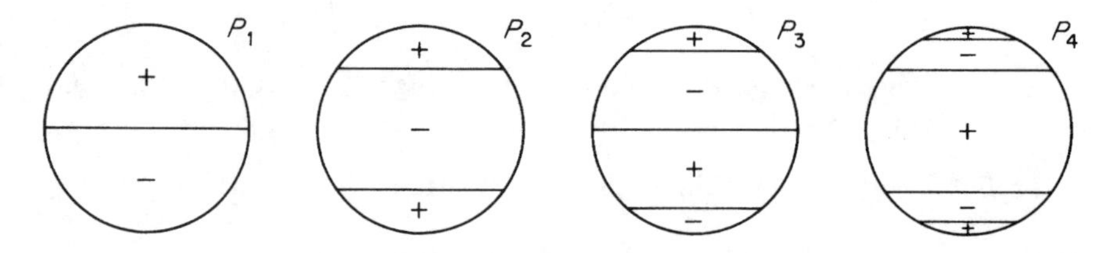

Figure 13: The nodal curves of the zonal harmonics are the parallels (ϕ = constant) on a sphere where $P_n(\cos(\phi)) = 0$. The nodal curves are shown in projection for $n = 1, 2, 3, 4$.

Exercises

1. Equation (4) may be solved by assuming

$$\Phi(\phi) = \frac{1}{2}a_0 + \sum_1^\infty a_k \cos(k\phi).$$

Find the relations among the coefficients a_k by computing the terms of the equation in the form of series. Use the identity

$$\sin(\phi)\sin(k\phi) = \frac{1}{2}\left[\cos(k-1)\phi - \cos(k+1)\phi\right].$$

Show that the coefficients are all zero after a_n if $\mu^2 = n(n+1)$.

2. Derive the formula for the coefficients b_n, as shown in Eq. (10).

3. Find $P_5(x)$, first from the formulas for the a's and second by using Eq. (9) with $n = 4$.

4. Verify Eqs. (6) and (7) for $n = 0, 1, 2$, and Eq. (9) for $n = 2, 3$.

5. One of the solutions of $(1-x^2)y'' - 2xy' = 0$ is $y(x) = 1(\mu^2 = 0)$. Find another independent solution of this differential equation.

6. Show that the orthogonality relation for the eigenfunctions $\Phi_n(\phi) = P_n(\cos(\phi))$ is

$$\int_0^\pi \Phi_n(\phi)\Phi_m(\phi)\sin(\phi)d\phi = 0, \qquad n \neq m.$$

7. Obtain the relation

$$P'_{n+1}(x) = (n+1)P_n(x) + xP'_n(x)$$

by differentiating Eq. (9) and eliminating P'_{n-1} between that and Eq. (8). Note that Eq. (16) follows from this relation.

8. Let $F = (x^2-1)^n$. Show that F satisfies the differential equation

$$(x^2-1)F' = 2nxF.$$

9. Differentiate both sides of the preceding equation $n+1$ times to show that the nth derivative of F satisfies Legendre's equation (5). Use Leibniz's rule for derivatives of a product.

10. Obtain Eq. (6) by these manipulations:

a. Multiply through Eq. (9) by P_{n+1}, integrate from -1 to 1, and use the orthogonality of P_{n+1} with P_{n-1}.

b. Replace $(2n+1)P_n$ by means of Eq. (8).

c. P_{n+1} is orthogonal to xP'_{n-1}, which is a polynomial of degree n.

d. Solve what remains for the desired integral.

11. Find the Legendre series for the function $f(x) = |x|$, $-1 < x < 1$.

12. Find the Legendre series for the function below. Note that $f(x) - 1/2$ is an odd function.

$$f(x) = \begin{cases} 0, & -1 < x < 0 \\ 1, & 0 < x < 1 \end{cases}$$

5.10 SOME APPLICATIONS OF LEGENDRE POLYNOMIALS

In this section we follow through the details involved in solving some problems in which Legendre polynomials are used. First, we complete the problem stated in the previous section.

A. Potential in a Sphere

We consider the axially symmetric potential equation — that is, with no variation in the longitudinal- or θ-direction. The unknown function u might represent an electrostatic potential, steady-state temperature, etc.

$$\frac{1}{\rho^2}\left\{\frac{\partial}{\partial\rho}\left(\rho^2\frac{\partial u}{\partial\rho}\right) + \frac{1}{\sin(\phi)}\frac{\partial}{\partial\phi}\left(\sin(\phi)\frac{\partial u}{\partial\phi}\right)\right\} = 0, \quad 0 < \rho < c, 0 < \phi < \pi \tag{1}$$

$$u(c,\phi) = f(\phi), \quad 0 < \phi < \pi. \tag{2}$$

Of course, the function u is to be bounded at the singular points $\phi = 0$, $\phi = \pi$, and $\rho = 0$. The assumption that u has the product form, $u(\rho, \phi) = \Phi(\phi)R(\rho)$ allows us to transform the partial differential equation into

$$\frac{(\rho^2 R'(r))'}{R(r)} + \frac{(\sin(\phi)\Phi'(\phi))'}{\sin(\phi)\Phi(\phi)} = 0.$$

From here we obtain equations for R and Φ individually,

$$(\rho^2 R')' - \mu^2 R = 0, \qquad 0 < \rho < c \tag{3}$$

$$(\sin(\phi)\Phi')' + \mu^2 \sin(\phi)\Phi = 0, \qquad 0 < \phi < \pi. \tag{4}$$

In Section 9 we found the eigenfunctions of Eq. (4), subject to the boundedness conditions at $\phi = 0$ and π, to be $\Phi_n(\phi) = P_n(\cos(\phi))$, corresponding to the eigenvalues $\mu_n^2 = n(n+1)$. We must still solve

Eq. (3) for R. After the differentiation has been carried out, the problem for R becomes

$$\rho^2 R_n'' + 2\rho R_n' - n(n+1)R_n = 0, \qquad 0 < \rho < c$$
$$R_n \text{ bounded at } \rho = 0.$$

The equation is of the Cauchy-Euler type, solved by assuming $R = \rho^\alpha$ and determining α. Two solutions, ρ^n and $\rho^{-(n+1)}$, are found, of which the second is unbounded at $\rho = 0$. Hence $R_n = \rho^n$, and our product solutions of the potential equation have the form

$$u_n(\rho, \phi) = R_n(\rho)\Phi_n(\phi) = \rho^n P_n(\cos(\phi)).$$

The general solution of the partial differential equation that is bounded in the region $0 < \rho < c$, $0 < \phi < \pi$ is thus the linear combination

$$u(\rho, \phi) = \sum_{n=0}^{\infty} b_n \rho^n P_n(\cos(\phi)). \tag{5}$$

At $\rho = c$, the boundary condition becomes

$$u(c, \phi) = \sum_{n=0}^{\infty} b_n c^n P_n(\cos(\phi)) = f(\phi), \quad 0 < \phi < \pi. \tag{6}$$

The coefficients b_n are then found to be

$$b_n = \frac{2n+1}{2c^n} \int_0^\pi f(\phi) P_n(\cos(\phi)) \sin(\phi) d\phi. \tag{7}$$

B. Heat Equation on a Spherical Shell

The temperature on a spherical shell satisfies the three-dimensional heat equation. If initially there is no dependence on θ, then there will never be such dependence. Furthermore, if the shell is thin (thickness much less than average radius R) we may also assume that temperature does not vary in the radial direction. The heat equation then becomes one-dimensional,

$$\frac{1}{\sin(\phi)} \frac{\partial}{\partial \phi}\left(\sin(\phi)\frac{\partial u}{\partial \phi}\right) = \frac{R^2}{k}\frac{\partial u}{\partial t}, \qquad 0 < \phi < \pi, \quad 0 < t \tag{8}$$

$$u(\phi, 0) = f(\phi), \qquad 0 < \phi < \pi. \tag{9}$$

Naturally, we require boundedness of u at $\phi = 0$ and $\phi = \pi$.

The assumption of a product form for the solution, $u(\phi, t) = \Phi(\phi)T(t)$, leads to the conclusion that

$$\frac{(\sin(\phi)\Phi'(\phi))'}{\sin(\Phi(\phi))} = \frac{R^2 T'(t)}{kT(t)} = -\mu^2.$$

Thus, we have the eigenvalue problem

$$(\sin(\phi)\Phi')' + \mu^2 \sin(\phi)\Phi = 0, \quad 0 < \phi < \pi$$

$$\Phi(0) \text{ and } \Phi(\pi) \text{ bounded.}$$

The solution of this problem was found in Section 9 to be $\mu^2 = n(n+1)$ and

$$\Phi_n(\phi) = P_n(\cos(\phi)), \quad n = 0, 1, 2, \cdots.$$

Obviously, the other factor in a product solution must be

$$T_n(t) = \exp(-n(n+1)kt/R^2).$$

Now a series of constant multiples of product solutions is the most general solution of our problem:

$$u(\phi, t) = \sum_{n=0}^{\infty} b_n P_n(\cos(\phi)) e^{-n(n+1)kt/R^2}. \tag{10}$$

The initial condition, Eq. (9), now takes the form of a Legendre series

$$\sum_{n=0}^{\infty} b_n P_n(\cos(\phi)) = f(\phi), \quad 0 < \phi < \pi. \tag{11}$$

From the information in Section 9, we know that the coefficients b_n must be chosen to be

$$b_n = \frac{2n+1}{2} \int_0^{\pi} P_n(\cos(\phi)) f(\phi) \sin(\phi) d\phi.$$

Then if $f(\phi)$ is sectionally smooth for $0 < \phi < \pi$, the series of Eq. (11) actually equals $f(\phi)$, and thus the function $u(\phi, t)$ in Eq. (10) satisfies the problem originally posed.

For instance, if $f(\phi) = T_0$ in the northern hemisphere $(0 < \phi < \pi/2)$ and $f(\phi) = -T_0$ in the southern $(\pi/2 < \phi < \pi)$, then the coefficients are

$$\begin{aligned} b_n &= \frac{2n+1}{2} \int_0^{\pi} f(\phi) P_n(\cos(\phi)) \sin(\phi) d\phi \\ &= \frac{2n+1}{2} \left[\int_{-1}^{1} f(\cos^{-1}(x)) P_n(x) dx \right] \\ &= T_0 \frac{2n+1}{n+1} P_{n-1}(0), \end{aligned}$$

as found in the previous section. Figure 14 shows graphs of $u(\phi, t)$ as a function of ϕ in the interval $0 < \phi < \pi$ for various times.

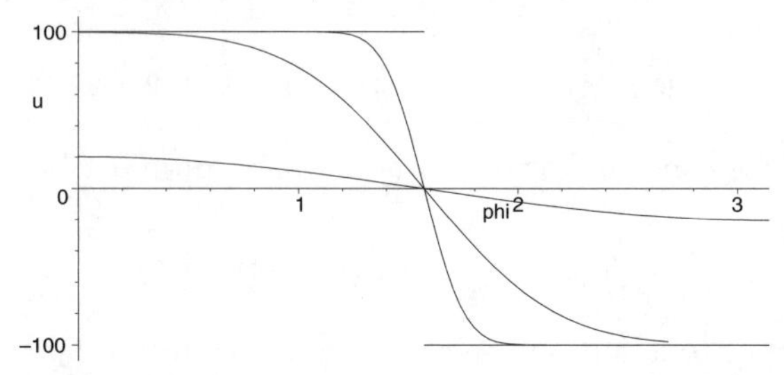

Figure 14: Graphs of the solution of the example problem, with $u(\phi, 0)$ positive in the north and negative in the south. The function $u(\phi, t)$ is shown as a function of ϕ in the range 0 to π for times chosen so that the dimensionless time kt/R^2 takes the values 0, 0.01, 0.1 and 1; for convenience, $T_0 = 100$.

C. Spherical Waves

In Section 8 of this chapter, we solved the wave equation in spherical coordinates for the case where the initial conditions depend only on the radial variable ρ. Now we consider the case where the variable ϕ is also present. A full statement of the problem is

$$\frac{1}{\rho^2} \frac{\partial}{\partial \rho} \left(\rho^2 \frac{\partial u}{\partial \rho} \right) + \frac{1}{\rho^2 \sin(\phi)} \frac{\partial}{\partial \phi} \left(\sin(\phi) \frac{\partial u}{\partial \phi} \right) = \frac{1}{c^2} \frac{\partial^2 u}{\partial t^2},$$

$$0 < \rho < a, \quad 0 < \phi < \pi, \quad 0 < t \qquad (12)$$

$$\begin{aligned} u(a, \phi, t) &= 0, & 0 < \phi < \pi, \quad & 0 < t \\ u(\rho, \phi, 0) &= f(\rho, \phi), & 0 < \rho < a, \quad & 0 < \phi < \pi \\ \frac{\partial u}{\partial t}(\rho, \phi, 0) &= g(\rho, \phi), & 0 < \rho < a, \quad & 0 < \phi < \pi. \end{aligned}$$

As usual, we require in addition that u be bounded as $\rho \to 0$ and as $\phi \to 0$ and $\phi \to \pi$.

First, we seek solutions in the product form $u(\rho, \phi, t) = R(\rho)\Phi(\phi)T(t)$. Inserting u in this form into the partial differential equation (12) and manipulating, we find

$$\frac{1}{\rho^2}\left(\frac{(\rho^2 R')}{R} + \frac{(\sin(\phi)\Phi')'}{\sin(\phi)\Phi}\right) = \frac{T''}{c^2 T}. \tag{13}$$

Both sides of this equation must have the same and constant value, say $-\lambda^2$. Thus, we must have

$$\frac{(\rho^2 R')'}{R} + \frac{(\sin(\phi)\Phi')'}{\sin(\phi)\Phi} = -\lambda^2\rho^2.$$

Again we see that the ratio containing Φ must be constant, say $-\mu^2$. Hence, we have two separate problems for the functions Φ and R.

$$(\sin(\phi)\Phi')' + \mu^2 \sin(\phi)\Phi = 0, \quad 0 < \phi < \pi$$
$$\Phi(\phi) \text{ bounded at } \phi = 0, \quad \pi.$$
$$(\rho^2 R')' - \mu^2 R + \lambda^2\rho^2 R = 0, \quad 0 < \rho < a$$
$$R(a) = 0,$$
$$R(\rho) \text{ bounded at } 0.$$

The first of these problems is now quite familiar, and we know its solution to be

$$\mu_n^2 = n(n+1), \quad \Phi_n(\phi) = P_n(\cos(\phi)), \quad n = 0, 1, 2, \cdots.$$

The second problem is less familiar. In standard form, the differential equation is

$$R'' + \frac{2}{\rho}R' - \frac{\mu^2}{\rho^2}R + \lambda^2 R = 0.$$

Comparison with the four-parameter form of Bessel's equation (Eq. (1) of Section 8) shows $\alpha = -1/2$, $\gamma = 1$ and $p^2 = \mu^2 + \alpha^2$. Since $\mu = n(n+1)$, $p^2 = n^2 + n + \frac{1}{4}$, and then $p = n + \frac{1}{2}$. Thus, the general solution of the differential equation is

$$R_n(\rho) = \rho^{-1/2}\left[AJ_{n+\frac{1}{2}}(\lambda\rho) + BY_{n+\frac{1}{2}}(\lambda\rho)\right].$$

The fact that the Bessel functions of the second kind, $Y_p(\lambda\rho)$, are unbounded at $\rho = 0$ allows us to discard them from the solution, leaving

$$R_n(\rho) = \rho^{-\frac{1}{2}} J_{n+\frac{1}{2}}(\lambda\rho)$$

as the bounded solution. These functions occur frequently in problems in spherical coordinates. Sometimes the functions

$$j_n(z) = \sqrt{\frac{\pi}{2z}} J_{n+\frac{1}{2}}(z)$$

called *spherical Bessel functions of the first kind of order n* are introduced. As noted in Section 8, there is a relation to sines and cosines:

$$j_0(z) = \sin(z)/z$$
$$j_1(z) = (\sin(z) - z\cos(z))/z^2$$
$$j_2(z) = ((3 - z^2)\sin(z) - 3z\cos(z))/z^3.$$

We have yet to satisfy the boundary condition $R_n(a) = 0$. This cannot be done by formula, except for $n = 0$. In this case, $R_0(a) = 0$ comes down to $\sin(\lambda a)/\lambda a = 0$, so that $\lambda_m = m\pi/a$ for $m = 1, 2, \cdots$. For other n's, solutions of $J_{n+1/2}(\lambda a) = 0$ must be found numerically. For instance, for $n = 1$ the equation is

$$\sin(\lambda a) - \lambda a \cos(\lambda a) = 0,$$

with solutions $\lambda a = 4.493,\ 7.725,\ 10.904, \cdots$. (See *Handbook of Mathematical Functions* by Abramowitz and Stegun listed in the Bibliography.)

Finally we can put together some product solutions. Clearly the factor function $T(t)$ will be a sine or cosine of λct. Thus our product solutions have the form

$$\rho^{-1/2} J_{n+1/2}(\lambda_{nm}\rho) P_n(\cos(\phi)) \sin(\lambda_{nm}ct),$$
$$\rho^{-1/2} J_{n+1/2}(\lambda_{nm}\rho) P_n(\cos(\phi)) \cos(\lambda_{nm}ct).$$

The solution $u(\rho, \phi, t)$ will be an infinite series of constant multiples of these functions. We will not write it out.

Let us summarize some of the information we have obtained. First, the frequencies of vibration of a sphere are $\lambda_{mn}c$ (radians per unit time) where λ_{mn} is the mth positive solution of

$$J_{n+1/2}(\lambda a) = 0.$$

Second, the nodal surfaces (loci of points where a product solution is 0 for all time) are the values of ρ and ϕ for which

$$J_{n+1/2}(\lambda_{mn}\rho) P_n(\cos(\phi)) = 0.$$

One or the other factor must be 0, so these surfaces are either concentric spheres, $\rho = const.$, determined by $J_{n+1/2}(\lambda_{mn}\rho) = 0$, or else cones $\phi = const$, determined by $P_n(\cos(\phi)) = 0$.

Finally, let us observe that, because $P_0(\cos(\phi)) \equiv 1$, the product solutions with $n = 0$ are precisely what we found as product solutions of the problem in Section 8, part B.

Exercises

1. Solve the potential equation in the sphere $0 < \rho < 1$, $0 < \phi < \pi$ with the boundary condition

$$u(1,\phi) = \begin{cases} 1, & 0 < \phi < \pi/2 \\ 0, & \pi/2 < \phi < \pi \end{cases}$$

together with appropriate boundedness conditions.

2. Solve the potential equation in a hemisphere, $0 < \rho < 1$, $0 < \phi < \pi/2$, subject to boundedness conditions at $\rho = 0$ and $\phi = 0$, and the boundary conditions

$$u(1,\phi) = 1, \quad 0 < \phi < \pi/2,$$

$$u(\rho,\pi/2) = 0, \quad 0 < \rho < 1.$$

Hint: use odd-order Legendre polynomials.

3. Solve this heat problem with convection on a spherical shell.

$$\frac{1}{R^2\sin(\phi)}\frac{\partial}{\partial\phi}\left(\sin(\phi)\frac{\partial u}{\partial\phi}\right) - \gamma^2(u - T) = \frac{1}{k}\frac{\partial u}{\partial t}, \quad 0 < \phi < \pi, \quad 0 < t,$$

$$u(\phi, 0) = 0, \quad 0 < \phi < \pi$$

4. Solve this heat problem on a hemispherical shell

$$\frac{1}{R^2\sin(\phi)}\frac{\partial}{\partial\phi}\left(\sin(\phi)\frac{\partial u}{\partial\phi}\right) = \frac{1}{k}\frac{\partial u}{\partial t}, \qquad 0 < \phi < \pi/2, \quad 0 < t,$$

$$\frac{\partial u}{\partial\phi}(\pi/2, t) = 0, \qquad 0 < t$$

$$u(\phi, 0) = \cos(\phi), \quad 0 < \phi < \pi/2.$$

5. Solve the eigenvalue problem

$$\frac{1}{\rho^2}\left[\frac{\partial}{\partial \rho}\left(\rho^2 \frac{\partial u}{\partial \rho}\right) + \frac{1}{\sin(\phi)}\frac{\partial}{\partial \phi}\left(\sin(\phi)\frac{\partial u}{\partial \phi}\right)\right] = -\lambda^2 u,$$
$$0 < \rho < R, 0 < \phi < \pi$$
$$u(R, \phi) = 0, \quad 0 < \phi < \pi$$

subject to boundedness conditions at $\rho = 0$ and at $\phi = 0, \pi$.

6. In Part C of this section we mention nodal surfaces (i.e., surfaces where the function is 0). Find the nodal surfaces of the function

$$\rho^{-1/2} J_{3/2}(\lambda\rho) P_1(\cos(\phi))$$

if λ is the second positive solution of $J_{3/2}(\lambda) = 0$.

7. Describe in words the nodal surfaces for

$$\rho^{-1/2} J_{5/2}(\lambda\rho) P_2(\cos(\phi))$$

if λ is the second positive solution of $J_{5/2}(\lambda) = 0$.

8. Solve the potential problem in the exterior of a sphere.

$$\frac{1}{\rho^2}\left[\frac{\partial}{\partial \rho}\left(\rho^2 \frac{\partial u}{\partial \rho}\right) + \frac{1}{\sin(\phi)}\frac{\partial}{\partial \phi}\left(\sin(\phi)\frac{\partial u}{\partial \phi}\right)\right] = 0,$$
$$R < \rho, 0 < \phi < \pi$$
$$u(R, \phi) = f(\phi), \quad 0 < \phi < \pi$$

9. L.M. Chiappetta and D.R. Sobel (Temperature distribution within a hemisphere exposed to a hot gas stream, SIAM Review 26 (1984), p. 575-577) analyze the steady-state temperature in the rounded tip of a combustion-gas sampling probe. The tip is approximately hemispherical in shape. Its outer surface is exposed to hot gases at temperature T_G, and its base is cooled by water at temperature T_W circulating inside the probe. If $T(\rho, \phi)$ is the temperature inside the tip, it should satisfy the conditions

$$\frac{1}{\rho^2}\left[\frac{\partial}{\partial \rho}\left(\rho^2 \frac{\partial u}{\partial \rho}\right) + \frac{1}{\sin(\phi)}\frac{\partial}{\partial \phi}\left(\sin(\phi)\frac{\partial u}{\partial \phi}\right)\right] = 0,$$
$$0 < \rho < R, 0 < \phi < \frac{\pi}{2}$$

$$T(\rho, \pi/2) = T_W, \qquad 0 < \rho < R$$
$$k\frac{\partial T}{\partial \rho}(R, \phi) = h[T_G - T(R, \phi)], \quad 0 < \phi < \frac{\pi}{2}$$

together with boundedness conditions at $\rho = 0$ and at $\phi = 0$.

The authors then change the variables to simplify the problem. Let $r = \rho/R$, $u(r, \phi) = T(\rho, \phi) - T_W$, and show that the problem for u becomes

$$\frac{\partial}{\partial \rho}\left(\rho^2 \frac{\partial u}{\partial \rho}\right) + \frac{1}{\sin(\phi)}\frac{\partial}{\partial \phi}\left(\sin(\phi)\frac{\partial u}{\partial \phi}\right) = 0, \quad 0 < r < 1, 0 < \phi < \pi/2,$$
$$u(r, \pi/2) = 0, \quad 0 < r < 1$$
$$K\frac{\partial u}{\partial r}(1, \phi) + u(1, \phi) = D, \quad 0 < \phi < \pi/2,$$

where $K = k/hR$ and $D = T_G - T_W$.

10. Solve the problem in Exercise 9. Hint: use odd-indexed Legendre polynomials to satisfy the boundary condition at $\phi = \pi/2$.

5.11 COMMENTS AND REFERENCES

We have seen just a few problems in two or three dimensions, but they are sufficient to illustrate the complications that may arise. A serious drawback to the solution by separation of variables is that double and triple series tend to converge slowly, if at all. Thus, if a numerical solution to a two- or three-dimensional problem is needed, it may be advisable to sidestep the analytical solution by using an approximate numerical technique from the beginning.

One advantage of using special coordinate systems is that some problems that are two-dimensional in Cartesian coordinates may be one-dimensional in another system. This is the case, for instance, when distance from a point (r in polar or ρ in spherical coordinates) is the only significant space variable. Of course, nonrectangular systems may arise naturally from the geometry of a problem.

As Sections 3 and 4 point out, solving the two-dimensional heat or wave equation in a region $\mathcal{R}$ of the plane depends on being able to solve the eigenvalue problem, $\nabla^2\phi = -\lambda^2\phi$ in $\mathcal{R}$ with $\phi = 0$ on the boundary. The solution of this problem in a region bounded by coordinate curves (that is, in a generalized rectangle) is known for many coordinate systems.

We have discussed the most common cases; others can be found in *Methods of Theoretical Physics* by Morse and Feshbach. Information about the special functions involved is available from the *Handbook of Mathematical Functions* by Abramowitz and Stegun, and also from *Special Functions for Engineers and Applied Mathematicians* by L.C. Andrews. Eigenfunctions and eigenvalues are known for a few regions that are not generalized rectangles. (See Miscellaneous Exercises 20 and 21 in the text that follows.)

Eigenvalues of the Laplacian in a region can be estimated by a Rayleigh quotient much as in Section 5 of chapter 3. Furthermore, we have theorems of the following type. Let λ_1^2 be the lowest eigenvalue of $\nabla\phi = -\lambda^2\phi$ in $\mathcal{R}$ with $\phi = 0$ on the boundary. Let $\bar{\lambda}_1^2$ have the same meaning for another region $\bar{\mathcal{R}}$. If $\bar{\mathcal{R}}$ fits inside $\mathcal{R}$, then $\bar{\lambda}_1^2 \geq \lambda_1^2$. (The smaller the region, the larger the first eigenvalue.) For further information, see *Methods of Mathematical Physics*, Vol. 1, by Courant and Hilbert. In the famous article "Can one hear the shape of a drum?", *American Mathematical Monthly*, 73 (1966, pp 1-23), Mark Kac shows that one can find the area, perimeter, and connectivity of a region from the eigenvalues of the Laplacian for that region. However, Kac's title question has been answered negatively. In the *Bulletin of the American Mathematical Society* 27 (1992), pp. 134-138, authors C. Gordon, D.L. Webb and S. Wolpert display two plane regions, or "drums," of different shapes, on which the Laplacian has exactly the same eigenvalues.

The nodal curves of a membrane shown in Fig. 9 can be realized physically. Photographs of such curves, along with an explanation of the physics of the vibrating membrane, will be found in "The Physics of Kettledrums" by Thomas D. Rossing, *Scientific American*, November, 1982.

MISCELLANEOUS EXERCISES

1. Solve the heat problem

$$\begin{aligned}
\nabla^2 u &= \frac{1}{k}\frac{\partial u}{\partial t}, && 0 < x < a, \quad 0 < y < b, \quad 0 < t \\
\frac{\partial u}{\partial x}(0, y, t) &= 0, \frac{\partial u}{\partial x}(a, y, t) = 0, && 0 < y < b, \quad 0 < t \\
u(x, 0, t) &= 0, u(x, b, t) = 0, && 0 < x < a, \quad 0 < t \\
u(x, y, 0) &= \frac{Tx}{a}, && 0 < x < a, \quad 0 < y < b.
\end{aligned}$$

2. Same as Exercise 1, but the initial condition is

$$u(x, y, 0) = \frac{Ty}{b}, \quad 0 < x < a, \quad 0 < y < b.$$

3. Let $u(x, y, t)$ be the solution of the heat equation in a rectangle as stated here. Find an expression for $u(a/2, b/2, t)$. Write out the first three nonzero terms for the case $a = b$.

$$\nabla^2 u = \frac{1}{k}\frac{\partial u}{\partial t}, \quad 0 < x < a, \quad 0 < y < b, \quad 0 < t$$

$$u = 0 \qquad \text{on all boundaries}$$

$$u(x, y, 0) = T, \qquad 0 < x < a, \quad 0 < y < b$$

4. Find the nodal lines of the square membrane. These are loci of points satisfying $\phi_{mn}(x, y) = 0$ where ϕ_{mn} satisfies $\nabla^2\phi = -\lambda^2\phi$ in the square and $\phi = 0$ on the boundary.

5. Find the solution of the boundary value problem

$$\frac{1}{r}\frac{d}{dr}\left(r\frac{du}{dr}\right) = -1, \quad 0 < r < a$$

$$u(0) \text{ bounded}, \quad u(a) = 0$$

both directly and by assuming that both $u(r)$ and the constant function 1 have Bessel series on the interval $0 < r < a$:

$$u(r) = \sum_{n=1}^{\infty} C_n J_0(\lambda_n r), \quad 0 < r < a$$

$$1 = \sum_{n=1}^{\infty} c_n J_0(\lambda_n r), \quad 0 < r < a.$$

$$\left(\text{Hint}: \frac{1}{r}\frac{d}{dr}\left(r\frac{dJ_0(\lambda r)}{dr}\right) = -\lambda^2 J_0(\lambda r).\right)$$

6. Suppose that $w(x, t)$ and $v(y, t)$ are solutions of the partial differential equations

$$\frac{\partial^2 w}{\partial x^2} = \frac{1}{k}\frac{\partial w}{\partial t}, \qquad \frac{\partial^2 v}{\partial y^2} = \frac{1}{k}\frac{\partial v}{\partial t}.$$

Show that $u(x, y, t) = w(x, t)v(y, t)$ satisfies the two-dimensional heat equation

$$\frac{\partial^2 u}{\partial x^2} + \frac{\partial^2 u}{\partial y^2} = \frac{1}{k}\frac{\partial u}{\partial t}.$$

7. Use the idea of Exercise 6 to solve the problem stated in Exercise 1.

8. Let $w(x, y)$ and $v(z, t)$ satisfy the equations

$$\frac{\partial^2 w}{\partial x^2} + \frac{\partial^2 w}{\partial y^2} = 0, \qquad \frac{\partial^2 v}{\partial z^2} = \frac{1}{c^2}\frac{\partial^2 v}{\partial t^2}.$$

Show that the product $u(x, y, z, t) = w(x, y)v(z, t)$ satisfies the three-dimensional wave equation

$$\frac{\partial^2 u}{\partial x^2} + \frac{\partial^2 u}{\partial y^2} + \frac{\partial^2 u}{\partial z^2} = \frac{1}{c^2}\frac{\partial^2 u}{\partial t^2}.$$

9. Find the product solutions of the equation

$$\frac{1}{r}\frac{\partial}{\partial r}\left(r\frac{\partial u}{\partial r}\right) = \frac{1}{k}\frac{\partial u}{\partial t}, \qquad 0 < r, \quad 0 < t$$

that are bounded as $r \to 0+$ and as $r \to \infty$.

10. Show that the boundary value problem

$$((1 - x^2)\phi')' = -f(x), \qquad -1 < x < 1$$

$$\phi(x) \text{ bounded at } x = \pm 1$$

has as its solution

$$\phi(x) = \int_0^x \frac{1}{1 - y^2}\int_y^1 f(z)dzdy$$

provided that the function f satisfies

$$\int_{-1}^1 f(z)dz = 0.$$

11. Suppose that the functions $f(x)$ and $\phi(x)$ in the preceding exercise have expansions in terms of Legendre polynomials

$$f(x) = \sum_{k=0}^{\infty} b_k P_k(x), \qquad -1 < x < 1$$

$$\phi(x) = \sum_{k=0}^{\infty} B_k P_k(x), \qquad -1 < x < 1.$$

What is the relation between B_k and b_k?

12. By applying separation of variables to the problem

$$\nabla^2 u = 0, \quad 0 < \rho < a, \quad 0 \le \phi < \pi$$

with u bounded at $\phi = 0$, π, and u periodic (2π) in θ, derive the following equation for the factor function $\Phi(\phi)$:

$$\sin(\phi)(\sin(\phi)\Phi')' - m^2\Phi + \mu^2 \sin^2(\phi)\Phi = 0,$$

where $m = 0, 1, 2, \cdots$ comes from the factor $\Theta(\theta)$.

13. Using the change of variables $x = \cos(\phi)$, $\Phi(\phi) = y(x)$ on the equation of Exercise 12, derive a differential equation for $y(x)$.

14. Solve the heat conduction problem

$$\frac{1}{r}\frac{\partial}{\partial r}\left(r\frac{\partial u}{\partial r}\right) = \frac{1}{k}\frac{\partial u}{\partial t}, \qquad 0 < r < a, \quad 0 < t$$

$$\frac{\partial u}{\partial r}(a, t) = 0, \qquad 0 < t$$

$$u(r, 0) = T_0 - \Delta\left(\frac{r}{a}\right)^2, \quad 0 < r < a.$$

15. Solve the following potential problem in a cylinder:

$$\frac{1}{r}\frac{\partial}{\partial r}\left(r\frac{\partial u}{\partial r}\right) + \frac{\partial^2 u}{\partial z^2} = 0, \qquad 0 < r < a, \quad 0 < z < b$$

$$u(a, z) = 0, \qquad 0 < z < b$$

$$u(r, 0) = 0, \quad u(r, b) = U_0, \quad 0 < r < a.$$

16. Find the solution of the heat conduction problem

$$\frac{1}{r}\frac{\partial}{\partial r}\left(r\frac{\partial u}{\partial r}\right) = \frac{1}{k}\frac{\partial u}{\partial t}, \quad 0 < r < a, \quad 0 < t$$

$$u(a, z) = T_0, \qquad 0 < t$$

$$u(r, 0) = T_1, \qquad 0 < r < a.$$

17. Find some frequencies of vibration of a cylinder by finding product solutions of the problem

$$\frac{1}{r}\frac{\partial}{\partial r}\left(r\frac{\partial u}{\partial r}\right) + \frac{\partial^2 u}{\partial z^2} = \frac{1}{c^2}\frac{\partial^2 u}{\partial t^2}, \quad 0 < r < a, \quad 0 < z < b, \quad 0 < t$$

$$u(r, 0, t) = 0, \quad u(r, b, t) = 0, \qquad 0 < r < a, \quad \alpha < t$$

$$u(a, z, t) = 0, \qquad 0 < z < b, \quad 0 < t.$$

18. Derive the given formula for the solution of the following potential equation in a spherical shell:

$$\nabla^2 u = 0, \quad a < \rho < b, \qquad 0 < \phi < \pi$$

$$u(a, \phi) = f(\cos(\phi)), \quad u(b, \phi) = 0, \qquad 0 < \phi < \pi$$

$$u(\rho, \phi) = \sum_{n=0}^{\infty} A_n \frac{b^{2n+1} - \rho^{2n+1}}{b^{2n+1} - a^{2n+1}} \left(\frac{a}{\rho}\right)^{n+1} P_n(\cos(\phi)),$$

$$A_n = \frac{2n+1}{2} \int_{-1}^{1} f(x) P_n(x) dx.$$

19. Show that the function $\phi(x, y) = \sin(\pi x)\sin(2\pi y) - \sin(2\pi x)\sin(\pi y)$ is an eigenfunction for the triangle $\mathcal{T}$ bounded by the lines $y = 0$, $y = x$, $x = 1$. That is,

$$\nabla^2 \phi = -\lambda^2 \phi \text{ in } \mathcal{T}$$

$$\phi = 0 \text{ on the boundary of } \mathcal{T}.$$

What is the eigenvalue λ^2 associated with ϕ?

20. Observe that the function ϕ in Exercise 19 is the difference of two different eigenfunctions of the 1×1 square (see Section 3) corresponding to the same eigenvalue. Use this idea to construct other eigenfunctions for the triangle $\mathcal{T}$ of Exercise 19.

21. Let $\mathcal{T}$ be the equilateral triangle in the xy-plane whose base is the interval $0 < x < 1$ of the x-axis and whose sides are segments of the lines $y = \sqrt{3}x$ and $y = \sqrt{3}(1 - x)$. Show that for $n = 1, 2, 3, \cdots$, the function

$$\phi_n(x, y) = \sin\left(4n\pi y/\sqrt{3}\right) + \sin(2n\pi(x - y/\sqrt{3})) - \sin(2n\pi(x + y/\sqrt{3}))$$

is a solution of the eigenvalue problem: $\nabla^2 \phi = -\lambda^2 \phi$ in $\mathcal{T}$, $\phi = 0$ on the boundary of $\mathcal{T}$. What are the eigenvalues λ_n^2 corresponding to the function ϕ_n that is given? (See "The eigenvalues of an equilateral triangle", SIAM Journal of Mathematical Analysis, 11 (1980), pp. 819-827, by Mark A. Pinsky.)

22. In Comments and References, Section 11, a theorem is quoted that relates the least eigenvalue of a region to that of a smaller region. Confirm the theorem by comparing the solution of Exercise 19 with the smallest eigenvalue of one-eighth of a circular disk of radius 1:

$$\frac{1}{r}\frac{\partial}{\partial r}\left(r\frac{\partial \phi}{\partial r}\right) + \frac{1}{r^2}\frac{\partial^2 \phi}{\partial \theta^2} = -\lambda^2 \phi, \quad 0 < \theta < \frac{\pi}{4}, \qquad 0 < r < 1$$

$$\phi(r, 0) = 0, \qquad \phi\left(r, \frac{\pi}{4}\right) = 0, \quad 0 < r < 1$$

$$\phi(1, \theta) = 0, \qquad 0 < \theta < \frac{\pi}{4}.$$

23. Same task as Exercise 22, but use the triangle of Exercise 21 and the smallest eigenvalue of one-sixth of a circular disk of radius 1.

24. Show that $u(\rho, t) = t^{-3/2}e^{-\rho^2/4t}$ is a solution of the three-dimensional heat equation $\nabla^2 u = \dfrac{\partial u}{\partial t}$, in spherical coordinates.

25. For what exponent b is $u(r, t) = t^b e^{-r^2/4t}$ a solution of the two-dimensional heat equation $\nabla^2 u = \dfrac{\partial u}{\partial t}$? (Use polar coordinates.)

26. Suppose that an estuary extends from $x = 0$ to $x = a$, where it meets the open sea. If the floor of the estuary is level but its width is proportional to x, then the water depth $u(x, t)$ satisfies

$$\frac{1}{x}\frac{\partial}{\partial x}\left(x\frac{\partial u}{\partial x}\right) = \frac{1}{gU}\frac{\partial^2 u}{\partial t^2}, \quad 0 < x < a, \quad 0 < t,$$

where g is the acceleration of gravity and U is mean depth. The tidal motion of the sea is represented by the boundary condition

$$u(a, t) = U + h\cos(\omega t).$$

Find a bounded solution of the partial differential equation that satisfies the boundary condition by setting

$$u(x, t) = U + y(x)\cos(\omega t).$$

(See Lamb, *Hydrodynamics*, pp. 275-276.)

27. Is there any combination of parameters for which the solution of Exercise 26 does not exist in the form suggested?

28. If the estuary of Exercise 26 has uniform width but variable depth $h = Ux/a$, then the equation for u is

$$\frac{\partial}{\partial x}\left(x\frac{\partial u}{\partial x}\right) = \frac{a}{gU}\frac{\partial^2 u}{\partial t^2}, \quad 0 < x < a, \quad 0 < t$$

subject to the same boundary condition as in Exercise 26. Find a bounded solution in the form suggested. (See Eq. (1) of Section 8.)

29. The equation for radially symmetric waves in n-dimensional space is

$$\frac{1}{r^{n-1}}\frac{\partial}{\partial r}\left(r^{n-1}\frac{\partial u}{\partial r}\right) = \frac{1}{c^2}\frac{\partial^2 u}{\partial t^2}$$

where r is distance to the origin. Find product solutions of this equation that are bounded at the origin.

30. Show that the equation of Exercise 29 has solutions of the form

$$u(r,t) = \alpha(r)\phi(r - ct)$$

for $n = 1$ and $n = 3$. (See "A simple proof that the world is three-dimensional " by Tom Morley, SIAM Review 27 (1985) pp. 69-71.)

31. A certain kind of chemical reactor contains particles of a solid catalyst and a liquid that reacts with a gas bubbled through it. M. Chidambaran (Catalyst mixing in bubble column slurry reactors, Canadian Journal of Chemical Engineering, 67 (1989), 503-506) uses the following problem to model the catalyst concentration C in a cylindrical reactor:

$$D_r\frac{1}{r}\frac{\partial}{\partial r}\left(r\frac{\partial C}{\partial r}\right) + D_z\frac{\partial^2 C}{\partial z^2} + U\frac{\partial C}{\partial z} = 0, \quad 0 < r < R, \quad 0 < z < L$$

$$\frac{\partial C}{\partial r}(R, z) = 0, \quad 0 < z < L.$$

Here, D_r and D_z are the diffusion constants in the radial and axial directions. The term containing $\partial C/\partial z$ represents physical movement of particles at speed U.

Show that the change of variables $\rho = r/R$, $\zeta = z/L$, $u(\rho, \zeta) = C(r, z)$ leads to the equivalent equations

$$b\frac{1}{\rho}\frac{\partial}{\partial \rho}\left(\rho\frac{\partial u}{\partial \rho}\right) + \frac{\partial^2 u}{\partial \zeta^2} + p\frac{\partial u}{\partial \zeta} = 0, \quad 0 < \rho < 1, \quad 0 < \zeta < 1,$$

$$\frac{\partial u}{\partial \rho}(1, \zeta) = 0, \quad 0 < \zeta < 1$$

and identify the parameters b and p.

32. When a boundedness condition at $\rho = 0$ is added, product solutions of the foregoing equation are found to have the form $u(\rho, \zeta) = R(\rho)Z(\zeta)$

$$R_0(\rho) = 1, \quad Z_0(\zeta) = \begin{cases} e^{-p\zeta} \\ 1 \end{cases}$$

$$R_n(\rho) = J_0(\lambda_n \rho), \quad Z_n(\zeta) = \begin{cases} e^{m_1 \zeta} \\ e^{m_2 \zeta} \end{cases}$$

where $m_1 < 0 < m_2$ are the roots of the equation $m^2 + pm - \lambda_n^2 b = 0$ and λ_n is chosen to satisfy $J_0'(\lambda_n) = 0$.

a. Check the details of the solution.

b. Show that the λ's satisfy also $J_1(\lambda_n) = 0$.

33. The solution of the problem in Exercise 31 has the form

$$u(\rho, \zeta) = a_0 e^{-p\zeta} + b_0 + \sum_{n=1}^{\infty} (a_n e^{m_1 \zeta} + b_n e^{m_2 \zeta}) J_0(\lambda_n \rho).$$

The coefficients would normally be found by applying boundary conditions

$$u(\rho, 0) = f(\rho), \quad u(\rho, 1) = g(\rho), \quad 0 < \rho < 1.$$

In this case, however, information is scarce. The author suggests discarding the solutions that do not approach 0 as $\zeta \to \infty$. The justification is that $g(\rho)$ is approximately 0. The solution then becomes

$$u(\rho, \zeta) = a_0 e^{-p\zeta} + \sum_{n=1}^{\infty} a_n e^{m_1 \zeta} J_0(\lambda_n \rho),$$

and the coefficients should be determined by

$$a_0 = 2 \int_0^1 f(\rho) \rho d\rho, \quad a_n = \frac{\int_0^1 f(\rho) J_0(\lambda_n \rho) \rho d\rho}{\int_0^1 J_0^2(\lambda_n \rho) \rho d\rho}.$$

The function f is known only roughly through experiment. Use the numbers in the table below to find a_0 and a_1 by the trapezoidal rule of numerical integration.

ρ	0	0.1	0.2	0.3	0.4
$f(\rho)$	8.8	8.9	9.2	9.8	10.3
$J_0(\lambda_1\rho)$	10	0.964	0.858	0.696	0.493

ρ	0.5	0.6	0.7	0.8	0.9	1.0
$f(\rho)$	11.2	12.0	13.1	14.1	14.8	15
$J_0(\lambda_1\rho)$	0.273	0.056	-0.135	-0.281	-0.373	-0.403

Chapter 6

Laplace Transform

6.1 DEFINITION AND ELEMENTARY PROPERTIES

The Laplace transform serves as a device for simplifying or mechanizing the solution of ordinary and partial differential equations. It associates a function $f(t)$ with a function of another variable $F(s)$ from which the original function can be recovered.

Let $f(t)$ be sectionally continuous in every interval $0 \leq t < T$. The Laplace transform of f, written $\mathcal{L}(f)$ or $F(s)$, is defined by the integral

$$\mathcal{L}(f) = F(s) = \int_0^\infty e^{-st} f(t) \; dt. \tag{1}$$

We use the convention that a function of t is represented by a lower case letter and its transform by the corresponding capital letter. The variable s may be real or complex, but in the computation of transforms by the definition, s is usually assumed to be real. Two simple examples are

$$\mathcal{L}(1) = \int_0^\infty e^{-st} \cdot 1 \; dt = \frac{1}{s}$$

$$\mathcal{L}(e^{at}) = \int_0^\infty e^{-st} e^{at} \; dt = \left. \frac{-e^{-(s-a)t}}{s-a} \right|_0^\infty = \frac{1}{s-a}.$$

Not every sectionally continuous function of t has a Laplace transform, for the defining integral may fail to converge. For instance, $\exp(t^2)$ has no transform. However, there is a simple sufficient condition, as expressed in the following theorem.

Theorem
Let $f(t)$ be sectionally continuous in every finite interval $0 \leq t < T$. If, for some constant k, it is true that

$$\lim_{t\to\infty} e^{-kt} f(t) = 0,$$

then the Laplace transform of f exists for $\mathrm{Re}(s) > k$.

A function that satisfies the limit condition in the hypotheses of the theorem is said to be of *exponential order.*

The Laplace transform inherits two important properties from the integral used in its definition:

$$\mathcal{L}\left(cf(t)\right) = c\mathcal{L}\left(f(t)\right), c \text{ constant} \tag{2}$$

$$\mathcal{L}\left(f(t) + g(t)\right) = \mathcal{L}\left(f(t)\right) + \mathcal{L}\left(g(t)\right). \tag{3}$$

By exploiting these properties, we easily determine that

$$\begin{aligned}\mathcal{L}(\cosh(at)) &= \mathcal{L}\left[\frac{1}{2}\left(e^{at} + e^{-at}\right)\right] \\ &= \frac{1}{2}\left(\frac{1}{s-a} + \frac{1}{s+a}\right) = \frac{s}{s^2 - a^2} \\ \mathcal{L}(\sin(\omega t)) &= \mathcal{L}\left[\frac{1}{2i}\left(e^{i\omega t} - e^{-i\omega t}\right)\right] \\ &= \frac{1}{2i}\left(\frac{1}{s - i\omega} - \frac{1}{s + i\omega}\right) = \frac{\omega}{s^2 + \omega^2}.\end{aligned}$$

Notice the use of linearity properties with complex constants and functions.

Because of the factor e^{-st} in the definition of the Laplace transform, exponential multipliers are easily handled by the "shifting theorem":

$$\begin{aligned}\mathcal{L}\left(e^{bt} f(t)\right) &= \int_0^\infty e^{-st} e^{bt} f(t) \ dt \\ &= \int_0^\infty e^{-(s-b)t} f(t) \ dt = F(s-b)\end{aligned}$$

where $F(s) = \mathcal{L}(f(t))$. For instance,

$$\mathcal{L}\left(e^{bt}\sin(\omega t)\right) = \frac{\omega}{(s-b)^2 + \omega^2} = \frac{\omega}{s^2 - 2sb + b^2 + \omega^2}$$

since $\mathcal{L}(\sin(\omega t)) = \omega/(s^2 + \omega^2)$.

The real virtue of the Laplace transform is revealed by its effect on derivatives. Suppose $f(t)$ is continuous and has a sectionally continuous derivative $f'(t)$. Then by definition

$$\mathcal{L}(f'(t)) = \int_0^\infty e^{-st} f'(t)\ dt.$$

Integrating by parts, we get

$$\mathcal{L}(f'(t)) = e^{-st} f(t)\Big|_0^\infty - \int_0^\infty (-s)e^{-st} f(t)\ dt.$$

If $f(t)$ is of exponential order, $e^{-st} f(t)$ must tend to 0 as t tends to infinity (for large enough s) so that the foregoing equation becomes

$$\begin{aligned}\mathcal{L}(f'(t)) &= -f(0) + s\int_0^\infty e^{-st} f(t)\ dt \\ &= -f(0) + s\mathcal{L}(f(t)).\end{aligned}$$

(If $f(t)$ has a jump at $t = 0$, $f(0)$ is to be interpreted as $f(0+)$.)

Similarly, if f and f' are continuous, f'' is sectionally continuous, and if all three functions are exponential order, then

$$\begin{aligned}\mathcal{L}(f''(t)) &= -f(0) + s\mathcal{L}(f'(t)) \\ &= -f(0) - sf(0) + s^2\mathcal{L}(f(t)).\end{aligned}$$

An easy generalization extends this formula to the nth derivative

$$\mathcal{L}[f^{(n)}(t)] = -f^{(n-1)}(0) - sf^{(n-2)}(0) - \cdots - s^{n-1}f(0) + s^n\mathcal{L}(f(t)) \quad (4)$$

on the assumption that f and its first $n-1$ derivatives are continuous, $f^{(n)}$ is sectionally continuous, and all are of exponential order.

We may apply Eq. (4) to the function $f(t) = t^k$, k being a nonnegative integer. Here we have

$$f(0) = f'(0) = \cdots = f^{(k-1)}(0) = 0, \quad f^{(k)}(0) = k!, \quad f^{(k+1)}(t) = 0.$$

Thus, Eq. (4) with $n = k+1$ yields

$$0 = -k! + s^{k+1}\mathcal{L}(t^k),$$

or

$$\mathcal{L}(t^k) = \frac{k!}{s^{k+1}}.$$

A different application of the derivative rule is used to transform integrals. If $f(t)$ is sectionally continuous, then $\int_0^t f(t')\ dt'$ is a continuous function, equal to zero at $t = 0$, and has derivative $f(t)$. Hence

$$\mathcal{L}\left(f(t)\right) = s\mathcal{L}\left[\int_0^t f(t')\ dt'\right]$$

or

$$\mathcal{L}\left[\int_0^t f(t')\ dt'\right] = \frac{1}{s}\mathcal{L}\left(f(t)\right). \tag{5}$$

Differentiation and integration with respect to s may produce transformations of previously inaccessible functions. We need the two formulas

$$-\frac{de^{-st}}{ds} = te^{-st}, \quad \int_s^\infty e^{-s't}\ ds' = \frac{1}{t}e^{-st}$$

to derive the results

$$\mathcal{L}\left(tf(t)\right) = -\frac{dF(s)}{ds}, \quad \mathcal{L}\left(\frac{1}{t}f(t)\right) = \int_s^\infty F(s')\ ds'. \tag{6}$$

(Note that, unless $f(0) = 0$, the transform of $f(t)/t$ will not exist.) Examples of the use of these formulas are

$$\mathcal{L}(t\sin(\omega t)) = -\frac{d}{ds}\left(\frac{\omega}{s^2+\omega^2}\right) = \frac{2s\omega}{(s^2+\omega^2)^2}$$

$$\mathcal{L}\left(\frac{\sin(t)}{t}\right) = \int_s^\infty \frac{ds'}{s'^2+1} = \frac{\pi}{2} - \tan^{-1}(s) = \tan^{-1}\left(\frac{1}{s}\right).$$

When a problem is solved by use of Laplace transforms, a prime difficulty is computation of the corresponding function of t. Methods for computing the "inverse transform" $f(t) = \mathcal{L}^{-1}(F(s))$ include integration in the complex plane, convolution, partial fractions (discussed in Section 2), and tables of transforms. The last method, which involves the least work, is the most popular. The transforms in Table 2 were all calculated from the definition or by use of formulas in this section.

Exercises

1. By using linearity and the transform of e^{at}, compute the transform of each of the following functions:

a.	$\sinh(at)$	**b.**	$\cos(\omega t)$	**c.**	$\cos^2(\omega t)$
d.	$\sin(\omega t - \phi)$	**e.**	$e^{2(t+1)}$	**f.**	$\sin^2(\omega t)$.

2. Use differentiation with respect to t to find the transform of

a. te^{at} from $\mathcal{L}(e^{at})$

b. $\sin(\omega t)$ from $\mathcal{L}(\cos(\omega t))$

c. $\cosh(at)$ from $\mathcal{L}(\sinh(at))$.

3. Compute the transform of each of the following directly from the definition:

a. $f(t) = \begin{cases} 0, & 0 < t < a \\ 1, & a < t \end{cases}$

b. $f(t) = \begin{cases} 0, & 0 < t < a \\ 1, & a < t < b \\ 0, & b < t \end{cases}$

c. $f(t) = \begin{cases} t, & 0 < t < a \\ a, & a < t. \end{cases}$

4. The *Heaviside step function* is defined by the formula

$$H_a(t) = \begin{cases} 1, & t > a \\ 0, & t < a. \end{cases}$$

Assuming $a \geq 0$, show that the Laplace transform of H_a is

$$\mathcal{L}(H_a(t)) = \frac{e^{-as}}{s}.$$

5. Use completion of square and the shifting theorem to find the inverse transform of

a. $\dfrac{1}{s^2 + 2s}$ b. $\dfrac{s}{s^2 + 2s}$ c. $\dfrac{1}{s^2 + 2as + b^2}$, $b > a$.

6. Find the Laplace transform of the square-wave function

$$f(t) = \begin{cases} 1, & 0 < x < a \\ 0, & a < x < 2a \end{cases}, \qquad f(x + 2a) = f(x).$$

(Hint: Break up the integral as shown in the following, evaluate the integrals, and add up a geometric series

$$F(s) = \sum_{n=0}^{\infty} \int_{2na}^{2(n+1)a} f(t)e^{-st}\,dt.)$$

7. Use any method to find the inverse transform of the following.

a. $\dfrac{1}{(s-a)(s-b)}$ **b.** $\dfrac{s}{(s^2-a^2)^2}$

c. $\dfrac{s^2}{(s^2+\omega^2)^2}$ **d.** $\dfrac{1}{(s-a)^3}$

e. $\dfrac{(1-e^{-s})}{s}$

8. Use any theorem or formula to find the transform of the following.

a. $\dfrac{1-\cos(\omega t)}{t}$ **b.** $\displaystyle\int_0^t \frac{\sin(at')}{t'}\,dt'$

c. $t^2 e^{-at}$ **d.** $t\cos(\omega t)$

e. $\sinh(at)\sin(\omega t)$

6.2 PARTIAL FRACTIONS AND CONVOLUTIONS

Because of the formula for the transform of derivatives, the Laplace transform finds important application to linear differential equations with constant coefficients, subject to initial conditions. In order to solve the simple problem

$$u' + au = 0, \quad u(0) = 1$$

we transform the entire equation, obtaining

$$\mathcal{L}(u') + a\mathcal{L}(u) = 0$$

or

$$sU - 1 + aU = 0$$

where $U = \mathcal{L}(u)$. The derivative has been "transformed out," and U is determined by simple algebra to be

$$U(s) = \frac{1}{s+a}.$$

Table 1: Properties of Laplace transform

$$\mathcal{L}(f) = F(s) = \int_0^\infty e^{-st} f(t)\ dt$$

$$\mathcal{L}\left(cf(t)\right) = c\mathcal{L}\left(f(t)\right)$$

$$\mathcal{L}\left(f(t) + g(t)\right) = \mathcal{L}\left(f(t)\right) + \mathcal{L}\left(g(t)\right)$$

$$\mathcal{L}\left(f'(t)\right) = -f(0) + sF(s)$$

$$\mathcal{L}\left(f''(t)\right) = -f'(0) - sf(0) + s^2F(s)$$

$$\mathcal{L}\left(f^{(n)}(t)\right) = -f^{(n-1)}(0) - sf^{(n-2)}(0) - \cdots - s^{n-1}f(0) + s^nF(s)$$

$$\mathcal{L}\left(e^{bt}f(t)\right) = F(s-b)$$

$$\mathcal{L}\left(\int_0^t f(t')\ dt'\right) = \frac{1}{s}F(s) \qquad \mathcal{L}\left(\frac{1}{t}f(t)\right) = \int_s^\infty F(s')\ ds'$$

$$\mathcal{L}\left(tf(t)\right) = \frac{dF}{ds}$$

By consulting the table we find that $u(t) = e^{-at}$.

Equations of higher order can be solved in the same way. When transformed, the problem

$$u'' + \omega^2 u = 0, \quad u(0) = 1, \quad u'(0) = 0$$

becomes

$$s^2U - s \cdot 1 - 0 + \omega^2 U = 0.$$

Note how both initial conditions have been incorporated into this one equation. Now we solve the transformed equation algebraically to find

$$U(s) = \frac{s}{s^2 + \omega^2},$$

the transform of $\cos(\omega t)$.

Table 2: Laplace transforms

$f(t)$	$F(s)$
0	0
1	$\frac{1}{s}$
e^{at}	$\frac{1}{s-a}$
$\cosh(at)$	$\frac{s}{s^2-a^2}$
$\sinh(at)$	$\frac{a}{s^2-a^2}$
$\cos(\omega t)$	$\frac{s}{s^2+\omega^2}$
$\sin(\omega t)$	$\frac{\omega}{s^2+\omega^2}$
t	$\frac{1}{s^2}$
t^k	$\frac{k!}{s^{k+1}}$
$e^{bt}\cos(\omega t)$	$\frac{s-b}{s^2-2bs+b^2+\omega^2}$
$e^{bt}\sin(\omega t)$	$\frac{\omega}{s^2-2bs+b^2+\omega^2}$
$e^{bt}t^k$	$\frac{k!}{(s-b)^{k+1}}$
$e^{at}-1$	$\frac{a}{s(s-a)}$
$t\cos(\omega t)$	$\frac{s^2-\omega^2}{(s^2+\omega^2)^2}$
$t\sin(\omega t)$	$\frac{2s\omega}{(s^2+\omega^2)^2}$

In general we may outline our procedure as follows:

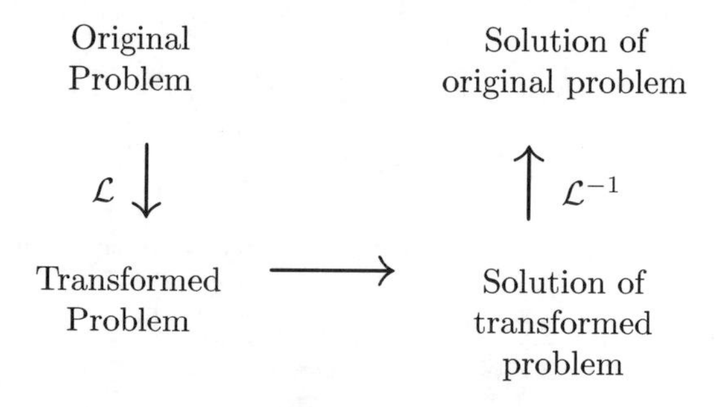

In the step marked $\mathcal{L}^{-1}$, we must compute the function of t to which the solution of the transformed problem corresponds. This is the difficult part of the process. The key property of the inverse transform is its linearity, as expressed by

$$\mathcal{L}^{-1}\left(c_1 F_1(s) + c_2 F_2(s)\right) = c_1 \mathcal{L}^{-1}\left(F_1(s)\right) + c_2 \mathcal{L}^{-1}\left(F_2(s)\right).$$

This property allows us to break down a complicated transform into a sum of simple ones.

A simple mass-spring-damper system leads to the initial value problem

$$u'' + au' + \omega^2 u = 0, \quad u(0) = u_0, \quad u'(0) = u_1$$

whose transform is

$$s^2 U - su_0 - u_1 + a(sU - u_0) + \omega^2 U = 0.$$

Determination of U gives it as the ratio of two polynomials

$$U(s) = \frac{su_0 + (u_1 + au_0)}{s^2 + as + \omega^2}.$$

Although this expression is not in our table, it can be worked around to a function of $s + a/2$ whose inverse transform is available. The shift theorem then gives $u(t)$. There is, however, a better way.

The inversion of a rational function of s (that is, the ratio of two polynomials) can be accomplished by the technique of "partial fractions." Suppose we wish to compute the inverse transform of

$$U(s) = \frac{cs + d}{s^2 + as + b}.$$

The denominator has two roots r_1 and r_2, which we assume for the moment to be distinct. Thus

$$s^2 + as + b = (s - r_1)(s - r_2)$$
$$U(s) = \frac{cs + d}{(s - r_1)(s - r_2)}.$$

The rules of elementary algebra suggest that U can be written as a sum

$$\frac{cs + d}{(s - r_1)(s - r_2)} = \frac{A_1}{s - r_1} + \frac{A_2}{s - r_2} \tag{1}$$

for some choice of A_1 and A_2. Indeed, by finding the common denominator form for the right-hand side and matching powers of s in the numerator, we obtain

$$\frac{cs + d}{(s - r_1)(s - r_2)} = \frac{A_1(s - r_2) + A_2(s - r_1)}{(s - r_1)(s - r_2)}$$
$$c = A_1 + A_2, \quad d = -A_1 r_2 - A_2 r_1.$$

When A_1 and A_2 are determined, the inverse transform of the right-hand side of Eq. (1) is easily found:

$$\mathcal{L}^{-1}\left(\frac{A_1}{s - r_1} + \frac{A_2}{s - r_2}\right) = A_1 \exp(r_1 t) + A_2 \exp(r_2 t).$$

For a specific example, suppose that

$$U(s) = \frac{s + 4}{s^2 + 3s + 2}.$$

The roots of the denominator are $r_1 = -1$ and $r_2 = -2$. Thus

$$\frac{s + 4}{s^2 + 3s + 2} = \frac{A_1}{s + 1} + \frac{A_2}{s + 2} = \frac{(A_1 + A_2)s + (2A_1 + A_2)}{(s + 1)(s + 2)}.$$

We find $A_1 = 3$, $A_2 = -2$. Hence

$$\mathcal{L}^{-1}\left(\frac{s + 4}{s^2 + 3s + 2}\right) = \mathcal{L}^{-1}\left(\frac{3}{s + 1} - \frac{2}{s + 2}\right) = 3e^{-t} - 2e^{-2t}.$$

A little calculus takes us much further. Suppose that U has the form

$$U(s) = \frac{q(s)}{p(s)}$$

where p and q are polynomials, and the degree of q is less than the degree of p. Assume that p has distinct roots $r_1, \cdots, r_k$

$$p(s) = (s - r_1)(s - r_2) \cdots (s - r_k).$$

We try to write U in the fraction form

$$U(s) = \frac{A_1}{s - r_1} + \frac{A_2}{s - r_2} + \cdots + \frac{A_k}{s - r_k} = \frac{q(s)}{p(s)}.$$

The algebraic determination of the A's is very tedious, but notice that

$$\frac{(s - r_1)q(s)}{p(s)} = A_1 + A_2 \frac{s - r_1}{s - r_2} + \cdots + A_k \frac{s - r_1}{s - r_k}.$$

If s is set equal to r_1, the right-hand side is just A_1. The left-hand side becomes $0/0$, but L'Hôpital's rule gives

$$\lim_{s \to r_1} \frac{(s - r_1)q(s)}{p(s)} = \lim_{s \to r_1} \frac{(s - r_1)q'(s) + q(s)}{p'(s)} = \frac{q(r_1)}{p'(r_1)}.$$

Therefore A_1 and all the other A's are given by

$$A_i = \frac{q(r_i)}{p'(r_i)}.$$

Consequently, our rational function takes the form

$$\frac{q(s)}{p(s)} = \frac{q(r_1)}{p'(r_1)} \frac{1}{s - r_1} + \cdots + \frac{q(r_k)}{p'(r_k)} \frac{1}{s - r_k}.$$

From this point we can easily obtain the inverse transform, as expressed in the conclusion of the following theorem.

Theorem 1
Let p and q be polynomials, q of lower degree than p, and let p have only simple roots, $r_1, r_2, \cdots, r_k$. Then

$$\mathcal{L}^{-1}\left(\frac{q(s)}{p(s)}\right) = \frac{q(r_1)}{p'(r_1)} \exp(r_1 t) + \cdots + \frac{q(r_k)}{p'(r_k)} \exp(r_k t). \tag{2}$$

(Equation (2) is known as *Heaviside's formula.*)

Let us apply the theorem to the example in which $q(s) = s + 4$, $p(s) = s^2 + 3s + 2$, $p'(s) = 2s + 3$. Then

$$\mathcal{L}^{-1}\left(\frac{s + 4}{s^2 + 3s + 2}\right) = \frac{-1 + 4}{2(-1) + 3} e^{-t} + \frac{-2 + 4}{2(-2) + 3} e^{-2t} = 3e^{-t} - 2e^{-2t}.$$

In nonhomogeneous differential equations also, the Laplace transform is a useful tool. To solve the problem

$$u' + au = f(t), \quad u(0) = u_0$$

we again transform the entire equation, obtaining

$$\begin{aligned} sU - u_0 + aU &= F(s) \\ U(s) &= \frac{u_0}{s+a} + \frac{1}{s+a}F(s). \end{aligned}$$

The first term in this expression is recognized as the transform of $u_0 e^{-at}$. If $F(s)$ is a rational function, partial fractions may be used to invert the second term. However, we can identify that term by solving the problem another way. For example,

$$\begin{aligned} e^{at}(u' + au) &= e^{at} f(t) \\ (ue^{at})' &= e^{at} f(t) \\ ue^{at} &= \int_0^t e^{at'} f(t') \ dt' + c \\ u(t) &= \int_0^t e^{-a(t-t')} f(t') \ dt' + ce^{-at}. \end{aligned}$$

The initial condition requires that $c = u_0$. On comparing the two results, we see that

$$\mathcal{L}\left[\int_0^t e^{-a(t-t')} f(t') \ dt'\right] = \frac{1}{s+a}F(s).$$

Thus the transform of the combination of e^{-at} and $f(t)$ on the left is the product of the transforms of e^{-at} and $f(t)$. This simple result can be generalized in the following way.

Theorem 2
If $g(t)$ and $f(t)$ have Laplace transforms $G(s)$ and $F(s)$, respectively, then

$$\mathcal{L}\left[\int_0^t g(t-t') f(t') \ dt'\right] = G(s)F(s). \tag{3}$$

(This is known as the convolution theorem.) The integral on the left is called the *convolution* of g and f, written

$$g(t) * f(t) = \int_0^t g(t-t') f(t') \ dt'.$$

It can be shown that the convolution follows these rules

$$g * f = f * g \tag{4a}$$

$$f * (g * h) = (f * g) * h \tag{4b}$$

$$f * (g + h) = f * g + f * h. \tag{4c}$$

The convolution theorem provides an important device for inverting Laplace transforms, which we shall apply to find the general solution of the nonhomogeneous problem

$$u'' - au = f(t), \quad u(0) = u_0, \quad u'(0) = u_1.$$

The transformed equation is readily solved, yielding

$$U(s) = \frac{su_0 + u_1}{s^2 - a} + \frac{1}{s^2 - a} F(s).$$

Because $1/(s^2 - a)$ is the transform of $\sinh(\sqrt{a}t)/\sqrt{a}$, we easily determine that u is

$$u(t) = u_0 \cosh(\sqrt{a}t) + \frac{u_1}{\sqrt{a}} \sinh(\sqrt{a}t) + \int_0^t \frac{\sinh(\sqrt{a}(t - t'))}{\sqrt{a}} f(t')\ dt'. \tag{5}$$

A slightly different problem occurs if the mass in a spring-mass system is struck while the system is in motion. The mathematical model of the system might be

$$u'' + \omega^2 u = f(t), \quad u(0) = u_0, \quad u'(0) = u_1$$

where $f(t) = F_0$ for $t_0 < t < t_1$ and $f(t) = 0$ for other values. The transform of u is

$$U(s) = \frac{su_0 + u_1}{s^2 + \omega^2} + \frac{1}{s^2 + \omega^2} F(s).$$

The inverse transform of $U(s)$ is then

$$u(t) = u_0 \cos(\omega t) + u_1 \frac{\sin(\omega t)}{\omega} + \int_0^t \frac{\sin(\omega(t - t'))}{\omega} f(t')\ dt'.$$

The convolution in this case is easy to calculate:

$$\int_0^t \frac{\sin(\omega(t - t'))}{\omega} f(t')\ dt' = \begin{cases} 0, & t < t_0 \\ F_0 \dfrac{1 - \cos(\omega(t - t_0))}{\omega^2}, & t_0 < t < t_1 \\ F_0 \dfrac{\cos(\omega(t - t_1)) - \cos(\omega(t - t_0))}{\omega^2}, & t_1 < t. \end{cases}$$

Exercises

1. Solve the initial value problems

a.	$u' - 2u = 0,$	$u(0) = 1$	
b.	$u' + 2u = 0,$	$u(0) = 1$	
c.	$u'' + 4u' + 3u = 0,$	$u(0) = 1,$	$u'(0) = 0$
d.	$u'' + 9u = 0,$	$u(0) = 0,$	$u'(0) = 1.$

2. Solve the initial value problem

$$u'' + 2au' + u = 0, \quad u(0) = u_0, \quad u'(0) = u_1$$

in the three cases: $0 < a < 1$, $a = 1$, $a > 1$.

3. Solve the inhomogeneous problems with zero initial conditions:

a.	$u' + au = 1$	**b.**	$u'' + u = t$
c.	$u'' + 4u = \sin(t)$	**d.**	$u'' + 4u = \sin(2t)$
e.	$u'' + 2u' = 1 - e^{-t}$	**f.**	$u'' - u = 1.$

4. Complete the square in the denominator and use the shift theorem $[F(s-a) = \mathcal{L}(e^{at}f(t))]$ to invert

$$U(s) = \frac{su_0 + (u_1 + 2au_0)}{s^2 + 2as + \omega^2}.$$

There are three cases corresponding to

$$\omega^2 - a^2 > 0, = 0, < 0.$$

5. Use partial fractions to invert the following transforms:

a.	$\dfrac{1}{s^2 - 4}$	**b.**	$\dfrac{1}{s^2 + 4}$
c.	$\dfrac{(s+3)}{s(s^2+2)}$	**d.**	$\dfrac{4}{s(s+1)}.$

6. Prove properties (4a) and (4c) of the convolution.

7. Compute the convolution $f * g$ for

a. $f(t) = 1$, $g(t) = \sin(t)$

b. $f(t) = e^t$, $g(t) = \cos(\omega t)$

c. $f(t) = t$, $g(t) = \sin(t)$.

8. Demonstrate the following properties of convolution either directly or by using Laplace transform:

a. $1 * f'(t) = f(t) - f(0)$

b. $(t * f(t))'' = f(t)$

c. $(f * g)' = f' * g = f * g'$, if $f(0) = g(0) = 0$.

6.3 PARTIAL DIFFERENTIAL EQUATIONS

In applying the Laplace transform to partial differential equations, we treat variables other than t as parameters. Thus, the transform of a function $u(x,t)$ is defined by

$$\mathcal{L}(u(x,t)) = \int_0^\infty e^{-st} u(x,t) \ dt = U(x,s).$$

For instance, we easily find the transforms

$$\begin{aligned} \mathcal{L}(e^{-at} \sin(\pi x)) &= \frac{1}{s+a} \sin(\pi x) \\ \mathcal{L}(\sin(x+t)) &= \frac{s \sin(x) + \cos(x)}{s^2+1}. \end{aligned}$$

The transform U naturally is a function not only of s but also of the "untransformed" variable x. We assume that derivatives or integrals with respect to the untransformed variable pass through the transform

$$\begin{aligned} \mathcal{L}\left(\frac{\partial u}{\partial x}\right) &= \int_0^\infty \frac{\partial u(x,t)}{\partial x} e^{-st} \ dt \\ &= \frac{\partial}{\partial x} \int_0^\infty u(x,t) e^{-st} \ dt = \frac{\partial}{\partial x}(U(x,s)). \end{aligned}$$

If we wish to focus on the role of x as a variable and keep s in the background as a parameter, we might use the symbol for the ordinary derivative

$$\mathcal{L}\left(\frac{\partial u}{\partial x}\right) = \frac{dU}{dx}.$$

The rule for transforming a derivative with respect to t can be found, as before, with integration by parts

$$\mathcal{L}\left(\frac{\partial u}{\partial t}\right) = s\mathcal{L}(u(x,t)) - u(x,0).$$

If the Laplace transform is applied to a boundary value-initial value problem in x and t, all time derivatives disappear, leaving an ordinary differential equation in x. We shall illustrate this technique with some trivial examples. Incidentally, we assume from here on that problems have been prepared (for example, by dimensional analysis) so as to eliminate as many parameters as possible.

Example 1.

$$\begin{aligned}
\frac{\partial^2 u}{\partial x^2} &= \frac{\partial u}{\partial t}, && 0 < x < 1, \quad 0 < t \\
u(0,t) &= 1, \quad u(1,t) = 1, && 0 < t \\
u(x,0) &= 1 + \sin(\pi x), && 0 < x < 1.
\end{aligned}$$

The partial differential equation *and the boundary conditions* (that is, everything that is valid for $t > 0$) are transformed, while the initial condition is incorporated by the transform

$$\begin{aligned}
\frac{d^2 U}{dx^2} &= sU - (1 + \sin(\pi x)), \quad 0 < x < 1 \\
U(0,s) &= \frac{1}{s}, \quad U(1,s) = \frac{1}{s}.
\end{aligned}$$

This boundary value problem is solved to obtain

$$U(x,s) = \frac{1}{s} + \frac{\sin(\pi x)}{s + \pi^2}.$$

We direct our attention now to U as a function of s. Because $\sin(\pi x)$ is a constant with respect to s, tables may be used to find

$$u(x,t) = 1 + \sin(\pi x) \exp(-\pi^2 t).$$

Example 2.

$$\begin{aligned}
\frac{\partial^2 u}{\partial x^2} &= \frac{\partial^2 u}{\partial t^2}, && 0 < x < 1, \quad 0 < t \\
u(0,t) &= 0, \quad u(1,t) = 0, && 0 < t \\
u(x,0) &= \sin(\pi x), && 0 < x < 1 \\
\frac{\partial u}{\partial t}(x,0) &= -\sin(\pi x), && 0 < x < 1.
\end{aligned}$$

Under transformation the problem becomes

$$\frac{d^2U}{dx^2} = s^2U - s\sin(\pi x) + \sin(\pi x), \quad 0 < x < 1$$

$$U(0,s) = 0, \quad U(1,s) = 0.$$

The function U is found to be

$$U(x,s) = \frac{s-1}{s^2+\pi^2}\sin(\pi x)$$

from which we find the solution,

$$u(x,t) = \sin(\pi x)\left(\cos(\pi t) - \frac{1}{\pi}\sin(\pi t)\right).$$

Example 3.
Now we consider a problem that we know to have a more complicated solution:

$$\begin{aligned} \frac{\partial^2 u}{\partial x^2} &= \frac{\partial u}{\partial t}, &&0 < x < 1, \quad 0 < t \\ u(0,t) &= 1, &&u(1,t) = 1, \quad 0 < t \\ u(x,0) &= 0, &&0 < x < 1. \end{aligned}$$

The transformed problem is

$$\begin{aligned} \frac{d^2U}{dx^2} &= sU, \quad 0 < x < 1 \\ U(0,s) &= \frac{1}{s}, \quad U(1,s) = \frac{1}{s}. \end{aligned}$$

The general solution of the differential equation is well known to be a combination of $\sinh(\sqrt{s}x)$ and $\cosh(\sqrt{s}x)$. Application of the boundary conditions yields

$$\begin{aligned} U(x,s) &= \frac{1}{s}\cosh(\sqrt{s}x) + \frac{(1-\cosh(\sqrt{s}))\sinh(\sqrt{s}x)}{s\sinh(\sqrt{s})} \\ &= \frac{\sinh(\sqrt{s}x) + \sinh(\sqrt{s}(1-x))}{s\sinh(\sqrt{s})}. \end{aligned}$$

This function rarely appears in a table of transforms. However, by extending the Heaviside formula, we can compute an inverse transform.

When U is the ratio of two transcendental functions (not polynomials) of s, we wish to write

$$U(x, s) = \sum A_n(x) \frac{1}{s - r_n}.$$

In this formula, the numbers r_n are values of s for which the "denominator" of U is zero or, rather, for which $|U(x, s)|$ becomes infinite; the A_n are functions of x but not s. From this form we expect to determine

$$u(x, t) = \sum A_n(x) \exp(r_n t).$$

This solution should be checked for convergence.

The hyperbolic sine (also the cosh, cos, sin, and exponential functions) is not infinite for any finite value of its argument. Thus $U(x, s)$ becomes infinite only where s or $\sinh(\sqrt{s})$ is zero. Because $\sinh(\sqrt{s}) = 0$ has no real root besides zero, we seek complex roots by setting $\sqrt{s} = \xi + i\eta$ (ξ and η real).

The addition rules for hyperbolic and trigonometric functions remain valid for complex arguments. Furthermore, we know that

$$\cosh(iA) = \cos(A), \quad \sinh(iA) = i \sin(A).$$

By combining the addition rule and these identities we find

$$\sinh(\xi + i\eta) = \sinh(\xi) \cos(\eta) + i \cosh(\xi) \sin(\eta).$$

This function is zero only if both the real and imaginary parts are zero. Thus ξ and η must be chosen to satisfy simultaneously

$$\sinh(\xi) \cos(\eta) = 0, \quad \cosh(\xi) \sin(\eta) = 0.$$

Of the four possible combinations, only $\sinh(\xi) = 0$ and $\sin(\eta) = 0$ produce solutions. Therefore $\xi = 0$ and $\eta = \pm n\pi (n = 0, 1, 2, \cdots)$; whence

$$\sqrt{s} = \pm in\pi, \quad s = -n^2\pi^2.$$

Recall that only the value of s, not the value of $\sqrt{s}$, is significant.

Finally then, we have located $r_0 = 0$, and $r_n = -n^2\pi^2 (n = 1, 2, \cdots)$. We proceed to find the $A_n(x)$ by the same method used in Section 2. The computations are done piecemeal and then the solution is assembled.

Part a.
($r_0 = 0$). In order to find A_0 we multiply both sides of our proposed partial fractions development

$$U(x,s) = \sum_{n=0}^{\infty} A_n(x)\frac{1}{s-r_n}$$

by $s - r_0 = s$ and take the limit as s approaches r_0. The right-hand side goes to A_0. On the left-hand side we have

$$\lim_{s\to\infty} s\frac{\sinh(\sqrt{s}x) + \sinh(\sqrt{s}(1-x)}{s\sinh(\sqrt{s})} = x + 1 - x = 1 = A_0(x).$$

Thus the part of $u(x,t)$ corresponding to $s = 0$ is $1 \cdot e^{0t} = 1$, which is easily recognized as the steady-state solution.

Part b.
($r_n = -n^2\pi^2$, $n = 1, 2, \cdots$). For these cases, we find

$$A_n = \frac{q(r_n)}{p'(r_n)}$$

where q and p are the obvious choices. We take $\sqrt{r}_n = +in\pi$ in all calculations

$$\begin{aligned}
p'(s) &= \sinh(\sqrt{s}) + s\frac{1}{2\sqrt{s}}\cosh(\sqrt{s}) \\
p'(r_n) &= \frac{1}{2}in\pi\cosh(in\pi) = \frac{1}{2}in\pi\cos(n\pi) \\
q(r_n) &= \sinh(in\pi x) + \sinh(in\pi(1-x)) \\
&= i[\sin(n\pi x) + \sin(n\pi(1-x))].
\end{aligned}$$

Hence the portion of $u(x,t)$ that arises from each r_n is

$$A_n(x)\exp(r_n t) = 2\frac{\sin(n\pi x) + \sin(n\pi(1-x))}{n\pi\cos(n\pi)}\exp(-n^2\pi^2 t).$$

Part c.
On assembling the various pieces of the solution, we get

$$u(x,t) = 1 + \frac{2}{\pi}\sum_{1}^{\infty}\frac{\sin(n\pi x) + \sin(n\pi(1-x))}{n\cos(n\pi)}\exp(-n^2\pi^2 t).$$

The same solution would be found by separation of variables but in a slightly different form.

Example 4.
Now consider the wave problem

$$\frac{\partial^2 u}{\partial x^2} = \frac{\partial^2 u}{\partial t^2}, \quad 0 < x < 1, \quad 0 < t$$
$$u(0,t) = 0, \qquad \frac{\partial u(1,t)}{\partial x} = 0, \quad 0 < t$$
$$u(x,0) = 0, \qquad \frac{\partial u(x,0)}{\partial t} = x, \quad 0 < x < 1.$$

The transformed problem is

$$\frac{d^2 U}{dx^2} = s^2\, U - x, \quad 0 < x < 1$$
$$U(0,s) = 0, \quad U'(1,s) = 0$$

and its solution (by undetermined coefficients or otherwise) gives

$$U(x,s) = \frac{sx\cosh(s) - \sinh(sx)}{s^3\cosh(s)}.$$

The numerator of this function is never infinite. The denominator is zero at $s = 0$ and $s = \pm i(2n-1)\pi/2$ $(n = 1, 2, \cdots)$. We shall again use the Heaviside formula to determine the inverse transform of U.

Part a. $(r_0 = 0)$.
The limit as s approaches zero of $sU(x,s)$ may be found by L'Hôpital's rule or by using the Taylor series for sinh and cosh. From the latter,

$$sU(x,s) = \frac{sx\left(1 + \frac{s^2}{2} + \cdots\right) - \left(sx + \frac{s^3x^3}{6} + \cdots\right)}{s^2\left(1 + \frac{s^2}{2} + \cdots\right)}$$
$$= \frac{s^3\left(\frac{x}{2} - \frac{x^3}{6} - \cdots\right)}{s^2\left(1 + \frac{s^2}{2} + \cdots\right)} \to 0.$$

Thus, in spite of the formidable appearance of s^3 in the denominator, $s = 0$ is not really a significant value and contributes nothing to $u(x,t)$.

Part b.
It is convenient to take the remaining roots in pairs. We label

$$\pm i(2n-1)\pi/2 = \pm i\rho_n.$$

The derivative of the denominator is

$$p'(s) = 3s^2 \cosh(s) + s^3 \sinh(s)$$

$$\begin{aligned} p'(\pm i\rho_n) &= \pm i^3 \rho_n^3 \sinh(\pm i\rho_n) \\ &= \rho_n^3 \sin(\rho_n) \end{aligned}$$

since $\sinh(i\rho) = i\sin(\rho)$, and $(\pm i)^4 = 1$. The contribution of these two roots together may be calculated using the exponential definition of sine

$$\begin{aligned} &\frac{q(i\rho_n)}{p'(i\rho_n)} \exp(i\rho_n t) + \frac{q(-i\rho_n)}{r'(-i\rho_n)} \exp(-i\rho_n t) \\ &\qquad = \frac{-\sinh(i\rho_n x)\exp(i\rho_n t) + \sinh(i\rho_n x)\exp(-i\rho_n t)}{\rho_n^3 \sin(\rho_n)} \\ &\qquad = \frac{\sin(\rho_n x)}{\rho_n^3 \sin(\rho_n)} i(-\exp(i\rho_n t) + \exp(-i\rho_n t)) \\ &\qquad = \frac{2\sin(\rho_n x)\sin(\rho_n t)}{\rho_n^3 \sin(\rho_n)}. \end{aligned}$$

Part c.
The final form of $u(x,t)$, found by adding up all the contributions from Part b, is the same as would be found by separation of variables

$$u(x,t) = 2\sum_1^\infty \frac{\sin(\rho_n x)\sin(\rho_n t)}{\rho_n^3 \sin(\rho_n)}.$$

Exercises

1. Find all values of s, real and complex, for which the following functions are zero:

a. $\cosh(\sqrt{s})$ **b.** $\cosh(s)$

c. $\sinh(s)$ **d.** $\cosh(s) - s\sinh(s)$

e. $\cosh(s) + s\sinh(s)$.

2. Find the inverse transforms of the following functions in terms of an infinite series:

a. $\dfrac{1}{s}\tanh(s)$ **b.** $\dfrac{\sinh(sx)}{s\cosh(s)}$.

3. Find the transform $U(x, s)$ of the solution of each of the following problems.

a.

$$\frac{\partial^2 u}{\partial x^2} = \frac{\partial u}{\partial t}, \quad 0 < x < 1, \quad 0 < t$$
$$u(0, t) = 0, \qquad u(1, t) = t, \quad 0 < t$$
$$u(x, 0) = 0, \qquad 0 < x < 1$$

b.

$$\frac{\partial^2 u}{\partial x^2} = \frac{\partial u}{\partial t}, \quad 0 < x < 1, \quad 0 < t$$
$$u(0, t) = 0, \qquad u(1, t) = e^{-t}, \quad 0 < t$$
$$u(x, 0) = 1, \qquad 0 < x < 1$$

4. Solve each of the problems in Exercise 3, inverting the transform by means of the extended Heaviside formula.

5. Solve each of the following problems by Laplace transform methods.

a.

$$\frac{\partial^2 u}{\partial x^2} = \frac{\partial u}{\partial t}, \quad 0 < x < 1, \quad 0 < t$$
$$u(0, t) = 0, \qquad u(1, t) = 1, \quad 0 < t$$
$$u(x, 0) = 0, \qquad 0 < x < 1$$

b.

$$\frac{\partial^2 u}{\partial x^2} = \frac{\partial u}{\partial t}, \quad 0 < x < 1, \quad 0 < t$$
$$u(0, t) = 0, \qquad u(1, t) = 0, \quad 0 < t$$
$$u(x, 0) = 1, \qquad 0 < x < 1$$

6.4 MORE DIFFICULT EXAMPLES

The technique of separation of variables, once mastered, seems more straightforward than the Laplace transform. However, when time-dependent boundary conditions or inhomogeneities are present, the Laplace transform offers a distinct advantage. Following are some examples that display the power of transform methods.

Example 1.
A uniform insulated rod is attached at one end to an insulated container of fluid. The fluid is circulated so well that its temperature is uniform and equal to that at the end of the rod. The other end of the rod is maintained at a constant temperature. A dimensionless initial value-boundary value problem that describes the temperature in the rod is

$$\frac{\partial^2 u}{\partial x^2} = \frac{\partial u}{\partial t}, \qquad 0 < x < 1, \quad 0 < t$$

$$\frac{\partial u(0,t)}{\partial x} = \gamma \frac{\partial u(0,t)}{\partial t}, \quad u(1,t) = 1, \quad 0 < t$$

$$u(x,0) = 0, \qquad 0 < x < 1.$$

The transformed problem and its solution are

$$\frac{d^2 U}{dx^2} = sU, \qquad 0 < x < 1$$

$$\frac{dU}{dx}(0,s) = s\gamma U(0,s), \quad U(1,s) = \frac{1}{s}$$

$$U(x,s) = \frac{\cosh(\sqrt{s}x) + \sqrt{s}\gamma \sinh(\sqrt{s}x)}{s(\cosh(\sqrt{s}) + \sqrt{s}\gamma \sinh(\sqrt{s}))} = \frac{q(s)}{p(s)}.$$

Aside from $s = 0$, the denominator has no real zeros. Thus we again search for complex zeros by employing $\sqrt{s} = \xi + i\eta$. The real and imaginary parts of the denominator are to be computed by using the addition formulas for cosh and sinh. The requirement that both real and imaginary parts be zero leads to the equation

$$(\cosh(\xi) + \xi\gamma \sinh(\xi))\cos(\eta) - \eta\gamma \cosh(\xi)\sin(\eta) = 0 \tag{1}$$

$$\eta\gamma \sinh(\xi)\cos(\eta) + (\sinh(\xi) + \xi\gamma\cosh(\xi))\sin(\eta) = 0. \tag{2}$$

We may think of these as simultaneous equations in $\sin(\eta)$ and $\cos(\eta)$. Because $\sin^2(\eta) + \cos^2(\eta) = 1$, the system has a solution only when its determinant is zero. Thus after some algebra, we arrive at the condition

$$(1+\xi^2\gamma^2+\eta^2\gamma^2)\sinh(\xi)\cosh(\xi)+\xi\gamma(\sinh^2(\xi)+\cosh^2(\xi))=0.$$

The only solution occurs when $\xi = 0$, for otherwise both terms of this equation have the same sign.

After setting $\xi = 0$, we find that Eq. (1) reduces to

$$\tan(\eta) = \frac{1}{\eta\gamma}$$

for which there are an infinite number of solutions. We shall number the positive solutions $\eta_1, \eta_2, \cdots$. Now we have found that the roots of $p(s)$ are $r_0 = 0$ and $r_k = (i\eta_k)^2 = -\eta_k^2$. The computation of the inverse transform follows.

Part a.
The limit of $sU(x,s)$ as s tends to zero is easily found to be 1. Thus this root contributes $1 \cdot e^{0t} = 1$ to $u(x,t)$.

Part b. $(r_k = -\eta_k^2)$.
First, we compute

$$p'(s) = \cosh(\sqrt{s}) + \sqrt{s}\gamma\sinh(\sqrt{s}) + \frac{1}{2}\sqrt{s}(1+\gamma)\sinh(\sqrt{s}) + \frac{1}{2}\gamma s\cosh(\sqrt{s}).$$

Using the fact that $\cosh(\sqrt{r_k}) + \sqrt{r_k}\gamma\sinh(\sqrt{r_k}) = 0$, we may reduce the foregoing to

$$p'(r_k) = \frac{-1}{2\gamma}(1+\gamma+\eta_k^2\gamma^2)\cos(\eta_k).$$

Hence the contribution to $u(x,t)$ of r_k is

$$\frac{q(r_k)}{p'(r_k)}\exp(r_k t) = -2\gamma\frac{\cos(\eta_k x) - \eta_k\gamma\sin(\eta_k x)}{(1+\gamma+\eta_k^2\gamma^2)\cos(\eta_k)}\exp(-\eta_k^2 t).$$

Part c.
The construction of the final solution is left to the reader. We note that an attempt to solve this problem by separation of variables would find difficulties, for the eigenfunctions are not orthogonal.

Example 2.
Sometimes one is interested not in the complete solution of a problem, but only in part of it. For example, in the problem of heat conduction in a semi-infinite solid with time-varying boundary conditions, we may seek that part of the solution that persists after a long time. (This may or may not be a steady-state solution.) Any initial condition that is bounded in x gives rise only to transient temperatures; these being of no interest, we assume a zero initial condition. Thus, the problem to be studied is

$$\frac{\partial^2 u}{\partial x^2} = \frac{\partial u}{\partial t}, \quad 0 < x < \quad 0 < t$$
$$u(0,t) = f(t), \quad 0 < t$$
$$u(x,0) = 0, \qquad 0 < x.$$

The transformed equation and its general solution are

$$\frac{d^2 U}{dx^2} = sU, \quad 0 < x, \quad U(0,s) = F(s)$$
$$U(x,s) = A\exp(-\sqrt{s}x) + B\exp(\sqrt{s}x).$$

We make two further assumptions about the solution: first, that $u(x,t)$ is bounded as x tends to infinity and second, that $\sqrt{s}$ means the square root of s that has a nonnegative real part. Under these two assumptions, we must choose $B = 0$ and $A = F(s)$, making

$$U(x,s) = F(s)\exp(-\sqrt{s}x).$$

In order to find the persistent part of $u(x,t)$, we apply the Heaviside inversion formula to those values of s having nonnegative real parts because a value of s with *negative* real part corresponds to a function containing a decaying exponential—a transient. To understand this fact, consider the pair

$$f(t) = 1 - e^{-\beta t} + \alpha \sin(\omega t)$$
$$F(s) = \frac{1}{s} - \frac{1}{s+\beta} + \frac{\alpha\omega}{s^2+\omega^2}.$$

Now we return to the original problem. The values of s for which $U(x,s) = F(s)\exp(-\sqrt{s}x)$ becomes infinite are 0, $\pm i\omega$, $-\beta$. The last value is discarded, as it is negative. Thus, the persistent part of the solution is given by

$$A_0 e^{0t} + A_1 e^{i\omega t} + A_2 e^{-i\omega t}$$

and the coefficients are found from

$$\begin{aligned}
A_0 &= \lim_{s\to 0}\left[sF(s)\exp(-\sqrt{s}x)\right] = 1\\
A_1 &= \lim_{s\to i\omega}\left[(s-i\omega)F(s)\exp(-\sqrt{s}x)\right] = \frac{\alpha}{2i}\exp(-\sqrt{i\omega}x)\\
A_2 &= \lim_{s\to -i\omega}\left[(s+i\omega)F(s)\exp(-\sqrt{s}x)\right] = -\frac{\alpha}{2i}\exp(-\sqrt{-i\omega}x).
\end{aligned}$$

We also need to know that the roots of $\pm i$ with positive real part are

$$\sqrt{i} = \frac{1}{\sqrt{2}}(1+i), \quad \sqrt{-i} = \frac{1}{\sqrt{2}}(1-i).$$

Thus the function we seek is

$$\begin{aligned}
1 + \frac{1}{2i}\exp\left[i\omega t - \sqrt{\frac{\omega}{2}}(1+i)x\right] - \frac{\alpha}{2i}\exp\left[-i\omega t - \sqrt{\frac{\omega}{2}}(1-i)x\right]\\
= 1 + \alpha\exp\left(-\sqrt{\frac{\omega}{2}}x\right)\sin\left(\omega t - \sqrt{\frac{\omega}{2}}x\right).
\end{aligned}$$

Example 3.
If a steel wire is exposed to a sinusoidal magnetic field, the boundary value-initial value problem that describes its displacement is

$$\begin{aligned}
&\frac{\partial^2 u}{\partial x^2} = \frac{\partial^2 u}{\partial t^2} - \sin(\omega t), && 0 < x < 1, \ \ 0 < t\\
&u(0,t) = 0, && u(1,t) = 0, \ \ 0 < t\\
&u(x,0) = 0, && \frac{\partial u}{\partial t}(x,0) = 0, \ \ 0 < x < 1.
\end{aligned}$$

The nonhomogeneity in the partial differential equation represents the effect of the force due to the field. The transformed equation and its solution are

$$\begin{aligned}
&\frac{d^2 U}{dx^2} = s^2 U - \frac{\omega}{s^2+\omega^2}, \qquad 0 < x < 1\\
&U(0,s) = 0, \qquad U(1,s) = 0,\\
&U(x,s) = \frac{\omega}{s^2(s^2+\omega^2)}\,\frac{\cosh(\frac{1}{2}s) - \cosh(s(\frac{1}{2}-x))}{\cosh(\frac{1}{2}s)}.
\end{aligned}$$

Several methods are available for the inverse transformation of U. An obvious one would be to compute

$$v(x,t) = \mathcal{L}^{-1}\left(\frac{\cosh(\frac{1}{2}s) - \cosh(s(\frac{1}{2}-x))}{s^2\cosh(\frac{1}{2}s)}\right)$$

and write $u(x,t)$ as a convolution

$$u(x,t) = \int_0^t \sin(\omega(t-t'))v(x,t') \ dt'.$$

The details of this development are left as an exercise.

We could also use the Heaviside formula. The application is now routine, except in the interesting case where $\cosh(i\omega/2) = 0$: that is, where $\omega = (2n-1)\pi$, one of the natural frequencies of the wire.

Let us suppose $\omega = \pi$, so that

$$U(x,s) = \frac{\pi}{s^2(s^2+\pi^2)} \frac{\cosh(\frac{1}{2}s) - \cosh(s(\frac{1}{2}-x))}{\cosh(\frac{1}{2}s)}.$$

At the points $s = 0$, $s = \pm i\pi$, $s = \pm(2n-1)i\pi$, $n = 2, 3, \cdots$, $U(x,s)$ becomes undefined. The computation of the parts of the inverse transform corresponding to the points other than $\pm i\pi$ is easily carried out. However, at these two troublesome points, our usual procedure will not work. Instead of expecting a partial-fraction decomposition containing

$$\frac{A_{-1}}{s+i\pi} + \frac{A_1}{s-i\pi}$$

and other terms of the same sort, we must seek terms like

$$\frac{A_{-1}(s+i\pi) + B_{-1}}{(s+i\pi)^2} + \frac{A_1(s-i\pi) + B_1}{(s-i\pi)^2}$$

whose contribution to the inverse transform of U would be

$$A_{-1}e^{-i\pi t} + B_{-1}te^{-i\pi t} + A_1 e^{i\pi t} + B_1 te^{i\pi t}.$$

One can compute A_1 and B_1, for example, by noting that

$$\begin{aligned} B_1 &= \lim_{s\to i\pi} [(s-i\pi)^2 U(x,s)] \\ A_1 &= \lim_{s\to i\pi} \left\{ (s-i\pi)\left[U(x,s) - \frac{B_1}{(s-i\pi)^2}\right]\right\} \end{aligned}$$

and similarly for A_{-1} and B_{-1}. The limit for B_1 is not too difficult. For example,

$$\begin{aligned}
B_1 &= \lim_{s\to i\pi}\left\{\frac{\pi}{s^2(s+i\pi)}\frac{\cosh(\frac{1}{2}s)-\cosh(s(\frac{1}{2}-x))}{(\cosh(\frac{1}{2}s))/(s-i\pi)}\right. \\
&= \frac{\pi}{-\pi^2(2i\pi)}\frac{\cosh(\frac{1}{2}i\pi)-\cosh(i\pi(\frac{1}{2}-x))}{\frac{1}{2}\sinh(\frac{1}{2}i\pi)} \\
&= \frac{-1}{\pi^2}\cos(\pi(\tfrac{1}{2}-x)) = \frac{-1}{\pi^2}\sin(\pi x).
\end{aligned}$$

The limit for A_1 is rather more complicated but may be computed by L'Hôpital's rule. Nevertheless, as $B_{-1} = B_1$, we already see that $u(x,t)$ contains the term

$$B_1te^{i\pi t} + B_{-1}te^{-i\pi t} = -\frac{2t}{\pi^2}\sin(\pi x)\cos(\pi t)$$

whose amplitude increases with time. This, of course, is the expected resonance phenomenon.

Exercises

1. Find the persistent part of the solution of the heat problem

$$\begin{aligned}
\frac{\partial^2 u}{\partial x^2} &= \frac{\partial u}{\partial t}, \quad 0 < x < 1, \quad 0 < t \\
\frac{\partial u}{\partial x}(0,t) &= 0, \quad \frac{\partial u}{\partial x}(1,t) = 1, \quad 0 < t \\
u(x,0) &= 0, \quad 0 < x < 1.
\end{aligned}$$

2. Verify that the persistent part of the solution to Example 2 actually satisfies the heat equation. What boundary condition does it satisfy?

3. Find the function $v(x,t)$ whose transform is

$$\frac{\cosh(\frac{1}{2}s)-\cosh(s(\frac{1}{2}-x))}{s^2\cosh(\frac{1}{2}s)}.$$

What boundary value-initial value problem does $v(x,t)$ satisfy?

4. Solve

$$\begin{aligned} &\frac{\partial^2 u}{\partial x^2} = \frac{\partial^2 u}{\partial t^2}, && 0 < x < 1, \quad 0 < t \\ &u(0,t) = 0, && u(1,t) = 0, \quad 0 < t \\ &u(x,0) = 0, && \frac{\partial u}{\partial t}(x,0) = 1, \quad 0 < x < 1. \end{aligned}$$

5. a. Solve for $\omega \neq \pi$

$$\begin{aligned} &\frac{\partial^2 u}{\partial x^2} = \frac{\partial^2 u}{\partial t^2} - \sin(\pi x)\sin(\omega t), && 0 < x < 1, \quad 0 < t \\ &u(0,t) = 0, \quad u(1,t) = 0, && 0 < t \\ &u(x,0) = 0, \quad \frac{\partial u}{\partial t}(x,0) = 0, && 0 < x < 1. \end{aligned}$$

b. Examine the special case $\omega = \pi$.

6. Obtain the complete solution of Example 1 and verify that it satisfies the boundary conditions and the heat equation.

7. a. Solve

$$\begin{aligned} &\frac{\partial^2 u}{\partial x^2} = \frac{\partial u}{\partial t}, && 0 < x < 1, \quad 0 < t \\ &u(0,t) = 0, && u(1,t) = 1 - e^{-at}, \quad 0 < t \\ &u(x,0) = 0, && 0 < x < 1. \end{aligned}$$

b. Examine the special case where $a = n^2\pi^2$ for some integer n.

6.5 COMMENTS AND REFERENCES

Laplace had virtually nothing to do with the Laplace transform, although a method of his for solving certain differential equations can be interpreted as an example of its use. The real development began in the late nineteenth century, when Oliver Heaviside invented a powerful, but unjustified, symbolic method for studying the ordinary and partial differential equations of mathematical physics. By the 1920's, Heaviside's method had been legitimatized and recast as the Laplace transform that we now use. Later generalizations are Schwartz's theory of distributions (1940's) and Mikusinski's operational calculus (1950's). The former seems to be the more general. Both theories give an interpretation of $F(s) = 1$, which is not the Laplace transform of any function, in the sense we use.

There are a number of other transforms, under the names of Fourier, Mellin, Hänkel, and others, similar in intent to the Laplace transform, in which some other function replaces e^{-st} in the defining integral. *Operational Mathematics* by Churchill, has more information about the applications of transforms. Extensive tables of transforms will be found in *Tables of Integral Transforms* by Erdelyi et al. (See the Bibliography.)

MISCELLANEOUS EXERCISES

1. Solve the heat conduction problem

$$\frac{\partial^2 u}{\partial x^2} - \gamma^2(u - T) = \frac{\partial u}{\partial t}, \quad 0 < x < 1, \quad 0 < t$$
$$\frac{\partial u}{\partial t}(0,t) = 0, \qquad \frac{\partial u}{\partial x}(1,t) = 0, \quad 0 < t$$
$$u(x,0) = T_0, \quad 0 < x < 1.$$

2. Find the "persistent part" of the solution of

$$\frac{\partial^2 u}{\partial x^2} = \frac{\partial u}{\partial t}, \quad 0 < x < 1, \quad 0 < t$$
$$\frac{\partial u}{\partial x}(0,t) = 0, \qquad u(1,t) = t, \quad 0 < t$$
$$u(x,0) = 0, \qquad 0 < x < 1.$$

3. Find the complete solution of the problem in Exercise 2.

4. A solid object and a surrounding fluid exchange heat by convection. The temperatures u_1 and u_2 are governed by the equations below. Solve them by means of Laplace transforms.

$$\frac{du_1}{dt} = -\beta_1(u_1 - u_2)$$
$$\frac{du_2}{dt} = -\beta_2(u_2 - u_1)$$
$$u_1(0) = 1, \quad u_2(0) = 0$$

5. Solve the following nonhomogeneous problem with transforms:

$$\frac{\partial^2 u}{\partial x^2} = \frac{\partial u}{\partial t} - 1, \quad 0 < x < 1, \quad 0 < t$$
$$u(0,t) = 0, \qquad u(1,t) = 0, \quad 0 < t$$
$$u(x,0) = 0, \qquad 0 < x < 1.$$

6. Find the transform of the solution of the problem

$$\frac{\partial^2 u}{\partial x^2} = \frac{\partial u}{\partial t}, \quad 0 < x < 1, \quad 0 < t$$
$$u(0,t) = 0, \qquad u(1,t) = 1, \quad 0 < t$$
$$u(x,0) = 0, \qquad 0 < x < 1.$$

7. Find the solution of the problem in Exercise 6 by using the extended Heaviside formula.

8. Solve the heat problem

$$\frac{\partial^2 u}{\partial x^2} = \frac{\partial u}{\partial t}, \qquad 0 < x, \quad 0 < t$$
$$u(0,t) = 0, \qquad 0 < t$$
$$u(x,0) = \sin(x), \quad 0 < x.$$

9. Find the transform of the solution of

$$\frac{\partial^2 u}{\partial x^2} = \frac{\partial u}{\partial t}, \qquad 0 < x, \quad 0 < t$$
$$u(0,t) = 0, \qquad 0 < t$$
$$u(x,0) = 1, \qquad 0 < x$$
$$u(x,t) \quad \text{bounded as} \quad x \to \infty.$$

10. At the end of Section 12, Chapter 2, the problem in Exercise 9 was solved by other means. Use this fact to identify

$$\frac{1}{s}\left(1 - e^{-\sqrt{s}x}\right) = \mathcal{L}\left[\operatorname{erf}\left(\frac{x}{\sqrt{4t}}\right)\right]$$

and

$$\frac{1}{s}e^{-\sqrt{s}x} = \mathcal{L}\left[1 - \operatorname{erf}\left(\frac{x}{\sqrt{4t}}\right)\right].$$

(The latter function is called the *complementary error function*, defined by $\operatorname{erfc}(q) \equiv 1 - \operatorname{erf}(q)$.)

11. Find the function of t whose Laplace transform is

$$F(s) = e^{-x\sqrt{s}}.$$

12. Using the definition of sinh in terms of exponentials and a geometric series, show that

$$\frac{\sinh(\sqrt{s}x)}{\sinh(\sqrt{s})} = \sum_{n=0}^{\infty} \left(e^{-\sqrt{s}(2n+1-x)} - e^{-\sqrt{s}(2n+1+x)} \right).$$

13. Use the series in Exercise 12 to find a solution of the problem in Exercise 6 in terms of complementary error functions.

14. Show the following relation by using Exercise 11 and differentiating with respect to s:

$$\mathcal{L}\left[\frac{1}{\sqrt{\pi t}} \exp\left(\frac{-k^2}{4t}\right)\right] = \frac{1}{\sqrt{s}} e^{-k\sqrt{s}}.$$

15. Find the Laplace transform of the odd periodic extension of the function

$$f(t) = \pi - t, \quad 0 < t < \pi$$

by transforming its Fourier series term-by-term.

16. Suppose that the function $f(t)$ is periodic with period $2a$. Show that the Laplace transform of f is given by the formula

$$F(s) = \frac{G(s)}{1 - e^{-2as}},$$

where

$$G(s) = \int_0^{2a} f(t) e^{-st} \, dt.$$

(Hint: See Section 1, Exercise 6.)

17. Apply the extended Heaviside method to the inversion of a transform with the form

$$F(s) = \frac{G(s)}{1 - e^{-2as}}$$

where $G(s)$ does not become infinite for any value of s.

18. Show that for a periodic function $f(t)$ the quantities

$$c_n = \frac{1}{2a} G\left(\frac{in\pi}{a}\right)$$

($G(s)$ is defined in Exercise 16) are the complex Fourier coefficients.

19. How is it possible to determine that a Laplace transform $F(s)$ corresponds to a periodic $f(t)$?

20. Is this function the transform of a periodic function?

$$F(s) = \frac{1}{s^2 + a^2}$$

21. Use the method of Exercise 16 to find the transform of the periodic extension of

$$f(t) = \begin{cases} 1, & 0 < t < \pi \\ -1, & \pi < t < 2\pi. \end{cases}$$

22. Same as Exercise 21, but use the function of Exercise 15.

23. Use the method of Exercise 16 to find the transform of

$$f(t) = |\sin(t)|.$$

24. Find the transform of the solution of the problem

$$\frac{\partial^2 u}{\partial x^2} = \frac{\partial^2 u}{\partial t^2}, \quad 0 < x, \quad 0 < t$$

$$u(0,t) = h(t), \quad 0 < t$$

$$u(x,0) = 0, \qquad \frac{\partial u}{\partial t}(x,0) = 0, \quad 0 < x$$

$$u(x,t) \text{ bounded as } x \to \infty.$$

Use the solution of the same problem as found in Chapter 3, Section 6, to verify the rule

$$\mathcal{L}^{-1}\left(e^{-sx}H(s)\right) = \begin{cases} h(t-x), & t > x \\ 0, & t < x. \end{cases}$$

25. Solve this wave problem with time-varying boundary condition, assuming $\omega \neq n\pi$, $n = 1, 2, \cdots$.

$$\frac{\partial^2 u}{\partial x^2} = \frac{\partial^2 u}{\partial t^2}, \quad 0 < x < 1, \quad 0 < t$$

$$u(0,t) = 0, \qquad u(1,t) = \sin(\omega t), \quad 0 < t$$

$$u(x,0) = 0, \qquad \frac{\partial u}{\partial t}(x,0) = 0, \quad 0 < x < 1.$$

26. Solve the problem in Exercise 25 in the special case $\omega = \pi$.

27. Certain techniques for growing a crystal from a solution or a melt may cause striations — variations in the concentration of impurities. Authors R.T. Gray, M.F. Larrousse and W.R. Wilcox (Diffusional decay of striations, Journal of Crystal Growth 92 (1988) 530-542) use a material balance on a slice of a cylindrical ingot to derive this boundary-value problem for the impurity concentration, C.

$$\frac{\partial}{\partial x}\left(D(x)\frac{\partial C}{\partial x}\right) - V\frac{\partial C}{\partial x} = \frac{\partial C}{\partial t}, \qquad 0 < x < \quad 0 < t$$

$$C(0,t) = C_a + A\sin\left(\frac{2\pi t}{t_C}\right), \qquad 0 < t$$

$$C(x,0) = C_a, \qquad 0 < x.$$

Here, V is the crystal growth rate, t_C is the striation period, C_a is the average concentration in the solid, and $D(x)$ is the diffusivity of the impurity at distance x from the growth face (which is located at $x = 0$). Of course, $C(x,t)$ is bounded as $x \to \infty$.

Next, the equations are made dimensionless by introducing new variables

$$\bar{C} = \frac{C - C_a}{A}, \qquad \bar{x} = \frac{V_x}{D(0)}, \qquad \bar{t} = \frac{V^2 t}{D(0)}.$$

The new problem is

$$\frac{\partial}{\partial \bar{x}}\left(\frac{D(\bar{x})}{D(0)}\frac{\partial \bar{C}}{\partial \bar{x}}\right) - \frac{\partial \bar{C}}{\partial \bar{x}} = \frac{\partial \bar{C}}{\partial \bar{t}}, \qquad 0 < \bar{x}, \quad 0 < \bar{t}$$

$$\bar{C}(0,\bar{t}) = \sin(\omega \bar{t}), \qquad 0 < \bar{t}$$

$$\bar{C}(\bar{x},0) = 0, \qquad 0 < \bar{x}$$

where $\omega = 2\pi D(0)/V^2 t_C$.

Because $D(\bar{x})$ depends in a complicated way on $\bar{x}$, a numerical solution was used. To check the numerical solution, the authors wished to find an analytical solution of the problem corresponding to constant diffusivity, $D(\bar{x}) = D(0)$. Let u be the solution of

$$\frac{\partial^2 u}{\partial \bar{x}^2} = \frac{\partial u}{\partial \bar{x}}\frac{\partial u}{\partial \bar{t}}, \qquad 0 < \bar{x}, \quad 0 < \bar{t}$$

$$u(0,\bar{t}) = \sin(\omega \bar{t}), \qquad 0 < \bar{t}$$

$$u(\bar{x},0) = 0, \qquad 0 < \bar{x}$$

$$u \text{ bounded as } x \to \infty.$$

Find the Laplace transform of the solution of this problem.

28. The authors of the paper mentioned in Exercise 27 were particularly interested in the persistent part of the solution. Use the methods of Section 4 to show that the persistent part of the solution is

$$u_1 = \frac{1}{2i}\left(f(i\omega) - f(-i\omega)\right)$$

where $f(i\omega) = \exp\left(\left(\frac{1}{2} - \sqrt{\frac{1}{4} + i\omega}\right)\bar{x} + i\omega\bar{t}\right)$.

29. Find the square root required in the foregoring expression by setting

$$\sqrt{\tfrac{1}{4} + i\omega} = \alpha + i\beta,$$

so that

$$\alpha^2 - \beta^2 + 2i\alpha\beta = \frac{1}{4} + i\omega,$$

or

$$\begin{cases} \alpha^2 - \beta^2 & = \dfrac{1}{4} \\ 2\alpha\beta & = \omega. \end{cases}$$

(To solve these equations: (i) solve the second for β; (ii) substitute the expression found into the first; (iii) solve the resulting biquadratic for α.)

30. Noting that the persistent part is $u_1 = Im(f(i\omega))$ (see Exercise 28), determine

$$u_1(\bar{x}, \bar{t}) = e^{(\frac{1}{2} - \alpha)\bar{x}} \sin(\omega\bar{t} - \beta\bar{x})$$

where α and β are as in Exercise 29.

Chapter 7

Numerical Methods

7.1 BOUNDARY VALUE PROBLEMS

More often than not, significant practical problems in partial—and even ordinary—differential equations cannot be solved by analytical methods. Difficulties may arise from variable coefficients, irregular regions, unsuitable boundary conditions, interfaces, or just overwhelming detail. Now that machine computation is cheap and easily accessible, numerical methods provide reliable answers to formerly difficult problems. In this chapter we examine a few methods that are simple and equally adaptable to machine or manual computation. In addition, we shall see how some of these methods can be carried out using a spreadsheet program.

If we cannot find a simple analytic formula for the solution of a boundary value problem, we may be satisfied with a table of (approximate) values of the solution. For instance, the solution of the problem

$$\frac{d^2u}{dx^2} - 12xu = -1, \qquad 0 < x < 1, \tag{1}$$

$$u(0) = 1, \qquad u(1) = -1 \tag{2}$$

may be written out in terms of Airy functions, but the values of u shown in Table 1 are more informative for most of us. One way to obtain such a table is to replace the original analytical problem by an arithmetical problem as described in what follows.

First, the values of x for the table will be uniformly spaced across the interval $0 \le x \le 1$, which we assume to be the interval of the boundary value problem:

Table 1: Approximate solution of equations (1) and (2)

x	$u(x)$
0.0	1.0
0.2	0.643
0.4	0.302
0.6	−0.026
0.8	−0.406
1.0	−1.0

$$x_i = i\Delta x, \qquad \Delta x = \frac{1}{n}.$$

These are called *meshpoints*. The numbers approximating the values of u are

$$u_i \cong u(x_i), \qquad i = 0, 1, \cdots, n.$$

These numbers are required to satisfy a set of equations obtained from the boundary value problem by making the replacements shown in Table 2. The entry $f(x)$ refers to any coefficient or inhomogeneity in the differential equation.

For example, the boundary-value problem in Eqs. (1) and (2) would be replaced by the algebraic equations

$$\frac{u_{i+1} - 2u_i + u_{i-1}}{(\Delta x)^2} - 12x_i u_i = -1 \tag{3}$$

$$u_0 = 1, \qquad u_n = -1. \tag{4}$$

Equation (3) holds for $i = 1, \cdots, n-1$ so that the unknowns $u_1, \cdots, u_{n-1}$ would be determined by this set of equations. The equations become specific when we choose n. Let us take $n = 5$ so that $\Delta x = 1/5$, and the

Table 2: Constructing replacement equations

Differential equation	Boundary condition
$u(x) \to u_i$	$u(0) \to u_0$
$\dfrac{d^2u}{dx^2}(x) \to \dfrac{u_{i+1} - 2u_i + u_{i-1}}{(\Delta x)^2}$	$\dfrac{du}{dx}(0) \to \dfrac{u_1 - u_{-1}}{2\Delta x}$
$\dfrac{du}{dx}(x) \to \dfrac{u_{i+1} - u_{i-1}}{2\Delta x}$	$u(1) \to u_n$
$f(x) \to f(x_i)$	$\dfrac{du}{dx}(1) \to \dfrac{u_{n_1} - u_{n-1}}{2\Delta x}$

four ($i = 1, 2, 3, 4$) versions of Eq. (3) are

$$\begin{aligned}
25(u_2 - 2u_1 + u_0) - \frac{12}{5}u_1 &= -1, \\
25(u_3 - 2u_2 + u_1) - \frac{24}{5}u_2 &= -1, \\
25(u_4 - 2u_3 + u_2) - \frac{36}{5}u_3 &= -1, \\
25(u_5 - 2u_4 + u_3) - \frac{48}{5}u_4 &= -1.
\end{aligned} \tag{5}$$

When we use the boundary conditions

$$u_0 = 1, \qquad u_5 = -1, \tag{6}$$

and collect coefficients, the foregoing equations become

$$\begin{aligned}
-52.4u_1 + 25u_2 \qquad\qquad\qquad &= -26 \\
25u_1 - 54.8u_2 + 25u_3 \qquad\quad &= -1 \\
25u_2 - 57.2u_3 + 25u_4 &= -1 \\
25u_3 - 59.6u_4 &= 24.
\end{aligned} \tag{7}$$

This system of four simultaneous equations can be solved manually by elimination or by software. The result will be a set of numbers giving the approximate values of u at the points $x_1 = 0.2, \cdots, x_4 = 0.8$. The numbers in Table 1 were obtained by a similar process, but using $n = 100$ instead of $n = 5$.

A slightly more complicated example is the problem

$$\frac{d^2u}{dx^2} - 10u = f(x), \quad 0 < x < 1 \tag{8}$$

$$u(0) = 1, \qquad \frac{du}{dx}(1) = -1, \tag{9}$$

$$f(x) = \begin{cases} 0, & 0 < x < \frac{1}{2} \\ -50, & x = \frac{1}{2} \\ -100, & \frac{1}{2} < x < 1. \end{cases}$$

The replacement equations for this problem are easily obtained by using Table 2. They are

$$\frac{u_{i+1} - 2u_i + u_{i-1}}{(\Delta x)^2} - 10u_i = f(x_i) \tag{10}$$

$$u_0 = 1, \qquad \frac{u_{n+1} - u_{n-1}}{2\Delta x} = -1. \tag{11}$$

We need to know $u_0, u_1, \cdots, u_n$. The derivative boundary condition at $x = 1$ forces us to include u_{n+1} among the unknowns, so we will need to use Eq. (10) for $i = 1, 2, \cdots, n$ in order to have enough equations to find all the unknowns. Since we have no use for u_{n+1}, the usual practice is to solve the boundary-condition replacement for u_{n+1},

$$u_{n+1} = u_{n-1} - 2\Delta x, \tag{12}$$

and then use this expression in the version of Eq. (10) that corresponds to $i = n$. Thus, the equation

$$\frac{u_{n+1} - 2u_n + u_{n-1}}{(\Delta x)^2} - 10u_n = f(x_n)$$

is combined with Eq. (12) to get

$$\frac{2u_{n-1} - 2\Delta x - 2u_n}{(\Delta x)^2} - 10u_n = f(x_n). \tag{13}$$

Then Eq. (10) for $i = 1, \cdots, n-1$ and Eq. (13) give n equations that determine unknowns $u_1, u_2, \cdots, u_n$.

To be specific, let us take $n = 4$, so that $\Delta x = 1/4$. The three $(i = 1, 2, 3)$ versions of Eq. (10) are

$$\begin{aligned} 16(u_2 - 2u_1 + u_0) - 10u_1 &= 0 & (i = 1) \\ 16(u_3 - 2u_2 + u_1) - 10u_2 &= -50 & (i = 2) \\ 16(u_4 - 2u_3 + u_2) - 10u_3 &= -100 & (i = 3) \end{aligned}$$

and Eq. (13) adapted to $n = 4$ is

$$16\left(2u_3 - \frac{1}{2} - 2u_4\right) - 10u_4 = -100.$$

When these equations are cleaned up and the boundary condition $u_0 = 1$ is applied, the result is the following system of four equations:

$$\begin{aligned} -42u_1 + 16u_2 \qquad\qquad &= -16 \\ 16u_1 - 42u_2 + 16u_3 \qquad &= -50 \\ 16u_2 - 42u_3 + 16u_4 &= -100 \\ 32u_3 - 42u_4 &= -92. \end{aligned} \tag{14}$$

In Table 3 are shown the values of u_i obtained by solving Eq. (14) and also more exact values found by using $n = 100$. Elimination is not the only

Table 3: Approximate solution of equations (8) and (9)

x	$u(n = 4)$	$u(n = 100)$
0	1	1
0.25	2.174	2.155
0.50	4.707	4.729
0.75	7.057	7.125
1	7.567	7.629

way to get the solution of a system like Eq. (7) or (14). An alternative is an iterative method, which generates a sequence of approximate solutions.

For one such method, we solve algebraically the ith equation for the ith unknown. In the resulting set of equations there are "circular references": the equation for u_2 refers to u_1 and u_3, while the equations for these refer to u_2, etc. We may start with some guessed values for the u's, feed them through the equations to get improved values for the u's, and repeat the process until the values settle down. This method requires a lot of arithmetic but no strategy, while elimination is just the reverse. It may also work with nonlinear equations, where elimination can not.

So far, we have given no justification for the procedure of constructing replacement equations. The explanation is not difficult; it depends on the fact that certain difference quotients approximate derivatives. If $u(x)$ is a function with several derivatives then

$$\frac{u(x_{i+1}) - u(x_{i-1})}{2\Delta x} = u'(x_i) + \frac{(\Delta x)^2}{24} u^{(3)}(\bar{x}_i) \tag{15}$$

$$\frac{u(x_{i+1}) - 2u(x_i) + u(x_{i-1})}{(\Delta x)^2} = u''(x_i) + \frac{(\Delta x)^2}{12} u^{(4)}(\bar{\bar{x}}_i) \tag{16}$$

where $\bar{x}_i$ and $\bar{\bar{x}}_i$ are points near x_i.

Now suppose that $u(x)$ is the solution of the boundary value problem

$$\frac{d^2u}{dx^2} + k(x)\frac{du}{dx} + p(x)u(x) = f(x), \quad 0 < x < 1 \tag{17}$$

$$\alpha u(0) - \alpha' u'(0) = a, \qquad \beta u(1) + \beta' u'(1) = b. \tag{18}$$

If $u(x)$ has enough derivatives, then at any point $x_i = i\Delta x$ it satisfies the differential equation (17) and thus also satisfies the equation

$$dd\frac{u(x_{i+1}) - 2u(x_i) + u(x_{i-1})}{(\Delta x)^2} + k(x_i)\frac{u(x_{i+1}) - u(x_{i-1})}{2\Delta x} + p(x_i)u(x_i)$$

$$= f(x_i) + \delta_i \tag{19}$$

where

$$\delta_i = \frac{(\Delta x)^2}{12} u^{(4)}(\bar{\bar{x}}_i) + k(x_i)\frac{(\Delta x)^2}{24} u^{(3)}(\bar{x}_i).$$

Becasue δ_i is proportional to $(\Delta x)^2$, it is very small when Δx is small.

The replacement equation for Eq. (17) is, according to Table 2,

$$\frac{u_{i+1} - 2u_i + u_{i-1}}{(\Delta x)^2} + k(x_i)\frac{u_{i+1} - u_{i-1}}{2\Delta x} + p(x_i)u_i = f(x_i). \tag{20}$$

Thus, the values of u at $x_0, x_1, \cdots, x_n$, which satisfy Eq. (19) exactly, will nearly satisfy Eq. (20); vice versa, the numbers $u_0, u_1, \cdots, u_n$, which satisfy the replacement equations (20), nearly satisfy Eq. (19). It can be proved that the calculated numbers $u_0, u_1, \cdots, u_n$ do indeed approach the appropriate values of $u(x_i)$ as Δx approaches 0 (under continuity and other conditions on $k(x)$, $p(x)$, $f(x)$).

Exercises

1. Set up and solve replacement equations with $n = 4$ for the problem

$$\frac{d^2u}{dx^2} = -1, \quad 0 < x < 1$$
$$u(0) = 0, \qquad u(1) = 1.$$

2. Solve the problem of Exercise 1 analytically. On the basis of Eqs. (15) and (16), explain why the numerical solution agrees exactly with the analytical solution.

3. Set up and solve replacement equations with $n = 4$ for the problem

$$\frac{d^2u}{dx^2} - u = -2x, \quad 0 < x < 1$$
$$u(0) = 0, \qquad u(1) = 1.$$

4. Solve the problem in Exercise 3 analytically and compare the numerical results with the true solution.

5. Set up and solve replacement equations with $n = 4$ for the problem

$$\frac{d^2u}{dx^2} = x, \quad 0 < x < 1$$
$$u(0) - \frac{du}{dx}(0) = 1, \quad u(1) = 0.$$

6. Solve the problem in Exercise 5 analytically and compare the numerical results with the true solution.

7. Set up and solve replacement equations for the problem

$$\frac{d^2u}{dx^2} + 10u = 0, \quad 0 < x < 1$$
$$u(0) = 0, \quad u(1) = -1.$$

Use $n = 3$ and $n = 4$. Sketch the results and explain why they vary so much.

In Exercises 8 to 11, set up and solve replacement equations for the problem stated and the given value of n. If a computer is available, also solve for n twice as large and compare results.

8. $\dfrac{d^2u}{dx^2} - 32xu = 0,\ 0 < x < 1$

$u(0) = 0,\ u(1) = 1 \quad (n = 4)$

9. $\dfrac{d^2u}{dx^2} - 25u = -25,\ 0 < x < 1$

$u(0) = 2,\ u(1) + u'(1) = 1 \quad (n = 5)$

10. $\dfrac{d^2u}{dx^2} + \dfrac{1}{1+x}\dfrac{du}{dx} = -1,\ 0 < x < 1$

$u(0) = 0,\ u(1) = 0 \quad (n = 3)$

11. $\dfrac{d^2u}{dx^2} + \dfrac{du}{dx} - u = -x$

$\dfrac{du}{dx}(0) = 0,\ u(1) = 1 \quad (n = 3)$

12. Use the Taylor series expansion

$$u(x+h) = u(x) + hu'(x) + \frac{h^2}{2}u''(x) + \frac{h^3}{6}u^{(3)}(x) + \frac{h^4}{24}u^{(4)}(x) + \cdots$$

with $x = x_i$ and $h = \pm\Delta x$, $(x_i + \Delta x = x_{i+1}, x_i - \Delta x = x_{i-1})$ to obtain representations similar to Eqs. (15) and (16).

7.2 HEAT PROBLEMS

In heat problems, we have two independent variables x and t assumed to be in the range $0 < x < 1$, $0 < t$. A table for a function $u(x,t)$ should give values at equally spaced points and times

$$x_i = i\Delta x, \qquad t_m = m\Delta t,$$

for $i = 0, 1, \cdots, n$ and $m = 0, 1, \cdots$. Here, $\Delta x = 1/n$ as before. We will use a subscript to denote position and a number in parentheses to denote the time level for the approximation to the solution of a problem. That is,

$$u_i(m) \cong u(x_i, t_m).$$

The spatial derivatives in a heat problem will be replaced by difference quotients as before

$$\frac{\partial^2 u}{\partial x^2}(x_i, t_m) \rightarrow \frac{u_{i+1}(m) - 2u_i(m) + u_{i-1}(m)}{(\Delta x)^2} \tag{1}$$

$$\frac{\partial u}{\partial x}(x_i, t_m) \rightarrow \frac{u_{i+1}(m) - u_{i-1}(m)}{2\Delta x}. \tag{2}$$

For the time derivative, there are several possible replacements. We limit ourselves to the forward difference

$$\frac{\partial u}{\partial t}(x_i, t_m) \rightarrow \frac{u_i(m+1) - u_i(m)}{\Delta t}, \tag{3}$$

which will yield explicit formulas for computing.

Now, to solve numerically the simple heat problem

$$\frac{\partial^2 u}{\partial x^2} = \frac{\partial u}{\partial t}, \quad 0 < x < 1, \quad 0 < t \tag{4}$$

$$u(0,t) = 0, \quad u(1,t) = 0, \quad 0 < t \tag{5}$$

$$u(x,0) = f(x), \quad 0 < x < 1 \tag{6}$$

we set up replacement equations according to Eqs. (1) to (3). Those equations are

$$\frac{u_{i-1}(m) - 2u_i(m) + u_{i+1}(m)}{(\Delta x)^2} = \frac{u_i(m+1) - u_i(m)}{\Delta t} \tag{7}$$

supposed valid for $i = 1, 2, \cdots, n-1$ and $m = 0, 1, 2, \cdots$.

The point of using a forward difference for the time derivative is that these equations may be solved for $u_i(m+1)$

$$u_i(m+1) = ru_{i-1}(m) + (1-2r)u_i(m) + ru_{i+1}(m) \tag{8}$$

where $r = \Delta t/(\Delta x)^2$. Thus each $u_i(m+1)$ is calculated from u's at the preceding time level. Because the initial condition gives each $u_i(0)$, the values of the u's at time level 1 can be calculated by setting $m = 0$ in Eq. (8):

$$u_i(1) = ru_{i-1}(0) + (1-2r)u_i(0) + ru_{i+1}(0).$$

Then the values of the u's at time level 2 can be found from these, and so on into the future. Of course, r has to be given a numerical value first, by choosing Δx and Δt.

It is convenient to display the numerical values of $u_i(m)$ in a table, making columns correspond to different meshpoints $x_0, x_1, \cdots, x_n$ and making rows correspond to the different time levels $t_0, t_1, \cdots$. See Table 4.

As an example, we take $\Delta x = 1/4$ and $r = 1/2$, making $\Delta t = 1/32$. The equations giving the u's at time level $m+1$ are

$$\begin{aligned} u_1(m+1) &= \tfrac{1}{2}\left(u_0(m) + u_2(m)\right) \\ u_2(m+1) &= \tfrac{1}{2}\left(u_1(m) + u_3(m)\right) \\ u_3(m+1) &= \tfrac{1}{2}\left(u_2(m) + u_4(m)\right). \end{aligned} \tag{9}$$

Recall that the boundary conditions of this problem specify $u_0(m) = 0$ and $u_4(m) = 0$ for $m = 1, 2, 3, \cdots$. Thus we fill in the columns of the table that correspond to points x_0 and x_4 with 0's (shown in italics in Table 4). Also the initial condition specifies $u_i(0) = f(x_i)$, so the top row of the table can be filled. In this example we take $f(x) = x$, and the corresponding values appear in italics in the top row of Table 4.

The initial condition, $u(x, 0) = x$, $0 < x < 1$, suggests that $u(1, 0)$ should be 1, while the boundary condition suggests that it should be 0. In fact, neither condition specifies $u(1, 0)$, nor is there a hard and fast rule telling what to do in case of conflict. Fortunately, it does not matter much, either. (See Exercise 1.)

The choice we made of $r = 1/2$ seems natural, perhaps, because it simplifies the computation. It might also seem desirable to take a larger value of r (signifying a larger time step) to get into the future more rapidly. For example, with $r = 1$ ($\Delta t = 1/16$) the replacement equations take the form

$$u_i(m+1) = u_{i-1}(m) - u_i(m) + u_{i+1}(m).$$

In Table 5 are values of $u_i(m)$ computed from this formula. No one can believe that these wildly fluctuating values approximate the solution to the heat problem in any sense. Indeed, they suffer from *numerical instability* due to using a time step too long relative to the mesh size. The analysis of instability requires familiarity with matrix theory, but there are some simple rules of thumb that guarantee stability.

First, write out the equations for each $u_i(m+1)$

$$u_i(m+1) = a_i u_{i-1}(m) + b_i u_i(m) + c_i u_{i+1}(m).$$

The coefficients must satisfy two conditions

1. no coefficient may be negative; and

Table 4: Numerical solution of equations (4) to (6)

m \ i	0	1	2	3	4
0	*0*	*0.25*	*0.5*	*0.75*	*1*
1	*0*	0.25	0.5	0.75	*0*
2	*0*	0.25	0.5	0.25	*0*
3	*0*	0.25	0.25	0.25	*0*
4	*0*	0.125	0.25	0.125	*0*
5	*0*	0.125	0.125	0.125	*0*

Table 5: Unstable solution

m \ i	0	1	2	3	4
0	*0*	*0.25*	*0.50*	*0.75*	*1*
1	*0*	0.25	0.50	0.75	*0*
2	*0*	0.25	0.50	−0.25	*0*
3	*0*	0.25	−0.50	0.75	*0*
4	*0*	−0.75	1.50	−1.25	*0*
5	*0*	2.25	−3.50	2.75	*0*

2. the sum of the coefficients is not greater than 1.

In the example, the replacement equations were

$$u_1(m+1) = ru_0(m) + (1-2r)u_1(m) + ru_2(m)$$
$$u_2(m+1) = ru_1(m) + (1-2r)u_2(m) + ru_3(m)$$
$$u_3(m+1) = ru_2(m) + (1-2r)u_3(m) + ru_4(m).$$

The second requirement is satisfied automatically, because $r+(1-2r)+r = 1$. But the first condition is satisfied only for $r \leq 1/2$. Thus the first choice of $r = 1/2$ corresponded to the longest stable time step.

Different problems give different maximum values for r. For the heat conduction problem

$$\frac{\partial^2 u}{\partial x^2} = \frac{\partial u}{\partial t}, \quad 0 < x < 1, \qquad 0 < t \tag{10}$$

$$u(0,t) = 1, \qquad \frac{\partial u}{\partial x}(1,t) + \gamma u(1,t) = 0, \qquad 0 < t \tag{11}$$

$$u(x,0) = 0, \qquad 0 < x < 1 \tag{12}$$

the replacement equations are found to be (for $n = 4$)

$$\begin{aligned}
u_1(m+1) &= ru_0(m) + (1-2r)u_1(m) + ru_2(m)\\
u_2(m+1) &= ru_1(m) + (1-2r)u_2(m) + ru_3(m)\\
u_3(m+1) &= ru_2(m) + (1-2r)u_3(m) + ru_4(m)\\
u_4(m+1) &= 2ru_3(m) + (1-2r-\tfrac{1}{2}r\gamma)u_4(m).
\end{aligned} \tag{13}$$

(Remember that $u(1,t)$, corresponding to u_4, is an unknown. The boundary condition has been incorporated into the equation for $u_4(m+1)$.) Again, the second stability requirement is satisfied automatically; but the first rule requires that

$$1 - 2r - \frac{1}{2}r\gamma \leq 0 \quad \text{or} \quad r \geq \frac{1}{(2+\frac{1}{2}\gamma)}. \tag{14}$$

Exercises

1. Solve Eqs. (4) to (6) numerically with $f(x) = x$, as in the text ($\Delta x = 1/4$, $r = 1/2$), but take $u_4(0) = 0$. Compare your results with Table 4.

2. Solve Eqs. (4) to (6) numerically with $f(x) = x$, $\Delta x = 1/4$, $u_4(0) = 1$ as in the text, but use $r = 1/4$. Compare your results with Table 4. Be sure to compare results at corresponding times.

3. For the problem in Eqs. (10) to (12), find the longest stable time step when $\gamma = 1$, and compute the numerical solution with the corresponding value of r.

4. Solve the problem in Eqs. (10) to (12) with $\Delta x = 1/4$, $r = 1/2$ and $\gamma = 0$, for m up to 5.

For each problem in the following exercises, set up the replacement equations for $n = 4$, compute the longest stable time step, and calculate the numerical solution for a few values of m.

5. $\dfrac{\partial^2 u}{\partial x^2} = \dfrac{\partial u}{\partial t}$, $\quad u(0,t) = u(1,t) = t, \quad u(x,0) = 0$

6. $\dfrac{\partial^2 u}{\partial x^2} - u = \dfrac{\partial u}{\partial t}, \quad u(0,t) = u(1,t) = 1, \quad u(x,0) = 0$

7. $\dfrac{\partial^2 u}{\partial x^2} = \dfrac{\partial u}{\partial t} - 1, \quad u(0,t) = u(1,t) = 0, \quad u(x,0) = 0$

8. $\dfrac{\partial^2 u}{\partial x^2} = \dfrac{\partial u}{\partial t}, \quad u(0,t) = 0, \quad \dfrac{\partial u}{\partial x}(1,t) + u(1,t) = 1, \quad u(x,0) = 0$

9. $\dfrac{\partial^2 u}{\partial x^2} = \dfrac{\partial u}{\partial t}, \quad \dfrac{\partial u}{\partial x}(0,t) = 0, \quad u(1,t) = 1, \quad u(x,0) = x$

7.3 WAVE EQUATION

The simple vibrating string problem that we studied in Chapter 3,

$$\frac{\partial^2 u}{\partial x^2} = \frac{\partial^2 u}{\partial t^2}, \quad 0 < x < 1, \qquad 0 < t \tag{1}$$

$$u(0,t) = 0, \quad u(1,t) = 0, \qquad 0 < t \tag{2}$$

$$u(x,0) = f(x), \quad \frac{\partial u}{\partial t}(x,0) = g(x), \qquad 0 < x < 1, \tag{3}$$

rarely needs treatment by numerical methods, because the D'Alembert solution provides a simple and direct means of calculating the solution $u(x,t)$ for arbitrary x and t. However, if the partial differential equation contains u or an inhomogeneity, or if the boundary conditions are more complex, a series solution or a solution of the D'Alembert type may not be practical. In many such cases, simple numerical techniques are quite rewarding.

In order to convert the wave equation (1) into a suitable difference equation, we first designate points $x_i = i\Delta x (\Delta x = 1/n)$ and times $t_m = m\Delta t$ for which the approximation to u will be found: $u(x_i, t_m) \cong u_i(m)$. Then the partial derivatives with respect to both x and t are replaced by central differences

$$\frac{\partial^2 u}{\partial x^2} \to \frac{u_{i+1}(m) - 2u_i(m) + u_{i-1}(m)}{(\Delta x)^2}$$

$$\frac{\partial^2 u}{\partial t^2} \to \frac{u_i(m+1) - 2u_i(m) + u_i(m-1)}{(\Delta t)^2}.$$

The wave equation (1) becomes this partial difference equation

$$\frac{u_{i+1}(m) + 2u_i(m) + u_{i-1}(m)}{(\Delta x)^2} = \frac{u_i(m+1) - 2u_i(m) + u_i(m-1)}{(\Delta t)^2}$$

or, with $\rho = \Delta t/\Delta x$,

$$u_i(m+1) - 2u_i(m) + u_i(m-1) = \rho^2(u_{i+1}(m) - 2u_i(m) + u_{i-1}(m)).$$

The replacement equations may be solved for the unknowns $u_i(m+1)$, yielding the equation

$$u_i(m+1) = \rho^2 u_{i-1}(m) + 2(1-\rho^2)u_i(m) + \rho^2 u_{i+1}(m) - u_i(m-1) \quad (4)$$

valid for $i = 1, 2, \cdots, n-1$. Naturally, the boundary conditions, Eq. (2), carry over as $u_0(m) = 0$, $u_n(m) = 0$. It is obvious that Eq. (4) requires us to know the approximate solution at time levels m and $m-1$ in order to find it at time level $m+1$. In other words, to get $u_i(1)$ we need $u_{i-1}(0)$, $u_i(0)$, $u_{i+1}(0)$—which are available from the initial condition—and also $u_i(-1)$! Of course, we have not yet applied the second initial condition,

$$\frac{\partial u}{\partial t}(x, 0) = g(x), \qquad 0 < x < 1.$$

If we replace the time derivative by a central difference approximation, this equation translates into

$$\frac{u_i(1) - u_i(-1)}{2\Delta t} = g(x_i) \quad (5)$$

for $i = 1, 2, \cdots, n-1$. Equation (5), together with a slightly modified version of Eq. (4) (with $m = 0$ and $u_i(0) = f(x_i)$), yields the system

$$\begin{aligned} u_i(1) + u_i(-1) &= \rho^2 f(x_{i-1}) + 2(1-\rho^2)f(x_i) + \rho^2 f(x_{i+1}) \\ u_i(1) - u_i(-1) &= 2\Delta t g(x_i), \end{aligned} \quad (6)$$

which we can easily solve for the u's at the first time level

$$u_i(1) = \frac{1}{2}\rho^2 f(x_{i-1}) + (1-\rho^2)f(x_i) + \frac{1}{2}\rho^2 f(x_{i+1}) + \Delta t g(x_i). \quad (7)$$

Thus, in order to solve the problem in Eqs. (1) to (3) numerically, we use the initial condition $u_i(0) = f(x_i)$ to fill the first line of our table, use the *starting equation* (7) to fill the next line, and continue with the *running equation* (4) to fill subsequent lines.

Let us now attempt to solve a simple problem. Suppose that $g(x) \equiv 0$ for $0 < x < 1$ and that $f(x)$ is given by

$$f(x) = \begin{cases} 2x, & 0 < x < \dfrac{1}{2} \\ 2(1-x), & \dfrac{1}{2} < x < 1. \end{cases} \quad (8)$$

Also, we shall choose $n = 4$ and $\rho = 1$ for convenience. (That is, $\Delta t = \Delta x = 1/4$.) Our rule for calculation, Eq. (4), is then

$$u_i(m+1) = u_{i-1}(m) + u_{i+1}(m) - u_i(m-1). \tag{9}$$

In Table 6 are the calculated values of $u_i(m)$. Entries in italics are given data. It is easy to check that this numerical solution is identical with the D'Alembert solution of this particular problem. (See Exercise 6.) However, if the initial velocity were not identically zero, the numerical solution would in general be only an approximation to the true solution. In our study of the heat equation, Section 2, we saw that the choice of Δx

Table 6: Numerical solution of equations (1) to (3)

m \ i	0	1	2	3	4
0	*0*	*0.5*	*1*	*0.5*	*0*
1	*0*	0.5	0.5	0.5	*0*
2	*0*	0	0	0	*0*
3	*0*	−0.5	−0.5	−0.5	*0*
4	*0*	−0.5	−1	−0.5	*0*
5	*0*	−0.5	−0.5	0.5	*0*
6	*0*	0	0	0	*0*

and Δt was not free. The same is true for the wave equation. Suppose we attempt to solve the same problem as above, but with $\rho^2 = (\Delta t/\Delta x)^2$ chosen to be 2. Then Eq. (4) becomes

$$u_i(m+1) = 2(u_{i-1}(m) - u_i(m) + u_{i+1}(m)) - u_i(m-1)$$

and the "solution" corresponding to this rule of calculation is shown in Table 7 (again, entries in italics are given data). Of course, the results bear no resemblance to the solution of the wave equation. They suffer from the same sort of instability as that observed in Section 2. There is a rule of thumb, similar to the one to be found there, applicable to the wave equation.

First, write out the equations for each $u_i(m+1)$ in terms of the u's at time levels m and $m-1$

$$u_i(m+1) = a_i u_{i-1}(m) + b_i u_i(m) + c_i u_{i+1}(m) - u_i(m-1).$$

The coefficients must satisfy two conditions:

1. none of the coefficients a_i, b_i, c_i may be negative; and
2. the sum of the coefficients is not greater than 2:

$$a_i + b_i + c_i \leq 2.$$

Of course, $u_i(m-1)$ appears with a coefficient of -1; nothing can be done about that, nor does it enter into the forementioned rules.

Table 7: Unstable numerical solution

m \ i	0	1	2	3	4
0	*0*	*0.5*	*1*	*0.5*	*0*
1	*0*	0.5	0	0.5	*0*
2	*0*	−1.5	1	−1.5	*0*
3	*0*	4.5	−8	4.5	*0*
4	*0*	−23.5	33	−23.5	*0*

In Eq. (4) we see that both conditions are met when $\rho = \Delta t / \Delta x$ is less than or equal to 1; in other words, the time step must not exceed the space step. However, using $\rho^2 = 1$ when acceptable often provides the best accuracy.

We conclude with one more example, illustrating how numerical results can be obtained easily in some cases that might be puzzling analytically. Suppose that we are to solve the problem

$$\frac{\partial^2 u}{\partial x^2} = \frac{\partial^2 u}{\partial t^2} - 16\cos(\pi t), \quad 0 < x < 1, \quad 0 < t \tag{10}$$

$$u(0,t) = 0, \quad u(1,t) = 0, \quad 0 < t \tag{11}$$

$$u(x,0) = 0, \quad \frac{\partial u}{\partial t}(x,0) = 0, \quad 0 < x < 1. \tag{12}$$

We replace the partial derivatives as before, obtaining

$$\frac{u_{i+1}(m) - 2u_i(m) + u_{i-1}(m)}{(\Delta x)^2} = \frac{u_i(m+1) - 2u_i(m) + u_i(m-1)}{(\Delta t)^2} - 16\cos(\pi t_m).$$

When this is solved for $u_i(m+1)$, we find

$$u_i(m+1) = (2-2\rho^2)u_i(m) + \rho^2 u_{i+1}(m) + \rho^2 u_{i-1}(m) - u_i(m-1) + 16(\Delta t)^2 \cos(\pi m \Delta t). \quad (13)$$

Let us take $\Delta x = \Delta t = 1/4$ again, so $\rho = 1$ and Eq. (13) simplifies to

$$u_i(m+1) = u_{i+1}(m) + u_{i-1}(m) - u_i(m-1) + \cos\left(\frac{m\pi}{4}\right). \quad (14)$$

This is our running equation. The starting equation comes from combining Eqs. (14) for $m = 0$,

$$u_i(1) = -u_i(-1) + 1$$

(note $u_i(0) = 0$), with the replacement initial condition

$$\frac{u_i(1) - u_i(-1)}{2\Delta t} = 0,$$

or

$$u_i(1) = u_i(-1) = \frac{1}{2}$$

for $i = 1, 2, 3$. Now we have the top two lines of Table 8, and the rest are filled using Eqs. (14) (with $\cos(\pi/4) \simeq 0.71$, and so forth). Entries in italics are given data.

Table 8: Numerical solution of equations (10) to (12)

i m	0	1	2	3	4
0	*0*	*0*	*0*	*0*	*0*
1	*0*	0.5	0.5	0.5	*0*
2	*0*	1.31	1.71	1.21	*0*
3	*0*	1.21	1.91	1.21	*0*
4	*0*	0.00	0.00	0.00	*0*
5	*0*	−2.21	−2.91	−2.21	*0*
6	*0*	−3.62	−5.12	−3.62	*0*
7	*0*	−2.91	−4.33	−2.91	*0*

The complete analytical solution of this problem is

$$u(x,t) = \frac{32}{\pi^2} t \sin(\pi t)\sin(\pi x) + \frac{32}{\pi^3}\sum_{n=3}^{\infty} \frac{1-\cos(n\pi)}{n(n^2-1)}(\cos(\pi t) - \cos(n\pi t))\sin(n\pi x).$$

At $x = 1/2$, the sum of the infinite series is 0, so

$$u\left(\frac{1}{2}, t\right) = \frac{32}{\pi^2} t \sin(\pi t).$$

Comparison of the values of this function at times t_m with the middle column of Table 8 shows the numerical solution off by a few percent.

Exercises

1. Obtain an approximate solution of Eqs. (1), (2), and (3) with $f(x) \equiv 0$ and $g(x) \equiv 1$. Take $\Delta x = 1/4$, $\rho = 1$.

2. Compare the results of Exercise 1 with the D'Alembert solution.

3. Obtain an approximate solution of Eqs. (1), (2), and (3) with $f(x) \equiv 0$ and $g(x) = \sin(\pi x)$. Take $\Delta x = 1/4$, $\rho = 1$.

4. Compare the results of Exercise 3 with the exact solution $u(x,t) = (1/\pi)\sin(\pi x)\sin(\pi t)$.

5. Obtain an approximate solution of Eqs. (1), (2) and (3) with $g(x) \equiv 0$ and $f(x)$ as in Eq. (8). Use $\Delta x = 1/4$ and $\rho^2 = 1/2$.

6. Compare the entries of Table 6 with the D'Alembert solution.

7. Obtain an approximate solution of this problem with a time-varying boundary condition, using $\Delta x = \Delta t = 1/4$.

$$\frac{\partial^2 u}{\partial x^2} = \frac{\partial^2 u}{\partial t^2}, \quad 0 < x < 1, \quad 0 < t$$

$$u(0,t) = 0, \quad u(1,t) = h(t), \quad 0 < t$$

$$u(x,0) = 0, \quad \frac{\partial u}{\partial t}(x,0) = 0, \quad 0 < x < 1$$

$$h(t) = \begin{cases} 1, & 0 < t < 1 \\ -1, & 1 < t < 2 \end{cases}$$

and $h(t+2) = h(t)$, $h(0) = h(1) = 0$.

8. Same task as Exercise 7 but $h(t) = \sin(\pi t)$. Use $\sin(\pi/4) \cong 0.7$ instead of $\sqrt{2}/2$.

9. Find starting and running equations for the following problem. Using $\Delta x = 1/4$, find the longest stable time step and compute values of the approximate solution for m up to 8.

$$\begin{aligned} &\frac{\partial^2 u}{\partial x^2} = \frac{\partial^2 u}{\partial t^2} + 16u, && && 0 < x < 1, \quad 0 < t \\ &u(0,t) = 0, && u(1,t) = 0, && 0 < t \\ &u(x,0) = f(x), && \frac{\partial u}{\partial t}(x,0) = 0, && 0 < x < 1 \end{aligned}$$

where $f(x)$ is given in Eq. (8).

10. Using $\Delta x = 1/4$ and $\rho^2 = 1/2$, compare the numerical solution of the problem in Exercise 9 with and without the $16u$ term in the partial differential equation.

7.4 POTENTIAL EQUATION

In this section, we will be concerned with approximate solutions of the potential equation and related equations in a region $\mathcal{R}$ of the xy-plane. For the sake of simplicity, we will limit ourselves to regions whose boundaries can be made to coincide with the lines on a sheet of graph paper with square divisions. Thus, we admit such shapes as rectangles, L's and T's, but not circles or triangles. The graph paper provides us with a ready-made mesh of points in the region $\mathcal{R}$ and on its boundary, at which we wish to know the solution of our problem. These points are to be numbered in some fashion — usually left to right and bottom to top.

On such a mesh, the replacement for the Laplacian operator is the following

$$\frac{\partial^2 u}{\partial x^2} + \frac{\partial^2 u}{\partial y^2} \to \frac{u_W - 2u_i + u_E}{(\Delta x)^2} + \frac{u_N - 2u_i + u_S}{(\Delta y)^2} \tag{1}$$

where the subscripts E, W stand for the indices of the mesh points to the left and right of point i, and N, S stand for those above and below (see Fig. 1). The result is sometimes called the *five-point approximation to the Laplacian.* Because we are assuming that $\Delta x = \Delta y$, we obtain a further simplification in the replacement:

$$\frac{\partial^2 u}{\partial x^2} + \frac{\partial^2 u}{\partial y^2} \to \frac{u_N + u_S + u_E + u_W - 4u_i}{(\Delta x)^2}. \tag{2}$$

Now, let us set up the replacement equations for this simple problem, which was solved analytically in Chapter 4.

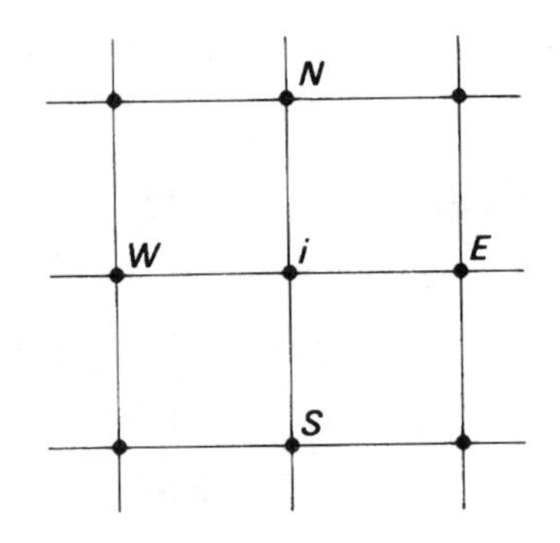

Figure 1: Point i on a square mesh and its four neighbors.

$$\frac{\partial^2 u}{\partial x^2} + \frac{\partial^2 u}{\partial y^2} = 0, \qquad 0 < x < 1, \qquad 0 < y < 1 \tag{3}$$

$$u(0, y) = 0, \qquad u(1, y) = 0, \qquad 0 < y < 1 \tag{4}$$

$$u(x, 0) = f(x), \quad u(x, 1) = f(x), \qquad 0 < x < 1 \tag{5}$$

$$f(x) = \begin{cases} 2x, & 0 < x < \frac{1}{2}, \\ 2(1 - x), & \frac{1}{2} \leq x < 1. \end{cases} \tag{6}$$

Let us take $\Delta x = \Delta y = 1/4$ and number the mesh points inside the 1×1 square as shown in Fig. 2.

At each of the nine mesh points, we will have the replacement equation

$$u_N + u_S + u_E + u_W - 4u_i = 0. \tag{7}$$

Together, these make up a system of nine equations in the nine unknowns $u_1, u_2, \cdots, u_9$. Referring to Fig. 2, where the values of u at boundary points are shown, we can write down the equations to be solved:

$$\begin{aligned}
u_2 + u_4 + \tfrac{1}{2} - 4u_1 &= 0 \\
u_1 + u_3 + u_5 + 1 - 4u_2 &= 0 \\
u_2 + u_6 + \tfrac{1}{2} - 4u_3 &= 0 \\
u_1 + u_5 + u_7 - 4u_4 &= 0 \\
u_2 + u_4 + u_6 + u_8 - 4u_5 &= 0 \\
u_3 + u_5 + u_9 - 4u_6 &= 0 \\
u_4 + u_8 + \tfrac{1}{2} - 4u_7 &= 0 \\
u_5 + u_7 + u_9 + 1 - 4u_8 &= 0 \\
u_6 + u_8 + \tfrac{1}{2} - 4u_9 &= 0.
\end{aligned} \tag{8}$$

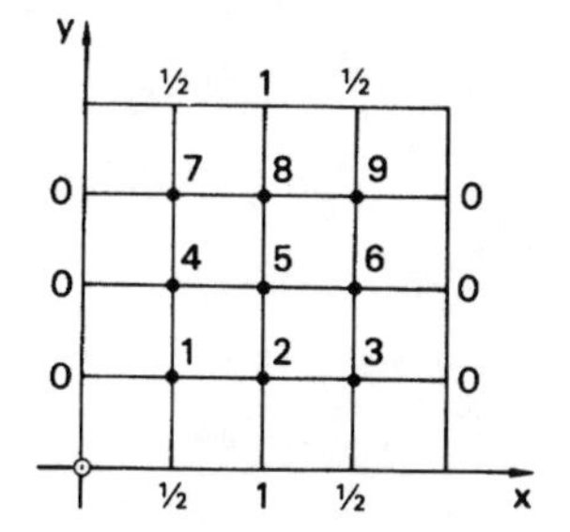

Figure 2: Numbering for mesh points, and values on boundary.

This is simply a system of simultaneous equations. It can be solved by elimination to obtain the results shown in Fig. 3. In this particular case, there are numerous symmetries in the problem, so that $u_1 = u_3 = u_7 = u_9$, $u_2 = u_8$ and $u_4 = u_6$. Thus, only u_1, u_2, u_4, and u_5 need to be found. The system can be reduced to four equations in these four unknowns, which can even be solved manually.

As a second example, we set up the replacement equations for the problem

$$\frac{\partial^2 u}{\partial x^2} + \frac{\partial^2 u}{\partial y^2} = 16(u-1), \quad 0 < x < 1, \qquad 0 < y < 1 \tag{9}$$

$$u(x,0) = 0, \qquad u(x,1) = 0, \qquad 0 < x < 1 \tag{10}$$

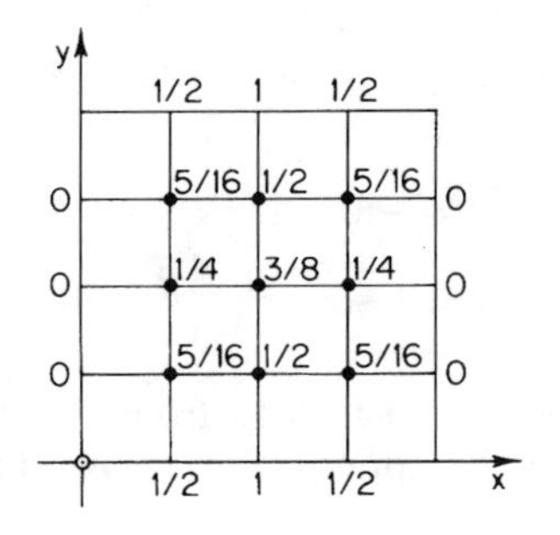

Figure 3: Numerical solution of Eqs. (3) to (6).

$$u(0, y) = 0, \qquad u(1, y) = 0, \qquad 0 < y < 1. \tag{11}$$

We may use the same numbering as in the first example (Fig. 2). At each mesh point, the replacement is

$$\frac{u_N + u_S + u_E + u_W - 4u_i}{(\Delta x)^2} = 16(u_i - 1). \tag{12}$$

Because $\Delta x = 1/4$, then $(1/\Delta x)^2 = 16$, and the typical replacement equation becomes

$$u_N + u_S + u_E + u_W - 4u_i = u_i - 1$$

or

$$u_N + u_S + u_E + u_W - 5u_i = -1. \tag{13}$$

Finally, we may write out the equations to be solved. The first four of the nine equations, corresponding to Eq. (13) with $i = 1, 2, 3, 4$ are

$$\begin{aligned} u_2 + u_4 - 5u_1 &= -1 \\ u_1 + u_3 + u_5 - 5u_2 &= -1 \\ u_2 + u_6 - 5u_3 &= -1 \\ u_1 + u_5 + u_7 - 5u_4 &= -1. \end{aligned} \tag{14}$$

The solution of this problem is left as an exercise.

On more complicated regions, the replacement for the Laplacian operator has exactly the same form, since we still use the "graph-paper mesh."

The system of equations to be solved will be rather less regular than that for a rectangle. As an example, consider the problem

$$\frac{\partial^2 u}{\partial x^2} + \frac{\partial^2 u}{\partial y^2} = -16 \text{ in } \mathcal{R} \tag{15}$$

$$u = 0 \text{ on the boundary of } \mathcal{R} \tag{16}$$

where $\mathcal{R}$ is an L-shaped region formed from a 1×1 square by removing a $1/4 \times 1/4$ square from the upper right corner. The general replacement equation is

$$u_N + u_S + u_E + u_W - 4u_i = -1. \tag{17}$$

With the numbering shown in Fig. 4, the eight equations to be solved are

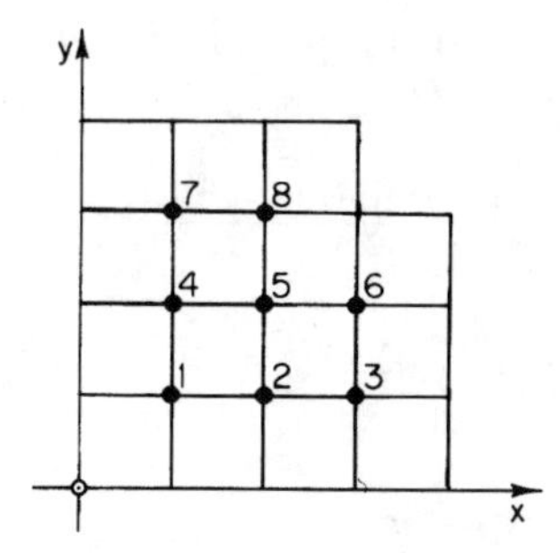

Figure 4: Mesh numbering for L-shaped region.

$$\begin{aligned}
u_2 + u_4 - 4u_1 &= -1 \\
u_1 + u_3 + u_5 - 4u_2 &= -1 \\
u_2 + u_6 - 4u_3 &= -1 \\
u_1 + u_5 + u_7 - 4u_4 &= -1 \\
u_2 + u_4 + u_6 + u_8 - 4u_5 &= -1 \\
u_3 + u_5 - 4u_6 &= -1 \\
u_4 + u_8 - 4u_7 &= -1 \\
u_5 + u_7 - 4u_8 &= -1.
\end{aligned} \tag{18}$$

The results, rounded to three digits, are shown in Eq. (19). Note the equalities, which arise from symmetries in the problem:

$$\begin{aligned} u_1 &= 0.656 \\ u_2 = u_4 &= 0.813 \\ u_3 = u_7 &= 0.616 \\ u_5 &= 0.981 \\ u_6 = u_8 &= 0.649. \end{aligned} \tag{19}$$

Systems of up to 10 equations, such as those in the foregoing examples, can readily be solved by elimination. It is easy to see, however, that we might well want a finer mesh to get better accuracy, and that a finer mesh will increase the number of equations dramatically. For example, if we use $\Delta x = \Delta y = 1/10$ in a numerical solution of Eqs. (3) to (5), the system to be solved contains 81 unknowns (or 25 if we use symmetry). Problems involving many thousands of unknowns are quite common. These large systems of simultaneous equations are almost always solved by *iterative methods*, which generate a sequence of approximate solutions.

Consider again the potential problem in Eqs. (3) to (6). Let us take a mesh with $\Delta x = \Delta y = 1/N$ and number the points of the mesh with a double index so that

$$u(x_i, y_j) \cong u_{i,j}. \tag{20}$$

Then the replacement equations for the potential equation are

$$\frac{u_{i+1,j} - 2u_{i,j} + u_{i-1,j}}{(\Delta x)^2} + \frac{u_{i,j+1} - 2u_{i,j} + u_{i,j-1}}{(\Delta y)^2} = 0,$$

or, using $\Delta x = \Delta y$ and some algebra,

$$u_{i,j} = \frac{1}{4}\left(u_{i+1,j} + u_{i-1,j} + u_{i,j+1} + u_{i,j-1}\right), \tag{21}$$

valid for i and j ranging from 1 to $N-1$. (This is the same as Eq. (7).) The boundary conditions, Eqs. (4) and (5), determine

$$u_{0,j} = 0, \qquad u_{N,j} = 0, \qquad j = 0, \cdots, N \tag{22}$$

$$u_{i,0} = f(x_i), \qquad u_{i,N} = f(x_i), \qquad i = 0, \cdots, N. \tag{23}$$

The simplest iterative method, called the Gauss-Seidel method, works this way. We sweep through the array of u's, replacing each $u_{i,j}$ by the combination of u's given on the right-hand side of Eq. (21). After several sweeps through the array, the numbers no longer change much. When the new and old values of $u_{i,j}$ at each point agree closely enough, we stop.

The result is a set of numbers that satisfy Eq. (21) approximately. Since the exact solution of the replacement equations is still just an approximation to the solution of the original problem in Eqs. (3) to (6), it is not urgent to get that exact solution of the replacement equations.

An iterative method such as the Gauss-Seidel method is very easy to implement on a spreadsheet without programming. (See the Appendix.)

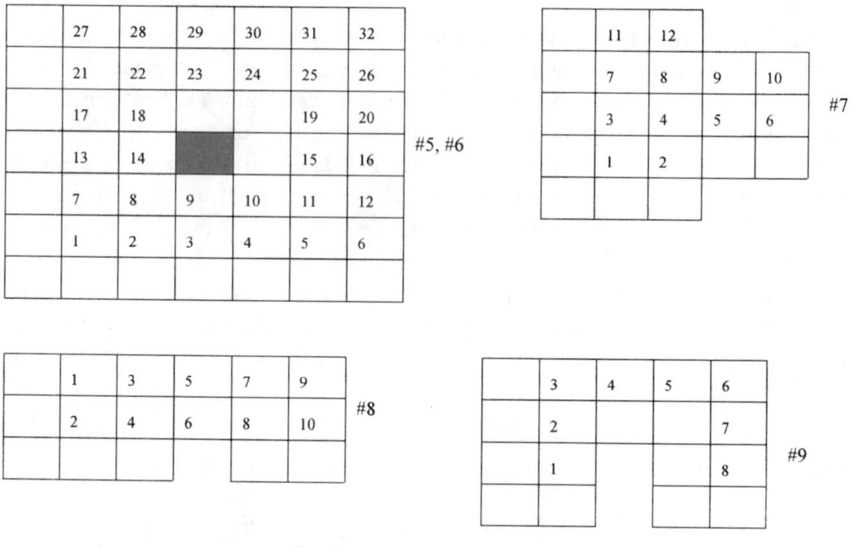

Figure 5: Regions and mesh numbering for Exercises 5 through 9.

Exercises Set up and solve replacement equations for each of the following problems. Use symmetry to reduce the number of unknowns.

1. $\nabla^2 u = -1$, $0 < x < 1$, $0 < y < 1$, $u = 0$ on the boundary. $\Delta x = \Delta y = 1/4$.

2. Same as Exercise 1 with $\Delta x = \Delta y = 1/8$. Compare the solutions.

3. $\nabla^2 u = 0$, $0 < x < 1$, $0 < y < 1$, $u(0, y) = 0$, $u(x, 0) = 0$, $u(1, y) = y$, $u(x, 1) = x$. $\Delta x = \Delta y = 1/4$.

4. Same as Exercise 3 with $\Delta x = \Delta y = 1/8$.

5. The region $\mathcal{R}$ is a square of side 1 from the center of which a similar square of side 1/7 has been removed; $\nabla^2 u = 0$ in $\mathcal{R}$, $u = 0$ on the outside boundary, and $u = 1$ on the inside boundary; $\Delta x = \Delta y = 1/7$. See Fig. 5.

6. Same as Exercise 5, but the partial differential equation is $\nabla^2 u = -1$, and the boundary condition is $u = 0$ on all boundaries. See Fig. 5.

7. The region $\mathcal{R}$ has the shape of a T, made by removing strips from the corners of a 1×1 square. The partial differential equationis $\nabla^2 u = -25$ in $\mathcal{R}$, and $u = 0$ on the boundary. Take $\Delta x = \Delta y = 1/5$. See Fig. 5 for numbering of mesh points.

8. The region is a rectangle, 2 units wide and 1 unit high. The potential equation holds in the interior; $u = 1$ on the upper half of the boundary (the top and the upper halves of the vertical sides), and $u = 0$ on the lower half. Take $\Delta x = \Delta y = 1/3$. See Fig. 5.

9. The region, as seen in Fig. 5, is shaped like an upsidedown U and is formed by removing a small (1×2) rectangle from the bottom of a larger (5×4) one. In the interior of the region, $\nabla^2 u = 0$. The boundary conditions are: $u = 1$ on the left and right sides and the top of the rectangle; $u = 0$ on the bottom and on the boundary formed by the removal of the small rectangle. Use $\Delta x = \Delta y = 1$.

7.5 TWO-DIMENSIONAL PROBLEMS

Separation of variables and other analytical methods produce satisfactory solutions to two-dimensional problems in only the nicest cases. However, simple numerical methods work quite well on two-dimensional problems. In this elementary exposition, we will limit ourselves to the heat and wave equations on two-dimensional regions that "fit on graph paper," as in Section 4.

We will compute an approximation to the solution of a problem, denoting space position with one or two subscripts and time level with an index in parentheses. Both heat and wave problems will require the replacement of the Laplacian operator. We use the same replacement as in Section 4,

$$\frac{\partial^2 u}{\partial x^2} + \frac{\partial^2 u}{\partial y^2} \to \frac{u_E(m) - 2u_i(m) + u_W(m)}{(\Delta x)^2} + \frac{u_N(m) - 2u_i(m) + u_S(m)}{(\Delta y)^2}.$$

Because we are using a square mesh, with $\Delta x = \Delta y$, the replacement simplifies to

$$\frac{\partial^2 u}{\partial x^2} + \frac{\partial^2 u}{\partial y^2} \to \frac{u_N(m) + u_S(m) + u_E(m) + u_W(m) - 4u_i(m)}{(\Delta x)^2} \tag{1}$$

where N, S, E, W stand for the indices of the four grid points adjacent to the point with the index i.

Now let us consider this heat problem on a rectangle

$$\frac{\partial^2 u}{\partial x^2} + \frac{\partial^2 u}{\partial y^2} = \frac{\partial u}{\partial t}, \quad 0 < x < 1.25, \qquad 0 < y < 1, \quad 0 < t \tag{2}$$

$$u(0, y, t) = 0, \quad u(1.25, y, t) = 0, \qquad 0 < y < 1, \quad 0 < t \tag{3}$$

$$u(x, 0, t) = 0, \quad u(x, 1, t) = 0, \qquad 0 < x < 1.25, \quad 0 < t \tag{4}$$

$$u(x, y, 0) = 1, \quad 0 < x < 1.25, \qquad 0 < y < 1. \tag{5}$$

We take $\Delta x = \Delta y = 1/4$ and number the interior points of the region as shown in Fig. 6. Then we will be computing the approximations

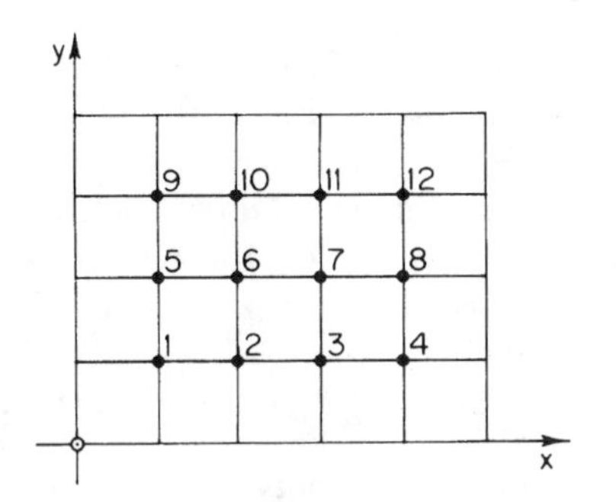

Figure 6: Mesh numbering for numerical solution of Eqs. (2)-(5).

$$u_1(m) \cong u\left(\tfrac{1}{4}, \tfrac{1}{4}, t_m\right), u_2(m) \cong u\left(\tfrac{1}{2}, \tfrac{1}{4}, t_m\right), u_3(m) \cong u\left(\tfrac{3}{4}, \tfrac{1}{4}, t_m\right), \cdots \tag{6}$$

and so forth, for $m = 1, 2, \cdots$. The replacement equations are obtained using Eq. (1) for the Laplacian and a forward difference to replace the time derivative. The typical equation is

$$\frac{u_N(m) + u_S(m) + u_E(m) + u_W(m) - 4u_i(m)}{(\Delta x)^2} = \frac{u_i(m+1) - u_i(m)}{\Delta t}. \tag{7}$$

When we solve this equation for $u_i(m+1)$, we obtain

$$u_i(m+1) = r\left[u_N(m) + u_S(m) + u_E(m) + u_W(m)\right] + (1-4r)u_i(m) \quad (8)$$

in which

$$r = \frac{\Delta t}{\Delta x^2} = \frac{\Delta t}{\Delta y^2} = 16\Delta t.$$

The stability considerations of Section 2 are still important and the rules of thumb are still valid. We must limit r by the requirement that $1-4r \geq 0$, or, in this case, $\Delta t \leq 1/64$. We shall take the longest acceptable time step, $\Delta t = 1/64$, $r = 1/4$, which makes the equations a little simpler.

At $m = 0$, all temperatures are given as 1. For $m \geq 1$, all the boundary temperatures are zero and the $u_i(m)$ are all found to equal 1. For $m = 2$, we calculate

$$\begin{aligned}
u_1(2) &= \frac{1}{4}\left(u_2(1) + u_5(1) + 0 + 0\right) = \frac{1}{2}\\
u_2(2) &= \frac{1}{4}\left(u_1(1) + u_3(1) + u_6(1) + 0\right) = \frac{3}{4}\\
&\vdots\\
u_5(2) &= \frac{1}{4}\left(u_1(1) + u_6(1) + u_9(1) + 0\right) = \frac{3}{4}\\
u_6(2) &= \frac{1}{4}\left(u_2(1) + u_5(1) + u_7(1) + u_{10}(1)\right) = 1.
\end{aligned}$$

The 0's in these equations stand for boundary temperatures.

An alert calculator will notice that only the unknowns u_1, u_2, u_5, u_6 need be calculated, since, in this example, the others will be given at each time step by symmetry

$$u_1(m) = u_4(m) = u_9(m) = u_{12}(m), \quad u_5(m) = u_8(m),$$
$$u_6(m) = u_7(m), \quad u_2(m) = u_3(m) = u_{10}(m) = u_{11}(m).$$

In Table 9 are computed values of the significant u's at a few times.

Now consider this heat problem, which is not solvable by separation of variables:

$$\frac{\partial^2 u}{\partial x^2} + \frac{\partial^2 u}{\partial y^2} = \frac{\partial u}{\partial t} \quad \text{in } \mathcal{R} \qquad (9)$$

$$u = f(t) \quad \text{on } \mathcal{C}, \qquad (10)$$

$$u = 0 \quad \text{in } \mathcal{R} \text{ at } t = 0. \qquad (11)$$

Table 9: Numerical solution of equations (2) to (5)

m \ i	1	2	5	6
0	1	1	1	1
1	$\frac{1}{2}$	$\frac{3}{4}$	$\frac{3}{4}$	1
2	$\frac{3}{8}$	$\frac{9}{16}$	$\frac{1}{2}$	$\frac{13}{16}$
3	$\frac{17}{64}$	$\frac{7}{16}$	$\frac{25}{64}$	$\frac{39}{64}$

Here, $\mathcal{R}$ is an L-shaped region and $\mathcal{C}$ is its boundary. The function f we take to be $f(t) = t$, but more complicated functions can be used.

To start the numerical solution, we set up a square grid, as shown in Fig. 7. The spacing is $\Delta x = \Delta y = 1/5$ and the numbering of the points is shown. The typical replacement equation is just as given in Eqs. (7) and (8). We must bear in mind, however, that some points are adjacent to boundary points where the temperature is given by $f(t)$.

Because $\Delta x = \Delta y = 1/5$, the parameter r in Eq. (8) is

$$r = \frac{\Delta t}{\Delta x^2} = 25\Delta t.$$

Clearly, the longest stable time step is $\Delta t = 1/100$, corresponding to $r = 1/4$. Using this value of r simplifies the typical replacement equation to

$$u_i(m+1) = \frac{1}{4}\left(u_N(m) + u_S(m) + u_E(m) + u_W(m)\right). \tag{12}$$

Specifically, we have

$$u_1(m+1) = \frac{1}{4}\left(u_2(m) + u_5(m) + 2f(t_m)\right)$$

$$u_2(m+1) = \frac{1}{4}\left(u_1(m) + u_3(m) + u_6(m) + f(t_m)\right)$$

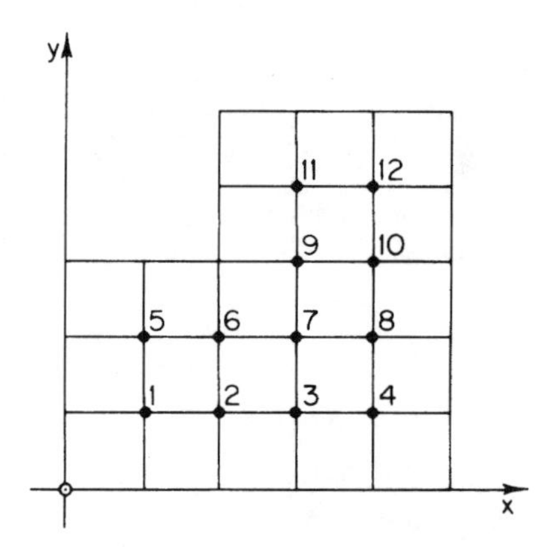

Figure 7: Mesh numbering for numerical solution of Eqs. (9) to (11).

and so on. The $f(t_m)$ terms enter because point 1 is adjacent to two boundary points and point 2 to one boundary point. Note that symmetry about the line through points 4 and 7 makes it unnecessary to compute $u_8(m), \cdots, u_{12}(m)$. Table 10 contains calculated values of u for the first four time levels.

Table 10: Numerical solution of equations (9) to (11). Entries are $100 \times u_i(m)$

i m	1	2	3	4	5	6	7	$f(t_m)$
0	0	0	0	0	0	0	0	0
1	0	0	0	0	0	0	0	1.0
2	0.5	0.25	0.25	0.5	0.5	0.25	0	2.0
3	1.22	0.75	0.69	0.75	1.22	0.69	0.25	3.0
4	1.99	1.40	1.19	1.84	1.98	1.30	0.69	4.0

In solving two-dimensional wave problems, we replace the Laplacian as in the foregoing and use a central difference for the time derivative, as we did in Section 3:

$$\frac{\partial^2 u}{\partial t^2} \to \frac{u_i(m+1) - 2u_i(m) + u_i(m-1)}{\Delta t^2}. \tag{13}$$

As an example, let us consider the vibrations of a square membrane, as described by the problem

$$\frac{\partial^2 u}{\partial x^2} + \frac{\partial^2 u}{\partial y^2} = \frac{\partial^2 u}{\partial t^2}, \quad 0 < x < 1, \quad 0 < y < 1, \quad 0 < t \tag{14}$$

$$u(x,0,t) = 0, \quad u(x,1,t) = 0, \quad 0 < x < 1, \quad 0 < t \tag{15}$$

$$u(0,y,t) = 0, \quad u(1,y,t) = 0, \quad 0 < y < 1, \quad 0 < t \tag{16}$$

$$u(x,y,0) = f(x,y), \quad 0 < x < 1, \quad 0 < y < 1 \tag{17}$$

$$\frac{\partial u}{\partial t}(x,y,0) = g(x,y), \quad 0 < x < 1, \quad 0 < y < 1. \tag{18}$$

A typical replacement for the wave equation (14) is constructed using Eq. (1) for the Laplacian and Eq. (13) for the time derivative:

$$\begin{aligned} &\frac{u_i(m+1) - 2u_i(m) + u_i(m-1)}{(\Delta t)^2} \\ &\qquad = \frac{u_N(m) + u_S(m) + u_E(m) + u_W(m) - 4u_i(m)}{(\Delta x)^2}. \end{aligned} \tag{19}$$

As usual we solve for $u_i(m+1)$, using the abbreviation $\rho = \Delta t / \Delta x$. The result is

$$\begin{aligned} u_i(m+1) = \rho^2 \left[u_E(m) + u_W(m) + u_N(m) + u_S(m) \right] \\ +(2 - 4\rho^2) u_i(m) - u_i(m-1). \end{aligned} \tag{20}$$

The stability rules given above still apply. Thus we must choose $\rho^2 \le 1/2$ in order to get a sensible solution.

Let us now be specific. We shall take $\Delta x = \Delta y = 1/4$, $\rho^2 = 1/2$ (that is, $\Delta t = 1/4\sqrt{2}$), and suppose that the initial data from Eqs. (11) and (12) are

$$f(x,y) = \begin{cases} 1 & \text{near } x = \frac{1}{4}, \quad y = \frac{1}{4}, \\ 0 & \text{elsewhere}, \end{cases}$$

$$g(x,y) \equiv 0.$$

The running equation is Eq. (20) which, with $\rho^2 = 1/2$, simplifies to

$$u_i(m+1) = \frac{1}{2} \left[u_E(m) + u_W(m) + u_N(m) + u_S(m) \right] - u_i(m-1). \tag{21}$$

To find the starting equation we solve Eq. (21) with $m = 0$ together with the replacement equation for the initial-velocity condition, Eq. (18). The equations are

$$u_i(1) + u_i(-1) = \frac{1}{2}\left[u_E(0) + u_W(0) + u_N(0) + u_S(0)\right]$$
$$u_i(1) - u_i(-1) = 2\Delta t g_i.$$

Because $g(x, y) = 0$ in this instance, we find

$$u_i(1) = \frac{1}{4}\left[u_E(0) + u_W(0) + u_N(0) + u_S(0)\right]$$

as the starting equation; the right-hand side contains known values of u only. In Fig. 8 are representations of the numerical solution at various time levels.

The simple numerical technique that we have developed can be adapted easily to treat inhomogeneities, boundary conditions involving derivatives of u, or time-varying boundary conditions. Even nonrectangular regions can be handled, provided that they fit neatly on a rectangular grid. Several exercises illustrate these points.

Exercises

In Exercises 1 to 5, set up replacement equations using the given space mesh and the numbering shown in the figure cited. Then find the $u_i(m)$ for a few values of m using the largest stable value of r. Let boundary conditions override the initial condition if there is a disagreement.

1. $\nabla^2 u = \dfrac{\partial u}{\partial t}, \quad 0 < x < 1, \quad 0 < y < 0.75, \quad 0 < t$

$u(0, y, t) = 0, \quad u(1, y, t) = 0, \quad 0 < y < 0.75, \quad 0 < t$

$u(x, 0, t) = 0, \quad u(x, 0.75, t) = 1, \quad 0 < x < 1, \quad 0 < t$

$u(x, y, 0) = 0, \quad 0 < x < 1, \quad 0 < y < 0.75$

$\Delta x = \Delta y = 1/4$ (See Fig. 9a.)

2. $\nabla^2 u = \dfrac{\partial u}{\partial t}$ in $\mathcal{R}$, $0 < t$

$u = 0$ on boundary, $0 < t$

$u = 1$ in $\mathcal{R}$, $t = 0$

The region $\mathcal{R}$ is an inverted T: starting with a rectangle of width 1 and height 3/4, remove a $1/4 \times 1/4$ square from the upper left and right corners. Take $\Delta x = \Delta y = 1/4$. (See Fig. 9b.)

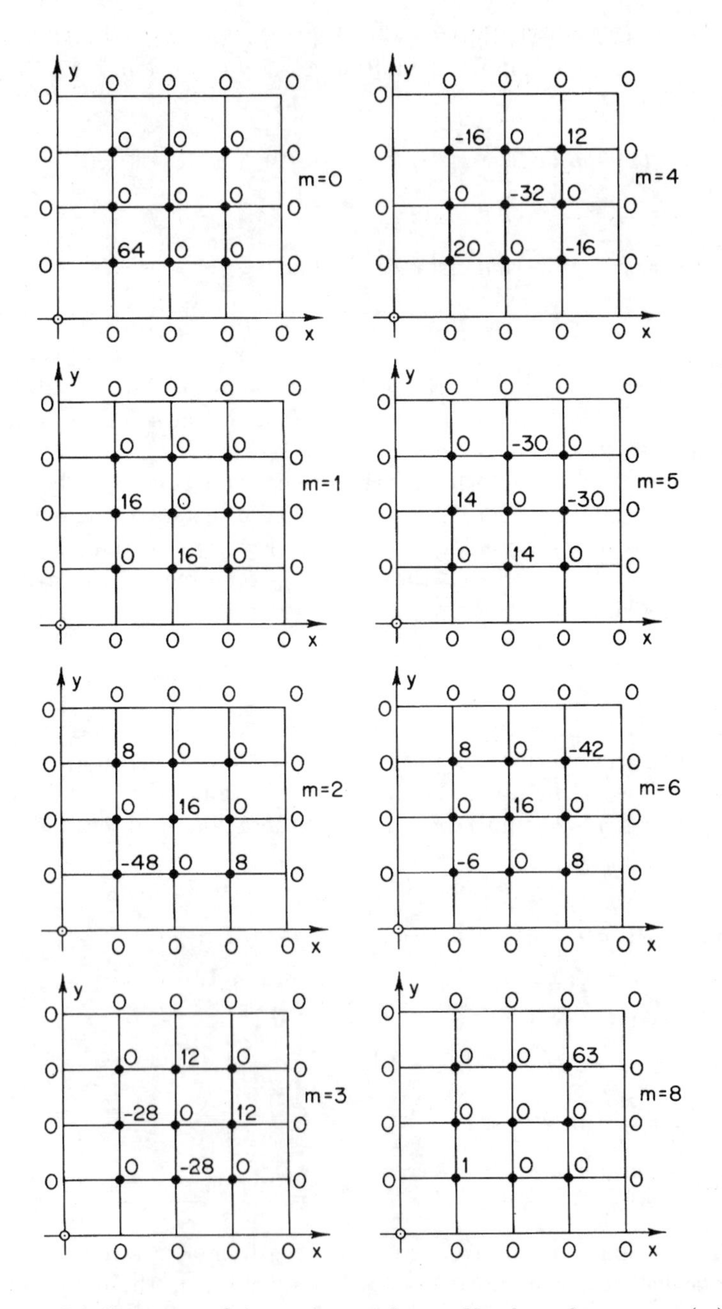

Figure 8: Displacements of the square membrane. Numbers shown are $u_i(m) \times 64$.

3. Same as Exercise 2, except that the region is a cross. (See Fig. 9c.)

4. Same as Eqs. (9) to (11), except that the boundary condition is $u = 1$ on the bottom ($y = 0$) and $u = 0$ elsewhere. (See Fig. 7.)

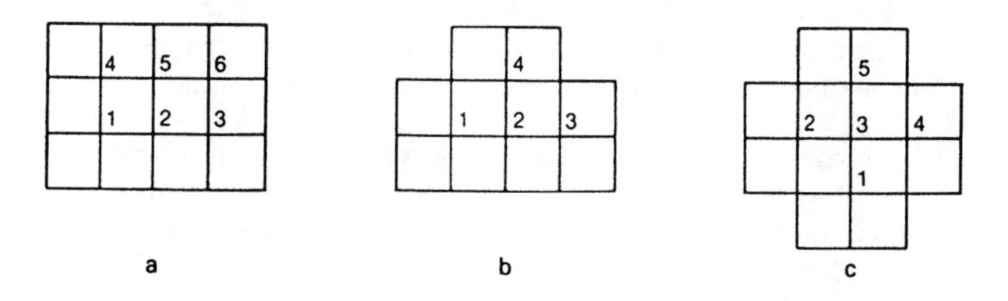

Figure 9: Regions for Exercises 1,2,3.

5. $\nabla^2 u = \dfrac{\partial u}{\partial t}, \quad 0 < x < 1, \quad 0 < y < 1, \quad 0 < t$

$$u(0, y, t) = 0, \quad u(1, y, t) = 1, \quad 0 < y < 1, \quad 0 < t$$
$$u(x, 0, t) = 0, \quad u(x, 1, t) = 1, \quad 0 < x < 1, \quad 0 < t$$
$$u(x, y, 0) = 0, \quad 0 < x < 1, \quad 0 < y < 1$$

$\Delta x = \Delta y = 1/4$ (See Fig. 2.)

6. Find a numerical solution of the heat problem on a 1×1 square with $\Delta x = \Delta y = 1/4$. Initially $u = 0$ and on the outside boundary $u = 0$. There is a tiny hole in the center of the square so that $u(1/2, 1/2, t) = 1$, $t > 0$. (Actually, the region is a punctured square.)

7. Solve numerically Eqs. (14) to (18) with $\Delta x = \Delta y = 1/4$, $\rho^2 = 1/2$. Take $f(x, y) \equiv 0$ and

$$g(x, y) = \begin{cases} 4 & \text{at} \quad \left(\frac{1}{2}, \frac{1}{2}\right) \\ 0 & \text{elsewhere.} \end{cases}$$

Physically, u describes the vibrations of a square membrane struck in the middle.

8. Obtain an approximate solution of Eqs. (14) to (18) with $f(x, y) \equiv 0$ and $g(x, y) = 4\sqrt{2}$. Take $\Delta x = \Delta y = 1/4$ and $\rho^2 = 1/2$.

9. Same as Exercise 8, but $f(x, y) \equiv 1$ and $g(x, y) \equiv 0$ in the square.

10. Obtain an approximate numerical solution to the wave equation on an L-shaped region (a 1×1 square with a $1/4 \times 1/4$ square removed from the upper right corner). Assume initial displacement $= 1$ in the lower right corner, initial velocity equal to 0 and zero displacement on the boundary. Take $\Delta x = \Delta y = 1/4$ and $\rho^2 = 1/2$.

11. Approximate the solution of the wave equation in a semi-infinite strip 3 units wide. Assume $u = 0$ on all boundaries, zero initial velocity, and an initial value for u that is 1 in a corner and 0 elsewhere. Take $\Delta x = \Delta y = 1$ and $\rho^2 = 1/2$.

7.6 COMMENTS AND REFERENCES

Our objective in this chapter has been to survey some elementary numerical methods for problems like those we attacked analytically in earlier chapters. We have only had enough space to touch on the central topics: obtaining replacement equations, solving linear systems of equations by direct and iterative methods, numerical stability, and order of error.

The methods we have introduced are satisfactory for a first introduction and for learning something about partial differential equations, but they are not adequate for any serious problem solving. New techniques for these problems are superior in speed, accuracy and stability but are also more complicated. Of the many texts available, two excellent ones are *Numerical Analysis* by Burden and Faires, for general methods, and *Numerical Solution of Partial Differential Equations* by Smith. (See the Bibliography.)

Almost all numerical methods for linear partial differential equations rely on the symbolism and theory of matrices. Two outstanding texts on matrix theory are *Applied Linear Algebra*, third edition, by Noble and Daniel, and *Matrices* by Barnett.

MISCELLANEOUS EXERCISES

1. Set up and solve replacement equations for this boundary value problem. Use $\Delta x = 1/3$.

$$\frac{d^2u}{dx^2} - \sqrt{24x}\ \ u = 0, \quad 0 < x < 1$$
$$\frac{du}{dx}(0) = 1, \quad u(1) = 1.$$

2. Use the change of variables $x = (r-a)/(b-a)$ and $v(r) = u(x)$ to convert the equation

$$\frac{1}{r}\frac{d}{dr}\left(r\frac{dv}{dr}\right) - q(r)v = f(r), \quad a < r < b$$

to an equation in u on the interval $0 < x < 1$.

3. By means of the transformation mentioned in Exercise 2, a heat problem on an annular ring is converted to

$$\frac{d^2u}{dx^2} + \frac{1}{1+x}\frac{du}{dx} = -(1+x), \quad 0 < x < 1$$
$$u(0) = 1, \qquad u(1) = 0.$$

Set up and solve replacement equations for this problem using $\Delta x = 1/4$.

4. The boundary value problem

$$\frac{1}{r}\frac{d}{dr}\left(r\frac{dv}{dr}\right) - \gamma^2 v = 0, \quad a < r < b$$
$$v(a) = 1, \quad v(b) = 0$$

can be transformed into the problem

$$\frac{d^2u}{dx^2} + \frac{1}{\alpha+x}\frac{du}{dx} - \gamma^2 L^2 u = 0, \quad 0 < x < 1$$
$$u(0) = 1, \quad u(1) = 0$$

where $L = b - a$ and $\alpha = a/L$. Set up and solve replacement equations using $\Delta x = 1/4$, $\alpha = 1$, $\gamma L = 1$.

5. Set up replacement equations for the heat problem in the following and solve for t up to $1/4$, using $\Delta x = 1/4$, $\Delta t = 1/32$

$$\frac{\partial^2 u}{\partial x^2} = \frac{\partial u}{\partial t}, \qquad 0 < x < 1, \quad 0 < t$$
$$u(0,t) = u(1,t) = 1 - e^{-t}, \quad 0 < t$$
$$u(x,0) = 0, \qquad 0 < x < 1$$

6. Same as Exercise 5, but use $u(0,t) = u(1,t) = 1 - e^{-32(\ln 2)t}$ so that $u(0,t_m) = 1 - (0.5)^m$.

7. Compare the numerical solution of the problem

$$\frac{\partial^2 u}{\partial x^2} = \frac{\partial u}{\partial t}, \quad 0 < x < 1, \quad 0 < t$$
$$u(0,t) = 0, \qquad u(1,t) = 0, \quad 0 < t$$
$$u(x,0) = 1, \qquad 0 < x < 1$$

with the solution of the problem consisting of the equation

$$\frac{\partial^2 u}{\partial x^2} - 16u = \frac{\partial u}{\partial t}, \qquad 0 < x < 1, \quad 0 < t$$

with the same initial and boundary conditions. Use $\Delta x = 1/4$, $\Delta t = 1/48$ in both cases.

8. In Exercise 7, what is the longest stable time step for each of the two problems?

9. Solve for several time levels using $\Delta x = 1/5$ and $r = 1/2$. What is Δt?

$$\frac{\partial^2 u}{\partial x^2} = \frac{\partial u}{\partial t}, \quad 0 < x < 1, \quad 0 < t$$
$$u(0,t) = 25t, \quad u(1,t) = 0, \quad 0 < t$$
$$u(x,0) = 0, \qquad 0 < x < 1$$

10. Same as Exercise 9, except that the second boundary condition is $\partial u/\partial x(1,t) = 0$.

11. The problem below describes the displacement of a string whose end is jerked.

$$\frac{\partial^2 u}{\partial x^2} = \frac{\partial^2 u}{\partial t^2}, \quad 0 < x < 1, \quad 0 < t$$
$$u(0,t) = 0, \qquad u(1,t) = 1, \quad 0 < t$$
$$u(x,0) = 0, \qquad \frac{\partial u}{\partial t}(x,0) = 0, \quad 0 < x < 1$$

Solve numerically through one period (until $t = 2$) with $\Delta x = \Delta t = 1/4$.

12. Same problems as Exercise 11, except that the right-hand boundary condition is $u(1,t) = h(t)$, $0 < t$, where

$$h(t) = \begin{cases} 1, & 0 < t \leq 1 \\ 0, & 1 < t \leq 2 \end{cases}$$

and $h(t+2) = h(t)$. Solve numerically with $\Delta x = \Delta t = 1/4$ for enough values of t so that resonance becomes noticeable.

13. Using $\Delta x = \Delta y = 1/4$, find a numerical solution of the problem

$$\frac{\partial^2 u}{\partial x^2} + \frac{\partial^2 u}{\partial y^2} = 0, \quad 0 < x < 1, \qquad 0 < y < 1$$

$$u(x,0) = 0, \quad u(x,1) = \frac{2}{\pi}\tan^{-1}\left(\frac{1}{x}\right), \quad 0 < x < 1$$

$$u(0,y) = 1, \quad u(1,y) = \frac{2}{\pi}\tan^{-1}(y), \qquad 0 < y < 1$$

14. The analytical solution of the problem in Exercise 13 is $u(x,y) = (2/\pi)\tan^{-1}(y/x)$. Compare your numerical results with the exact solution.

15. Using $\Delta x = \Delta y = 1/4$ and $r = 1/4$, find a numerical solution for the problem

$$\nabla^2 u = \frac{\partial u}{\partial t}, \qquad 0 < x < 1, \quad 0 < y < 1, \quad 0 < t$$

$$u(x,0,t) = u(x,1,t) = 0, \quad 0 < x < 1, \quad 0 < t$$

$$u(0,y,t) = u(1,y,t) = 0, \quad 0 < y < 1, \quad 0 < t$$

$$u(x,y,0) = 1, \qquad 0 < x < 1, \quad 0 < y < 1$$

16. The analytical solution of the problem in Exercise 15 is

$$u(x,y,t) = \sum_{n=1}^{\infty}\sum_{n=1}^{\infty} \frac{4(1-\cos(n\pi))(1-\cos(m\pi))}{\pi^2 mn} \sin(n\pi x)\sin(n\pi y)e^{-(m^2+n^2)\pi^2 t}.$$

Using just the term $m = n = 1$ of this solution, compare the ratio

$$R = \frac{u\left(\frac{1}{2}, \frac{1}{2}, t_{m+1}\right)}{u\left(\frac{1}{2}, \frac{1}{2}, t_m\right)}$$

and the ratio of the corresponding u's computed in Exercise 15.

Appendix: Using a Spreadsheet for Numerical Methods

Several of the numerical methods discussed in this chapter can be implemented easily using an electronic spreadsheet program. The explanations below were based on QuattroPro 6 for Windows. Modifications for other spreadsheets, if necessary at all, will be straightforward.

Only a few of the spreadsheet program's facilities are needed. We refer to them in the text with a capital letter.

Fill. Fills a specified block of cells with a sequence of numbers. The user specifies the starting number and the step.

Copy. Copies a formula (or value) from a specified cell to other cells.

Recalculate. Calculates all the formulas in the spreadsheet. Options may include Mode (Automatic/Manual), Number of iterations, Order (Natural/ Column-wise/Row-wise) and others. Not all spreadsheet programs have all options. In Manual mode, recalculation takes place when you press the Recalculation key (typically F9).

If you are not familiar with these facilities, use the Help screens to find them and learn about their properties.

1. Boundary value problems.

Suggested Layout

Use row 1 to register values of the independent variable x.
Use row 2 for values of and formulas for the approximate solution.
Use row 3 to store any constants necessary, such as Δx or n.
Use column A for labels.

Preparation

Set the Recalculation Mode to Manual, and Iterations to 25.

Setting up the Equations

- Fill a string of cells in row 1 with values of x_i for $i = 0, \cdots, n$.
- Write out the generic replacement equation at x_i, and solve it (on paper) algebraically for u_i.
- Insert the equivalent formula, using cell addresses, into a cell of row 2.
- Copy that formula into the cells of row 2 that correspond to the unknowns $u_2, \cdots, u_{n-1}$.
- Insert appropriate boundary conditions:

For conditions specifying values, such as $u(0) = T$ or $u(1) = U$, where T and U are numbers, insert the numbers T or U into the appropriate cells.
For conditions involving the derivative at 0 or 1, the replacement of the boundary condition becomes the equation for a cell outside the x-range.

Example 1. Find an approximate solution, with $n = 5$, of the problem (see Eqs. (1) to (7) in Section 7.1)

$$u'' - 12xu = -1, \qquad 0 < x < 1,$$

$$u(0) = 1, \qquad u(1) = -1.$$

First, follow the instructions for Preparation, setting the Recalculation Mode to Manual and Iterations to 25.

We are solving with $n = 5$, $\Delta x = 0.2$. Therefore, in row 1, fill cells *B1* to *G1* with the values 0, 0.2, 0.4, 0.6, 0,8, 1.0. The easiest way to do this is to use the Fill command with initial value 0 and step 0.2 in cells *B1..G1*. The spreadsheet looks like this:

	A	*B*	*C*	*D*	*E*	*F*	*G*	*H*	*I*
1		0	0.2	0.4	0.6	0.8	1.0		
2									
3									

The generic replacement equation is given in Eq. (3) of Section 7.1. With $\Delta x = 0.2$, it becomes

$$25\,(u_{i+1} - 2u_i + u_{i-1}) - 12x_i u_i = -1 \tag{1}$$

for $i = 1, \cdots, n-1$. When this equation is solved algebraically for u_i we find

$$u_i = (1 + 25(u_{i+1} + u_{i-1}))/(50 + 12x_i). \tag{2}$$

Now, we record appropriate translations of this equation in the spreadsheet. Cells $B2..G2$ are going to carry values $u_0, \cdots, u_5$, so cell $C2$ carries the value u_1 (just below cell $C1$, which carries x_1). Equation (2), with $i = 1$, is $u_1 = (1 + 25(u_2 + u_0))/(50 + 12x_1)$. Cells $D2$ and $B2$ carry $u_{1+1} = u_2$ and $u_{1-1} = u_0$, respectively. Thus, we record this formula in cell $C2$:

$$C2 : (1 + 25 * (D2 + B2))/(50 + 12 * C1).$$

Now Copy the formula from $C2$ to the block $C2..F2$. Because the spreadsheet uses relative addresses, it interprets $B2$ as "the contents of the cell to the left" and $C1$ as "the contents of the cell above," so it puts the correct formula into each cell. For instance, cell $D2$ acquires the formula

$$D2 : (1 + 25 * (E2 + C2))/(50 + 12 * D1).$$

Next, insert the boundary conditions: set cell $B2$ to value 1 and cell $G2$ to value -1.

The last step is to perform the calculations. Press the Recalculation key. The numbers in the cells can be seen changing while the 25 iterations are being calculated. Press the Recalculation key again. If you see any changes, press it again. When no more changes occur, the numbers in row 2 satisfy the replacement equations. The spreadsheet looks like this:

	A	*B*	*C*	*D*	*E*	*F*	*G*	*H*	*I*
1		0	0.2	0.4	0.6	0.8	1.0		
2		1	0.634	0.290	-0.039	-0.419	-1		
3									

The numbers in row 2 can be used with the graphing facility to produce a graph of the approximate solution.

Example 2. Solve the problem consisting of the same differential equation and the same boundary condition at $x = 0$, but with the boundary condition $u'(1) = -1$.

The replacement equation for the differential equation remains the same as the foregoing, but now it applies for $i = n$ as well. The second boundary condition is replaced by

$$(u_{n+1} - u_{n-1})/2\Delta x = -1. \tag{3}$$

Simplifying and using $n = 5$ and $\Delta x = 0.2$, we obtain the condition $u_6 = u_4 - 0.4$.

Now, in the spreadsheet, the replacement equation that applies to cells $C2..F2$ must be copied into cell $G2$ as well. That equation will contain a reference to the contents of cell $H2$, which corresponds to u_{n+1} (that is, u_6, since $n = 5$). The boundary condition requires that $u_6 = u_4 - 0.4$. Thus the formula for cell $H2$ is

$$H2 : +F2 - 0.4.$$

(Remember that the spreadsheet program interprets an entry beginning with a letter as being a label. The leading plus sign tells the program that $F2$ means a cell address.)

We are ready to calculate the solution of the replacement equations. Simply press the Recalculation key several times, until no changes are seen in the numbers in the cells. The solution is shown in the following. Of course, we are interested only in the values in cells $B2..G2$.

	A	B	C	D	E	F	G	H	I
1		0	0.2	0.4	0.6	0.8	1.0		
2		1	0.720	0.468	-0.267	0.103	-0.062	-0.297	
3									

2. One-dimensional heat problems.

Suggested Layout

Use row 1 to register values of the independent variable x.
Use row 2 for the initial values, $u(x_i, 0)$.
Use column A for values of the independent variable t.

Preparation

Set the Recalculation Order to Row-wise.

Setting up the Equations

- Fill a string of cells in row 1 with values of x_i.
- Fill a string of cells in column A with the values of t_m.
- Apply appropriate boundary conditions.

For conditions specifying values, such as $u(0,t) = T(t)$ or $u(1,t) = U(t)$, where U, T are given functions of t, fill the column under $x = 0$ or $x = 1$ with the appropriate formula. Because values of t appear in column A, the formulas will be easy to write.

For conditions involving the derivative at 0 or 1, find the replacement for the boundary condition. Solve algebraically for $u_{-1}(m)$ or $u_{n+1}(m)$ and use as the replacement equation for the corresponding cell.

- Apply the initial condition $u(x, 0) = f(x)$, by filling the first row of u's with the formula for $f(x)$, appropriately translated to spreadsheet formulas.
- Write out the generic replacement equation and solve it (on paper) for $u_i(m + 1)$.
- Insert the equivalent formula, using cell addresses, into a cell of row 3.
- Copy the formula to the cells corresponding to unknown values of $u_1(m)$.

Example 1. Find a numerical solution of the heat problem

$$\frac{\partial^2 u}{\partial x^2} = \frac{\partial u}{\partial t}, \quad 0 < x < 1, \qquad 0 < t \tag{1}$$

$$u(0, t) = 0, \quad u(1, t) = 0, \qquad 0 < t \tag{2}$$

$$u(x, 0) = x, \quad 0 < x < 1 \tag{3}$$

(Eqs. (4) to (7), Section 7.2). Use $\Delta x = 1/4$ and $r = 1/2$.

First, follow the instructions for Preparation. Next, fill cells $C1..G1$ with the x-values 0, 0.25, 0.5, 0.75, 1, and Fill some cells in column A (say, $A2..A30$) with t-values. Because $\Delta t = r(\Delta x)^2$, we must have $\Delta t = 1/32 = 0.03125$. When these steps are completed, the top of the spreadsheet looks like this:

	A	*B*	*C*	*D*	*E*	*F*	*G*	*H*	
1			0	0.25	0.5	0.75	1		
2	0								
3	0.03125								
4	0.0625								
5	0.09375								
6	0.125								

Next, fill in the initial condition, $u(x, 0) = x$, in cells $C2..G2$, and the boundary conditions, $u(0, t) = 0$, $u(1, t) = 0$ in cells $C3..C30$ and $G3..G30$. Note that we have made the initial condition override the boundary condition. Now the spreadsheet looks like this:

	A	B	C	D	E	F	G	H	I
1			0	0.25	0.5	0.75	1		
2	0		0	0.25	0.5	0.75	1		
3	0.03125		0				0		
4	0.0625		0				0		
5	0.09375		0				0		
6	0.125		0				0		

The generic replacement equation for the simple heat equation is Eq. (8), Section 7.2. With $r = 0.5$, it becomes

$$u_i(m+1) = 0.5(u_{i-1}(m) + u_{i+1}(m)).$$

In words, the value of u in any cell is the average of the values of u to the left and right in the row above it. Thus, we put in cell $D3$ the formula

$$D3 : 0.5 * (C2 + E2) \tag{5}$$

and copy this into cells $D3..F30$. The calculations are carried out automatically, and the spreadsheet is filled with approximations to the solution of the original problem. The first few lines are the same as Table 4.

	A	B	C	D	E	F	G	H	I
1			0	0.25	0.5	0.75	1		
2	0		0	0.25	0.5	0.75	1		
3	0.03125		0	0.25	0.5	0.25	0		
4	0.0625		0	0.25	0.25	0.25	0		
5	0.09375		0	0.125	0.25	0.125	0		
6	0.125		0	0.125	0.125	0.125	0		

Example 2. Find a numerical solution of the heat problem

$$\frac{\partial^2 u}{\partial x^2} = \frac{\partial u}{\partial t}, \quad 0 < x < 1, \qquad 0 < t \tag{6}$$

$$u(0,t) = 100t, \quad \frac{\partial u}{\partial x}(1,t) = 0, \qquad 0 < t \tag{7}$$

$$u(x,0) = 0, \quad 0 < x < 1. \tag{8}$$

Let us choose $\Delta x = 1/4$ and $r = 1/2$. We follow the same procedure as in the first example for filling in x's in the first row and t's in the first column. In row 2, we record the initial values of 0. In column C, record the left boundary condition $u(0, t_m) = 100t_m$. This can be done in several ways, but the one closest to the spirit of the problem is to treat t as a function of t (which it certainly is) by inserting in cell $C2$ the formula $C2 : +100 * A2$ and Copying to the block $C2..C30$. The top of the spreadsheet now looks like this:

	A	B	C	D	E	F	G	H	I
1			0	0.25	0.5	0.75	1		
2	0		0	0	0	0	0		
3	0.03125		3.125						
4	0.0625		6.25						
5	0.09375		9.375						
6	0.125		12.5						
7	0.15625		15.625						

The right-hand boundary condition requires zero gradient at $x = 1$. The replacement equation for this condition is

$$(u_{n+1}(m) - u_{n-1}(m))/2\Delta x = 0, \tag{9}$$

or $u_5(m) = u_3(m)$, because $n = 4$ here. Therefore, in cell $H2$ we place the formula $H2 : +F2$, and Copy it down column H.

Finally, the replacement equation is the same as in previous cases,

$$u_i(m + 1) = 0.5(u_{i-1}(m) + u_{i+1}(m)). \tag{10}$$

The equivalent of this equation is inserted into, say, cell $D3$ (see Eq. (5)) and Copied to all relevant cells in columns D through G. Numbers appear immediately throughout the block containing the formulas. The resulting spreadsheet looks like this (only two decimals are shown).

	A	B	C	D	E	F	G	H	I
1			0	0.25	0.5	0.75	1		
2	0		0	0	0	0	0	0	
3	0.03125		3.125	0	0	0	0	0	
4	0.0625		6.25	1.56	0	0	0	0	
5	0.09375		9.375	3.13	0.78	0	0	0	
6	0.125		12.5	5.08	1.56	0.39	0	0.39	
7	0.15625		15.625	7.03	2.73	0.78	0.39	0.78	

3. One-dimensional wave problems.

Suggested Layout

Use row 1 to register values of the independent variable x.
Use rows 2 and 3 for the initial values, $u(x_i, 0)$, $u(x_i, t_1)$.
Use column A for values of the independent variable t.

Preparation

Set the Recalculation Order to Row-wise.

Setting up the Equations

- Fill a string of cells in row 1 with values of x_i.
- Fill a string of cells in column A with the values of t_m.
- Fill row 2 with the initial condition, $u_i(0) = f(x_i)$.
- Fill row 3 using the starting equation (see Eq. (7), Section 7.3).
- Insert appropriate boundary conditions.

For conditions specifying boundary values, such as $u(0,t) = T(t)$ or $u(1,t) = U(t)$, fill the column under $x = 0$ or $x = 1$ with the appropriate formula. Because values of t appear in column A, the formulas will be easy to write.
For conditions involving the derivative at 0 or 1, find the replacement for the boundary condition. Solve algebraically for $u_{-1}(m)$ or $u_{n+1}(m)$ and use as the replacement equation for the corresponding cell.

- Write out the generic running equation. Insert the equivalent formula, using cell addresses, into a cell of row 4. Copy the formula to the cells corresponding to unknown values of $u_i(m)$.

Example 1. Find a numerical solution of this wave problem

$$\frac{\partial^2 u}{\partial x^2} = \frac{\partial^2 u}{\partial t^2}, \qquad 0 < x < 1, \qquad 0 < t \tag{1}$$

$$u(0,t) = 0, \qquad u(1,t) = 0, \qquad 0 < t, \tag{2}$$

$$u(x,0) = 1 - |2x - 1|, \qquad \frac{\partial u(x,0)}{\partial x} = 0, \qquad 0 < x < 1. \tag{3}$$

(See Eqs. (1) to (3), Section 7.3.) Use $\Delta x = \Delta t = 1/4$.

First, follow the instructions for Preparation. Next, Fill cells $C1..G1$ with the x-values 0, 0.25, 0.5, 0.75, 1, and Fill cells $A2..A30$ with t-values. When these steps are completed, the top of the spreadsheet looks like this:

	A	B	C	D	E	F	G	H	I
1			0	0.25	0.5	0.75	1		
2	0								
3	0.25								
4	0.5								
5	0.75								

Next, fill in the initial condition in cells $C2..G2$, and the boundary conditions, $u(0,t) = 0$, $u(1,t) = 0$ in cells $C3..C30$ and $G3..G30$. Note that there is no conflict between initial and boundary conditions. Now the spreadsheet looks like this:

	A	B	C	D	E	F	G	H	I
1			0	0.25	0.5	0.75	1		
2	0		0	0.5	1	0.5	0		
3	0.25		0				0		
4	0.5		0				0		
5	0.75		0				0		

The starting equation, Eq. (7) of Section 7.3, with $g(x) = 0$, and $\rho = 1$ becomes

$$u_i(1) = 0.5(f(x_{i-1}) + f(x_{i+1})). \tag{4}$$

We can put the equivalent equation into cell $D2$,

$$D2 : 0.5 * (C1 + E1) \tag{5}$$

and Copy it to cells $D2..F2$. Calculation takes place immediately, and the top of the spreadsheet is

	A	*B*	*C*	*D*	*E*	*F*	*G*	*H*	*I*
1			0	0.25	0.5	0.75	1		
2	0		0	0.5	1	0.5	0		
3	0.25		0	0.5	0.5	0.5	0		
4	0.5		0				0		
5	0.75		0				0		

The general running equation for $\rho = 1$ is given by Eq. (9), Section 7.3,

$$u_i(m+1) = u_{i-1}(m) + u_{i+1}(m) - u_i(m-1). \tag{6}$$

Thus, we put in cell $D4$ the formula

$$D4 : +C3 + E3 - D2 \tag{7}$$

and Copy this into cells $D4..F30$. The calculations are carried out automatically, and the spreadsheet is filled with approximations to the solution of the original problem. The first few lines are the same as Table 6.

	A	*B*	*C*	*D*	*E*	*F*	*G*	*H*	*I*
1			0	0.25	0.5	0.75	1		
2	0		0	0.5	1	0.5	0		
3	0.25		0	0.5	0.5	0.5	0		
4	0.5		0	0	0	0	0		
5	0.75		0	-0.5	-0.5	-0.5	0		
6	1.0		0	-0.5	-1	-0.5	0		
7	1.25		0	-0.5	-0.5	-0.5	0		

Example 2. Obtain a numerical solution of the following problem for t up to 4 using $\Delta x = \Delta t = 1/5$.

$$\frac{\partial^2 u}{\partial x^2} = \frac{\partial^2 u}{\partial t^2}, \quad 0 < x < 1, \quad 0 < t \tag{8}$$

$$u(0,t) = \sin(\pi t), \quad u(1,t) = 0, \quad 0 < t \tag{9}$$

$$u(x,0) = 0, \quad \frac{\partial u(x,0)}{\partial t} = 0, \quad 0 < x < 1. \tag{10}$$

As Δx and Δt are both 0.2, we must construct anew the top row and left column of the spreadsheet. The boundary condition on the right, $u(1,t) = 0$, is satisfied by putting 0's into column H. To satisfy the boundary condition on the left, we insert this formula into cell $C2$,

$$C2 : @\sin(@PI * A2), \tag{11}$$

and Copy it into cells $C2..C22$. The sign @ tells the spreadsheet program that a function name follows. The same convention (or a similar one) applies to the constant π.

Because both initial conditions are 0, the starting equation provides zero values for time t_1. Thus, cells $D2..G2$ and $D3..G3$ are filled with 0's. The top of the spreadsheet is

	A	*B*	*C*	*D*	*E*	*F*	*G*	*H*	*I*
1			0	0.2	0.4	0.6	0.8	1	
2	0		0	0	0	0	0	0	
3	0.2		0.588	0	0	0	0	0	
4	0.4		0.951					0	
5	0.6		0.951					0	
6	0.8		0.588					0	
7	1		0					0	
8	1.2		-0.588					0	

Finally, the block $D3..G22$ must be filled with the running equation. Because $\Delta x = \Delta t$, the running equation is the same as in Example 1. (See Eq. (7).) When the block receives the correct formulas, calculation takes place immediately, and the first few rows of the spreadsheet are:

	A	B	C	D	E	F	G	H	I
1			0	0.2	0.4	0.6	0.8	1	
2	0		0	0	0	0	0	0	
3	0.2		0.588	0	0	0	0	0	
4	0.4		0.951	0.588	0	0	0	0	
5	0.6		0.951	0.951	0.588	0	0	0	
6	0.8		0.588	0.951	0.951	0.588	0	0	
7	1		0	0.588	0.951	0.951	0.588	0	
8	1.2		-0.588	0	0.588	0.951	0.951	0	
9	1.4		-951	-0.588	0	0.588	0.363	0	
10	1.6		-.951	-.951	-0.588	0	-0.363	0	

Inspection of a longer piece of the spreadsheet clearly shows that the function u grows steadily in amplitude. Figure A.1 shows calculated values of $u(0.4, t)$ for $0 \leq t \leq 6$.

4. Two-dimensional potential problems.

Solving the potential equation and other related equations in two-dimensional regions is one of the most satisfying applications of spreadsheets to numerical methods. The problem can be set up in a simple natural way, and the numerical solution is displayed within the region where the problem is to be solved.

For simplicity, we start with problems on a rectangle. Then it will be easy to modify our procedures for nonrectangular regions.

Suggested layout

Use row 1 and column 1 to register values of the coordinates x and y.

Use a rectangular region for values of and formulas for the approximate solution.

Preparation

Set the Default Recalculation Mode to Manual and Iterations to 25.

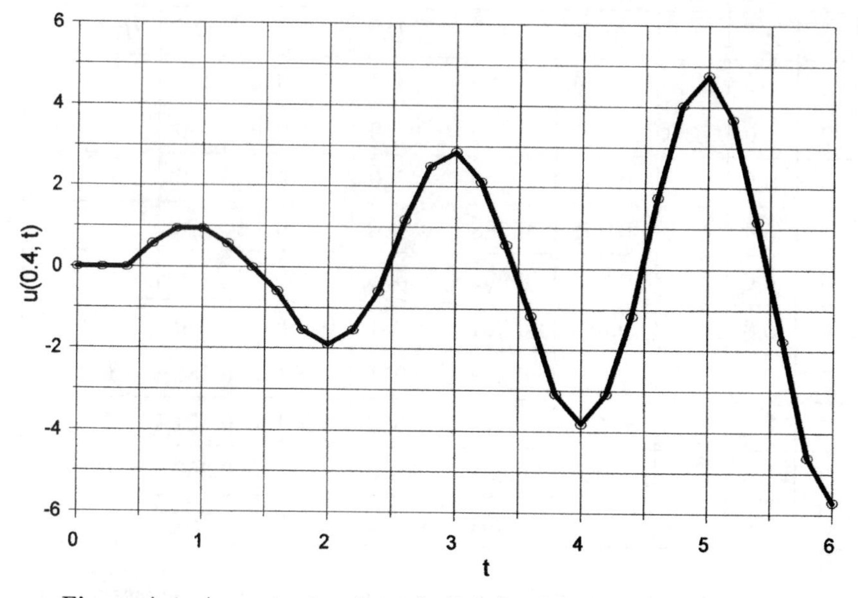

Figure A.1: Approximate values of $u(0.4, t)$, solution of Eqs. (8) to (10).

Setting up

- Fill strings of cells in row 1 and column 1 with values of the independent variables x_i and y_j. These define a rectangular block that will contain the approximate solution.

- Write out the generic replacement equation and solve it algebraically (on paper) for the unknown $u_{i,j}$.

- Insert the equivalent formula, using cell addresses, into a cell of the region and Copy it into the rest of the cells of the region.

- Fill in boundary conditions:

  For Dirichlet conditions (values of u given at boundary points) simply fill the appropriate cells with the given values.
  For conditions involving the normal derivative at boundary points, the replacement of the boundary condition becomes the equation for a cell outside the region.

Example 1. Solve this potential equation in the square numerically, using $\Delta x = \Delta y = 1/4$. (See Eqs. (3) to (6), Section 7.4.)

$$\frac{\partial^2 u}{\partial x^2} + \frac{\partial^2 u}{\partial x^2} = 0, \qquad 0 < x < 1, \qquad 0 < y < 1 \tag{1}$$

$$u(0, y) = 0, \qquad u(1, y) = 0, \qquad 0 < y < 1 \tag{2}$$

$$u(x, 0) = f(x), \quad u(x, 1) = f(x), \qquad 0 < x < 1 \tag{3}$$

where $f(x) = 1 - |2x - 1|$. (This function's graph is an isosceles triangle.)

First, follow the instructions for Preparation. Next, place values of the independent variables into row 1 and column 1. It is convenient to leave some empty space around the block that will contain the solution. Thus let us specify cell $C3$ (you might prefer $D4$) as the upper left corner of the region. Thus, we put the x-values into row 1 starting with cell $C1$ and y-values into column 1, starting with cell $A3$. Because $\Delta x = \Delta y = 1/4$, the spreadsheet now looks like this:

	A	*B*	*C*	*D*	*E*	*F*	*G*	*H*	*I*
1			0	0.25	0.5	0.75	1		
2									
3	1								
4	0.75								
5	0.5								
6	0.25								
7	0								

Now we fill in columns C and G with the zero boundary conditions. In rows 3 and 7, we fill in the nonzero boundary conditions. In cell $C3$ place the formula

$$C3 : 1 - @ABS(2 * C\$1 - 1)$$

and copy it to cells $C3..G3$, then to cells $C7..G7$. The @ signals the spreadsheet that a function follows. The form of address $C\$1$ means that the *column* is a relative address, but the *row* is an absolute address. Thus, when this formula is copied, cell $D3$ contains the formula

$$D3 : 1 - @ABS(2 * D\$1 - 1)$$

and cell $D7$ contains exactly the same formula. Computation of values takes place immediately and the spreadsheet looks like this.

	A	B	C	D	E	F	G	H	I
1			0	0.25	0.5	0.75	1		
2									
3	1		0	0.5	1	0.5	0		
4	0.75		0				0		
5	0.5		0				0		
6	0.25		0				0		
7	0		0	0.5	1	0.5	0		

Finally, the generic replacement for the potential equation is

$$u_N + u_S + u_E + u_W - 4u_i = 0 \tag{4}$$

(See Eq. (7), Section 7.4.) Solved algebraically for u_i it becomes

$$u_i = (u_N + u_S + u_E + u_W)/4. \tag{5}$$

The equation must be translated into an equivalent, using cell addresses. We insert into cell $D4$ the formula

$$D4 : 0.25 * (D3 + D5 + C4 + E4)$$

and copy this into the entire block $D4..F6$. Numbers appear immediately. Press the Recalculation key. In this case, the correct approximate solution appears after one cycle of recalculations. The spreadsheet looks like this (compare Fig. 5):

	A	B	C	D	E	F	G
1			0	0.25	0.5	0.75	1
2							
3	1		0	0.5	1	0.5	0
4	0.75		0	0.3125	0.5	0.3125	0
5	0.5		0	0.25	0.375	0.25	0
6	0.25		0	0.3125	0.5	0.3125	0
7	0		0	0.5	1	0.5	0
8							

Example 2. Solve the problem made from the one in Example 1 by changing the boundary condition at $y = 0$ to

$$\frac{\partial u(x,0)}{\partial y} = 0, \qquad 0 < x < 1.$$

The same setup can be reused with two changes. First, the replacement for the potential equation must be applied to cells $D7..F7$, which comprise the bottom boundary of the region. Second, we must find and apply the replacement of the boundary condition. If point number i is a point on the bottom boundary, then the boundary condition there is

$$(u_N - u_S)/2\Delta x = 0, \quad or \quad u_N = u_S \tag{6}$$

Thus, we put into cell $D8$ the formula $D8 : +D6$ and Copy it into cells $D8..F8$. (Recall that a formula must not begin with a letter, so we use the leading plus sign to indicate that the letter D is a cell address.)

Numbers appear (or change) immediately in the rectangle of unknowns. After a few Recalculation cycles, these numbers stabilize to the correct approximate solution, and the spreadsheet looks like this (only three decimals are shown here):

	A	*B*	*C*	*D*	*E*	*F*	*G*	*H*	*I*
1			0	0.25	0.5	0.75	1		
2									
3	1		0	0.5	1	0.5	0		
4	0.75		0	0.268	0.436	0.268	0		
5	0.5		0	0.138	0.206	0.138	0		
6	0.25		0	0.077	0.112	0.077	0		
7	0		0	0.060	0.086	0.060	0		
8				0.077	0.112	0.077			

The numbers in cells $D8..F8$ have no interpretation in terms of the original problem.

Example 3. Torsion stresses in a prismatic beam can be found from the solution of Poisson's equation, $\nabla^2 u = -H$, on a region that is the cross section of the beam. The boundary condition is $u = 0$ on the boundary. (The region must be simply connected. Stresses are partial derivatives of u.) Appropriate scaling allows us to replace the constant H by 1 or by any other convenient constant. To find torsion stresses in a beam whose cross section is a particular L-shaped region $\mathcal{R}$, we would solve this problem

$$\frac{\partial^2 u}{\partial x^2} + \frac{\partial^2 u}{\partial x^2} = -H \text{ in } \mathcal{R} \tag{7}$$

$$u(x, y) = 0 \text{ for points } (x, y) \text{ on the boundary.} \tag{8}$$

Assume that the L-shaped region is a 1×1 square from which a $1/2 \times 1/2$ corner has been removed. Solve numerically with $\Delta x = \Delta y = 1/10$.

It will be convenient for us to assume that the constant H is 100. It turns out that the values of u will lie between 0 and 10. Thus the display will not contain many leading 0's. The replacement for the partial differential equation is

$$(u_N + u_S + u_E + u_W - 4u_i)/(\Delta x)^2 = -100. \tag{9}$$

Solved for u_i, this equation becomes

$$\begin{aligned} u_i &= (100(\Delta x)^2 + u_N + u_S + u_E + u_W)/4, \quad \text{or} \\ u_i &= (1 + u_N + u_S + u_E + u_W)/4. \end{aligned} \tag{10}$$

Here, we have used the fact that $\Delta x = 1/10$ and $100(\Delta x)^2 = 1$.

In the spreadsheet, set up values of x and y in row 1 and column 1. We will assume that the removed quarter of the region is in the lower right corner.

The jobs to be done are: set the boundary conditions and fill internal cells with the equivalent of Eq. (10). Of course there are many ways to accomplish these, but the shape of the region suggests these steps.

(1) Fill the block $C3..M13$ with 0's.

(2) Put the equivalent of Eq. (10) into cell $D4$, the upper left cell of the interior of the L-shaped region. The formula is

$$D4 : 0.25 * (1 + D3 + D5 + C4 + E4).$$

(3) Copy this formula into the block $D4..L12$. At this point, we are ready to find the numerical solution of the Poisson equation in a square.

(4) Fill the "removed corner," $H8..M13$, with 0's.

Now the spreadsheet is ready for the computation. After a few Recalculation cycles, the spreadsheet contains the values of the approximate solution, as shown below. The graph in Fig. A.2 was made from this solution.

	A	B	C	D	E	F	G	H	I	J	K	L	M
1			0	0.1	0.2	0.3	0.4	0.5	0.6	0.7	0.8	0.9	1
2													
3	0		0	0	0	0	0	0	0	0	0	0	0
4	0.1		0	1.08	1.66	1.94	2.03	2.00	1.90	1.74	1.46	0.96	0
5	0.2		0	1.66	2.62	3.07	3.18	3.07	2.87	2.59	2.14	1.37	0
6	0.3		0	1.94	3.07	3.56	3.54	3.23	2.93	2.61	2.15	1.37	0
7	0.4		0	2.03	3.18	3.54	3.20	2.36	2.02	1.77	1.47	0.96	0
8	0.5		0	2.00	3.07	3.23	2.36	0	0	0	0	0	0
9	0.6		0	1.90	2.87	2.93	2.02	0	0	0	0	0	0
10	0.7		0	1.74	2.59	2.61	1.77	0	0	0	0	0	0
11	0.8		0	1.46	2.14	2.15	1.47	0	0	0	0	0	0
12	0.9		0	0.96	1.37	1.37	0.96	0	0	0	0	0	0
13	1		0	0	0	0	0	0	0	0	0	0	0
14													
15													

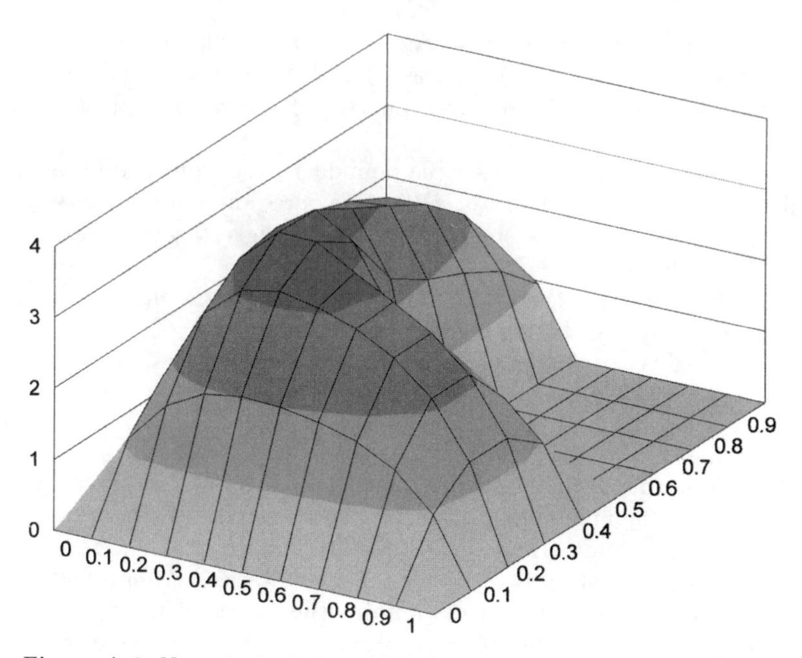

Figure A.2: Numerical solution of Poisson equation on an L-shaped region.

Appendix Exercises

1. A current-carrying wire is exposed in its middle section to convection but is encased by ceramic insulation near its ends. (See Fig. A.3.) The temperature $u(x)$ in the wire is governed by different differential equations in different parts of the wire:

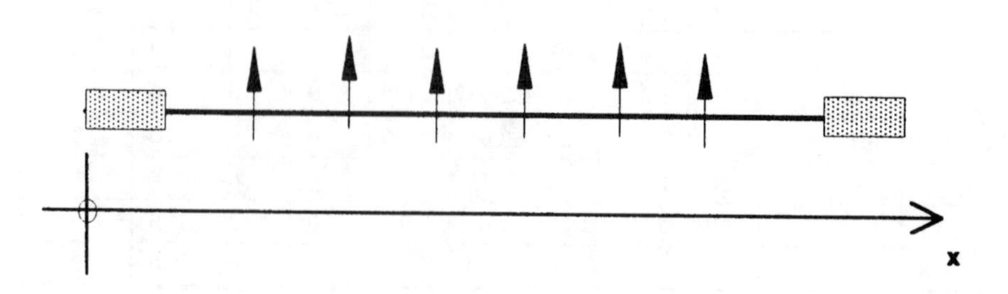

Figure A.3: Current-carrying wire exposed to convection in the middle, with ceramic insulators at the ends.

$$\frac{d^2u}{dx^2} = -H, \qquad 0 < x < aL,$$
$$\frac{d^2u}{dx^2} = -H - b^2(T-u), \quad aL < x < (1-a)L$$
$$\frac{d^2u}{dx^2} = -H, \qquad (1-a)L < x < L.$$

Note that u'' jumps in value at aL and $(1-a)L$, but u and u' are continuous there.

In these equations, H is a constant that represents resistance heating, b^2 contains a heat transfer coefficient and geometric parameters, and a is the fraction of the wire that is covered at each end by the ceramic insulation. For boundary conditions, we will assume that the temperature at the endpoints of the wire is the same as that of the surrounding air,

$$u(0) = T, \qquad u(L) = T.$$

Because of the obvious symmetry with respect to the midpoint $x = 1/2L$, we can simplify the problem to

$$\frac{d^2u}{dx^2} = -H, \qquad 0 < x < aL,$$
$$\frac{d^2u}{dx^2} = -H - b^2(T-u), \quad aL < x < \tfrac{1}{2}L$$
$$u(0) = T, \qquad \frac{du(\frac{1}{2}L)}{dx} = 0.$$

Now, remove the dimensions by defining $X = x/L$, $K = b^2L^2$, $U = (u-T)(b^2/H)$, and the problem becomes

$$\frac{d^2U}{dX^2} = -K, \qquad 0 < X \le a,$$
$$\frac{d^2U}{dX^2} = -K(1-U), \qquad a < X < 0.5$$
$$U(0) = 0, \qquad \frac{dU(0.5)}{dX} = 0.$$

Solve this problem numerically, taking $\Delta X = 1/20$, $a = 0.125$, and $K = 0.25/\Delta X^2$. Graph the solution. Other interesting values for K are $0.5\Delta X^2$ and $0.1\Delta X^2$.

2. In Chapter 0, we derived the following nonlinear problem as a model of the displacement of a hanging cable loaded by its own weight.

$$\begin{aligned} \frac{d^2u}{dx^2} &= (w/T)\sqrt{1+\left(\frac{du}{dx}\right)^2}, \qquad 0 < x < a \\ u(0) &= h_0, \qquad u(a) = h_1. \end{aligned}$$

Remove the dimensions by changing variables to $X = x/a$, $U = (u - h_0)/a$. Then the problem becomes

$$\begin{aligned} \frac{d^2u}{dx^2} &= K\sqrt{1+\left(\frac{du}{dx}\right)^2}, \qquad 0 < x < 1, \\ U(0) &= 0, \qquad U(1) = H, \end{aligned}$$

where $K = wa/T$ and $H = (h_1 - h_0)/a$. Set up the replacement equations for the numerical solution of this problem.

3. Use a spreadsheet program to solve numerically the problem in Exercise 2. Take $n = 10$, $K = 1$ and $H = 0$. Compare the numerical solution to the exact solution found in Exercises 4 and 5 of Chapter 0, Section 3.

4. Problems in cylindrical coordinates often present difficulties even when it is possible to write out the solution in terms of Bessel functions. One such problem describes the steady-state temperature $U(r)$ in an annulus:

$$\frac{1}{r}\frac{d}{dr}\left(r\frac{dU}{dr}\right) - k^2(U-T) = 0, \qquad a < r < b$$

$$U(a) = T_0, \qquad U(b) = T_1.$$

(Compare Chapter 5, Section 5, Exercise 9.)

Introduce dimensionless variables $x = (r-a)/(b-a)$ and $u = (U-T)/M$, where M is the largest of the three quantities $T_0 - T$, $T_1 - T$, $T_1 - T_0$. Then the problem becomes

$$\frac{1}{s+x}\frac{d}{dx}\left((s+x)\frac{du}{dx}\right) - AU = 0, \qquad 0 < x < 1$$

$$u(0) = h_0, \qquad u(1) = h_1$$

where $A = k^2(b-a)^2$, and $s = a/(b-a)$ is a shape parameter.

a. If $M = 0$, what is the solution of the dimensional problem?

b. If $M \neq 0$, find h_0 and h_1 in terms of other parameters, and show that they are both between -1 and 1.

5. Obtain a numerical solution of the problem in Exercise 4 using $\Delta x = 1/10$, $A = 20$, $h_0 = 1$, $h_1 = 0$. Try the two values $s = 0.1$ and $s = 1.0$; sketch annuli for these shape parameters.

6. A problem of diffusion in a material with variable properties may be expressed as

$$D(x)\frac{\partial^2 u}{\partial x^2} = \frac{\partial u}{\partial t}, \quad 0 < x < 1, \quad 0 < t$$
$$u(0,t) = 0, \qquad u(1,t) = 0, \quad 0 < t$$
$$u(x,0) = 100, \quad 0 < x < 1$$

with $D(x) = 1 + \alpha x$. If $\alpha = 0$, this problem can be solved analytically as in Chapter 2, Section 3. An important question is this: if $\alpha \neq 0$, how different in the solution? In particular, how noticeable is the induced asymmetry?

Study these questions by solving numerically with $\alpha = 0, 0.5, 1$. Use $\Delta x = 1/10$, $\Delta t = 1/400$ ($r = 1/4$).

7. Miscellaneous Exercises 18 to 20 of Chapter 3 give equations that model the phenomenon of water hammer. Initially, water is flowing steadily through a pipe from a large reservoir to open air. At time 0, a valve is closed suddenly at the end of the pipe, causing violent changes in pressure and velocity in the water. If the equations are combined, we can obtain this problem for dimensionless pressure:

$$\frac{\partial^2 p}{\partial x^2} = \frac{\partial^2 p}{\partial t^2}, \qquad 0 < x < 1, \quad 0 < t$$
$$p(0,t) = 0, \quad \frac{\partial p}{\partial x}(1,t) = 0, \qquad 0 < t,$$
$$p(x,0) = -x, \quad \frac{\partial p}{\partial t}(x,0) = 0, \quad 0 < x < 1.$$

Solve this problem numerically using $\Delta x = \Delta t = 1/10$ for time 0 to 2.

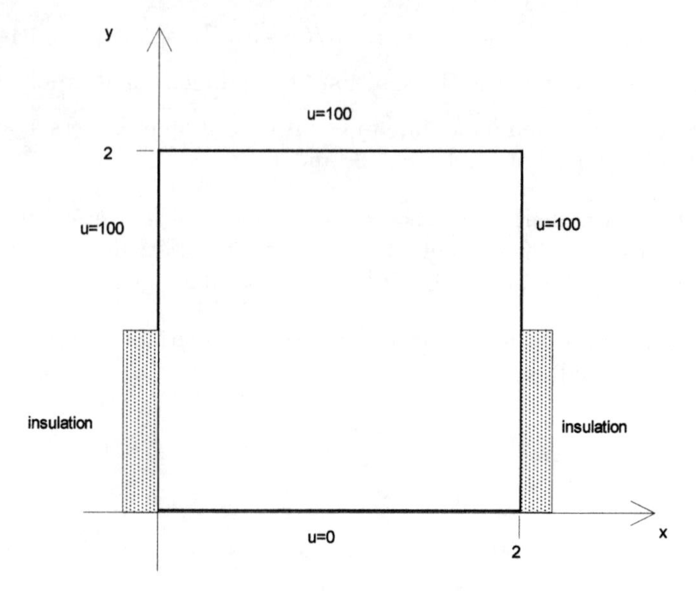

Figure A.4: Partially insulated plate.

8. Find the steady-state temperature in a square plate that is insulated along the lower half of each vertical side and has controlled temperature on the remaining parts of its sides, as shown in Fig. A.4. Because of symmetry, we can solve the problem in just one-half of the plate and use the symmetry condition

$$\frac{\partial u}{\partial x}(1, y) = 0$$

along the center line. The full problem is

$$\begin{aligned}
&\frac{\partial^2 u}{\partial x^2} + \frac{\partial^2 u}{\partial y^2} = 0, && 0 < x < 2, \quad 0 < y < 2, \\
&u(x, 2) = 100, \quad u(x, 0) = 0, && 0 < x < 2, \\
&u(0, y) = 100, \quad u(2, y) = 100, && 1 \leq y < 2, \\
&\frac{\partial u}{\partial x}(0, y) = 0, \quad \frac{\partial u}{\partial x}(2, y) = 0, && 0 < y < 1.
\end{aligned}$$

9. Part of a certain structure is a beam subjected to torsion. The engineer designing the structure must choose one of the two cross sections shown in Fig. A.5. The criterion is to be torsional rigidity,

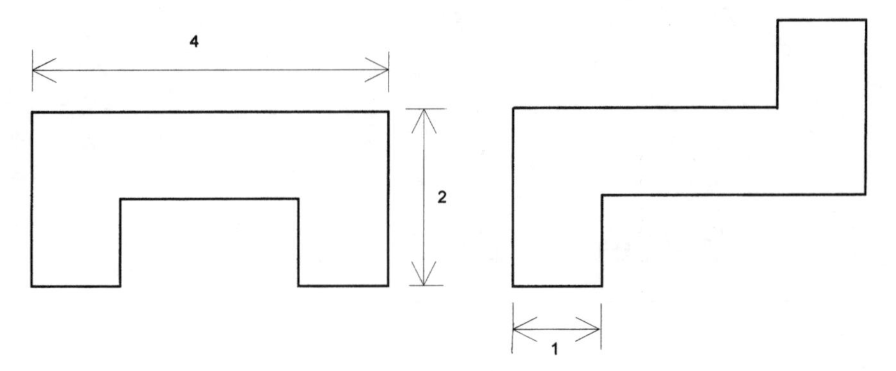

Figure A.5: Beam sections.

which is proportional to the integral over the cross section of the function u that satisfies Poisson's equation $\nabla^2 = -1$ in the cross section and $u = 0$ on the boundary.

For each cross section: (1) solve the Poisson equation numerically using $\Delta x = \Delta y = 1/4$; (2) approximate the required integral as $\Delta x \Delta y$ times the sum of the values of u.

Which section gives greater rigidity?

Bibliography

Abramowitz, M., and I. Stegun (eds). *Handbook of Mathematical Functions*, 10th ed. Washington, D.C., National Bureau of Standards, 1972.

Andrews, L.C. *Special Functions for Engineers and Applied Mathematicians.* New York, MacMillan, 1985.

Barnett, S. *Matrices: Methods and Applications.* New York, Oxford University Press, 1990.

Brigham, E.O. *The Fast Fourier Transform and its Applications.* Englewood Cliffs NJ, Prentice-Hall, 1988.

Burden, R.L., and J.D. Faires. *Numerical Analysis*, 6th ed. Belmont CA, Brooks/Cole, 199.

Carslaw, H.S., and J.C. Jaeger. *Conduction of Heat in Solids*, 2nd ed. New York, Oxford University Press, 1986.

Churchill, S.W. *Viscous Flows.* Boston, MA, Butterworths, 1988.

Churchill, R.V., and J.W. Brown. *Fourier Series and Boundary Value Problems*, 5th ed. New York, McGraw-Hill, 1992.

Churchill, R.V. *Operational Mathematics*, 3rd ed. New York, McGraw-Hill, 1972.

Courant, R., and D. Hilbert, *Methods of Mathematical Physics*, Vol. I. New York, Wiley-Interscience, 1953/1989.

Crank, J. *The Mathematics of Diffusion*, 2nd ed. New York, Oxford University Press, 1980.

Davis, P.J., R. Hersh and E.A. Marchisotto, *The Mathematical Experience.* Cambridge MA, Birkhauser, 1995.

Duff, G.F.D., and D. Naylor. *Differential Equations of Applied Mathematics.* New York, John Wiley & Sons, 1966.

Erdelyi, A., W. Magnus, F. Oberhettinger, and F. Tricomi. *Tables of Integral Transforms,* Vols. 1 and 2. New York, McGraw-Hill, 1954.

Feller, W. *Introduction to Probability Theory and its Applications,* Vol. I, 3rd ed. New York, John Wiley & Sons, 1968.

Isenberg, C. *The Science of Soap Films and Soap Bubbles.* Avon, U.K., Tieto Ltd, 1978. (Reprinted by Dover, 1992.)

Jerri, A.J. *Integral and Discrete Transforms with Applications and Error Analysis.* New York, Marcel Dekker, 1991.

Jones, D.S., and B.D. Sleeman. *Differential Equations and Mathematical Biology.* London/Boston, George Allen & Unwin, 1983.

Kirchhoff, R.J. *Potential Flows: Computer Graphic Solutions.* New York/Basel, Marcel Dekker, 1998.

Lamb, H. *Hydrodynamics,* 6th ed. Cambridge, Cambridge University Press, 1932 (Reprinted by Dover, New York, 1945).

Main, I.G. *Vibrations and Waves in Physics,* 3d ed., Cambridge, Cambridge University Press, 1993.

Morse, P.M., and H. Feshbach. *Methods of Theoretical Physics.* New York, McGraw-Hill, 1953.

Murray, J.D. *Mathematical Biology.* Berlin/New York, Springer, 1993.

Noble, B., and J.W. Daniel. *Applied Linear Algebra,* 3rd ed. Englewood Cliffs, NJ, Prentice-Hall, 1988.

O'Neil, P.V. *Advanced Engineering Mathematics,* 4th ed., Boston MA, Brooks/Cole, 1995.

The Physics of Music, San Francisco, Freeman, 1978.

Protter, M.H., and H.F. Weinberger. *Maximum Principles in Differential Equations.* New York, Springer-Verlag, 1984.

Sagan, H. *Boundary and Eigenvalue Problems in Mathematical Physics.* New York, John Wiley & Sons, 1966.

Smith, G.D. *Numerical Solution of Partial Differential Equations*, 3rd ed. New York, Oxford University Press, 1986.

Street, R.L. *Analysis and Solution of Partial Differential Equations.* Monterey, CA, Brooks/Cole, 1973.

Tolstov, G.P. *Fourier Series.* Englewood Cliffs, NJ, Prentice-Hall, 1962 (Reprinted by Dover, New York, 1976).

Wan, F.Y.M. *Mathematical Models and Their Analysis.* New York, Harper & Row, 1989.

Widder, D.V. *The Heat Equation.* New York, Academic Press, 1975.

Appendix: Mathematical References

Trigonometric Functions

$$\begin{aligned}
\sin(A \pm B) &= \sin(A)\cos(B) \pm \cos(A)\sin(B) \\
\cos(A \pm B) &= \cos(A)\cos(B) \mp \sin(A)\sin(B) \\
\sin(A) + \sin(B) &= 2\sin\left(\frac{A+B}{2}\right)\cos\left(\frac{A-B}{2}\right) \\
\sin(A) - \sin(B) &= 2\cos\left(\frac{A+B}{2}\right)\sin\left(\frac{A-B}{2}\right) \\
\cos(A) + \cos(B) &= 2\cos\left(\frac{A+B}{2}\right)\cos\left(\frac{A-B}{2}\right) \\
\cos(A) - \cos(B) &= 2\sin\left(\frac{A+B}{2}\right)\sin\left(\frac{A-B}{2}\right) \\
\sin(A)\sin(B) &= \frac{1}{2}\left(\cos(A-B) - \cos(A+B)\right) \\
\sin(A)\cos(B) &= \frac{1}{2}\left(\sin(A-B) + \sin(A+B)\right) \\
\cos(A)\cos(B) &= \frac{1}{2}\left(\cos(A-B) + \cos(A+B)\right)
\end{aligned}$$

$$\cos(A) = \frac{1}{2}\left(e^{iA} + e^{-iA}\right), \quad \sin(A) = \tfrac{1}{2i}\left(e^{iA} - e^{-iA}\right)$$

$$\cos^2(A) + \sin^2(A) = 1, \qquad 1 + \tan^2(A) = \sec^2(A)$$

Hyperbolic Functions

$$\cosh(A) = \frac{1}{2}\left(e^{A} + e^{-A}\right), \quad \sinh(A) = \frac{1}{2}\left(e^{A} - e^{-A}\right)$$

$$d\cosh(u) = \sinh(u)du, \quad d\sinh(u) = \cosh(u)du$$

$$\sinh(A \pm B) = \sinh(A)\cosh(B) \pm \cosh(A)\sinh(B)$$

$$\cosh(A \pm B) = \cosh(A)\cosh(B) \pm \sinh(A)\sinh(B)$$

$$\sinh(A) + \sinh(B) = 2\sinh\left(\frac{A+B}{2}\right)\cosh\left(\frac{A-B}{2}\right)$$

$$\sinh(A) - \sinh(B) = 2\cosh\left(\frac{A+B}{2}\right)\sinh\left(\frac{A-B}{2}\right)$$

$$\cosh(A) + \cosh(B) = 2\cosh\left(\frac{A+B}{2}\right)\cosh\left(\frac{A-B}{2}\right)$$

$$\cosh(A) - \cosh(B) = 2\sinh\left(\frac{A+B}{2}\right)\sinh\left(\frac{A-B}{2}\right)$$

$$\sinh(A)\sinh(B) = \frac{1}{2}\left(\cosh(A+B) - \cosh(A-B)\right)$$

$$\sinh(A)\sinh(B) = \frac{1}{2}\left(\cosh(A+B) + \sinh(A-B)\right)$$

$$\cosh(A)\cosh(B) = \frac{1}{2}\left(\sinh(A+B) + \cosh(A-B)\right)$$

$$\cosh^2(A) - \sinh^2(A) = 1, \qquad 1 - \tanh^2(A) = \text{sech}^2(A)$$

Calculus

1. Derivative of a product

$$(uv)' = u'v + uv'$$

$$(uv)'' = u''v + 2u'v' + uv''$$

$$(uv)^{(n)} = u^{(n)}v + \binom{n}{1} u^{(n-1)}v' + \cdots + \binom{n}{n-1} uv^{(n-1)} + uv^{(n)}$$

In this formula, $\binom{n}{k} = \dfrac{n!}{(n-k)!k!}$ is a binomial coefficient.

2. Rules of integration

a. $\int_a^b (c_1 f_1(x) + c_2 f_2(x))\, dx = c_1 \int_a^b f_1(x)dx + c_2 \int_a^b f_2(x)dx$

b. $\int_a^a f(x)dx = 0$

c. $\int_a^b f(x)dx = -\int_b^a f(x)dx$

d. $\int_a^b f(x)dx = \int_a^c f(x)dx + \int_c^b f(x)dx$

3. Derivatives of integrals

a. $\frac{d}{dt}\int_a^b f(x,t)dx = \int_a^b \frac{\partial f}{\partial t}(x,t)dx$

b. $\frac{d}{dt}\int_a^t f(x)dx = f(t)$ (Fundamental theorem of calculus; a is constant)

c. $\frac{d}{dt}\int_{u(t)}^{v(t)} f(x,t)dx = f(v(t),t)v'(t) - f(u(t),t)u'(t) + \int_{u(t)}^{v(t)} \frac{\partial f}{\partial t}(x,t)dx$ (Leibniz's rule)

4. Integration by parts

a. $\int uv'dx = uv - \int vu'dx$

b. $\int uv''dx = v'u - vu' + \int vu''dx$

5. Functions defined by integrals

a. Natural logarithm

$$\ln(x) = \int_1^x \frac{dz}{z}$$

b. Since-integral function

$$\mathrm{Si}(x) = \int_0^x \frac{\sin(z)}{z}dz$$

c. Normal probability distribution function

$$\Phi(x) = \frac{1}{\sqrt{2\pi}}\int_{-\infty}^x e^{-z^2/2}dz$$

d. Error function

$$\mathrm{erf}(x) = \frac{2}{\sqrt{\pi}} \int_0^x e^{-z^2} dz. \text{ Note: } \mathrm{erf}(x) = 2\Phi(\sqrt{2}x) - 1$$

e. Integrated Bessel function

$$IJ(x) = \int_0^x J_0(z) dz$$

Table of Integrals Any letter except x represents a constant. The integration constants have been left off.

1. Rational functions

1.1 $\displaystyle\int \frac{dx}{h + kx} = \frac{1}{k} \ln |h + kx|$

1.2 $\displaystyle\int \frac{dx}{x^2 + a^2} = \frac{1}{a} \tan^{-1}\left(\frac{x}{a}\right)$

1.3 $\displaystyle\int \frac{xdx}{x^2 + a^2} = \frac{1}{2} \ln(x^2 + a^2)$

1.4 $\displaystyle\int \frac{dx}{x^2 - a^2} = \frac{1}{2a} \ln \left|\frac{x - a}{x + a}\right|$

1.5 $\displaystyle\int \frac{xdx}{x^2 - a^2} = \frac{1}{2} \ln |x^2 - a^2|$

2. Radicals

2.1 $\displaystyle\int \frac{dx}{\sqrt{x^2 + a^2}} = \ln\left(x + \sqrt{x^2 + a^2}\right) \text{ or } \sinh^{-1}\left(\frac{x}{a}\right)$

2.2 $\displaystyle\int \frac{xdx}{\sqrt{x^2 + a^2}} = \sqrt{x^2 + a^2}$

2.3 $\displaystyle\int \frac{dx}{\sqrt{x^2 - a^2}} = \ln\left(x + \sqrt{x^2 - a^2}\right) \quad (x > a)$

2.4 $\displaystyle\int \frac{xdx}{\sqrt{x^2 - a^2}} = \sqrt{x^2 - a^2}$

2.5 $\displaystyle\int \frac{xdx}{\sqrt{x^2 - a^2}} = \sin^{-1}\left(\frac{x}{a}\right) \quad |x| < a$

2.6 $\displaystyle\int \frac{xdx}{\sqrt{x^2 - a^2}} = -\sqrt{x^2 - a^2} \quad |x| < a$

3. Exponentials and hyperbolic functions

3.1 $\displaystyle\int e^{kx} dx = \frac{e^{kx}}{k}$

3.2 $\displaystyle\int xe^{kx}dx = \frac{kx-1}{k^2}e^{kx}$

3.3 $\displaystyle\int \sinh(kx)dx = \frac{\cosh(kx)}{k}$

3.4 $\displaystyle\int \cosh(kx)dx = \frac{\sinh(kx)}{k}$

3.5 $\displaystyle\int x\sinh(kx)dx = \frac{x\cosh(kx)}{k} - \frac{\sinh(kx)}{k^2}$

3.6 $\displaystyle\int x\cosh(kx)dx = \frac{x\sinh(kx)}{k} - \frac{\cosh(kx)}{k^2}$

4. Sines and cosines

4.1 $\displaystyle\int \sin(\lambda x)dx = \frac{-\cos(\lambda x)}{\lambda}$

4.2 $\displaystyle\int \cos(\lambda x)dx = \frac{\sin(\lambda x)}{\lambda}$

4.3 $\displaystyle\int x\sin(\lambda x)dx = \frac{\sin(\lambda x)}{\lambda^2} - \frac{x\cos(\lambda x)}{\lambda}$

4.4 $\displaystyle\int x\cos(\lambda x)dx = \frac{\cos(\lambda x)}{\lambda^2} + \frac{x\sin(\lambda x)}{\lambda}$

4.5 $\displaystyle\int x^2\sin(\lambda x)dx = \frac{2x\sin(\lambda x)}{\lambda^2} + \frac{(2-\lambda^2x^2)\cos(\lambda x)}{\lambda^3}$

4.6 $\displaystyle\int x^2\cos(\lambda x)dx = \frac{2x\cos(\lambda x)}{\lambda^2} + \frac{(\lambda^2x^2-2)\sin(\lambda x)}{\lambda^3}$

4.7 $\displaystyle\int \sin(\lambda x)\sin(\mu x)dx = \frac{\sin(\mu-\lambda)x}{2(\mu-\lambda)} - \frac{\sin(\mu+\lambda)x}{2(\mu+\lambda)} \quad (\lambda \neq \mu)$

4.8 $\displaystyle\int \sin(\lambda x)\cos(\mu x)dx = \frac{\cos(\mu-\lambda)x}{2(\mu-\lambda)} - \frac{\cos(\mu+\lambda)x}{2(\mu+\lambda)} \quad (\lambda \neq \mu)$

4.9 $\displaystyle\int \cos(\lambda x)\cos(\mu x)dx = \frac{\sin(\mu-\lambda)x}{2(\mu-\lambda)} + \frac{\sin(\mu+\lambda)x}{2(\mu+\lambda)} \quad (\lambda \neq \mu)$

4.10 $\displaystyle\int \sin^2(\lambda x)dx = \frac{x}{2} - \frac{\sin(2\lambda x)}{4\lambda}$

4.11 $\displaystyle\int \sin(\lambda x)\cos(\lambda x)dx = \frac{\sin^2(\lambda x)}{2\lambda}$

4.12 $\displaystyle\int \cos^2(\lambda x)dx = \frac{x}{2} + \frac{\sin(2\lambda x)}{4\lambda}$

4.13 $\displaystyle\int e^{kx}\sin(\lambda x)dx = \frac{e^{kx}(k\sin(\lambda x) - \lambda\cos(\lambda x))}{k^2+\lambda^2}$

4.14 $\displaystyle\int e^{kx}\cos(\lambda x)dx = \frac{e^{kx}(k\cos(\lambda x) + \lambda\sin(\lambda x))}{k^2+\lambda^2}$

4.15 $\displaystyle\int \sinh(kx)\sin(\lambda x)dx = \frac{k\cosh(kx)\sin(\lambda x) - \lambda\sinh(kx)\cos(\lambda x)}{k^2+\lambda^2}$

4.16 $\displaystyle\int \sinh(kx)\cos(\lambda x)dx = \frac{k\cosh(kx)\cos(\lambda x) + \lambda\sinh(kx)\sin(\lambda x)}{k^2+\lambda^2}$

4.17 $\displaystyle\int \cosh(kx)\sin(\lambda x)dx = \frac{k\sinh(kx)\sin(\lambda x) - \lambda\cosh(kx)\cos(\lambda x)}{k^2+\lambda^2}$

4.18 $\displaystyle\int \cosh(kx)\cos(\lambda x)dx = \frac{k\sinh(kx)\cos(\lambda x) + \lambda\cosh(kx)\sin(\lambda x)}{k^2+\lambda^2}$

5. Bessel functions

5.1 $\displaystyle\int xJ_0(\lambda x)dx = \frac{xJ_1(\lambda x)}{\lambda}$

5.2 $\displaystyle\int x^2J_0(\lambda x)dx = \frac{x^2J_1(\lambda x)}{\lambda} + \frac{xJ_0(\lambda x)}{\lambda^2} - \frac{1}{\lambda^3}IJ(\lambda x)^*$

5.3 $\displaystyle\int J_1(\lambda x)dx = -\frac{J_0(\lambda x)}{\lambda}$

5.4 $\displaystyle\int x^{n+1}J_n(\lambda x)dx = \frac{x^{n+1}J_{n+1}(\lambda x)}{\lambda}$

5.5 $\displaystyle\int J_n(\lambda x)\frac{dx}{x^{n-1}} = -\frac{J_{n-1}(\lambda x)}{\lambda x^{n-1}}$

5.6 $\displaystyle\int J_0^2(\lambda x)xdx = \frac{x^2}{2}\left[J_0^2(\lambda x) + J_1^2(\lambda x)\right]$

5.7 $\displaystyle\int J_n^2(\lambda x)xdx = \frac{x^2}{2}\left[J_n^2(\lambda x) - J_{n-1}(\lambda x)J_{n+1}(\lambda x)\right]$

$$= \frac{x^2}{2}\left[J_n'(\lambda x)\right]^2 + \left(\frac{x^2}{2} - \frac{n^2}{2\lambda^2}\right)\left[J_n(\lambda x)\right]^2$$

6. Legendre Polynomials

6.1 $\displaystyle\int P_n(x)dx = -\frac{-(1-x^2)}{n(n+1)}P_n'(x)$

6.2 $\displaystyle\int xP_n(x)dx = \frac{(1-x^2)}{(n+2)(n-1)}\left(P_n(x) - xP_n'(x)\right).$

*See above: Calculus 5e.

Answers to Odd-Numbered Exercises

Chapter 0

Section 0-1, P. 10

1. $\phi(x) = c_1 \cos(\lambda x) + c_2 \sin(\lambda x)$

3. The equation has constant coefficients $k = 0$, $p = 0$; $u(t) = c_1 + c_2 t$.

5. $w(r) = c_1 r^{\lambda} + c_2 r^{-\lambda}$

7. Integrate, solve for dv/dx, and integrate again;
   $v(x) = c_1 + c_2 \ln|h + kx|$.

9. $u(x) = c_1 + c_2/x^2$

11. $u(r) = c_1 + c_2 \ln(r)$

13. Characteristic polynomial $m^4 + \lambda^4 = 0$; roots $m = \pm(1 \pm i)\lambda/\sqrt{2}$. General solution $u(x) = e^{\mu x}(c_1 \cos(\mu x) + c_2 \sin(\mu x)) + e^{-\mu x}(c_3 \cos(\mu x) + c_4 \sin(\mu x))$, $\mu = \lambda/\sqrt{2}$.

15. Characteristic polynomial $(m^2 + \lambda^2)^2 = 0$; roots $m = \pm i\lambda$ (double). General solution $u(x) = (c_1 + c_2 x)\cos(\lambda x) + (c_3 + c_4 x)\sin(\lambda x)$.

17. $v(t) = \ln(t)$ and $u_2(t) = t^b \ln(t)$

19. $u'' + \lambda^2 u = 0$; $R(\rho) = (a\cos(\lambda\rho) + b\sin(\lambda\rho))/\rho$

21. $t^2 d^2u/dt^2 = v'' - v'$; $t\,du/dt = v'$; $v'' + (k-1)v' + pv = 0$ (constant coefficients)

23. The relation is $u'' - p^2 u = 0$; general solution is $u(t) = c_1 e^{pt} + c_2 e^{-pt}$. If $u(t) = 0$, then $c_2 = -c_1 e^{2pt}$; if $u'(t) = 0$, then $c_2 = c_1 e^{2pt}$. If the c's have opposite signs, $u(t) = 0$ sometime, but $u'(t)$ is never 0. If the c's have the same sign, $u'(t) = 0$ sometime, but $u(t)$ is never 0. If one c is 0, then $u(t)$ and $u'(t)$ are never 0.

Section 0-2, P. 20

1. $u(t) = T + ce^{-at}$

3. $u(t) = te^{-at} + ce^{-at}$

5. $u(t) = \frac{1}{2} t \sin(t) + c_1 \cos(t) + c_2 \sin(t)$

7. $u(t) = \frac{1}{12} e^t + \frac{1}{2} te^{-t} + c_1 e^{-t} + c_2 e^{-2t}$

9. $u(\rho) = -\frac{1}{6}\rho^2 + \frac{c_1}{\rho} + c_2$

11. $h(t) = -320t + c_1 + c_2 e^{-0.1t}$, $c_1 = h_0 + 3200$, $c_2 = -3200$

13. $v(t) = t, u_p(t) = te^{-at}$

15. $v_1(x) = \sin(x) - \ln|\sec(x) + \tan(x)|, v_2 = -\cos(x)$;
    $u_p(x) = -\cos(x) \ln|\sec(x) + \tan(x)|$

17. $v_1(t) = t^2/2, v_2(t) = -t; u_p(t) = -t^2/2$

19. $v_1(t) = -1/2t, v_2(t) = -t/2, u_p(t) = -1.$

Section 0-3, P. 31

1. a. $u(x) = B\sin(x)$, arbitrary

   b. $u(x) = 1 - \cos(x) - \dfrac{1 - \cos(1)}{\sin(1)} \sin(x)$ (unique)

   c. No solution exists.

3. a. and b. $\lambda = (2n-1)\dfrac{\pi}{2a}$, $n = 1, 2, \cdots$

   c. $\lambda = \dfrac{n\pi}{a}$, $n = 0, 1, 2, \cdots$

5. $c = -a/2$, $c' = h - \dfrac{1}{\mu}\cosh\left(\dfrac{\mu a}{2}\right)$

7. $u(x) = T + c_1 \cosh(\gamma x) + c_2 \sinh(\gamma x)$, where $\gamma = \sqrt{\dfrac{hC}{\kappa A}}$ and

$$c_1 = T_0 - T, c_2 = -\frac{\kappa\gamma \sinh(\gamma a) + h\cosh(\gamma a)}{\kappa\gamma\cosh(\gamma a) + h\sinh(\gamma a)} c_1$$

9. $u(x) = T + T^* \left(1 - \cosh(\gamma x) - \dfrac{1 - \cosh(\gamma a)}{\sinh(\gamma a)} \sinh(\gamma x)\right)$

where $T^* = \dfrac{I^2 R}{hC}$ and $\gamma = \sqrt{\dfrac{hC}{\kappa A}}$

11. $u(y) = y(y - L)g/2\mu$

13. $P = EI(n\pi/L)^2$, $n = 1, 2, \cdots$

15. $u(x) = T + A\left(1 - \cosh(\gamma x) - \dfrac{1 - \cosh(\gamma a)}{\sinh(\gamma a)} \sinh(\gamma x)\right)$

$A = g/\kappa\gamma^2$, and $\gamma = \sqrt{\dfrac{hC}{\kappa A}}$

17. $u(r) = c_1 \ln(r/a) + c_2$, $c_1 = h_0 h_1 (T_a - T_W)/D$, $c_2 = [h_0(\kappa/b + h_1 \ln(b/a))T_W + (\kappa/a)h_1 T_a]/D$, $D = h_1\kappa/a + h_0\kappa/b + h_0 h_1 \ln(b/a)$

Section 0-4, P. 39

1. a. $u'' + \dfrac{1}{r}u' - u = 0$, $r = 0$

   b. $u'' - \dfrac{2x}{1 - x^2}u' = 0$, $x = \pm 1$

   c. $u'' + \cot(\phi)u' - u = 0$, $\phi = 0, \pm\pi, \pm 2\pi, \cdots$

   d. $u'' + \dfrac{2}{\rho}u' + \lambda^2 u = 0$, $\rho = 0$

3. $u(0)$ bounded; $u(\rho) = \dfrac{H}{6\kappa}(c^2 - \rho^2) + \dfrac{Hc}{3h} + T$

5. $u(\rho) = \dfrac{1}{\rho}(A\cos(\mu\rho) + B\sin(\mu\rho))$.

$u(\rho) \equiv 0$ unless $\mu a = \pi, 2\pi, \cdots$. The critical radius is $a = \dfrac{\pi}{\mu}$.

7. $u(r) = 325 + 10^4(0.25 - r^2)/4$; $u(0) = 950$

9. $u(x) = T_0 + AL^2(1 - e^{-x/L})$

Section 0-5, P. 47

1. $$G(x,z) = \begin{cases} z(a-x)/(-a), & 0 < z \le x \\ x(a-z)/(-a), & x \le z < a \end{cases}$$

3. $$G(x,z) = \begin{cases} \cosh(\gamma z)\sinh(\gamma(a-x))/(-\gamma\cosh(a)), & 0 < z \le x \\ \cosh(\gamma x)\sinh(\gamma(a-z))/(-\gamma\cosh(a)), & x \le z < a \end{cases}$$

5. $$G(\rho,z) = \begin{cases} \dfrac{(c-\rho)/\rho}{-c/z^2}, & 0 \le z < \rho \\ \dfrac{(c-z)/z}{-c/z^2}, & \rho \le z < c \end{cases}$$

7. $$G(x,z) = \begin{cases} \dfrac{\sinh(\gamma z)e^{-\gamma x}}{-\gamma}, & 0 < z \le x \\ \dfrac{\sinh(\gamma x)e^{-\gamma z}}{-\gamma}, & x \le z \end{cases}$$

9. $u(\rho) = (\rho^2 - c^2)/6$

11. $$u(x) = \int_0^a G(x,z)f(z)dz = \int_0^x \frac{z(a-x)}{-a}f(z)dz + \int_x^a \frac{x(a-z)}{-a}f(z)dz.$$

 There are two cases:

 (i) $x \le a/2$, so $u(x) = \displaystyle\int_{a/2}^a \frac{x(a-z)}{-a}dz$;

 and

 (ii) $x > a/2$, so $u(x) = \displaystyle\int_x^a \frac{x(a-z)}{-a}dz$.

 Results: $$u(x) = \begin{cases} -ax/8, & 0 < x < a/2 \\ -x(a-x)^2/2a, & a/2 < x < a \end{cases}$$

13. (i) At the left boundary, $x = l < z$, so the second line of Eq. (17) holds. The boundary condition (2) is satisfied by v because it is satisfied by u_1. At the right boundary, use the first line of Eq. (17).

 (ii) At $x = z$, both lines of Eq. (17) give the same value.

 (iii) $$v'(z+h) - v'(z-h) = \frac{u_1(z)u_2'(z+h) - u_1'(z-h)u_2(z)}{W(z)}$$

 As h approaches 0, the numerator approaches $W(z)$.

(iv) This is true because $u_1(x)$ and $u_2(x)$ are solutions of the homogeneous equation.

Chapter 0 Miscellaneous Exercises, P. 49

1. $u(x) = T_0 \cosh(\gamma x) + (T_1 - T_0 \cosh(\gamma a))\dfrac{\sinh(\gamma x)}{\sinh(\gamma a)}$

3. $u(x) = T_0$

5. $u(r) = p(a^2 - r^2)/4$

7. $u(\rho) = H(a^2 - \rho^2)/6 + T_0$

9. $u(x) = T + (T_1 - T)\cosh(\gamma x)/\cosh(\gamma a)$

11. $u(x) = T_0 + (T - T_0)e^{-\gamma x}$

13. $h(x) = \sqrt{ex(a-x) + h_0^2 + (h_1^2 - h_0^2)(x/a)}$

15. $u(x) = w(1 - e^{-\gamma x}\cos(\gamma x))EI/k$ where $\gamma = (k/4EI)^{1/4}$

17. $u(x) = \begin{cases} T_0 + Ax, & 0 < x < \alpha a \\ T_1 - B(a - x), & \alpha a < x < a \end{cases}$

$$A = \frac{\kappa_2}{\kappa_1(1-\alpha) + \kappa_2\alpha}\frac{T_1 - T_0}{a}, \quad B = \frac{\kappa_1}{\kappa_2}A$$

19. $u(x) = \dfrac{1}{2}(1 - e^{-2x}) - \dfrac{1}{2}(1 - e^{-2a})\dfrac{1 - e^{-x}}{1 - e^{-a}}$

21. a. $u(x) = \sinh(px)/\sinh(pa)$

    b. $u(x) = \cosh(px) - \dfrac{\cosh(pa)}{\sinh(pa)}\sinh(px) = \sinh(p(a-x))/\sinh(pa)$

    c. $u(x) = \cosh(px)/\cosh(pa)$

    d. $u(x) = \cosh(p(a-x))/\cosh(pa)$

    e. $u(x) = -\cosh(p(a-x))/p\sinh(pa)$

    f. $u(x) = \cosh(px)/p\sinh(pa)$

23. a. $u(x) = \dfrac{x}{2}\ln\left|\dfrac{1+x}{1-x}\right| - 1$

    b. $u(x) = 1 + (1-x)\ln\left|\dfrac{x}{1-x}\right|$

25. Multiply by u' and integrate: $\frac{1}{2}(u')^2 = \frac{1}{5}\gamma^2 u^5 + c_1$. Since $u(x) \to 0$ as $x \to \infty$, also $u'(x) \to 0$; thus $c_1 = 0$. Now $u' = -\sqrt{2\gamma^2/5}u^{5/2}$ or $u^{-5/2}u' = -\sqrt{2\gamma^2/5}$ (the negative root makes u decrease) can be integrated to result in $(-2/3)u^{-3/2} = -\sqrt{2\gamma^2/5}x + c_2$. The condition at $x = 0$ gives $c_2 = (-3/2)U^{-3/2}$.
Finally $u(x) = (U^{-3/2} + (3/2)\sqrt{2\gamma^2/5}x)^{-2/3}$.

27. $u(x) = \dfrac{w_0}{EI}\left(\dfrac{x^4}{24} - \dfrac{ax^3}{6} + \dfrac{a^2x^2}{2}\right)$

29. 459.77 rad/sec

31. $u(x) = C_0 e^{-ax}$

Chapter 1

Section 1-1, P. 61

1. a. $2\left(\sin(x) - \frac{1}{2}\sin(2x) + \frac{1}{3}\sin(3x) - + \cdots\right)$

   b. $\frac{\pi}{2} - \frac{4}{\pi}\left(\cos(x) + \frac{1}{9}\cos(3x) + \frac{1}{25}\cos(5x) + \cdots\right)$

   c. $\frac{1}{2} + \frac{2}{\pi}\left(\sin(x) + \frac{1}{3}\sin(3x) + \frac{1}{5}\sin(5x) + \cdots\right)$

   d. $\frac{2}{\pi} - \frac{4}{\pi}\left(\frac{1}{3}\cos(2x) + \frac{1}{15}\cos(4x) + \frac{1}{35}\cos(6x) + \cdots\right)$.

3. $f(x+p) = 1 = f(x)$ for any p and all x.

5. If c is a multiple of p, the graph of $f(x)$ between c and $c+p$ is the same as between 0 and p. Otherwise, let k be the integer such that kp lies between c and $c+p$:

$$\int_c^{c+p} f(x)dx = \int_c^{kp} f(x)dx + \int_{kp}^{c+p} f(x)dx = \int_{c^*}^{p} f(x)dx + \int_0^{c^*} f(x)dx$$

where $c^* = c - (k-1)p$.

7. a. $\cos^2(x) = \frac{1}{2} + \frac{1}{2}\cos(2x)$

   b. $\sin\left(x - \frac{\pi}{6}\right) = \cos(\frac{\pi}{6})\sin(x) - \sin(\frac{\pi}{6})\cos(x)$

   c. $\sin(x)\cos(2x) = -\frac{1}{2}\sin(x) + \frac{1}{2}\sin(3x)$

Section 1-2, P. 70

1. a. $\frac{1}{2} - \frac{4}{\pi^2}\left[\cos(\pi x) + \frac{1}{9}\cos(3\pi x) + \frac{1}{25}\cos(5\pi x) + \cdots\right]$

   b. $\frac{4}{\pi}\left[\sin\left(\frac{\pi x}{2}\right) + \frac{1}{3}\sin\left(\frac{3\pi x}{2}\right) + \frac{1}{5}\sin\left(\frac{5\pi x}{2}\right) + \cdots\right]$

   c. $\frac{1}{12} - \frac{1}{\pi^2}\left[\cos(2\pi x) - \frac{1}{4}\cos(4\pi x) + \frac{1}{9}\cos(6\pi x) - + \cdots\right].$

3. $\bar{f}(x) = f(x - 2na), \quad 2na < x < 2(n+1)a$

   $$f(x) \sim a_0 + \sum_1^\infty a_n \cos(n\pi x/a) + b_n \sin(n\pi x/a)$$

   $$a_0 = \frac{1}{2a}\int_0^{2a} f(x)dx, \quad a_n = \frac{1}{a}\int_0^{2a} f(x)\cos(n\pi x/a)dx$$

   $$b_n = \frac{1}{a}\int_0^{2a} f(x)\sin(n\pi x/a)dx$$

5. Odd: (a), (d), (e); even: (b), (c).

7. a. $\frac{2}{\pi}\left(\sin(\pi x) - \frac{1}{2}\sin(2\pi x) + - \cdots\right)$

   b. This function is its own Fourier series.

   c. $\frac{4}{\pi^2}\left(\sin(\pi x) - \frac{1}{9}\sin(3\pi x) + \frac{1}{25}\sin(5\pi x) - + \cdots\right).$

9. If $f(-x) = -f(x)$ and $f(x) = f(a - x)$ for $0 < x < a$, sine coefficients with even indices are zero. Example: square wave.

11. a. $f(x) = 1 = \frac{2}{\pi}\sum_1^\infty \frac{1 - \cos(n\pi)}{n}\sin\left(\frac{n\pi x}{a}\right)$

    b. $f(x) = \frac{a}{2} - \frac{2a}{\pi^2}\sum_1^\infty \frac{1 - \cos(n\pi)}{n^2}\cos\left(\frac{n\pi x}{a}\right)$

    $$= \frac{2a}{\pi}\sum_1^\infty \frac{-\cos(n\pi)}{n}\sin\left(\frac{n\pi x}{a}\right)$$

    d. $f(x) = \frac{2}{\pi}\left[1 - \sum_1^\infty \frac{1 + \cos(n\pi)}{n^2 - 1}\cos(nx)\right] = \sin(x)$.

13. Even, yes. Odd, yes only if $f(0) = f(a) = 0$.

Section 1-3, P. 77

1. a. Use $x = 0$; b. $x = \frac{1}{2}$; c. $x = 0$.

3. To $f(x)$ everywhere.

Section 1-4, P. 82

1. (c), (d), (f), (g) have uniformly convergent Fourier series.

3. All of the cosine series converge uniformly. The sine series converges uniformly only in case (b).

5. (a), (c).

Section 1-5, P. 88

1. $\sum_{n=1}^{\infty} \frac{1}{n^2} = \frac{\pi^2}{6}$

3. $f'(x) = 1$, $0 < x < \pi$. The sine series cannot be differentiated, because the odd periodic extension of f is not continuous. But the cosine series can be differentiated.

5. For the sine series: $f(0+) = 0$ and $f(a-) = 0$. For the cosine series no additional condition is necessary.

7. No. The function $\ln \left| 2\cos(\frac{x}{2}) \right|$ is not even sectionally continuous.

9. Since f is odd, periodic, and sectionally smooth, (c) follows, and also $b_n \to 0$ as $n \to \infty$. Then $\sum_{n=1}^{\infty} \left| n^k b_n e^{-n^2 t} \right|$ converges for all integers k $(t > 0)$ by the comparison test and ratio test:

$$\left| n^k b_n e^{-n^2 t} \right| \le M n^k e^{-n^2 t} \text{ for some} M$$

and

$$\frac{M(n+1)^k e^{-(n+1)^2 t}}{M n^k e^{-n^2 t}} = \left(\frac{n+1}{n} \right)^k e^{-(2n+1)t} \to 0$$

as $n \to \infty$. Then by Theorem 7, (a) is valid. Property (b) follows by direction substitution.

Section 1-6, P. 93

1. $\dfrac{1}{\pi}\displaystyle\int_{-\pi}^{\pi}\left(\ln\left|2\cos\left(\frac{x}{2}\right)\right|\right)^2 dx = \sum_{n=1}^{\infty}\frac{1}{n^2} = \frac{\pi^2}{6}$

3. a. Coefficients tend to zero.

   b. Coefficients tend to zero, although $\displaystyle\int_{-1}^{1}|x|^{-1}dx$ is infinite.

5. The integral must be infinite, because $\displaystyle\sum_{n=1}^{\infty} a_n^2 + b_n^2 = \infty$.

Section 1-7, P. 100

1. The equality to be proved is

$$2\sin\left(\tfrac{1}{2}y\right)\left(\frac{1}{2}+\sum_{n=1}^{N}\cos(ny)\right) = \sin\left(\left(N+\tfrac{1}{2}\right)y\right).$$

The left-hand side is transformed as follows

$$\begin{aligned}
2\sin\left(\tfrac{1}{2}y\right)&\left(\frac{1}{2}+\sum_{n=1}^{N}\cos(ny)\right)\\
&= \sin\left(\tfrac{1}{2}y\right)+\sum_{n=1}^{N}2\sin\left(\tfrac{1}{2}y\right)\cos(ny)\\
&= \sin\left(\tfrac{1}{2}y\right)+\sum_{n=1}^{N}\left(\sin\left(\left(n+\tfrac{1}{2}\right)y\right)-\sin\left(\left(n-\tfrac{1}{2}\right)y\right)\right)\\
&= \sin\left(\tfrac{1}{2}y\right)+\sum_{n=1}^{N}\sin\left(\left(n+\tfrac{1}{2}\right)y\right)-\sum_{n=0}^{N-1}\sin\left(\left(n+\tfrac{1}{2}\right)y\right)\\
&= \sin\left(\left(N+\tfrac{1}{2}\right)y\right)
\end{aligned}$$

because all other terms cancel.

3. $\phi(0+) = 1$, $\phi(0-) = -1$

5. a. $f'(x) = \frac{3}{4}x^{-1/4}$ for $0 < x < \pi$ (and f' is an odd function).

   Thus, f has a vertical tangent at $x = 0$, although it is continuous there.

   b. $\phi(y) = \dfrac{|y|^{3/4}}{2\sin(\frac{1}{2}y)}\cos(\tfrac{1}{2}y)\ \ -\pi < y < \pi$

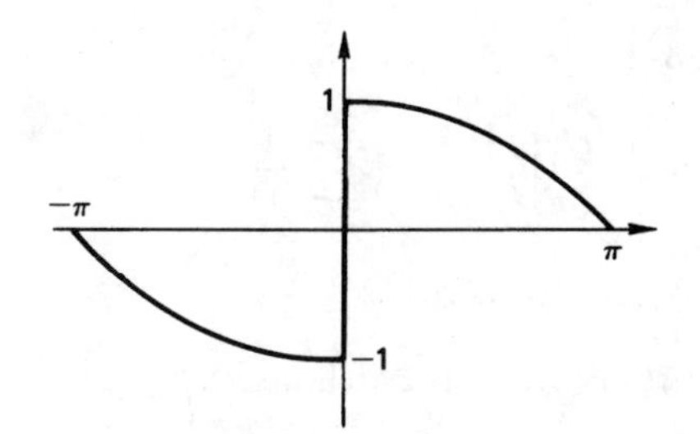

is a product of continuous functions and is therefore continuous, except perhaps where the denominator is 0. At $y = 0$, $\cos(\frac{1}{2}y) \cong 1$, $2\sin(\frac{1}{2}y) \cong y$, so $\phi(y) \cong |y|^{3/4}/y = \pm|y|^{-1/4}$ near $y = 0$.

c. Now, $\int_{-\pi}^{\pi} \phi^2(y)dy$ is finite, so the Fourier coefficients of ϕ approach zero.

Section 1-8, P. 106

1. $\hat{a}_6 = -0.00701$, $a_6 = -0.00569$

3. $\hat{a}_0 = \ \ 1.367$

$$\begin{array}{ll}
\hat{a}_1 = -0.844 & \hat{b}_1 = -0.043 \\
\hat{a}_2 = \ \ 0.208 & \hat{b}_2 = -0.115 \\
\hat{a}_3 = \ \ 0.050 & \hat{b}_3 = -0.050 \\
\hat{a}_4 = \ \ 0.042 & \hat{b}_4 = \ \ 0.00 \\
\hat{a}_5 = -0.0064 & \hat{b}_5 = \ \ 0.043 \\
\hat{a}_6 = \ \ 0.0167 &
\end{array}$$

Section 1-9, P. 113

1. Each function has the representations (for $x > 0$)

$$f(x) = \int_0^\infty A(\lambda)\cos(\lambda x)d\lambda = \int_0^\infty B(\lambda)\sin(\lambda x)d\lambda.$$

a. $A(\lambda = 2/\pi(1+\lambda^2)$, $B(\lambda) = 2\lambda/\pi(1+\lambda^2)$

b. $A(\lambda) = 2\sin(\lambda)/\pi\lambda$, $B(\lambda) = 2(1-\cos(\lambda))/\pi\lambda$

c. $A(\lambda) = 2(1-\cos(\lambda\pi))/\lambda^2\pi$, $B(\lambda) = 2(\pi\lambda - \sin(\lambda\pi))/\pi\lambda^2$

3. a. $\dfrac{1}{1+x^2} = \displaystyle\int_0^\infty e^{-\lambda}\cos(\lambda x)d\lambda$

b. $\dfrac{\sin(x)}{x} = \displaystyle\int_0^\infty A(\lambda)\cos(\lambda x)d\lambda$ where $A(\lambda) = \begin{cases} 1, & 0 < x < 1 \\ 0, & 1 < x \end{cases}$

5. The cosine and sine coefficient functions of $f'(x)$ are, respectively, $\lambda B(\lambda)$ and $-\lambda A(\lambda)$.

7. Change the variable of integration from x to λx.

9. a. $A(\lambda) \equiv 0, \quad B(\lambda) = \dfrac{2\sin(\lambda\pi)}{\pi(1-\lambda^2)}$

   b. $A(\lambda = \dfrac{1+\cos(\lambda\pi)}{\pi(1-\lambda^2)}, \quad B(\lambda) = \dfrac{\sin(\lambda\pi)}{\pi(1-\lambda^2)}$

   c. $A(\lambda) = \dfrac{2(1+\cos(\lambda\pi))}{\pi(1-\lambda^2)}, \quad B(\lambda) \equiv 0.$

11. Since $\int_{-\infty}^{\infty} |f(x)|dx$ is finite, the limits for $A(\lambda)$ and $B(\lambda)$ follow from the definition of the improper integral. Also

$$|a_0| \leq \frac{1}{2a}\int_{-\infty}^{\infty} |f(x)|dx \to 0.$$

Section 1-10, P. 118

1. a. $1 + \sum_{n=1}^{\infty} r^n \cos(nx) = \text{Re}\sum_{0}^{\infty}(re^{ix})^n = \text{Re}\dfrac{1}{1-re^{ix}}$

   b. $\sum_{n=1}^{\infty} \dfrac{\sin(nx)}{n!} = \text{Im}\sum_{n=1}^{\infty}\dfrac{e^{inx}}{n!} = \text{Im}\exp(e^{ix})$

3. $e^{\alpha x} = 2\dfrac{\sinh(\alpha\phi)}{\pi}\left(\dfrac{1}{2\alpha} + \sum_{n=1}^{\infty}\dfrac{(-1)^n}{\alpha^2+n^2}(\alpha\cos(nx) - n\sin(nx))\right)$

5. $f(x) = \int_{-\infty}^{\infty} C(\lambda)e^{i\lambda x}d\lambda$

   a. $C(\lambda) = \dfrac{1}{2\pi(1+i\lambda)}$, b. $C(\lambda) = \dfrac{1+e^{-i\lambda\pi}}{2\pi(1-\lambda^2)}$

Chapter 1 Miscellaneous Exercises, P. 124

1. $f(x) = \sum_{n=1}^{\infty} b_n \sin(nx)$

$$b_n = \begin{cases} 0, & n \text{ even} \\ \dfrac{4\sin(n\alpha)}{\pi\alpha n^2}, & n \text{ odd} \end{cases}$$

3. Yes. As $\alpha \to 0$, $\sin(n\alpha)/n\alpha \to 1$.

5. $f(x) = \sum_{n=1}^{\infty} b_n \sin(n\pi x/a)$

$$b_n = \frac{2h}{\pi^2}\frac{\sin(n\pi\alpha)}{n^2}\left(\frac{1}{\alpha} + \frac{1}{1-\alpha}\right)$$

7. a. $b_n = 0$, $a_n = 0$, $a_0 = 1$

b. $\sum_{n=1}^{\infty} b_n \sin(n\pi x/a)$, $b_n = \dfrac{2(1-\cos(n\pi))}{n\pi}$

c. and d. same as a.

e. same as b.

f. $a_0 + \sum_{n=1}^{\infty} a_n \cos(n\pi x/a) + b_n \sin(n\pi x/a)$

$a_0 = \dfrac{1}{2}$, $a_n = 0$, $b_n = \dfrac{1-\cos(n\pi)}{n\pi}$

9. $f(x) = a_0 + \sum_{n=1}^{\infty} a_n \cos(n\pi x/a) + b_n \sin(n\pi x/a)$

$a_0 = \dfrac{1}{2}a$, $a_n = -\dfrac{2a(1-\cos(n\pi))}{n^2\pi^2}$, $b_n = -\dfrac{2a\cos(n\pi)}{n\pi}$

$x = -a, \quad -a/2, \quad 0, \quad a, \quad 2a$

$\text{sum} = a, \quad 0, \quad 0, \quad a, \quad 0$

11. $f(x) = a_0 + \sum_{n=1}^{\infty} a_n \cos(nx)$

$a_0 = \dfrac{3}{4}$, $a_n = \dfrac{\sin(n\pi/2)}{n\pi}$

$x = 0, \quad \pi/2, \quad \pi, \quad 3\pi/2, \quad 2\pi$

$\text{sum} = 1, \quad \dfrac{3}{4}, \quad \dfrac{1}{2}, \quad \dfrac{3}{4}, \quad 1$

13. $f(x) = \sum_{n=1}^{\infty} b_n \sin(n\pi x)$, $b_n = 2(1+\cos(n\pi))/n\pi$

15. $f(x) = \sum_{n=1}^{\infty} b_n \sin(nx)$

$b_2 = \frac{1}{2}$, other $b_n = \frac{4\sin(n\pi/2)}{\pi(4-n^2)}$

17. $\sum_1^N \cos(nx) = \text{Re}\sum_1^N e^{inx} = \text{Re}\frac{e^{ix} - e^{iNx}}{1-e^{ix}} = \text{Re}\frac{e^{ix/2} - e^{i(2N-1)x/2}}{e^{-ix/2} - e^{ix/2}}$

The denominator is now $-2i\sin(x/2)$.

19. $f(x) = \sum_{n=1}^{\infty} b_n \sin(nx)$, $b_n = \frac{2a\sin(na+\pi)}{n^2a^2 - \pi^2}$

21. $f(x) = \int_0^{\infty} \left(\frac{\sin(\lambda a)}{\lambda\pi}\cos(\lambda x) + \frac{1-\cos(\lambda a)}{\lambda\pi}\sin(\lambda x)\right) d\lambda$

23. $f(x) = \int_0^{\infty} \frac{2\sin(\lambda\pi)}{\pi(1-\lambda^2)}\sin(\lambda x)d\lambda \quad (x > 0)$

29. Use $\int_0^{\infty} \frac{\sin(\lambda t)}{\lambda} d\lambda = \frac{\pi}{2}$

31. These answers are not unique.

a. $\sum_{n=1}^{\infty} b_n \sin(n\pi x)$, $b_n = 2/n\pi$

b. $a_0 + \sum_{n=1}^{\infty} a_n \cos(n\pi x)$, $a_0 = \frac{1}{2}$, $a_n = 2(1-\cos(n\pi))/n^2\pi^2$

c. $\int_0^{\infty} B(\lambda)\sin(\lambda x)d\lambda$, $B(\lambda) = 2(\lambda - \sin(\lambda))/(\pi\lambda^2)$

d. $\int_0^{\infty} A(\lambda)\cos(\lambda x)d\lambda$, $A(\lambda) = 2(1-\cos(\lambda))/(\pi\lambda^2)$

The integrals of parts c. and d. converge to 0 for $x > 1$.

33. Use $s = 6$ in Eq. (7) of Sect. 8.

$\hat{a}_0 = 0.78424 \quad \hat{a}_4 = -0.00924$

$\hat{a}_1 = 0.22846 \quad \hat{a}_5 = 0.00744$

$\hat{a}_2 = -0.02153 \quad \hat{a}_6 = -0.00347$

$\hat{a}_3 = 0.01410$

35. $a_0 = \frac{a}{6},\ a_n = \frac{2a}{n^2\pi^2}\left(\cos\left(\frac{2n\pi}{3}\right) - \cos\left(\frac{n\pi}{3}\right)\right)$

37. $a_0 = \frac{5}{8},\ a_n = \frac{2}{n^2\pi^2}\left(3\cos\left(\frac{n\pi}{2}\right) - 2 - \cos(n\pi)\right)$

39. $a_0 = \frac{1}{2},\ a_n = \frac{2}{n^2\pi^2}(1 - \cos(n\pi))$

41. $a_0 = \frac{a^2}{6},\ a_n = \frac{-2a^2}{n^2\pi^2}(1 + \cos(n\pi))$

43. $a_0 = \frac{1}{2},\ a_n = \frac{-1}{n\pi}2\sin\left(\frac{n\pi}{2}\right)$

45. $b_n = \dfrac{1 + \cos\left(\frac{n\pi}{2}\right) - 2\cos(n\pi)}{n\pi}$

47. $b_n = a\left(\frac{2\sin(n\pi/2)}{n^2\pi^2} - \frac{\cos(n\pi)}{n\pi}\right)$

49. $b_n = \frac{2}{n\pi}\left(\cos\left(\frac{n\pi}{4}\right) - \cos\left(\frac{3n\pi}{4}\right)\right)$

51. $b_n = 2n\pi\frac{(1 - e^{ka}\cos(n\pi))}{(a^2k^2 + n^2\pi^2)}$

53. $A(\lambda) = \frac{2}{\pi(1 + \lambda^2)}$

55. $A(\lambda) = \frac{2\sin(\lambda b)}{\pi\lambda}$

57. $A(\lambda) = \frac{2(1 - \cos(\lambda))}{\pi\lambda^2}$

59. $B(\lambda) = \frac{2\lambda}{\pi(1 + \lambda^2)}$

61. $B(\lambda) = \frac{2(1 - \cos(\lambda b))}{\lambda\pi}$

63. $B(\lambda) = \frac{2(\lambda - \sin(\lambda))}{\lambda^2\pi}$

65. The term $a_n\cos(nx) + b_n\sin(nx)$ appears in $S_n, S_{n+1}, \cdots, S_N$ and thus $N + 1 - n$ times in σ_N.

67. Use Eq. (13) of Section 7 and the identity in Exercise 66.

Chapter 2

Section 2-1, P. 141

1. One possibility: $u(x,t)$ is the temperature in a rod of length a whose lateral surface is insulated. The temperature at the left end is held constant at T_0. The right end is exposed to a medium at temperature T_1. Initially the temperature is $f(x)$.

3. $A\ \Delta x\ \ g = hC\Delta x(U - u(x,t))$, where h is a constant of proportionality and C is the circumference. Eq. (4) becomes

$$\frac{\partial^2 u}{\partial x^2} + \frac{hC}{\kappa A}(U - u) = \frac{1}{k}\frac{\partial u}{\partial t}.$$

5. If $\dfrac{\partial x}{\partial u}(0,t)$ is positive, then heat is flowing to the left, so $u(0,t)$ is greater than $T(t)$.

7. The second factor is approximately constant if T is much larger than u, or if T and u are approximately equal.

Section 2-2, P. 147

1. $v'' - \gamma^2(\nu - U) = 0,\ 0 < x < a$
   $v(0) = T_0,\ v(a) = T_1$
   $v(x) = U + A\cosh(\gamma x) + B\sinh(\gamma x),$
   $A = T_0 - U,\ B = \dfrac{(T_1 - U) - (T_0 - U)\cosh(\gamma a)}{\sinh(\gamma a)}$

   One interpretation: u as the temperature in a rod, with convective heat transfer from the cylindrical surface to a medium at temperature U.

3. $v(x) = T$. Heat is being generated at a rate proportional to $u - T$. If $\gamma = \pi/a$, the steady-state problem does not have a unique solution.

5. $v(x) = A\ln(\kappa_0 + \beta x) + B,\ A = (T_1 - T_0)/\ln(1 + a\beta/\kappa_0),$
   $B = T_0 - A\ln b$

7. a. $v(x) = T_0 + r(2a - x)x/2$
   b. $v(x) = A + B\sinh(\beta x) + C\sinh(\beta(a - x))$
   $A = \alpha/\beta^2,\ B = (T_1 - \alpha/beta^2)/\sinh(\beta a),\ C = (T_0 - \alpha/beta^2)/\sinh(\beta a).$

9. $Du'' - Su' = 0,\ 0 < x < a;\ u(0) = U_1,\ u(a) = 0.$
   $u(x) = U(e^{-Sx/D} - e^{-Sa/D})/(1 - e^{-Sa/D})$

Section 2-3, P. 154

1.

$$w(x,t) = -\frac{2}{\pi}(T_0 + T_1)\sin\left(\frac{\pi x}{a}\right)\exp\left(-\frac{\pi^2 kt}{a^2}\right)$$
$$-\frac{2}{\pi}\left(\frac{T_0 - T_1}{2}\right)\sin\left(\frac{2\pi x}{a}\right)\exp\left(-\frac{4\pi^2 kt}{a^2}\right)$$
$$-\cdots$$

3. The partial differential equation is

$$\frac{\partial^2 U}{\partial \xi^2} = \frac{\partial U}{\partial \tau}, \qquad 0 < \xi < 1, \qquad 0 < \tau$$

5. $w(x,t) = \sum_{n=1}^{\infty} b_n \sin\left(\frac{n\pi x}{a}\right)\exp(-n^2\pi^2 kt/a^2),\ b_n = T_0\frac{2(1-\cos(n\pi))}{\pi n}$

7. $w(x,t)$ as in 5. Above, with $b_n = \frac{2\beta a}{\pi} \cdot \frac{1}{n}$

9. a. $v(x) = c_1$

b. $\frac{\partial w}{\partial t} = D\frac{\partial^2 w}{\partial x^2}, \quad 0 < x < a,\ 0 < t$

$w(0,t) = 0, \quad w(a,t) = 0,\ 0 < t$

$w(x,0) = C_0 - C_1$

(c) $C(x,t) = C_1 + \sum_{n=1}^{\infty} b_n \sin\left(\frac{n\pi x}{a}\right)\exp(-n^2\pi^2 kt/a^2),$

$b_n = (C_0 - C_1)\frac{2(1-\cos(n\pi))}{\pi n}$

(d) $t = \frac{-a^2}{D\pi^2}\ln\left(\frac{\pi}{40}\right)$

(e) $t = 6444$ sec $= 107.4$ min

Section 2-4, P. 160

1. $a_0 = T_1/2,\ a_n = 2T_1(\cos(n\pi) - 1)/(n\pi)^2$

3. $u(x,t)$ as given in Eq. (9), with $\lambda_n = n\pi/a$, $a_0 = T_0/2$ and $a_n = 4T_0(2\cos(n\pi/2) - 1 - \cos(n\pi))/n^2\pi^2$.

5. (a) The general solution of the steady-state equation is $v(x) = c_1 + c_2 x$. The boundary conditions are $c_2 = S_0$, $c_2 = S_1$; thus there is a solution if $S_0 = S_1$. If heat flux is different at the ends, the temperature cannot approach a steady state. If $S_0 = S_1$, then $v(x) = c_1 + S_0 x$, c_1 undefined.

   (c) $A = (S_1 - S_0)/a$, $B = S_0$. If $S_0 \neq S_1$, then $\dfrac{\partial u}{\partial t} = kA$ for all t.

7. $\phi'' + \lambda^2 \phi = 0$, $0 < x < a$,

   $\phi(0) = 0$, $\phi(a) = 0$

   Solution: $\phi_n = \sin(\lambda_n x)$, $\lambda_n = n\pi/a (n = 1, 2, \cdots)$.

9. The series $\sum_{n=1}^{\infty} |A_n(t_1)|$ converges.

Section 2-5, P. 166

1. $v(x, t) = T_0$

3. The graph of G in the interval $0 < x < 2a$ is made by reflecting the graph of g in the line $x = a$. (Like an even extension.)

5. $g(x) \sim \sum_{n=1}^{\infty} b_n \sin\left(\dfrac{(2n-1)\pi x}{2a}\right)$, $0 < x < a$

   a. $b_n = \dfrac{8a(-1)^{n+1}}{\pi^2(2n-1)^2}$, b. $b_n = \dfrac{4T}{\pi(2n-1)}$

7. $u(x, t) = T_0 + \sum_{n=1}^{\infty} b_n \sin(\lambda_n x) \exp(-\lambda_n^2 kt)$, $\lambda_n = (2n-1)\pi/2a$,

$$b_n = \frac{8T(-1)^{n+1}}{\pi^2(2n-1)^2} - \frac{4T_0}{\pi(2n-1)}$$

9. The steady-state solution is $v(x) = T_0 - Tx(x - 2a)/2a^2$. The transient satisfies Eqs. (5)-(8) with

$$g(x) = T_0 - v(x) = \frac{Tx(x-2a)}{2a^2}.$$

11. $u(x,t) = T_0 + \sum_{n=1}^{\infty} c_n \cos(\lambda_n x) \exp(-\lambda_n^2 kt),$

$$\lambda_n = (2n-1)\pi/2a, \qquad c_n = \frac{4(T_1 - T_0)(-1)^{n+1}}{\pi(2n-1)}$$

13. (a) $u(x,t) = T_0 + \sum_{n=1}^{\infty} b_n \sin(\lambda_n x) \exp(-\lambda_n^2 kt)$, $\lambda_n = (2n-1)\pi/2a$,

$$b_n = \frac{1}{a}\int_0^{2a} g(x) \sin\left(\frac{n\pi x}{2a}\right) dx$$

In the integral for b_n, break the interval of integration at a; in the second integral, make the change of variable $y = 2a - x$. The two integrals cancel if n is even, and the coefficient is the same as Eq. (18) if n is odd.

(b) In the solution of Eqs. (1)-(4), the eigenfunction $\phi(x) = \sin((2n-1)\pi x/2a)$ has the property $\phi(2a-x) = \phi(x)$, so the sum of the series has the same property. This implies 0 derivative at $x = a$.

15. (a) $\frac{\partial u}{\partial t} = D\frac{\partial^2 u}{\partial x^2}$, $0 < x < L$, $0 < t$

$\frac{\partial u}{\partial x}(0,t) = 0$, $u(L,t) = S_0$, $0 < t$

$u(0,t) = 0$, $0 < x < L$

(b) $u(x,t) = S_0 + \sum_{n=1}^{\infty} c_n \cos(\lambda_n x) \exp(-\lambda_n^2 Dt),$

$c_n = 4S_0(-1)^n/(2n-1)$

Section 2-6, P. 175

1. The graph of $v(x)$ is a straight line from T_0 at $x = 0$ to T^* at $x = a$, where

$$T^* = T_0 + \frac{ha}{k+ha}(T_1 - T_0).$$

In all cases, T^* is between T_0 and T_1.

3. Negative solutions provide no new eigenfunctions.

7. $b_m = \dfrac{2(1-\cos(\lambda_m a))}{\lambda_m[a + (\kappa/h)\cos^2(\lambda_m a)]}$

9. $b_m = \dfrac{-2(\kappa + ah)\cos(\lambda_m a)}{\lambda_m(ah + \kappa\cos^2(\lambda_m a))}$

Section 2-7, P. 180

1. $\lambda_n = n\pi/\ln 2$, $\phi_n = \sin(\lambda_n \ln(x))$

3. a. $\sin(\lambda_n x)$, $\lambda_n = (2n-1)\pi/2a$

   b. $\cos(\lambda_n x)$, $\lambda_n = (2n-1)\pi/2a$

   c. $\sin(\lambda_n x)$, λ_n a solution of $\tan(\lambda a) = -\lambda$

   d. $\lambda_n \cos(\lambda_n x) + \sin(\lambda_n x)$, λ_n a solution of $\cot(\lambda a) = \lambda$

   e. $\lambda_n \cos(\lambda_n x) + \sin(\lambda_n x)$, λ_n a solution of $\tan(\lambda a) = 2\lambda/(\lambda^2 - 1)$

5. The weight functions in the orthogonality relations, and limits of integration are:

   a. $1 + x$, 0 to a; b. e^x, 0 to a; c. $\dfrac{1}{x^2}$, 1 to 2; d. e^x, 0 to a.

7. Because λ appears in a boundary condition.

9. The negative value of μ does not contradict Theorem 2 because the coefficient α_2 is not positive.

Section 2-8, P. 184

1. $x = \sum_{n=1}^{\infty} c_n\phi_n$, $1 < x < b$; $c_n = 2n\pi\dfrac{1 - b\cos(n\pi)}{n^2\phi^2 + \ln^2(b)}$

3. $1 = \sum_{n=1}^{\infty} c_n\phi_n$, $0 < x < a$; $c_n = 2n\pi\dfrac{1 - e^{a/2}\cos(n\pi)}{n^2\pi^2 + a^2/4}$

   (Hint: find the sine series of $e^{x/2}$.)

5. $b_n = \int_l^r f(x)\psi_n(x)p(x)dx$

7. 1 and $\sqrt{2}\cos(n\pi x)$, $n = 1, 2, \cdots$

Section 2-9, P. 189

1. a. $v(x) = \text{constant}$; b. $v(x) = AI(x) + B$

3. If $\partial u/\partial x = 0$ at both ends, then the steady-state problem is indeterminate, but Eqs. (1)-(3) are homogeneous, so separation of variables applies directly. Note that $\lambda_0 = 0$ and $\phi_0 = 1$. The constant term in the series for $u(x, t)$ is

$$a_0 = \frac{\int_l^r p(x) f(x) dx}{\int_l^r p(x) dx}.$$

Section 2-10, P. 193

1. The solution is as in Eq. (9), with $B(\lambda) = 2T(\cos(a) - \cos(b))/\lambda\pi$

3. $u(x, t)$ is given by Eq. (6) with $B(\lambda) = \dfrac{2T_0\lambda}{\pi(\alpha^2 + \lambda^2)}$.

5. $u(x, t) = \displaystyle\int_0^\infty A(\lambda) \cos(\lambda x) \exp(-\lambda^2 kt) d\lambda$; $A(\lambda) = \dfrac{2}{\pi\lambda} \sin(\lambda b)$

7. $u(x, t) = T_0 + \displaystyle\int_0^\infty B(\lambda) \sin(\lambda x) \exp(-\lambda^2 kt) d\lambda$;

   $B(\lambda) = \dfrac{2}{\pi} \displaystyle\int_0^\infty (f(x) - T_0) \sin(\lambda x) dx$

9. (a) $v(x) = C_0 e^{-ax}$

   (b) $$\frac{\partial w}{\partial t} = D\left(\frac{\partial^2 w}{\partial x^2} - a^2 w\right), \quad 0 < x, \quad 0 < t$$

   $w(0, t) = 0, \quad 0 < t$

   $w(x, 0) = -c_0 e^{-ax}, \quad 0 < x$

   (c) $w(x, t) = e^{-a^2 Dt} \displaystyle\int_0^\infty B(\lambda) \sin(\lambda x) e^{-\lambda^2 Dt} d\lambda$

   $B(\lambda) = -2C_0\lambda/(\pi(\lambda^2 + a^2))$

Section 2-11, P. 199

1. Break the interval of integration at $x' = 0$.

3. $B(\lambda) = 0, \quad A(\lambda) = \dfrac{2T_0 a}{(1 + \lambda^2 a^2)}$

5. The function $u(x,t)$, as a function of x, is the famous "bell-shaped" curve. The smaller t is, the more sharply peaked the curve.

7. In Eq. (3) replace both $f(x')$ and $u(x,t)$ by 1.

9. Using the integral given, obtain

$$u(x,t) = \frac{2}{\pi}\int_0^\infty \frac{1}{\lambda}\sin(\lambda x)e^{-\lambda^2 kt}d\lambda.$$

Note, however, that $B(\lambda) = 2/\lambda\pi$ is *not* found using the usual formulas for Fourier coefficient functions.

Section 2-12, P. 204

5. As $t \to 0+$, $x/\sqrt{4\pi kt} \to \begin{cases} +\infty & \text{if} \quad x > 0 \\ -\infty & \text{if} \quad x < 0 \end{cases}$

so $\operatorname{erf}(x/\sqrt{4\pi kt}) \to \begin{cases} +1 & \text{if} \quad x > 0 \\ -1 & \text{if} \quad x < 0 \end{cases}$

7. Make the substitution $x = y^2$. Then $I(x) = \sqrt{\pi}\,\operatorname{erf}(\sqrt{x}) + c$.

9. Let z be defined by $\operatorname{erf}(z) = -U_b/(U_i - U_b)$. Then $x(t) = z\sqrt{4kt}$.

Chapter 2 Miscellaneous Exercises, P. 208

1. SS: $v(x) = T_0$, $0 < x < a$

   EVP: $\phi'' + \lambda^2\phi = 0$, $\phi(0) = 0$, $\phi(a) = 0$, $\lambda_n = n\pi/a$, $\phi_n = \sin(\lambda_n x)$, $n = 1, 2, \cdots$

$$u(x,t) = T_0 + \sum_1^\infty b_n \sin(\lambda_n x)e^{-\lambda_n^2 kt}$$

$$b_n = \frac{2}{a}\int_0^a (T_1 - T_0)\sin\left(\frac{n\pi x}{a}\right)dx$$

3. SS: $v(x) = T_0 + \frac{r}{2}x(x-a)$, $0 < x < a$

   EVP: $\phi'' + \lambda^2\phi = 0$, $\phi(0) = 0$, $\phi(a) = 0$, $\lambda_n = n\pi/a$, $\phi_n = \sin(\lambda_n x)$, $n = 1, 2, \cdots$

$$u(x,t) = T_0 + \frac{r}{2}x(x-a) + \sum_1^\infty b_n \sin(\lambda_n x)\exp(-\lambda_n^2 kt)$$

$$b_n = \frac{2}{a}\int_0^a \left[T_1 - T_0 - \frac{r}{2}x(x-a)\right]\sin\left(\frac{n\pi x}{a}\right)dx$$

5. SS: $v(x) = 0$, $0 < x < a$

   Hint: put $-\gamma^2 u$ on the other side of the equation. Separation of variables gives $\phi''/\phi = \gamma^2 + T'/kT = -\lambda^2$.

   EVP: $\phi'' + \lambda^2\phi = 0$, $\phi'(0) = 0$, $\phi'(a) = 0$, $\lambda_0 = 0$, $\phi_0 = 1$; $\lambda_n = n\pi/a$, $\phi_n = \cos(\lambda_n x)$, $n = 1, 2, \cdots$

   $$u(x,t) = e^{-\gamma^2 kt}\left(a_0 + \sum a_n \cos(\lambda_n x)\exp(-\lambda_n^2 kt)\right)$$

   $a_0 = T_1/2$, $a_n = 2T_1(1-\cos(n\pi))/n^2\pi^2$

7. $u(x,t) = T_0$

9. $$u(x,t) = T_0 + \sum_{n=1}^{\infty} c_n \sin(\lambda_n x)\exp(-\lambda_n^2 kt),$$

   $$\lambda_n = \frac{(2n-1)\pi}{2a}, \qquad c_n = \frac{(T-1-T_0)\cdot 4}{(2n-1)\pi}$$

11. $$u(x,t) = T_0 + \int_0^\infty B(\lambda)\sin(\lambda x)\exp(-\lambda^2 kt)d\lambda,\ B(\lambda) = \frac{-2\lambda T_0}{\pi(\alpha^2+\lambda^2)}$$

13. $$u(x,t) = \int_0^\infty A(\lambda)\cos(\lambda x)\exp(-\lambda^2 kt)d\lambda,\ A(\lambda) = \frac{2T_0\sin(\lambda a)}{\pi\lambda}$$

15. $$u(x,t) = \int_0^\infty (A(\lambda)\cos(\lambda x) + B(\lambda)\sin(\lambda x))\exp(-\lambda^2 kt)d\lambda,$$

    $$A(\lambda) = \frac{T_0\sin(\lambda a)}{\pi\lambda}, \qquad B(\lambda) = \frac{T_0(1-\cos(\lambda a))}{\pi\lambda}$$

    or

    $$u(x,t) = \frac{T_0}{\sqrt{4\pi kt}}\int_0^a \exp\left(-\frac{(x'-x)^2}{4kt}\right)dx'$$

    $$= \frac{T_0}{2}\left[\operatorname{erf}\left(\frac{a-x}{\sqrt{4kt}}\right) + \operatorname{erf}\left(\frac{x}{\sqrt{4kt}}\right)\right]$$

17. Interpretation: u is the temperature in a rod with insulation on the cylindrical surface and on the left end. At the right end, heat is being forced into the rod at a constant rate (because $q(a,t) = -\kappa\frac{\partial u}{\partial x}(a,t) = -\kappa S$, so heat is flowing to the left, into the rod).

The accumulation of heat energy accounts for the steady increase of temperature.

19. $(1/6ka)u_3 - (a/6k)u_1$ satisfies the boundary conditions.

21. $w(x,t) = -\dfrac{2}{u}\dfrac{\partial u}{\partial x}$ where $u(x,t) = a_0 + \sum a_n \cos(n\pi x)\exp(-n^2\pi^2 t)$
where $a_0 = 2(1 - e^{-1/2})$ and $a_n = \dfrac{1 - e^{-1/2}\cos(n\pi)}{\frac{1}{4} + (n\pi)^2}$

23. $u_2 = \dfrac{\beta_2 V}{\beta_1 + \beta_2}$, $u_1 = 1 - \dfrac{\beta_1 V}{\beta_1 + \beta_2}$
where $V = 1 - \exp(-(\beta_1 + \beta_2)t)$ and $\beta_i = h/c_i$

25. $u(\rho, t) = \dfrac{1}{\rho}\sum_{n=1}^{\infty} b_n \sin(\lambda_n \rho)\exp(-\lambda_n^2 kt)$,

$\lambda_n = n\pi/a$, $b_n = \dfrac{2}{a}\displaystyle\int_0^a \rho T \sin(\lambda_n \rho)d\rho$

27. $v(x) = T_0 + Sx - S\dfrac{\sinh(\lambda x)}{\gamma\cosh(\gamma a)}$

29. If $\lambda = 0$, the differential equation is $\phi'' = 0$ with general solution $\phi(x) = c_1 + c_2 x$. The boundary conditions require $c_2 = 0$ but allow $c_1 \neq 0$. Thus, this value of λ permits the existence of a nonzero solution, and therefore $\lambda = 0$ is an eigenvalue.

31. Choose $B(\omega) = \dfrac{2}{\pi}\displaystyle\int_0^{\infty} f(t)\sin(\omega t)dt$. If f has a Fourier integral representation, then this choice of B will make $u(0,t) = f(t)$, $0 < t$.

33. (a) $v(x) = -Ix/aK + c_1 + c_2(1 - e^{-aKx/T})$
$c_1 = h_1, \quad c_2 = (h_2 - h_1 + IL/aK)/(1 - e^{-aKL/T})$
(b) $\dfrac{\partial^2 w}{\partial x^2} + \mu\dfrac{\partial w}{\partial x} = \dfrac{1}{k}\dfrac{\partial w}{\partial t}, \quad 0 < x < L, \quad 0 < t$
$w(0,t) = 0, \quad w(L,t) = 0, \quad 0 < t$
$w(x,0) = h_0(x) - v(x), \quad 0 < x < L$
where $\mu = aK/T$, $k = T/S$.
(c) $w(x,t) = \sum c_n \phi_n(x) e^{-\lambda_n^2 kT}, \quad \phi_n(x) = e^{-\mu x/2}\sin(n\pi x/L)$,
$\lambda_n^2 = \left(\dfrac{n\pi}{L}\right)^2 + \dfrac{\mu^2}{4}$
(d) $\lambda_n^2 = (7.30n^2 + .0133) \times 10^{-4} m^{-1}$

Chapter 3

Section 3-1, P. 220

1. $[u] = L$, $[c] = L/t$

3. $v(x) = \dfrac{(x^2 - ax)g}{2c^2}$

Section 3-2, P. 226

1. $u(x,t) = \sum_{n=1}^{\infty} b_n \sin\left(\frac{n\pi x}{a}\right) \sin\left(\frac{n\pi ct}{a}\right)$

$$b_n = \frac{2a(1-\cos(n\pi))}{n^2\pi^2 c}$$

3. $u(x,t) = \sum_{n=1}^{\infty} a_n \cos\left(\frac{n\pi ct}{a}\right) \sin\left(\frac{n\pi x}{a}\right), \quad a_n = 2U_0 \frac{1-\cos(n\pi/2)}{n\pi}$

5. a. $\sin\left(\frac{n\pi x}{a}\right)$ b. $\sin\left(\frac{2n-1}{2}\frac{\pi x}{a}\right)$

7. Product solutions are $\phi_n(x)T_n(t)$ where

$$\phi_n(x) = \sin(\lambda_n x), \quad T_n(t) = \exp(-kc^2t/2) \times \begin{cases} \sin(\mu_n t) \\ \cos(\mu_n t) \end{cases}$$

$$\lambda_n = \frac{n\pi}{a}, \quad \mu_n = \sqrt{\lambda_n^2 c^2 - \frac{1}{4}k^2c^4}$$

9. Product solutions are $\phi_n(x)T_n(t)$ where

$$\phi_n(x) = \sin\left(\frac{n\pi x}{a}\right)$$
$$T_n(t) = \sin \ \text{or} \ \cos\left(\frac{n^2\pi^2 ct}{a^2}\right).$$

Frequencies $n^2\pi^2c/a^2$.

11. The general solution of the differential equation is $\phi(x) = A\cos(\lambda x) + B\sin(\lambda x) + C\cosh(\lambda x) + D\sinh(\lambda x)$. Boundary conditions at $x = 0$ require $A = -C$, $B = -D$; those at $x = a$ lead to $C/D = -(\cosh(\lambda a) + \cos(\lambda a))/(\sinh(\lambda a) - \sin(\lambda a))$ and $1 + \cos(\lambda a)\cosh(\lambda a) = 0$. The first eigenvalues are $\lambda_1 = 1.875/a$, $\lambda_2 = 4.693/a$, and the eigenfunctions are similar to the functions shown in the figure.

13. $u(x,t) = \sum_{n=1}^{\infty}(a_n \cos(\mu_n t) + b_n(\sin \mu_n t)) \sin(\lambda_n x)$: $\lambda_n = n\pi/a, \mu_n = \sqrt{\lambda_n^2 + \gamma^2}c,$

$a_n = 2h(1 - \cos(n\pi))/n\pi, b_n = 0,$

$n = 1, 2, \cdots.$

15. Convergence is uniform because $\sum |b_n|$ converges.

Section 3-3, P. 234

1. Table shows $u(x,t)/h$.

t	0	$0.2a/c$	$0.4a/c$	$0.8a/c$	$1.4a/c$
$x = 0.25a$	0.5	0.5	0.2	-0.5	-0.2
$x = 0.5a$	1.0	0.6	0.2	-0.6	-0.2

3. $u(0, 0.5a/c) = 0$; $u(0.2a, 0.6a/c) = 0.2\alpha a$; $u(0.5a, 1.2a/c) = -0.2\alpha a$. (Hint: $G(x) = \alpha x$, $0 < x < a$.)

5. $$G(x) = \begin{cases} 0, & 0 < x < 0.4a \\ 5(x - 0.4a), & 0.4a < x < 0.6a \\ a, & 0.6a < x < a \end{cases}$$

Notice that G is a continuous function whose graph is composed of line segments.

7.

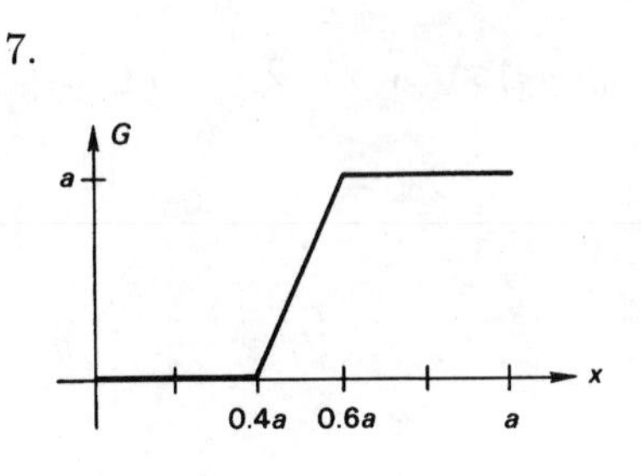

9.

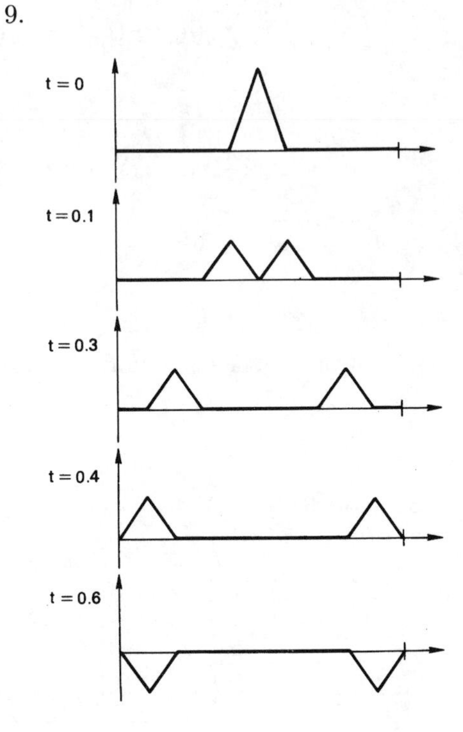

11. $u(x,t) = -c^2 \cos(t) + \phi(x - ct) + \psi(x + ct)$

Section 3-4, P. 238

1. If f and g are sectionally smooth, and f continuous.

3. The frequency is $c\lambda_n$ rads/sec, and the period is $2\pi/c\lambda_n$ sec.

5. Separation of variables leads to the following in place of Eqs. (11) and (12):

$$T'' + \gamma T' + \lambda^2 c^2 T = 0 \tag{11'}$$

$$(s(x)\phi')' - q(x)\phi + \lambda^2 p(x)\phi = 0. \tag{12'}$$

The solutions of Eq. (11') all approach 0 as $t \to \infty$, if $\gamma > 0$.

7. The period of $T_n(t) = a_n \cos(\lambda_n ct) + b_n \sin(\lambda_n ct)$ is $2\pi/\lambda_n c$. All T_n's have a common period p if and only if for each n there is an integer

m such that $m(2\pi/\lambda_n c) = p$, or $m = (pc/2\pi)\lambda_n$ is an integer. For λ_n as shown and $\beta = q/r$ where q and r are integers, this means

$$m = (pc/2\pi)\alpha(n + q/r)$$

or

$$m = \left(\frac{pc}{2\pi}\right)\frac{\alpha}{r}(rn + q).$$

Given α, p can be adjusted so that m is an integer whenever n is an integer.

Section 3-5, P. 242

1. If $q \geq 0$, the numerator in Eq. (3) must also be greater than or equal to 0, since $\phi_1(x)$ cannot be identically 0.

3. $2\pi^2/3$ is one estimate from $y = \sin(\pi x)$.

5. $\displaystyle\int_1^2 (y')^2 dx = \frac{1}{3}, \int_1^2 \frac{y^2}{x^4} dx = \frac{25}{6} - 6\ln 2;$

   $N(y)/D(y) = 42.83; \lambda_1 \leq 6.54.$

Section 3-6, P. 247

1. $u(x,t) = \frac{1}{2}[f_e(x+ct) + G_o(x+ct)] + \frac{1}{2}[f_e(x-ct) - G_o(x-ct)]$, where f_e is the even extension of f and G_o the odd extension of G.

3. 5.

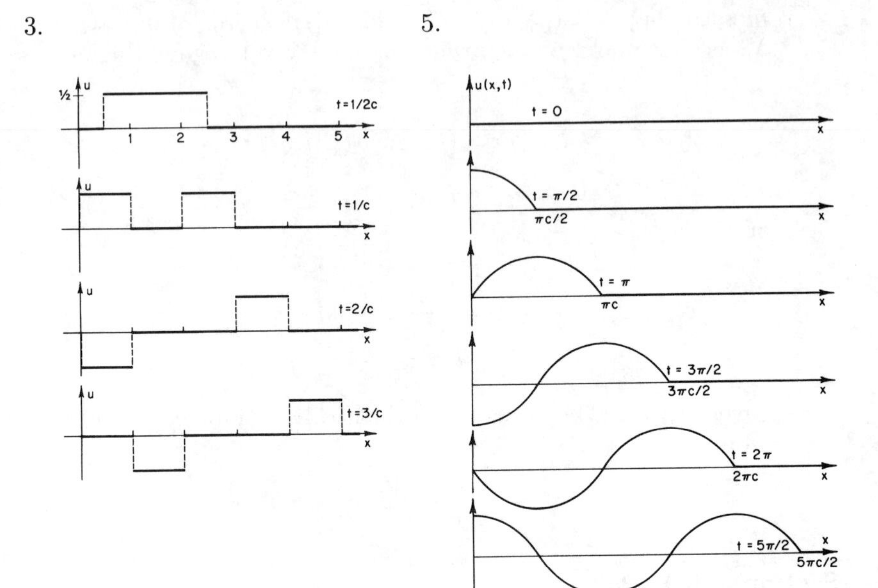

7. $u(x,t) = \dfrac{1}{2}[f(x+ct) + f(x-ct)] + \dfrac{1}{2c}\displaystyle\int_{x-ct}^{x+ct} g(y)dy$

Chapter 3 Miscellaneous Exercises, P. 251

1. $u(x,t) = \displaystyle\sum_{1}^{\infty} b_n \sin(\lambda_n x)\cos(\lambda_n ct), b_n = 2(1-\cos(n\pi))/n\pi, \lambda_n = n\pi/a.$

3.

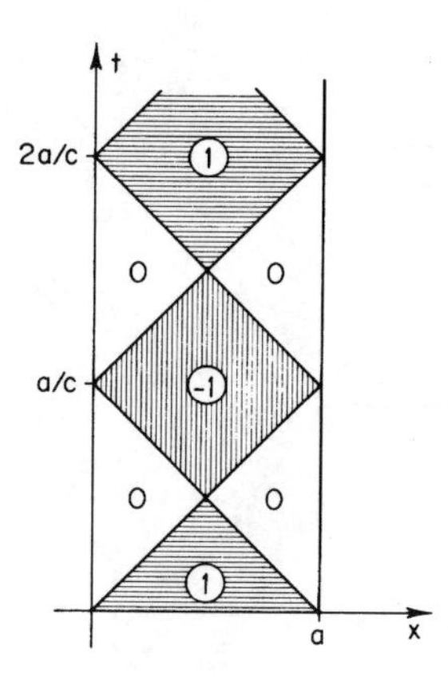

t
2a/c
1
0
0
a/c
-1
0
0
1
a
x

5.

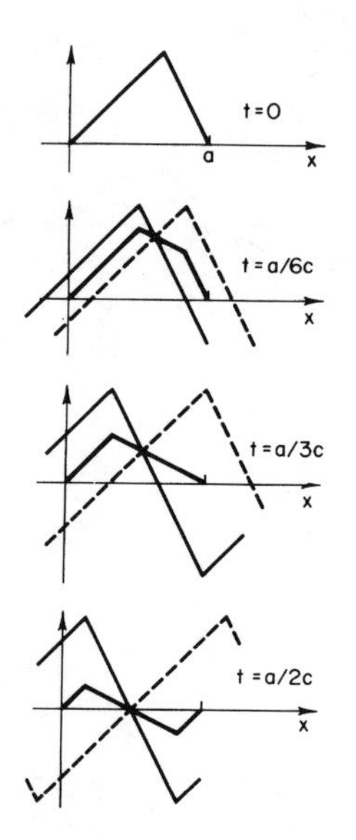

t=0
a
x
t=a/6c
x
t=a/3c
x
t=a/2c
x

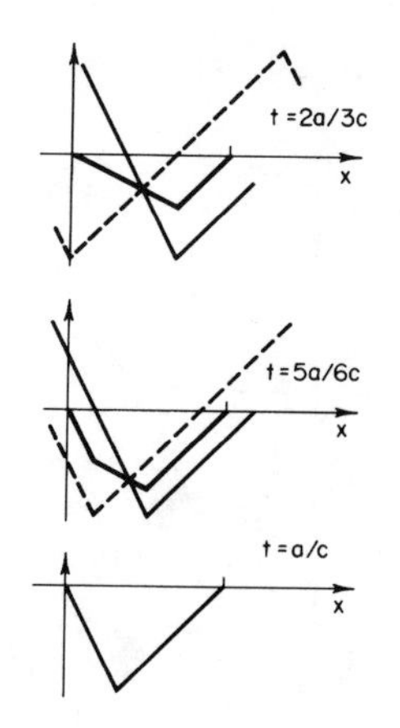

t=2a/3c
x
t=5a/6c
x
t=a/c
x

7.

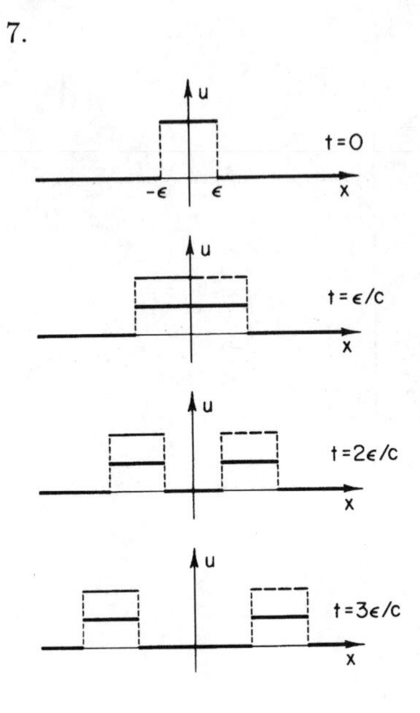

9.

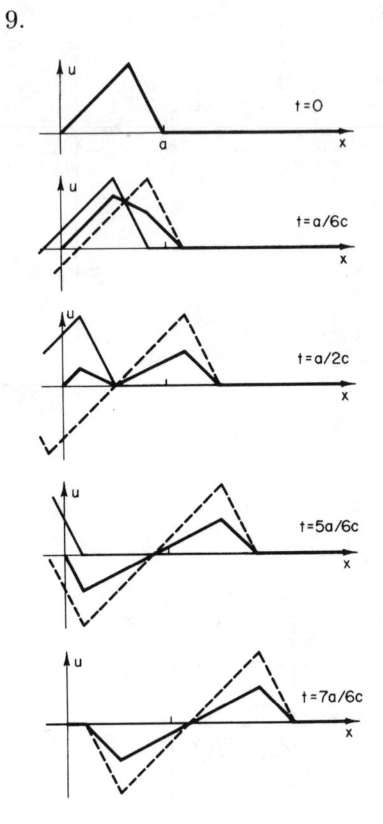

11.

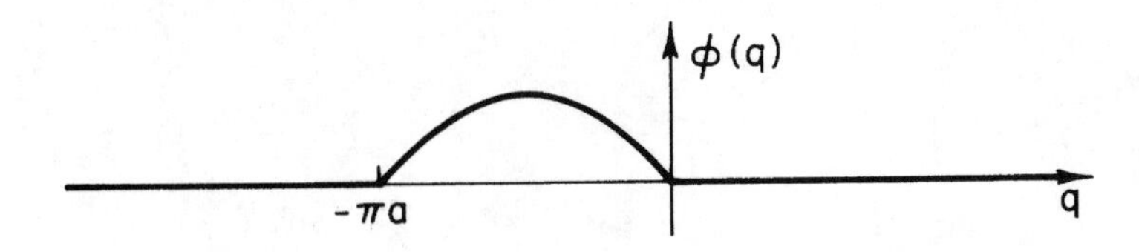

13.

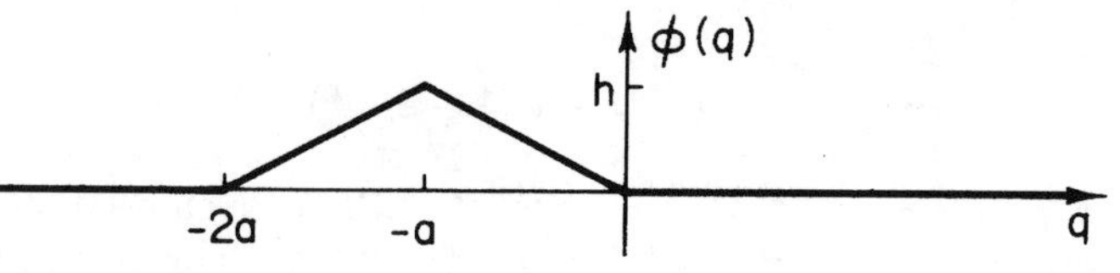

15. $\lambda_1^2 \leq 8.324$, using $y(x) = (x - 1/4)(5/4 - x)$. Also $\lambda_1^2 \leq 8.265$ from $y(x) = \sin(\pi(x - 1/4))$. For this case, one integral had to be calculated numerically (trapezoidal rule).

17. $f(q) = 12a^2\text{sech}^2(aq), c = 4a^2$

21. $v(x,t) = \sum_{n=1}^{\infty}(a_n \cos(\lambda_n ct) + b_n \sin(\lambda_n ct)) \sin(\lambda_n x)$

$\lambda_n = (2n-1)\pi/2a$

$a_n = \dfrac{8aU_0(-1)^{n+1}}{\pi^2(2n-1)^2}, \quad b_n = \dfrac{16a^2k(-1)^n}{\pi^3(2n-1)^3}$

23. $\dfrac{Y''}{Y} = \dfrac{2V}{k}\dfrac{\psi'}{\psi}$. The function $\phi(x - Vt)$ cancels from both sides.

25. $\phi_n(-Vt) = T_0 \exp(\lambda_n^2 kt/2)b_n, \quad t > 0$

$\phi_n(x) = T_1 \exp(\lambda_n^2 kx/2V)b_n, \quad x > 0$

where $\sum_{n=1}^{\infty} b_n \sin(\lambda_n y) = 1,\ 0 < y < b$

27. $\dfrac{\partial^2 i}{\partial x^2} = CR\dfrac{\partial i}{\partial t}, \quad \dfrac{\partial^2 e}{\partial x^2} = CR\dfrac{\partial e}{\partial t}$

29. $e(x,t) = \sum_{n=1}^{\infty} a_n \cos(\lambda_n ct) \sin(\lambda_n x)$,

$a_n = \dfrac{2V(1 - \cos(n\pi))}{n\pi}, \quad \lambda_n = n\pi/a$

31. $e(x,t) = V_0\dfrac{(a-x)}{a} + \sum_{n=1}^{\infty} b_n \sin(\lambda_n x) \exp(-\lambda_n^2 kt)$

$b_n = -\dfrac{2V_0 \cos(n\pi)}{n\pi}, \quad \lambda_n = n\pi/a,\ k = 1/Rc$

33. $e(x,t) = V + \frac{1}{2}(f_o(x+ct) + f_o(x-ct))$, where f_o is the odd extension of
$$f(x) = -V(1 - e^{-\alpha x}), \quad 0 < x$$

35. $\phi(x-ct) = e^{-c(x-ct)/k} = e^{(c^2t-cx)/k}$. The given c satisfies $c^2 = i\omega k$, so $\phi(x-ct) = e^{i\omega t-(1+i)px} = e^{-px}e^{i(\omega t-px)}$. Now form $\frac{1}{2}(\phi(x-ct) + \phi(x-ct)) = e^{-px}\cos(\omega t - px)$ and so forth.

37. Differentiate and substitute.

Chapter 4

Section 4-1, P. 261

1. $f + d = 0$

3. $Y(y) = A\sinh(\pi y)$, $A = 1/\sinh(\pi)$

5. $v(r) = a\ln(r) + b$

7. $$\frac{\partial u}{\partial x} = \frac{\partial v}{\partial r}\cos(\theta) - \frac{\partial v}{\partial \theta}\frac{\sin(\theta)}{r}$$
$$\frac{\partial u}{\partial y} = \frac{\partial v}{\partial r}\sin(\theta) + \frac{\partial v}{\partial \theta}\frac{\cos(\theta)}{r}$$

Section 4-2, P. 267

1. Show by differentiating and substituting that both are solutions of the differential equation. The Wronskian of the two functions is
$$\begin{vmatrix} \sinh(\lambda y) & \sinh(\lambda(b-y)) \\ \lambda\cosh(\lambda y) & -\lambda\cosh(\lambda(b-y)) \end{vmatrix} = -\lambda\sinh(\lambda b) \neq 0$$

3. In the case $b = a$, use two terms of the series: $u(a/2, a/2) = 0.32$.

5. $$u(x,y) = \sum_1^\infty b_n \sin\left(\frac{n\pi x}{a}\right)\frac{\sinh(n\pi y/a)}{\sinh(n\pi b/a)}, \quad b_n = \frac{8}{n^2\pi^2}\sin\left(\frac{n\pi}{2}\right)$$

7. a. $u(x,y) = 1$, but the form found by applying the methods of this section is
$$u(x,y) = \sum_{n=1}^\infty a_n \frac{\sinh(\lambda_n y) + \sinh(\lambda_n(b-y))}{\sinh(\lambda_n b)}\cos(\lambda_n x)$$

$$+\sum_{n=1}^{\infty} b_n \frac{\cosh(\mu_n x)}{\cosh(\mu_n a)} \sin(\mu_n y)$$

where

$$\lambda_n = \frac{(2n-1)\pi}{2a}, \quad a_n = \frac{4\sin\left(\frac{(2n-1)\pi}{2}\right)}{\pi(2n-1)},$$

$$\mu_n = \frac{n\pi}{b}, \quad b_n = \frac{2(1-\cos(n\pi))}{n\pi}.$$

b. $u(x,y) = y/b$, and this is found by the methods of this section. In this case, 0 is an eignevalue.

c. $\frac{4}{\pi}\sum_{1}^{\infty} \frac{(-1)^{n+1}\cos(\lambda_n y)}{(2n-1)} \frac{\sinh(\lambda_n(a-x))}{\sinh(\lambda_n a)}, \ \lambda_n = \left(\frac{2n-1}{2}\frac{\pi}{b}\right)$

Section 4-3, P. 270

1. $a_n = \frac{2}{a}\int_0^a f(x)\sin\left(\frac{n\pi x}{a}\right)dx$

3. $A(\mu) = \frac{2}{\pi}\int_0^\infty g_2(y)\sin(\mu y)dy$

5. a. $u(x,y) = \sum c_n \cos(\lambda_n x)\exp(-\lambda_n y), \ \lambda_n = (2n-1)\pi/2a$
   $c_n = 4(-1)^{n+1}/\pi(2n-1)$

   b. $u(x,y) = \int_0^\infty B(\lambda)\cosh(\lambda x)\sin(\lambda y)d\lambda, \ B(\lambda) = \frac{2\lambda}{\pi(\lambda^2+1)\cosh(\lambda a)}$

   c. $u(x,y) = \int_0^\infty A(\lambda)\cos(\lambda y)\sinh(\lambda x)d\lambda, \ A(\lambda) = \frac{2\sin(\lambda b)}{\pi\lambda\sinh(\lambda a)}$

7. $u(x,y) = \sum_{1}^{\infty} b_n \sin(\lambda_n x)\exp(-\lambda_n y)$

$$+\int_0^\infty \left(A(\mu)\frac{\sinh(\mu x)}{\sinh(\mu a)} + B(\mu)\frac{\sinh(\mu(a-x))}{\sinh(\mu a)}\right)\sin(\mu y)d\mu$$

   $\lambda_n = n\pi/a, \ b_n = 2(1-\cos(n\pi))/n\pi, \ A(\mu) = B(\mu) = 2\mu/\pi(\mu^2+1)$

   Also see Exercise 8.

9. a. $u(x,y) = \frac{2}{\pi}\int_0^\infty \frac{1-\cos(\lambda a)}{\lambda}\sin(\lambda x)\frac{\sinh(\lambda y)}{\sinh(\lambda b)}d\lambda$

   b. $u(x,y) = \frac{2}{\pi}\int_0^\infty \frac{\lambda}{1+\lambda^2}\sin(\lambda x)\frac{\sinh(\lambda(b-y))}{\sinh(\lambda b)}d\lambda$

11. $u(x,y) = \int_0^\infty \frac{2}{\pi(1+\lambda^2)} \frac{\sinh(\lambda x)}{\sinh(\lambda a)} \cos(\lambda y) d\lambda.$

13. $e^{-\lambda y}\sin(\lambda x)$, $\lambda > 0$.

15. $e^{-\lambda y}\sin(\lambda x)$, $e^{-\lambda y}\cos(\lambda x)$, $\lambda > 0$.

17. $u(x,y) = \frac{1}{\pi}\left[\frac{\pi}{2} + \tan^{-1}(x/y)\right]$

19. This solution is unbounded as x tends to infinity and cannot be found by the method of this section.

Section 4-4, P. 277

1. $v(r,\theta)$ is given by Eq. (10) with $b_n = 0$, $a_0 = \pi/2$, $a_n = -2\,(1-\cos(n\pi))/\pi n^2 c^n$.

3. The solution is as in Eq. (10) with $b_n = 0$, $a_0 = 1/\pi$, $a_1 = 1/2$, and $a_n = \frac{2\sin((n-1)\pi/2)}{\pi(n^2-1)}$ for $n \neq 1$.

5. Convergence is uniform in θ.

7. $a_0 = \frac{1}{2\pi}\int_{-\pi}^{\pi} f(\theta)d\theta$, $a_n = \frac{c^n}{\pi}\int_{-\pi}^{\pi} f(\theta)\cos(n\theta)d\theta$,

   $b_n = \frac{c^n}{\pi}\int_{-\pi}^{\pi} f(\theta)\sin(n\theta)d\theta$

9. $\frac{2}{\pi}\sum_{n=1}^{\infty} \frac{1-\cos(n\pi)}{nc^{2n}} r^{2n}\sin(2n\theta) = v(r,\theta)$

11. $v_n(r,\theta) = r^{n/\alpha}\sin(n\theta/\alpha)$ has $\partial v/\partial r$ unbounded as $r \to 0+$, if $n = 1$.

Section 4-5, P. 281

1. Hyperbolic (a) and (e); elliptic (b) and (c); parabolic (d).

3. Only (e).

5. a. $u(x,y) = \sum_1^\infty a_n \sin(n\pi x)e^{-n\pi y}$

   b. $u(x,y) = \sum_1^\infty a_n \sin(n\pi x)\cos(n\pi y)$

c. $u(x,y) = \sum_{1}^{\infty} a_n \sin(n\pi x) \exp(-n^2\pi^2 y)$

$$a_n = 2\int_0^1 f(x)\sin(n\pi x)dx$$

Chapter 4 Miscellaneous Exercises, P. 284

1. $u(x,y) = \sum_{1}^{\infty} \frac{\sinh(\lambda_n(a-x))}{\sinh(\lambda_n a)} \sin(\lambda_n y)$

   $\lambda_n = n\pi/b$, $b_n = 2(1-\cos(n\pi))/n\pi$

3. $u(x,y) = 1$. Note that 0 is an eigenvalue.

5. $u(x,y) = \sum_{n=1}^{\infty} \frac{a_n \sinh(\lambda_n x) + b_n \sinh(\lambda_n(a-x))}{\sinh(\lambda_n a)} \cos(\lambda_n y)$

   $\lambda_n = (2n-1)\pi/2b$, $a_n = b_n = 4(-1)^{n+1}/\pi(2n-1)$

7. $u(x,y) = w(x,y) + w(y,x)$, where

   $w(x,y) = \sum_{n=1}^{\infty} b_n \frac{\sinh(\lambda_n(a-y))}{\sinh(\lambda_n a)} \sin(\lambda_n x)$

   $\lambda_n = n\pi/a$, $b_n = \frac{8h}{n^2\pi^2} \sin\left(\frac{n\pi}{2}\right)$

9. $u(x,y) = \int_0^{\infty} A(\lambda) \frac{\sinh(\lambda(b-y))}{\sinh(\lambda b)} \cos(\lambda x) d\lambda$

   $A(\lambda) = 2\sin(\lambda a)/\lambda\pi$

11. $u(x,y) = \int_0^{\infty} A(\lambda)\cos(\lambda x)e^{-\lambda y} d\lambda$, $A(\lambda) = 2\alpha/\pi(\alpha^2 + \lambda^2)$

13. $u(x,y) = \frac{-1}{\pi} \tan^{-1}\left(\frac{x-x'}{y}\right)\Big|_{-\infty}^{\infty} = \frac{1}{\pi}\left[\frac{\pi}{2} - (-\frac{\pi}{2})\right]$

15. $u(r,\theta) = a_0 + \sum_{n=1}^{\infty} \left(\frac{r}{c}\right)^n (a_n\cos(n\theta) + b_n\sin(n\theta))$

    $a_0 = \frac{1}{2}$, $a_n = 0$, $b_n = \frac{(1-\cos(n\pi))}{n\pi}$

17. Same form as Exercise 15, but $a_0 = 2/\pi$,

    $a_n = 2(1+\cos(n\pi))/(1-n^2)$, $b_n = 0$ (and $a_1 = 0$).

19. $u(r,\theta) = (\ln(r) - \ln(b))/(\ln(a) - \ln(b))$

21. $u(r,\theta) = \sum_{1}^{\infty} b_n \left(\frac{r}{c}\right)^{n/2} \sin(n\theta/2)$, $b_n = \frac{1}{\pi}\int_0^{2\pi} f(\theta)\sin(n\theta/2)d\theta$

23. $u(x,y) = \sum c_n \sinh(\lambda_n y)\sin(\lambda_n x)$, $\lambda_n = (2n-1)\pi/2a$,
$c_n = 2\sin(\lambda_n a)/(a\lambda_n^2 \sinh(\lambda_n b))$

25. w satisfies the potential equation in the rectangle with boundary conditions
$w(0,y) = 0$, $w_x(a,y) = ay/b$, $0 < y < b$,
$w(x,0) = 0$, $w(x,b) = 0$, $0 < x < a$.
$w(x,y) = \sum_{n=1}^{\infty} b_n \sin(\lambda_n y)\cosh(\lambda_n x)$, $\lambda_n = n\pi/b$,
$b_n = 2a(-1)^{n+1}/n^2\pi^2\cosh(\lambda_n a)$

27. The equations become

$$\frac{\partial^2\phi}{\partial y\partial x} = \frac{\partial^2\phi}{\partial x\partial y}, \quad (1-M^2)\frac{\partial^2\phi}{\partial x^2} + \frac{\partial^2\phi}{\partial y^2} = 0.$$

29. $\phi(x,y) = \int_0^{\infty} (A(\alpha)\cos(\alpha x) + B(\alpha)\sin(\alpha x))e^{-\beta y}d\alpha + c$ where $\beta = \alpha\sqrt{1-M^2}$, c is an arbitrary constant, and

$$\left.\begin{matrix} A(\alpha) \\ B(\alpha) \end{matrix}\right\} = -\frac{U_0}{\beta\pi}\int_{-\infty}^{\infty} f'(x)\left\{\begin{matrix} \cos(\alpha x) \\ \sin(\alpha x) \end{matrix}\right\} dx$$

31. If $(x(s), y(s))$ is the parametric representation for the boundary curve $\mathcal{C}$, then the vector $y'\mathbf{i} - x'\mathbf{j}$ is normal to $\mathcal{C}$, and

$$\int_{\mathcal{C}} \frac{\partial u}{\partial n} ds = \int_{\mathcal{C}} \frac{\partial u}{\partial x} dy - \frac{\partial u}{\partial y} dx.$$

Now by Green's theorem,

$$\int_{\mathcal{C}} \frac{\partial u}{\partial x} dy - \frac{\partial u}{\partial y} dx = \iint_{\mathcal{R}} \left(\frac{\partial^2 u}{\partial x^2} + \frac{\partial^2 u}{\partial y^2}\right) dA$$

which is 0 since u satisfies the potential equation in $\mathcal{R}$.

33. Substitute directly.

35. $D + F = -\frac{1}{2}$

37. $u(x, y) = \frac{y(b-y)}{2} + \sum_{1}^{\infty} \frac{a_n \sinh(\mu_n x) + b_n \sinh(\mu_n(a-x))}{\sinh(\mu_n a)} \sin(\mu_n y)$

$\mu_n = \frac{n\pi}{b}, \quad a_n = b_n = \frac{2b^2}{\pi^3} \frac{1-\cos(n\pi)}{n^3}$

39. a. $v(x) = \frac{1}{2} x(a-x)$

b. w satisfies the potential equation in the rectangle with boundary conditions

$$\begin{aligned} w(0, y) &= 0, & w(a, y) &= 0, & 0 < y < b \\ w(x, 0) &= v(x), & w(x, b) &= v(x), & 0 < x < a. \end{aligned}$$

c. $w(x, y) = \sum (a_n \cosh(\lambda_n y) + b_n \sinh y(\lambda_n y)) \sin(\lambda_n x)$,

$\lambda_n = \frac{n\pi}{a}, \quad a_n = \frac{2a^2(1-(-1)^n)}{n^3\pi^3}, \; b_n = \frac{a_n(1-\cosh(\lambda_n b))}{\sinh(\lambda_n b)}$

d. The numerator is the integral of the first term of the series in c; the denominator is the integral of $v(x)$, minus the numerator.

Numerator integral: $(16a^4/\pi^5)\tanh(\pi b/2a)$
Integral of $v(x)$: $ba^3/12$

$b/a =$	1	2	4
ratio =	1.35	.455	.186

Chapter 5

Section 5-1, P. 296

1. $\frac{\partial^2 u}{\partial x^2} + \frac{\partial^2 u}{\partial y^2} = \frac{1}{c^2} \frac{\partial^2 u}{\partial t^2}, \quad 0 < x < a, \; 0 < y < b, \; 0 < t$

$u(x, 0, t) = 0, \; u(x, b, t) = 0, \; 0 < x < a, \; 0 < t$

$u(0, y, t) = 0, \; u(a, y, t) = 0, \; 0 < y < b, \; 0 < t$

$u(x, y, 0) = f(x, y), \; \frac{\partial u}{\partial t}(x, y, 0) = g(x, y), \; 0 < x < a, \; 0 < y < b$

3. $\dfrac{\partial^2 u}{\partial x^2} + \dfrac{\partial^2 u}{\partial y^2} + \dfrac{\partial^2 u}{\partial z^2} = \dfrac{1}{c^2}\dfrac{\partial^2 u}{\partial t^2}$

Section 5-2, P. 301

1. $\displaystyle\int_0^C \frac{\partial^2 u}{\partial z^2} dz = \frac{\partial u}{\partial z}\bigg|_0^C = 0$ by Eq. (12).

3. $W'' + (2h/b\kappa)(T_2 - W) = 0, \quad 0 < x < a, \quad W(0) = T_0, \quad W(a) = T_1$
   $W(x) = T_2 + A\cosh(\mu x) + B\sinh(\mu x)$, where $\mu^2 = 2h/b\kappa$,
   $A = T_0 - T_2, \quad B = (T_1 - T_2 - A\cosh(\mu a))/\sinh(\mu a)$

5. $\nabla^2 u = \dfrac{1}{k}\dfrac{\partial u}{\partial t}, \quad 0 < x < a, \quad 0 < y < b, \quad 0 < t$
   $\dfrac{\partial u}{\partial x}(0, y, t) = 0, \quad u(a, y, t) = T_0, \quad 0 < y < b, \quad 0 < t$
   $u(x, 0, t) = T_0, \quad \dfrac{\partial u}{\partial y}(x, b, t) = 0, \quad 0 < x < a, \quad 0 < t$
   $u(x, y, 0) = f(x, y), \quad 0 < x < a, \quad 0 < y < b,$

Section 5-3, P. 306

1. If $a = b$, the lowest eigenvalues are those with indices (m, n) in this order: $(1, 1)$; $(1, 2) = (2, 1)$; $(2, 2)$; $(3, 1) = (1, 3)$: $(3, 2) = (2, 3)$; $(1, 4) = (4, 1)$; $(3, 3)$.

3. Frequencies are $\lambda_{mn}c/2\pi$ (Hz) where λ_{mn}^2 are the eigenvalues found in the text.

5. $\lambda_{mn}^2 = (m\pi/a)^2 + (n\pi/b)^2$, for $m = 0, 1, 2, \cdots, n = 1, 2, 3, \cdots$.

7. a. $u(x, y, t) = 1$

   For b. and c. the solution has the form:

$$u(x, y, t) = \sum_{m,n} a_{mn} \cos\left(\frac{m\pi x}{a}\right) \cos\left(\frac{n\pi y}{b}\right) \exp(-\lambda_{mn}^2 kt)$$

   where $\lambda_{mn}^2 = (m\pi/a)^2 + (n\pi/b)^2$, and m and n run from 0 to ∞.

   b. $a_{00} = \dfrac{(a+b)}{2}, \quad a_{m0} = -\dfrac{2b(1 - \cos(m\pi))}{m^2\pi^2}$

   $a_{0n} = -\dfrac{2a(1 - \cos(n\pi))}{n^2\pi^2}, \quad a_{mn} = 0$ otherwise

c. $a_{00} = \dfrac{ab}{4}$, $a_{m0} = -\dfrac{ab(1-\cos(m\pi))}{m^2\pi^2}$, $a_{0n} = -\dfrac{ab(1-\cos(n\pi))}{n^2\pi^2}$

$$a_{mn} = \frac{4ab(1-\cos(n\pi))(1-\cos(m\pi))}{m^2n^2\pi^4}$$

if m and n are greater than zero.

9. The choice of a positive constant for either X''/X or Y''/Y, under the boundary conditions in Eqs. (9) and (10), will lead to the trivial solution.

Section 5-4, P. 309

1. The partial differential equations are the same, the boundary conditions become homogeneous, and in the initial conditions $g(r,\theta)$ is replaced by $g(r,\theta) - v(r,\theta)$.

3. In the heat problem, $T' + \lambda^2 kT = 0$. In the wave problem, $T'' + \lambda^2 c^2 T = 0$.

5. The boundary conditions Eqs. (10) and (11) would be replaced by
$$\Theta(0) = 0, \quad \Theta(\pi) = 0$$
Solutions are $\Theta(\theta) = \sin(n\theta)$, $n = 1, 2, \cdots$.

7. Taking the hint, and using the fact that $\nabla^2\phi = -\lambda^2\phi$, the left-hand side becomes
$$(\lambda_k^2 - \lambda_m^2)\iint_{\mathcal{R}} \phi_k\phi_m$$
while the right-hand side is zero, because of the boundary condition.

Section 5-5, P. 314

1. $\lambda_n = \alpha_n/a$, where α_n is the nth zero of the Bessel function J_0. The solutions are $\phi_n(r) = J_0(\lambda_n r)$, or any constant multiple thereof.

3. This is just the chain rule.

5. Rolle's theorem says that if a differentiable function is zero in two places, its derivative is zero somewhere between. From Exercise 4 it is clear that J_1 must be zero between consecutive zeros of J_0. Check Fig. 7 and Table 1.

7. Use the second formula of Exercise 6, after replacing μ by $\mu + 1$ on both sides.

9. $u(r) = T + (T_1 - T)I_0(\gamma r)/I_0(\gamma a)$.

Section 5-6, P. 320

1. $v(0,t)/T_0 \cong 1.602\exp(-5.78\tau) - 1.065\exp(-30.5\tau)$ where $\tau = kt/a^2$

3. $v(r,t) = \sum_{n=1}^{\infty} a_n J_0(\lambda_n r)\exp\left(-\lambda_n^2 kt\right)$, $\lambda_n = \alpha_n/a$. Use Eq. (13) and others to find $a_n = T_0 J_1(\alpha_n/2)/\alpha_n J_1^2(\alpha_n)$.

5. Integration leads to the equality

$$\int_0^a (r\phi'^2)' dr + \lambda^2 \int_0^a r^2(\phi^2)' dr = 0.$$

The first integral is evaluated directly. The second must be integrated by parts.

Section 5-7, P. 327

1. Use

$$\frac{1}{r}\frac{d}{dr}\left(r\frac{d}{dr}J_0(\lambda r)\right) = -\lambda^2 J_0(\lambda r).$$

3. The frequencies of vibration are $\lambda_{mn}c = \alpha_{mn}c/a$. The five lowest values of α_{mn}, in order, have subscripts (0,1), (1,1), (2,1), (0,2), and (3,1). See Table 1 in Section 6.

5. $\phi(a,\theta) = 0$ and $\phi(r,-\pi) = \phi(r,\pi)$, $\frac{\partial\phi}{\partial\theta}(r,-\pi) = \frac{\partial\phi}{\partial\theta}(r,\pi)$

7. Set $J_m(\lambda_{mn}r) = \phi_n$. Then $(r\phi_n')' = -\lambda_{mn}r\phi_n$ and $(r\phi_q')' = -\lambda_{mq}r\phi_q$ are the equations satisfied by the functions in the integrand. Follow the proof in Chapter 2, Section 7.

Section 5-8, P. 333

1. $\phi(x) = x^\alpha[AJ_p(\lambda x) + BY_p(\lambda x)]$, where $\alpha = (1-n)/2$, $p = |\alpha|$.

3. For $\lambda^2 = 0$, $Z = A + Bz$.

5. $\phi(\rho + ct) = \bar{F}_o(\rho + ct) + \bar{G}_e(\rho + ct)$

   $\psi(\rho - ct) = \bar{F}_o(\rho - ct) - \bar{G}_e(\rho - ct)$

   where $\bar{F}_o(x)$ is the odd periodic extension with period $2a$ of $xf(x)/2$, and $\bar{G}_e(x)$ is the even periodic extension with period $2a$ of $\int(x/2c)g(x)dx$.

7. The weight function is ρ^2 and the interval is 0 to a.

9. $v(x) = (b - x)(x - a)/(a + b)x^2$

11. No. The idea is to find a solution of the partial differential equation that depends on only one variable. That is impossible if f depends on both x and y.

13. $$a_n = -\frac{\int_a^b v(x)X_n(x)x^3dx}{\int_a^b X_n^2(x)x^3dx}$$

Section 5-9, P. 343

1. $[k(k+1) - \mu^2]a_{k+1} - [k(k-1) - \mu^2]a_{k-1} = 0$, valid for $k = 1, 2, \cdots$.

3. $P_5 = \frac{1}{8}(63x^5 - 70x^3 + 15x)$

5. $y = A \ln\left(\frac{1+x}{1-x}\right)$

7. Differentiate Eq. (9) and add to it n times Eq. (8).

9. Leibniz's rule states that
$$(uv)^{(k)} = \sum_{r=0}^{k} \binom{k}{r} u^{(k-r)}v^{(r)}.$$
(A superscript k in parentheses means kth derivative.) The right-hand side looks like the binomial theorem. In the case at hand, at most two terms are not zero.

11. $b_n = 0$ (n odd), $= \dfrac{-(-1)^{n/2}(2n+1)}{(n+2)(n-1)} \dfrac{1 \cdot 3 \cdot 5 \cdots (n-1)}{2 \cdot 4 \cdot 6 \cdots n}$ (n even)

Section 5-10, P. 351

1. The solution is as given in Eq. (5) with coefficients as shown in Eq. (7). The integration yields (see Section 5-9)

$b_0 = \frac{1}{2}$; $b_n = 0$ for n even; $b_1 = 3/4$ and

$$b_n = \frac{(-1)^{(n-1)/2}}{2 \cdot c^n} \frac{1 \cdot 3 \cdot 5 \cdots (n-2)}{2 \cdot 4 \cdot 6 \cdots (n-1)} \cdot \frac{2n+1}{n+1}, \; n = 3, 5, 7, \cdots.$$

3. $u(\phi, t) = T - \sum b_n P_n(\cos(\phi)) \exp(-(\mu^2 + n(n+1))kt/R^2)$, where $\mu^2 = \gamma^2 R^2$, n is odd, and b_n is as at the end of Part B, with $T_0 = T$.

5. The eigenfunctions are as in Part C, except that n must be odd in order to satisfy the boundary condition.

7. The nodal surfaces are: a sphere at $\rho = 0.634$ and two nappes of a cone given by $\phi = 0.304\pi$ and $\rho = 0.696\pi$.

Chapter 5 Miscellaneous Exercises, P. 354

1. $$u(x, y, t) = \sum_{m=1}^{\infty} a_m \sin(\mu_m y) \exp(-\mu_m^2 kt)$$

$$+ \sum_{n=1}^{\infty} a_{mn} \cos(\lambda_n x) \sin(\mu_m y) \exp(-(\mu_m^2 + \lambda_n^2)kt)$$

$$\mu_m = m\pi b, \quad \lambda_n = n\pi/a$$

$$a_m = T\frac{1 - \cos(m\pi)}{m\pi}$$

$$a_{mn} = \frac{4T}{\pi^3} \frac{(\cos(n\pi) - 1)(1 - \cos(m\pi))}{n^2 m}$$

3. $$u(a/2, b/2, t) = \sum_{n=1}^{\infty} b_{mn} \sin\left(\frac{n\pi}{2}\right) \sin\left(\frac{m\pi}{2}\right) \exp(-(\lambda_n^2 + \mu_m^2)kt)$$

where $\lambda_n = n\pi/a$, $\mu_m = m\pi/b$, and

$$b_{mn} = \frac{4T}{\pi^2} \frac{(1 - \cos(m\pi))(1 - \cos(n\pi))}{mn}$$

The first three nonzero terms are, for $a = b$, those with $(m, n) =$ (1,1), (1,3) = (3,1), (3,3). All terms with an even index are 0.

$$u(a/2, a/2, t) \cong \frac{16t}{\pi^2}\left(e^{-2\tau} - \frac{2}{3}e^{-10\tau} + \frac{1}{9}e^{-18\tau}\right)$$

where $\tau = kt\pi^2/a^2$.

5. $u(r) = (a^2 - r^2)/2$ and $u(r) = \sum_{1}^{\infty} C_n J_0(\lambda_n r)$, with $C_n = \dfrac{2a^2}{\alpha_n^3 J_1(\alpha_n)}$

7. $w(x,t) = a_0 + \sum_{n=1}^{\infty} a_n \cos(\lambda_n x) \exp(-\lambda_n^2 kt)$

$v(y,t) = \sum b_m \sin(\mu_m y) \exp(-\mu_m^2 kt)$

where $\mu_m = m\pi/b$, $\lambda_n = n\pi/a$, and initial conditions are

$$v(y,0) = 1, \quad 0 < y < b; \quad w(x,0) = Tx/a, \quad 0 < x < a.$$

9. $J_0(\lambda r) \exp(-\lambda^2 kt)$

11. $B_k = b_k/k(k+1)$ for $k = 1, 2, \cdots$; b_0 must be 0, and B_0 is arbitrary.

13. $((1-x^2)y')' - \dfrac{m^2}{1-x^2}y + \mu^2 y = 0$

15. $u(r,z) = \sum_{n=1}^{\infty} a_n \dfrac{\sinh(\lambda_n z)}{\sinh(\lambda_n b)} J_0(\lambda_n r)$

where $\lambda_n = \alpha_n/a$ and $a_n = \dfrac{2U_0}{\alpha_n J_1(\alpha_n)}$

17. $u(r,z,t) = \sin(\mu z) J_0(\lambda r) \sin(\nu ct)$ is a product solution if $\mu = m\pi/b$, $\lambda = \alpha_n/a$, and $\nu = \sqrt{\mu^2 + \lambda^2}$. The frequencies of vibration are therefore νc or

$$c\sqrt{\left(\frac{m\pi}{b}\right)^2 + \left(\frac{\alpha_n}{a}\right)^2}.$$

19. Each of the two terms satisfies $\nabla^2\phi = -(5\pi^2)\phi$. On $y = 0$ and $x = 1$, both terms are 0; on $y = x$ they are obviously equal in value, opposite in sign.

21. Each term satisfies $\nabla^2\phi = -(16\pi^2/3)\phi$.

On $y = 0$, $\phi = \sin(2n\pi x) - \sin(2n\pi x)$;

on $y = \sqrt{3}x$, $\phi = \sin(4n\pi x) + 0 - \sin(2n\pi \cdot 2x)$;

on $y = \sqrt{3}(1-x)$, $\phi = \sin(4n\pi x) + \sin(2n\pi(1-2x)) - \sin 2n\pi$.

23. For a sextant, $\phi_n = J_{3n}(\lambda r)\sin(3n\theta)$ and $J_{3n}(\lambda) = 0$. Thus $\lambda_1 = 6.380$, which is less than $\sqrt{16\pi^2/3} = 7.255$.

25. $b = -1$.

27. Since $y(x) = hJ_0(kx)/J_0(ka)$, where $k = \omega/\sqrt{gU}$, the solution cannot have this form if $J_0(ka) = 0$.

29. $u(r,t) = R(t)T(t)$; $R(r) = r^{-m}J_m(\lambda r)$, where $m = (n-2)/2$; $T(t) = a\cos(\lambda ct) + b\sin(\lambda ct)$.

31. $b = \dfrac{D_r L^2}{D_z R^2}, \quad \rho = \dfrac{UL}{D_z}$

33. $a_0 = 12.77, \quad a_1 = -4.88.$

Chapter 6

Section 6-1, P. 366

1. c. $\dfrac{s^2 + 2\omega^2}{s(s^2 + 4\omega^2)}$ d. $\dfrac{\omega\cos(\phi) - s\sin(\phi)}{s^2 + \omega^2}$

   e. $\dfrac{e^2}{s - 2}$ f. $\dfrac{2\omega^2}{s(s^2 + 4\omega^2)}$

3. a. $\dfrac{e^{-as}}{s}$ b. $\dfrac{(e^{-as} - e^{-bs})}{s}$ c. $\dfrac{1 - e^{-as}}{s^2}$

5. a. $e^{-t}\sinh(t)$ b. e^{-2t} c. $e^{-at}\dfrac{\sin(\sqrt{b^2 - a^2}t)}{\sqrt{b^2 - a^2}}$

7. a. $\dfrac{(e^{at - e^{bt}})}{(a - b)}$ b. $\dfrac{t}{2a}\sinh(at)$ d. $\dfrac{t^2 e^{at}}{2}$

   e. $f(t) = 1,\ 0 < t < 1;\ = 0,\ 1 < t$

Section 6-2, P. 376

1. a. e^{2t} b. e^{-2t} c. $\dfrac{3e^{-t} - e^{-3t}}{2}$ d. $\sin(3t)/3$

3. a. $\dfrac{(1 - e^{-at})}{a}$ b. $t - \sin(t)$ c. $\left(\sin(t) - \dfrac{1}{2}\sin(2t)\right)/3$

   d. $(\sin(2t) - 2t\cos(t))/8$ e. $-\dfrac{3}{4} + \dfrac{1}{2}t + e^{-t} - \dfrac{1}{4}e^{-2t}$

5. a. $\dfrac{(e^{2t} - e^{-2t})}{4}$ b. $\dfrac{1}{2}\sin(2t)$

   c. $\dfrac{3}{2} + \dfrac{i\sqrt{2} - 3}{4}\exp(-i\sqrt{2}t) - \dfrac{i\sqrt{2} - 3}{4}\exp(i\sqrt{2}t)$ d. $4(1 - e^{-t})$

7. a. $1 - \cos(t)$ b. $\dfrac{e^t - \cos(\omega t) + \omega\sin(\omega t)}{\omega^2 + 1}$ c. $t - \sin(t)$

Section 6-3, P. 383

1. a. $s = -\left(\frac{2n-1}{2}\pi\right)^2, \quad n = 1, 2, \cdots$

   b. $s = \pm i\frac{2n-1}{2}\pi,\ n = 1, 2, \cdots$

   c. $s = \pm in\pi,\ n = 0, 1, 2, \cdots$

   d. $s = i\eta$, where $\tan\eta = \frac{-1}{\eta}$

   e. $s = i\eta$, where $\tan\eta = \frac{1}{\eta}$

3. a. $\frac{\sinh(\sqrt{s}x)}{s^2\sinh(\sqrt{s})}$ b. $\frac{1}{s} - \frac{\cosh(\sqrt{s}(\frac{1}{2}-x))}{s(s+1)\cosh(\sqrt{s}/2)}$

5. a. $u(x,t) = x + \sum_{n=1}^{\infty}\frac{2\sin(n\pi x)}{n\pi\cos(n\pi)}\exp(-n^2\pi^2 t)$

   b. $u(x,t)$ is 1 minus the solution of Example 3.

Section 6-4, P. 390

1. $t + \frac{x^2}{2}$

3. $v(x,t) = \frac{4}{\pi^2}\sum_{1}^{\infty}\frac{\cos((2n-1)(\frac{1}{2}-x))\sin((2n-1)\pi t)}{(2n-1)^2\sin\left(\frac{2n-1}{2}\pi\right)}$

5. a. $\frac{\omega}{\omega^2-\pi^2}\left(\frac{1}{\pi}\sin(\pi t) - \frac{1}{\omega}\sin(\omega t)\right)\sin(\pi x)$

   b. $\frac{1}{2\pi^2}(\sin(\pi t) - \pi t\cos(\pi t))\sin(\pi x)$

7. a. $u(x,t) = x - \frac{\sin(\sqrt{a}x)}{\sin(\sqrt{a})}e^{-at} + \frac{2a}{\pi}\sum_{1}^{\infty}\frac{\sin(n\pi x)\exp(-n^2\pi^2 t)}{n(a-n^2\pi^2)\cos(n\pi)}$

   b. The term $-\frac{x\cos(n\pi x)}{\cos(n\pi)}\exp(-n^2\pi^2 t)$ arises.

Chapter 6 Miscellaneous Exercises, P. 392

1. $U(s) = \dfrac{T_0}{\gamma^2 + s} + \dfrac{\gamma^2 T}{s(\gamma^2 + s)}$

   $u(x,t) = T_0 \exp(-\gamma^2 t) + T(1 - \exp(-\gamma^2 t))$

3. $U(s) = \dfrac{\cosh(\sqrt{s}x)}{s^2 \cosh(\sqrt{s})}$

   $$u(x,t) = t - \frac{1 - x^2}{2} - \sum_{n=1}^{\infty} \frac{2\cos(\rho_n x)}{\rho_n \sin(\rho_n)} \exp(-\rho_n^2 t)$$

   where $\rho_n = (2n-1)\pi/2$

5. $$u(x,t) = \frac{x(1-x)}{2} - \sum_{n=1}^{\infty} \frac{4\cos(\rho_n(x - \frac{1}{2}))}{\rho_n \sin(\rho_n/2)} \exp(-\rho_n^2 t)$$

   where $\rho_n = (2n-1)\pi$

7. $$u(x,t) = x + \sum_{1}^{\infty} \frac{2\sin(n\pi x)}{n\pi\cos(n\pi)} \exp(-n^2\pi^2 t)$$

9. $U(x,s) = \dfrac{1}{s}(1 - \exp(-\sqrt{s}x))$

11. $f(t) = \dfrac{x}{\sqrt{4\pi t^3}} \exp(-x^2/4t)$

13. $$u(x,t) = \sum_{n=0}^{\infty} \left[\operatorname{erfc}\left(\frac{2n+1-x}{\sqrt{4t}}\right) - \operatorname{erfc}\left(\frac{2n+1+x}{\sqrt{4t}}\right) \right]$$

15. $$F(s) = 2\sum_{n=1}^{\infty} \frac{1}{s^2 + n^2}$$

17. $$f(t) = \sum_{-\infty}^{\infty} \frac{1}{2a} G\left(\frac{in\pi}{a}\right) e^{in\pi t/a}$$

19. $F(s)$ must be of the form $F(s) = G(s)/H(s)$ where $G(s)$ is never infinite. The solutions of $H(s) = 0$ must form an arithmetic sequence of purely imaginary numbers, and $H'(s) \neq 0$ if $H(s) = 0$.

21. $$F(s) = \frac{(1 - e^{-\pi s})^2}{s(1 - e^{-2\pi s})} = \frac{1 - e^{-\pi s}}{s(1 + e^{-\pi s})}$$

23. $F(s) = \dfrac{1+e^{-\pi s}}{(s^2+1)(1-e^{-\pi s})}$

25. $u(x,t) = \dfrac{\sin(\omega x)\sin(\omega t)}{\sin(\omega)} + \sum_{n=1}^{\infty} \dfrac{(-1)^{n+1}2\omega}{\omega^2 - n^2\pi^2} \sin(n\pi x)\sin(n\pi t)$

27. $U(x,s) = \dfrac{\omega}{s^2+\omega^2} e^{mx}$, $m = \frac{1}{2} - \sqrt{\frac{1}{4}+s}$

29. $\alpha^2 = \frac{1}{2}\left(\frac{1}{4} \pm \sqrt{(\frac{1}{4})^2 + \omega^2}\right)$. Since α must be real, take the + sign.

Chapter 7

Section 7-1, P. 404

1. $16(u_{i+1} - 2u_i j + u_{i-1}) = -1$, $i = 1,2,3$, $u_0 = 0$, $u_4 = 1$. Solution: $u_1 = 11/32$, $u_2 = 5/8$, $u_3 = 27/32$.

3. $16(u_{i+1} - 2u_i + u_{i-1}) - u_i = -\frac{1}{2}i$, $i = 1,2,3$, $u_0 = 0$, $u_4 = 1$. Solution: $u_1 = 0.285$, $u_2 = 0.556$, $u_3 = 0.800$.

5. $16(u_{i+1} - 2u_i + u_{i-1}) = \frac{1}{4}i$, $i = 0,1,2,3$, $u_0 - 2(u_1 - u_{-1}) = 1$, $u_4 = 0$. Solution: $u_0 = 0.422$, $u_1 = 0.277$, $u_2 = 0.148$, $u_3 = 0.051$.

7. $n = 3$: $u_1 = 4.76$, $u_2 = 4.24$; $n = 4$: $u_1 = 6.65$, $u_2 = 9.14$, $u_3 = 5.92$. The actual solution, $u(x) = -\sin(\sqrt{10}x)/\sin(\sqrt{10})$, has a maximum of about 50. The boundary value problem is nearly singular.

9. $25(u_{i+1} - 2u_i + u_{i-1}) - 25u_i = -25$, $i = 1,2,3,4,5$; $u_0 = 2$, $u_5 + (u_6 - u_4)/(2/5) = 1$. When the equation for $i = 5$ and the boundary condition are combined, they become $2u_4 - 3.4u_5 = -1.4$. Solution: $u_1 = 1.382$, $u_2 = 1.146$, $u_3 = 1.057$, $u_4 = 1.023$, $u_5 = 1.014$.

11. $9(u_{i+1} - 2u_i + u_{i-1}) + (3/2)(u_{i+1} - u_{i-1}) - u_i = -(1/3)i$, $i = 0,1,2$; $u_3 = 1$, $(u_1 - u_{-1})/(2/3) = 0$. When u_{-1} is eliminated and coefficients collected, the equations to solve are

    $-19u_0 + 18u_1 = 0$

    $7\frac{1}{2}u_0 - 19u_1 + 10\frac{1}{2}u_2 = -\frac{1}{3}$

    $7\frac{1}{2}u_1 - 19u_2 = -11\frac{1}{6}$.

    Solution: $u_0 = 0.795, u_1 = 0.839, u_2 = 0.919$.

Section 7-2, P. 409

1. Line m of the solution should be exactly the same as line $m+1$ of Table 4.

3. $r = 2/5$, $\Delta t = 1/40$

m \ i	0	1	2	3	4
0	*0*	*0*	*0*	*0*	*0*
1	*1*	0.	0.	0.	0.
2	*1*	0.4	0.	0.	0.
3	*1*	0.48	0.16	0.	0.
4	*1*	0.56	0.224	0.064	0.
5	*1*	0.6016	0.2944	0.1024	0.0512

5. $\Delta t = 1/32$. Remember that $u_4(m) = u_0(m) = m\Delta t$. All numbers in the table should be multiplied by Δt.

m \ i	0	1	2	3	4
0	*0*	*0*	*0*	*0*	*0*
1	*1*	0	0	0	*1*
2	*2*	1/2	0	1/2	*2*
3	*3*	1	1/2	1	*3*
4	*4*	7/4	1	7/4	*4*
5	*5*	5/2	7/4	5/2	*5*

7. $\Delta t = 1/32$. All numbers in this table should be multiplied by Δt.

m \ i	0	1	2	3	4
0	*0*	*0*	*0*	*0*	*0*
1	*0*	1	1	1	*0*
2	*0*	3/2	2	3/2	*0*
3	*0*	2	5/2	2	*0*
4	*0*	9/4	3	9/4	*0*
⋮					
∞	*0*	3	4	3	*0*

9. $\Delta t = \frac{1}{32}$. Remember $u_{-1} = u_1$. All entries in this table should be multiplied by $\Delta x = \frac{1}{4}$.

m \ i	0	1	2	3	4
0	*0*	*1*	*2*	*3*	*4*
1	1	1	2	3	*4*
2	1	3/2	2	3	*4*
3	3/2	3/2	9/4	3	*4*
4	3/2	15/8	9/4	25/8	*4*
5	15/8	15/8	5/2	25/8	*4*

Section 7-3, P. 415

1.

m \ i	0	1	2	3	4
0	*0*	*0*	*0*	*0*	*0*
1	*0*	1/4	1/4	1/4	*0*
2	*0*	1/4	1/2	1/4	*0*
3	*0*	1/4	1/4	1/4	*0*
4	*0*	0	0	0	*0*
5	*0*	−1/4	−1/4	−1/4	*0*

3. In this table, $\alpha = 1/\sqrt{2}$.

m \ i	0	1	2	3	4
0	*0*	*0*	*0*	*0*	*0*
1	*0*	$\alpha/4$	1/4	$\alpha/4$	*0*
2	*0*	1/4	$\alpha/2$	1/4	*0*
3	*0*	$\alpha/4$	1/4	$\alpha/4$	*0*
4	*0*	0	0	0	*0*
5	*0*	$-\alpha/4$	−1/4	$-\alpha/4$	*0*

5.

t_m	m \ i	0	1	2	3	4
0	0	*0*	*1/2*	*1*	*1/2*	*0*
0.177	1	*0*	1/2	3/4	1/2	*0*
0.354	2	*0*	3/8	1/4	3/8	*0*
0.530	3	*0*	0	−1/8	0	*0*
0.707	4	*0*	−7/16	−3/8	−7/16	*0*
0.884	5	*0*	−5/8	−11/16	−5/8	*0*

7.

m \ i	0	1	2	3	4
0	*0*	*0*	*0*	*0*	*0*
1	*0*	0	0	0	*1*
2	*0*	0	0	1	*1*
3	*0*	0	1	1	*1*
4	*0*	1	1	1	*0*
5	*0*	1	1	0	*−1*
6	*0*	0	0	−1	*−1*
7	*0*	−1	−2	−1	*−1*
8	*0*	−2	−2	−2	*0*

9. Run: $u_i(m+1) = (2 - 2\rho^2 - 16\Delta t^2)u_i(m) + \rho^2 u_{i-1}(m) + \rho^2 u_{i+1}(m) - u_i(m-1)$. Start: $u_i(1) = \frac{1}{2}((2 - 2\rho^2 - 16\Delta t^2)u_i(0) + \rho^2 u_{i-1}(0) + \rho^2 u_{i+1}(0))$. Longest stable time step: $\Delta t = 1/\sqrt{24}$ ($\rho^2 = 2/3$).

m \ i	0	1	2	3	4
0	*0*	*0.50*	*1.00*	*0.50*	*0*
1	*0*	0.33	0.33	0.33	*0*
2	*0*	−0.28	−0.56	−0.28	*0*
3	*0*	−0.70	−0.70	−0.70	*0*
4	*0*	−0.19	−0.38	−0.19	*0*
5	*0*	0.45	0.45	0.45	*0*
6	*0*	0.49	0.98	0.49	*0*
7	*0*	0.21	0.21	0.21	*0*
8	*0*	−0.35	−0.71	−0.35	*0*

Section 7-4, P. 422

1. At $(1/4, 1/4)$, 11/256; at $(1/2, 1/4)$, 14/256; at $(1/2, 1/2)$, 18/256.

3. In both 3. and 4. the exact solution is $u(x, y) = xy$, and the numerical solutions are exact.

5. Coordinates and values of the corresponding u_i are: $(1/7, 1/7)$, 5α; $(2/7, 1/7)$, 10α; $(3/7, 1/7)$, 14α; $(1/7,\ 2/7)$, 21α; $(2/7,\ 2/7)$, 32α. Here $\alpha = 19/1159$.

7. $u_1 = .670$, $u_2 = .721$, $u_3 = .961$, $u_4 = 1.212$, $u_5 = .954$, $u_6 = .651$. The remaining values are found by symmetry.

9. $u_1 = .386$, $u_2 = .542$, $u_3 = .784$, $u_4 = .595$. The remaining values are found by symmetry.

Section 7-5, P. 429

1. Use Eq. (8) with $r = 1/4$

m \ i	1	2	3	4	5	6
0	*0*	*0*	*0*	*0*	*0*	*0*
1	0	0	0	1/4	1/4	1/4
2	1/16	1/16	1/16	5/16	3/8	5/16
3	3/32	1/8	3/32	23/64	27/64	23/64

3. Note that $u_1 = u_2 = u_4 = u_5$; replacement equations become

$u_1(m + 1) = u_3(m)/4$, $u_3(m + 1) = u_1(m)$.

m \ i	1	3
0	*1*	*1*
1	1/4	1
2	1/4	1/4
3	1/16	1/4
4	1/16	1/16

5. Use Eq. (8) with $r = 1/4$. Note that $u_4 = u_2$, $u_7 = u_3$, $u_8 = u_6$.

m \ i	1	2	3	5	6	9
0	*0*	*0*	*0*	*0*	*0*	*0*
1	0	0	1/4	0	1/4	1/2
2	0	1/16	5/16	1/8	7/16	5/8
3	1/32	7/64	3/8	1/4	17/64	23/32

7. Use the same numbering as for Exercise 5. Note that $u_1 = u_3 = u_7 = u_9$ and $u_2 = u_4 = u_6 = u_8$. The running equations become

$$u_1(m+1) = \frac{1}{2}u_2(m) - u_1(m-1),$$

$$u_2(m+1) = \frac{1}{2}u_1(m) + \frac{1}{4}u_5(m) - u_2(m-1),$$

$$u_5(m+1) = u_2(m) - u_5(m-1)$$

$m =$	0	1	2	3	4	5	6	7
u_1	*0*	0	0	1/8	0	−5/16	0	15/32
u_2	*0*	0	1/4	0	−3/8	0	5/16	0
u_5	*0*	1	0	−3/4	0	3/8	0	−1/16

9.

m \ i	1	2	5
0	*1*	*1*	*1*
1	1/2	3/4	1
2	−1/4	0	1/2
3	−1/2	−3/4	−3/2
4	−1/2	−5/4	−2
5	−3/4	−3/4	−1

11. See figure below for numbering of points.

m \ i	1	2	3	4	5	6
0	*1*	*0*	*0*	*0*	*0*	*0*
1	0	1/4	1/4	0	0	0
2	−3/4	0	0	1/4	1/8	0
3	0	−1/2	−7/16	0	0	3/16

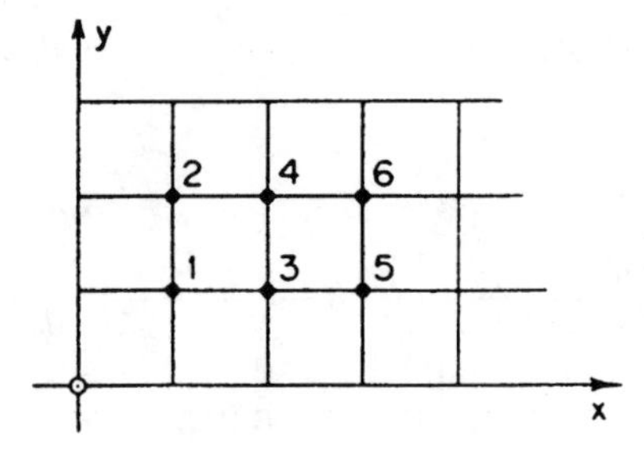

Chapter 7, Miscellaneous Exercises, P. 432

1. $\dfrac{u_{i+1} - 2u_i - u_{i-1}}{(\Delta x)^2} - \sqrt{24x_i}u_i = 0$, $i = 0, 1, 2,$

$\dfrac{u_1 - u_{-1}}{2\Delta x} = 1$, $u_3 = 1;$

$-18u_0 + 18u_1 = 6$

$9u_0 - 20.83u_1 + 9u_2 = 0$

$9u_1 - 22u_2 = -9$

$u_0 = -0.248, u_1 = 0.08, u_2 = 0.44$

3. $\dfrac{u_{i+1} - 2u_i + u_{i-1}}{(\Delta x)^2} + \dfrac{1}{1+x_i}\dfrac{u_{i+1} - u_{i-1}}{2\Delta x} = -(1 + x_i),$

$i = 1, 2, 3; u_0 = 1, u_4 = 0.$

$-32u_1 + 17.60u_2 = -15.65$

$14.67u_1 - 32u_2 + 17.33u_3 = -1.5$

$14.86u_2 - 32u_3 = -1.75$

$u_1 = 0.822, u_2 = 0.606, u_3 = 0.335$

5. $u_i(m+1) = (u_{i-1}(m) + u_{i+1}(m))/2$. Note that $u_3(m) = u_1(m)$ and $u_4(m) = u_0(m)$.

m \ i	0	1	2
0	*0*	*0*	*0*
1	0.03	0	0
2	0.06	0.015	0
3	0.09	0.03	0.15
4	0.12	0.053	0.03
5	0.14	0.075	0.053
6	0.17	0.1	0.075
7	0.20	0.122	0.10
8	0.22	0.15	0.122

7. First problem: $u_i(m+1) = (u_{i+1}(m) + u_i(m) + u_{i-1}(m))/3$; second problem $u_i(m+1) = (u_{i+1}(m) + u_{i-1}(m))/3$

	First Problem					Second Problem				
m \ i	0	1	2	3	4	0	1	2	3	4
0	*0*	*1*	*1*	*1*	*0*	*0*	*1*	*1*	*1*	*0*
1	*0*	2/3	1	2/3	*0*	*0*	1/3	2/3	1/3	*0*
2	*0*	5/9	7/9	5/9	*0*	*0*	2/9	2/9	2/9	*0*
3	*0*	14/27	17/27	14/27	*0*	*0*	2/27	4/27	2/27	*0*
4	*0*	31/81	45/81	31/81	*0*	*0*	4/81	4/81	4/81	*0*

9. $u_i(m+1) = (u_{i+1}(m) + u_{i-1}(m))/2$

m \ i	0	1	2	3	4	5
0	*0*	*0*	*0*	*0*	*0*	*0*
1	*1/2*	0	0	0	0	*0*
2	*1*	1/4	0	0	0	*0*
3	*3/2*	1/2	1/8	0	0	*0*
4	*2*	13/16	1/4	1/16	0	*0*
5	*5/2*	9/8	7/16	1/8	1/32	*0*
6	*3*	87/32	5/8	15/64	1/16	*0*

11.

i / m	0	1	2	3	4
0	*0*	*0*	*0*	*0*	*0*
1	*0*	0	0	0	*1*
2	*0*	0	0	1	*1*
3	*0*	0	1	1	*1*
4	*0*	1	1	1	*1*
5	*0*	1	1	1	*1*
6	*0*	0	1	1	*1*
7	*0*	0	0	1	*1*
8	*0*	0	0	0	*1*

13. Let $u_{ij} \cong u(x_i, y_j)$. Then $u_{11} = u_{22} = u_{33} = 0.5$, $u_{12} = 0.698$, $u_{13} = 0.792$, $u_{21} = 0.302$, $u_{23} = 0.624$, $u_{21} = 0.209$, $u_{32} = 0.376$.

15. Number as in Fig. 7.4. Then $u_1 = u_3 = u_7 = u_9$ and $u_2 = u_4 = u_6 = u_8$.

$m =$	0	1	2	3	4	5
u_1	*1*	1/2	3/8	1/4	3/16	1/8
u_2	*1*	3/4	1/2	3/8	1/4	3/16
u_5	*1*	1	3/4	1/2	3/8	1/4

Appendix Solutions, P. 454

1. Solution for $K\Delta x^2 = 0.25$;

$X:$	0	0.05	0.1	0.15	0.2	0.25	0.3	0.35	0.4	0.45	0.5
$U:$	0	0.555	0.86	0.914	0.948	0.968	0.98	0.988	0.992	0.994	0.995

3. The numerical solution is shown below. The exact solution differs by about .003

$X:$	0	0.1	0.2	0.3	0.4	0.5	0.6	0.7	0.8	0.9	1
$U:$	0	−0.045	−0.08	−0.105	−0.12	−0.125	−0.12	−0.105	−0.08	−0.045	0

5. Solution for $s = 0.1$

$x:$	0	0.1	0.2	0.3	0.4	0.5	0.6	0.7	0.8	0.9	1
$u:$	1	0.4921	0.2661	0.1503	0.087	0.051	0.0299	0.0173	0.0094	0.004	0

Solution for $s = 1$

$x:$	0	0.1	0.2	0.3	0.4	0.5	0.6	0.7	0.8
$u:$	1	0.6132	0.3773	0.2327	0.1437	0.0886	0.0542	0.0323	0.018

7.

$x \rightarrow$ t	0	0.1	0.2	0.3	0.4	0.5	0.6	0.7	0.8	0.9	
0	0	−0.1	−0.2	−0.3	−0.4	−0.5	−0.6	−0.7	−0.8	−0.9	[illegible]
0.1	0	−0.1	−0.2	−0.3	−0.4	−0.5	−0.6	−0.7	−0.8	−0.9	[illegible]
0.2	0	−0.1	−0.2	−0.3	−0.4	−0.5	−0.6	−0.7	−0.8	−0.8	[illegible]
0.3	0	−0.1	−0.2	−0.3	−0.4	−0.5	−0.6	−0.7	−0.7	−0.7	[illegible]
0.4	0	−0.1	−0.2	−0.3	−0.4	−0.5	−0.6	−0.6	−0.6	−0.6	[illegible]
0.5	0	−0.1	−0.2	−0.3	−0.4	−0.5	−0.5	−0.5	−0.5	−0.5	[illegible]
0.6	0	−0.1	−0.2	−0.3	−0.4	−0.4	−0.4	−0.4	−0.4	−0.4	[illegible]
0.7	0	−0.1	−0.2	−0.3	−0.3	−0.3	−0.3	−0.3	−0.3	−0.3	[illegible]
0.8	0	−0.1	−0.2	−0.2	−0.2	−0.2	−0.2	−0.2	−0.2	−0.2	[illegible]
0.9	0	−0.1	−0.1	−0.1	−0.1	−0.1	−0.1	−0.1	−0.1	−0.1	[illegible]
1	0	0	0	0	0	0	0	0	0	0	
1.1	0	0.1	0.1	0.1	0.1	0.1	0.1	0.1	0.1	0.1	[illegible]
1.2	0	0.1	0.2	0.2	0.2	0.2	0.2	0.2	0.2	0.2	[illegible]
1.3	0	0.1	0.2	0.3	0.3	0.3	0.3	0.3	0.3	0.3	[illegible]
1.4	0	0.1	0.2	0.3	0.4	0.4	0.4	0.4	0.4	0.4	[illegible]
1.5	0	0.1	0.2	0.3	0.4	0.5	0.5	0.5	0.5	0.5	[illegible]
1.6	0	0.1	0.2	0.3	0.4	0.5	0.6	0.6	0.6	0.6	[illegible]
1.7	0	0.1	0.2	0.3	0.4	0.5	0.6	0.7	0.7	0.7	[illegible]
1.8	0	0.1	0.2	0.3	0.4	0.5	0.6	0.7	0.8	0.8	[illegible]
1.9	0	0.1	0.2	0.3	0.4	0.5	0.6	0.7	0.8	0.9	[illegible]
2	0	0.1	0.2	0.3	0.4	0.5	0.6	0.7	0.8	0.9	

9. The solution for the left half of the channel section is shown below. The numbers (to three decimals) are identical for the dogleg section, so there is no difference in rigidity at this level of accuracy.

	0	0.25	0.5	0.75	1	1.25	1.5	1.75	2
0	0	0	0	0					
0.25	0	0.051	0.066	0.051	0				
0.5	0	0.075	0.099	0.076	0				
0.75	0	0.087	0.117	0.09	0				
1	0	0.093	0.129	0.105	0	0	0	0	0
1.25	0	0.095	0.139	0.139	0.109	0.099	0.096	0.095	0.095
1.5	0	0.086	0.129	0.14	0.135	0.13	0.128	0.127	0.126
1.75	0	0.059	0.087	0.097	0.098	0.096	0.095	0.095	0.095
2	0	0	0	0	0	0	0	0	0

Index